This new edition of a highly popular biochemistry text introduces the student to the basic principles of techniques that are used routinely in modern biochemistry and molecular biology. Its coverage is comprehensive, giving greatest attention to those techniques that undergraduate students of biosciences can expect to encounter in their practical classes. Principle and theory are discussed alongside details of equipment used and applications of the techniques.

The fourth edition contains new chapters on spectroscopic and mass spectrometric techniques. The chapter on enzyme techniques has been enlarged to cover protein techniques in general; the chapter on molecular biology techniques has been expanded to cover the polymerase chain reaction, that on electrophoretic techniques restructured to include the electrophoresis of nucleic acids and that on chromatographic techniques expanded to give greater coverage to the theory of chromatography and the practical details of high-performance liquid chromatography. The chapters on immunological techniques, radiobiological techniques and electrochemical techniques have been fully updated.

The book is essential reading for all bioscience students for whom practical biochemistry forms part of the syllabus. The book will also be of interest to postgraduates undertaking projects of a biochemical or molecular biological nature.

Principles and techniques of practical biochemistry

Principles and techniques of practical biochemistry

FOURTH EDITION

EDITED BY

Keith Wilson and John M. Walker

University of Hertfordshire, Hatfield

CAMBRIDGE
UNIVERSITY PRESS

Published by the Press Syndicate of the University of Cambridge
The Pitt Building, Trumpington Street, Cambridge CB2 1RP
40 West 20th Street, New York, NY 10011-4211, USA
10 Stamford Road, Oakleigh, Melbourne 3166, Australia

First published by Edward Arnold 1975
as *A biologist's guide to principles and techniques of practical biochemistry*
Second edition 1981
Third edition 1986
Third edition first published by Cambridge University Press 1992
Reprinted 1993
Fourth edition published by Cambridge University Press 1994 as
Principles and techniques of practical biochemistry

Printed in Great Britain at Butler & Tanner Ltd, Somerset

A catalogue record for this book is available from the British Library

Library of Congress cataloguing in publication data
Principles and techniques of practical biochemistry / edited by Keith
Wilson and John M. Walker. – 4th ed.
      p.   cm.
   Updated ed. of: A Biologist's guide to principles and techniques
of practical biochemistry. 3rd ed. 1986.
   Includes biographical references and index.
   ISBN 0-521-41769-4. – ISBN 0-521-42809-2 (pbk)
   1. Biochemistry – Methodology.   I. Wilson, Keith, 1936–   .
II. Walker, John M., 1948–   . III. Biologist's guide to principles
and techniques of practical biochemistry.
QP519.7.P75   1994
574.19′2 – dc20   93-6823 CIP

ISBN 0 521 41769 4 hardback (fourth edition)
ISBN 0 521 42809 2 paperback (fourth edition)
ISBN 0 521 42808 4 (third edition)

Contents

Contributors

DR S. BOFFEY
Division of Biosciences
University of Hertfordshire
Hatfield Campus
College Lane
Hatfield
Herts AL10 9AB, UK

DR D. BURRIN
Department of Biological Sciences
University of Coventry
Priory Street
Coventry CV1 5FB, UK

PROFESSOR D.B. GORDON
Department of Biological Sciences
Metropolitan University of Manchester
Chester Street
All Saints
Manchester M15 6BH, UK

PROFESSOR K.H. GOULDING
Academic Director
University of Middlesex
Bramley Road
London N14 4XS, UK

DR A. GRIFFITHS
Department of Biological Sciences
Oxford Brookes University
Headington
Oxford OX3 0BP, UK

DR P.K. ROBINSON
Department of Applied Biology
University of Central Lancashire
Preston
Lancs PR1 2HE, UK

DR I. SIMPKINS
Division of Biosciences
University of Hertfordshire
Hatfield Campus
College Lane
Hatfield
Herts AL10 9AB, UK

DR R.J. SLATER
Division of Biosciences
University of Hertfordshire
Hatfield Campus
College Lane
Hatfield
Herts AL10 9AB, UK

PROFESSOR J.M. WALKER
Division of Biosciences
University of Hertfordshire
Hatfield Campus
College Lane
Hatfield
Herts AL10 9AB, UK

PROFESSOR K. WILSON
School of Natural Sciences
University of Hertfordshire
Hatfield Campus
College Lane
Hatfield
Herts AL10 9AB, UK

Preface to the fourth edition

The spectacular progress over the past two decades in our understanding of biological processes at the molecular level has been made possible by the coming together of chemical, biochemical and genetic approaches to the investigation of those biological processes and by the availability of a wide range of analytical techniques. The development of routine methods for the isolation of proteins and nucleic acids and for the determination of their structure has led to the elucidation of the mechanisms of gene expression, cell–cell signalling and metabolic control. This in turn has led to important advances in the diagnosis and control of human disorders and for the introduction of commercially attractive characteristics into animal, plant and microbial cells. Such possibilities were little more than optimistic biochemical speculation 16 years ago, when the first edition of this book was published. In the preface to that edition we identified the need for increased emphasis on practical aspects of biochemistry in all biology courses and referred to some of the logistical problems this presented to the design of the undergraduate curriculum. This need for biochemistry to occupy a central position in the biological and life sciences is now accepted, but as the sciences expand so too do the pressures on the curriculum so that the logistical problems identified in 1976 remain as acute as before.

Our approach to the design of this fourth edition remains unchanged from that of the previous editions. Our aim has been to produce a text that gives students a basic appreciation of the underlying principles and practical strategy of the analytical and preparative techniques that are fundamental to the study and understanding of biochemistry. Greatest attention has been paid to those techniques that most students in the biosciences can expect to encounter in their practical classes. Less detailed, but essential, basic coverage is given to the more 'specialist' techniques, such as fast atom bombardment mass spectrometry, which students are unlikely to meet in the laboratory but which certainly will be referred to in lectures and which are likely to have a significant impact on the future development of biochemistry. The book is designed to be complementary to standard biochemistry textbooks and should be of value to students enrolled on degree and higher national certificate and higher national diploma course of BTEC in the biological, medical, paramedical and veterinary sciences of which biochemistry is a component. It is hoped that the book will also be of value to postgraduate students embarking on a programme involving biochemical techniques.

Major changes have been made to the content and structure of the new edition. Chapter 1 covers the rationale and methodology of biochemical experimentation. It now makes reference to the important issue of biological variability and to the need for careful data analysis as well as to the need for safe experimental design. Chapter 2 covers immunological techniques and Chapter 3 molecular biology techniques, making reference to new developments such as the polymerase chain reaction. Chapter 4 on enzyme techniques has been expanded to include protein techniques. Chapters 5 and 6 cover radioisotope techniques and centrifugation techniques, respectively. Chapters 7 and 8 relate to spectroscopic techniques and mass spectrometric techniques and have emerged from the development and division of the chapter on spectroscopic techniques in the three previous editions. Recent developments in mass spectrometry, specifically ion desorption methods and ion evaporation methods, will undoubtedly have a big impact in biochemistry and we believe justify greater space being devoted to them in this new edition. Chapter 9 on electrophoretic techniques has been redesigned and is now approached from the point of view of the study of proteins and nucleic acids but concomitantly it identifies the broad potential of such new techniques as capillary zone electrophoresis. Chapter 10 on chromatographic techniques has also been redesigned to include a greater appreciation than in previous editions of the underlying theoretical principles and growing importance of high performance liquid chromatography (HPLC) and of techniques such as chiral chromatography, hydrophobic interaction chromatography and lectin affinity chromatography. Chapter 11 on electrochemical techniques has been rewritten to give greater attention to topics such as ion-selective electrodes, electrochemical detectors and biosensors.

The fourth edition involves a new joint editor, John Walker, who has taken over from Ken Goulding, who is now heavily involved in the leadership of the new University of Middlesex. Several new contributors have been recruited including Robert Slater, who with John Walker has been intimately involved in the highly successful Molecular Biology Workshops held at Hatfield over many years and which, by means of the generous support of the British Council, have been taken to many European and Asian universities; Professor Derek Gordon, from the Metropolitan University of Manchester, who has an established reputation for his work on the application of mass spectrometry to biological problems; and Peter Robinson, from the University of Central Lancashire, who has taken over the chapter on electrochemical techniques.

We also acknowledge the support received from the staff of Edward Arnold (Publishers) Limited over the past 16 years and welcome the new opportunities for future collaboration with Cambridge University Press. As with previous editions, we thank numerous colleagues for their helpful discussions and permission to reproduce published data. Finally, it is a pleasure to thank students and academics from around the world who have written to us about aspects of previous editions. We continue to welcome comments from all those who use this book as part of their studies.

Keith Wilson
John M. Walker

Abbreviations and SI units

The following abbreviations have been used throughout this book without definition:

AMP   adenosine 5′-monophosphate
ADP   adenosine 5′-diphosphate
ATP   adenosine 5′-triphosphate
c.p.m.   counts per minute
DDT   2,2-bis-(p-chlorophenyl)-1,1,1-trichlorethane
DNA   deoxyribonucleic acid
d.p.m.   disintegrations per minute
e^- electron
EDTA   ethylenediaminetetra-acetate
e.m.f.   electromotive force
FAD   flavin adenine dinucleotide
FMN   flavin mononucleotide
kb   kilobase-pairs
M_r relative molecular mass
min   minute
NAD^+ nicotinamide adenine dinucleotide (oxidised)
NADH   nicotinamide adenine dinucleotide (reduced)
$NADP^+$ nicotinamide adenine dinucleotide phosphate (oxidised)
$NADPH^+$ nicotinamide adenine dinucleotide phosphate (reduced)
P_i inorganic phosphate

PP_i	inorganic pyrophosphate
p.p.m.	parts per million
RNA	ribonucleic acid
s.t.p.	standard temperature and pressure
Tris	2-amino-2-hydroxymethyl propane-1,3-diol

Basic SI Units (Système International D'Unités)

Physical quantity	Name of SI unit	Symbol
Length	metre	m
Mass	kilogram	kg
Time	second	s
Electric current	ampere	A
Thermodynamic temperature	kelvin	K
Amount of substance	mole	mol
Force	newton	N
Pressure	pascal	Pa
Power	watt	W
Electric charge	coulomb	C
Electric potential difference	volt	V
Electric resistance	ohm	Ω
Frequency	hertz	Hz
Magnetic flux density	tesla	T
Area	square metre	m^2
Volume	cubic metre	m^3
Velocity	metre per second	$m\ s^{-1}$
Acceleration	metre per second squared	$m\ s^{-2}$
Density	kilogram per cubic metre	$kg\ m^{-3}$
Electric field strength	volt per metre	$V\ m^{-1}$
Concentration	mole per cubic metre	$mol\ m^{-3}$
Magnetic field strength	ampere per metre	$A\ m^{-1}$
Dipole moment	coulomb metre	$C\ m$
Entropy	joule per kelvin	$J\ K^{-1}$

Volume

The SI unit of volume is the cubic metre, m^3. The litre has been redefined as being exactly equal to the cubic decimetre. Although the term litre still remains in common usage, it is recommended that both the litre and fractions of it (e.g. millilitre) are abandoned in exact scientific work.

$$
\begin{aligned}
1 \text{ litre (l)} &= 1 \text{ dm}^3 &= 10^{-3} \text{ m}^3 \\
1 \text{ millilitre (ml)} &= 1 \text{ cm}^3 &= 10^{-6} \text{ m}^3 \\
1 \text{ microlitre (}\mu\text{l)} &= 1 \text{ mm}^3 &= 10^{-9} \text{ m}^3 \\
1 \text{ nanolitre (nl)} &= 1 \text{ } \mu\text{m}^3 &= 10^{-12} \text{ m}^3
\end{aligned}
$$

Powers of units – prefixes

Multiple	Prefix	Symbol
10^{12}	tera	T
10^{9}	giga	G
10^{6}	mega	M
10^{3}	kilo	k
10^{2}	hecto	h
10	deca	da
10^{-1}	deci	d
10^{-2}	centi	c
10^{-3}	milli	m
10^{-6}	micro	μ
10^{-9}	nano	n
10^{-12}	pico	p
10^{-15}	femto	f

Conversion table for common units to SI equivalents

Unit	SI equivalent
ångström (Å)	$100 \text{ pm} = 10^{-10} \text{ m } (10^{-1} \text{ nm})$
atmosphere (standard) (760 mmHg at s.t.p.)	101 325 Pa
calorie	4.186 J
centigrade (°C)	$(t\,°\text{C} + 273) \text{ K}$

Curie, Ci	3.7×10^{10} Bq
erg	10^{-7} J
gauss (G)	10^{-4} T
micron, μ	1 μm
millimetre mercury (mmHg)	133.322 Pa
pound-force/sq.in. (lbf in.$^{-2}$) (p.s.i.)	6894.76 Pa
ln x	2.303 $\log_{10}x$

Values of some physical constants in SI units

Molar or universal gas constant (R)	8.314 J K^{-1}mol^{-1}
Planck constant (h)	6.63×10^{-34} J s
Molar volume of ideal gas at s.t.p.	22.41 dm^3 mol^{-1}
Faraday constant (F)	9.648×10^4 C mol^{-1}
Speed of light in a vacuum (C)	2.997×10^8 m s^{-1}

1

General principles of biochemical investigations

1.1 Introduction

Biochemistry is both a pure and an applied science, having developed, relatively recently, as a separate discipline combining chemistry, physics, botany, zoology and physiology. In many ways it retains an interdisciplinary perspective. In the pure, academic sense, biochemistry is concerned with advancing knowledge of the relationship between structure and function at the molecular and subcellular level and, as such, is highly theoretical and research orientated. Increasingly, however, biochemistry is being applied in a more pragmatic context to provide and improve goods and services in a wide variety of industries, e.g. food and drinks, pharmaceuticals, environmental protection, biotechnology and the health service. In such industries it is quite common for specialist biochemistry departments to be set up to support research and development. It is important to stress at the outset, however, that routine biochemical analyses also form an essential part of the quality control aspects in many manufacturing and most service industries and is not research orientated.

Biochemistry is particularly important within the life sciences disciplines, e.g. physiology, pharmacology, microbiology and ecology, because it tries to explain living systems in terms of their ability to transform energy from their environment. In this context, an understanding of the relationship between Gibbs free energy changes, $\Delta G^{o'}$, in relation to redox potential and the equilibrium constants of chemical reactions (Section 11.3.1) is essential for an appreciation of metabolism.

Biochemical studies have substantiated the cell as the fundamental unit of life, since it, alone, possesses all the characteristics needed for independent energy transformation and replication. This is realised mainly through the juxtaposition of redox carriers in membranes to facilitate phosphorylation of adenosine diphosphate (ADP) to the energy-rich adenosine triphosphate (ATP) and the evolution of the genetic code in DNA. Within cells, metabolism depends on the enzymically coupled turnover of a relatively few energy-rich group transfer molecules (such as certain acyl phosphates, nucleoside diphosphates and triphosphates, and enoyl phosphates) and strongly reducing substances (such as reduced pyridine nucleotides and flavin nucleotides), generated in catabolism, being used to overcome thermodynamic barriers in biosynthesis. Nutritional classification of organisms is based on both the external source of electrons for reduction purposes and the energy source. Organisms that rely on inorganic electron donors are said to be lithotrophic and those that rely on organic sources are said to be organotrophic. To each of these classes may be added the prefix photo, if energy is provided by light within the visible and far-red region of the spectrum, or chemo, if energy is provided by oxidation of either organic or inorganic compounds.

Since the inception of life, therefore, natural selection has operated on the basis of different environments selecting out the most efficient cells for energy transformation and self-duplication. In this evolutionary process two types of cell have emerged: the prokaryotic cell, which has neither organised nucleus nor organelles; and the eukaryotic cell, which has both. The greater biochemical compartmentalisation of eukaryotes has permitted much of the more recent stages of evolution to be concerned with natural selection for physiological complexity based on differentiation of alternative cell types. From this it follows that eukaryotes exhibit a much narrower range of nutritional types than do prokaryotes but display great heterogeneity in the differentiation of cells, tissues and organs that perform special physiological functions. Each different cell type in a multicellular organism must reflect accompanying biochemical and physiological differences operating within these cells and invoke mutual cooperation of cells in physiological processes. A large part of developmental biochemistry is concerned, therefore, with elucidating, at the molecular level, the mechanisms of selective gene expression leading to differentiation.

Despite phylogenic differences, a unifying factor of all cells is that they contain many identical chemical constituents, metabolic pathways and mechanisms of cellular recognition. This allows for a mode of biochemical deduction based on extrapolation of results obtained in one species (usually of lower phylogenic order) to another. Thus, microorganisms, cultures of animal tissue, or laboratory animals are used frequently for monitoring biochemical, physiological, pharmacological or toxicological responses to foreign exogenous compounds (xenobiotics) as a prelude to use of the latter in humans. However, this approach must be treated with caution because biological variation between cell types or species is possible and there may be gross physiological differences, particularly between unicellular and multicellular species.

1.2   General approaches to biochemical investigations

Biochemical investigations frequently require the purification of a particular compound from a complex mixture. In analytical separations, the objective is to identify and estimate small amounts of the compounds, and frequently it is not necessary to recover the compound after the separation process. In preparative separations, the main aim is to isolate and recover as large an amount as possible of the compound in a high degree of purity, in order, subsequently, to study its chemistry and/or its biological properties. Whether an analytical or preparative approach is being adopted may well dictate the choice of separation and purification techniques, mainly because a preparative approach requires much larger amounts of starting material and need not employ techniques that give a high percentage recovery.

Analytical separations may be used on a qualitative basis as a short cut to identifying the components present in a mixture. A compound may be suspected of being present in a mixture; then, if its behaviour during analytical separation is identical with the behaviour of a known compound (a standard), the two compounds are likely to be identical. In assessing the behaviour of the standard and the unknown, as many high resolution techniques as possible should be used in which the standard should undergo exactly the same treatment, preferably at the same time, as the unknown compound. This technique removes the need for analysing the compound by physical methods, after separation, in order to establish its structure.

It is not always necessary to carry out preliminary isolation and purification of molecules before estimating their concentration in biological material if they possess some unique property that may be used as a basis for their quantitative determination. Such unique features demonstrate biological specificity and are illustrated in the assay of enzymes in crude extracts, in intact cells and in cellular organelles. Some spectrophotometric methods by which enzyme assays may be carried out are given in Section 4.8.2. In cases where such specificity is not exhibited, then some degree of purification of the molecule is required before it can be analysed. Any purification procedures adopted will inevitably involve some loss of material. It is essential that the number of purification stages should be kept to a minimum and therefore the techniques employed should be those that are capable of giving the greatest resolution with small amounts of material. Some of the techniques, described in later chapters, such as chromatography (Chapter 10) and electrophoresis (Chapter 9) are particularly suitable.

High resolution methods of separation normally have a limited capacity and cannot be used to separate large amounts of material. The normal procedure in preparative separations, therefore, involves a first stage with a high separation capacity but which may have low resolution. Techniques such as precipitation, dialysis, adsorption, liquid–liquid chromatography and partition column chromatography, countercurrent distribution and continuous electrophoresis, all of which are discussed in subsequent chapters, are important because, although they may not achieve as good a resolution as other methods, they can be scaled up to cope with large amounts of material. Some of the high resolution methods, for example ion-exchange, exclusion and affinity chromatography, preparative gas–liquid chromatography (GLC) and high-performance liquid chromatography (HPLC) and preparative gel electrophoresis may then be employed for the final purification steps. Sequential methods used for separation should be based on as many different properties of the molecule as possible, as comparatively little further purification is likely to be obtained by using another method that exploits the same physical property for separation.

When one is confronted with the problem of choosing a particular method (or methods) to purify a compound, it is obvious that, although several techniques may be theoretically applicable, in practice certain techniques may be superior. Several factors must be taken into consideration when one decides which techniques to adopt. These include:

(i) Whether the separation is to be carried out on an analytical or preparative basis so that the resolution and/or capacity of the techniques are considered.

(ii) The physical properties of the compound, for example solubility, volatility, relative molecular mass and charge properties.

(iii) The stability of the compound during a particular treatment.

(iv) A consideration of comparable successful separations documented in the literature.

(v) The availability of equipment and expertise.

(vi) The cost of the procedure.

An important decision is often to evaluate a range of model systems and choose the one that is both appropriate and available to provide answers to the aims of the investigation. Essentially two types of experimental model are available, *in vivo* and *in vitro*. However, the distinction between these models is sometimes difficult. The *in vivo* technique applies to multicellular organisms and uses whole plants, whole animals or parts of animals in which isolated organs are perfused with nutrients.

1.2.1   *In vivo* studies

The use of whole animals and plants for biochemical experiments *in vivo* is very extensive, as alternative model techniques such as cell culture are often considered physiologically too simplistic to meet the needs of the investigation. Animals are used for fundamental research, for medical and clinical research, for veterinary and agricultural research, in the manufacture of vaccines, antibodies and hormones, and in testing the potency of drugs or other biological products where chemical determination is not feasible; they are also very widely used in toxicological testing. Much animal work is directed towards the study of the metabolism of foreign substances, xenobiotics (drugs, food additives), and extrapolation of results to humans as a prerequisite to clinical trials. Mice, rats and guinea-pigs are most commonly used because of their lower cost and ease of handling, but rabbits, cats, dogs, marmosets, monkeys and baboons are also used. The animals are bred specifically for experimental purposes and in the UK their husbandry is subject to scrutiny by the Home Office Inspectorate, who issue licences for work involving live vertebrates. Age, sex, nutritional state, strain, circadian rhythms and stress state of animals have to be closely regulated to minimise variability in results. Experiments should be performed, therefore, on small groups of animals, as physiologically similar as possible, and the collated results analysed statistically. When metabolic responses to xenobiotics are being monitored, the results obtained must be compared with a control group which has received a placebo (a completely inert substance such as lactose). In a human subject or a laboratory animal the fate of an administered compound can be traced by monitoring the concentrations of the compound and/or its metabolites in blood, urine, faeces, bile, expired air, sweat or saliva. Excretion patterns of small laboratory animals are best studied by placing the animal in a metabolic cage for the duration of the experiment. Such cages allow urine and faeces to be collected separately and expired carbon dioxide to be trapped. If ^{14}C-labelled compounds are used, expired $^{14}CO_2$ may be absorbed in sodium hydroxide solution. The assay of $^{14}CO_2$ expired allows an estimate of the extent of complete degradation of the compound to be made. Since many excretion products in urine and faeces are conjugated (i.e. linked to polar molecules such as glucuronic acid, sulphate and glycine), it is essential that samples are hydrolysed, enzymically or chemically, before the free metabolites are extracted, and identified by established analytical techniques such as GLC (Section 10.10) and HPLC (Section 10.4).

The technique of whole-body autoradiography (Section 5.2.3) has proved very useful as a method of visualising the distribution of radioactive materials administered to small animals. Experimental information on distribution, relative accumulation in tissues, rates of excretion and ability to cross biological membranes is obtainable by this technique. At a suitable time interval after injection with a radioactively labelled compound, the animal is killed by deep anaesthesia and rapidly cooled in an acetone–solid CO_2 mixture at $-78\,°C$ or in liquid N_2. The frozen animal is embedded in an aqueous solution of resin (gum acacia) at low temperature. After the resin has set, the animal may either be machined to a suitable level with a cutter attached to an electric drill, or serial sections made using a microtome that has a special tungsten carbide blade. The prepared animal surface is placed in close contact with X-ray film at a low temperature for one to two weeks, after which time the film is developed (see Fig. 5.12).

The fate of a compound in a living laboratory animal may also be studied through the vascular perfusion of an organ such as the liver or kidney *in situ*. This procedure involves infusing the compound through a fine hollow needle inserted into the artery carrying blood to the organ and performing subsequent analysis on the blood being transported from that organ by the corresponding vein.

The cost and ethics of whole-animal experimentation is leading to increasing effort being directed towards finding alternatives to whole-animal experiments and, coincidentally, to reducing the number of animals used for essential experimentation. The main approach being increasingly adopted is to change to *in vitro* methods where appropriate.

Higher plants

Traditionally, higher plants have been used to further basic understanding of the biochemistry of fundamental processes such as pathways for photosynthesis, respiration, photorespiration and nitrogen assimilation, which are aspects of primary metabolism relating to ATP and pyridine nucleotide turnover. Increasingly, however, plants are being used for 'applied' research in fields such as plant breeding, plant pathology, food research, environmental pollution, xenobiotic testing and in the search for so-called 'secondary products'. These secondary products are part of biochemical pathways not concerned primarily with energy production and conservation but which have been shown occasionally to have considerable economic importance as pharmaceuticals, pigments, perfumes and natural agrochemicals. Although much of the future of plant biochemistry is now clearly associated with the exploitation of plant tissue culture and molecular biology (Sections 1.5.3 and Chapter 3), progress with this has been held back by the relative paucity of knowledge of secondary metabolism and its regulation in plants.

Methods of studying the metabolism of whole plants depend on the objective of the experiment. Whole plants may be grown either outside in field plots or in more controlled environments, i.e. inside glasshouses or environmental cabinets where factors such as temperature, light, relative humidity and gaseous environment can be controlled within specified limits. Any test compound may be added to either relatively ill-defined compost or to sand/rockwool mixtures when plants are grown hydroponically in chemically defined nutrient solutions. Test compounds may be injected into roots, included in the rhizosphere, or applied to foliage by painting or spraying. Dilute organic solvents and mild detergents promote uptake through the waxy cuticle. Gaseous uptake proceeds through open stomata under conducive relative humidity conditions. The subsequent distribution of the compound proceeds either locally by plasmodesmata or over longer distances through the vascular system. Radiolabelling techniques may be used conveniently for metabolic investigations. Major difficulties in studying plant metabolism are that plant organs do not contain enzymes in such high concentrations as are found in animal organs, and the metabolic rate of plants is usually lower.

1.2.2 *In vitro* studies

In vitro methods involve the incubation of biologically derived material in artificial physical and chemical environments. The term *in vitro* is equally applicable to enzyme preparations, to isolated organelles, to intact microorganisms and to excised parts of animals or plants. Conditions may be chosen to promote a limited degree of growth, differentiation and development, as for instance in cell, tissue and organ culture of animals and plants. The specific advantage of cell and tissue culture methods is that they reduce the physiological and biochemical constraints imposed by contiguous cells. The approach has found widespread application in the biosciences. In its most fundamental sense, cell culture facilitates the investigation of the developmental potential (totipotency) of cells, i.e. the capacity, within the limits of its genetic constitution, of one cell to form any other type of cell in the organism, given an appropriate artificial chemical and physical environment. A general criticism of experimentation *in vitro* is that extrapolation to the situation *in vivo* may be unjustified, i.e. that the methodology *in vitro* may be the study of artefacts.

1.2.3 Organ and tissue slice techniques

In perfusion of isolated organs, the liver, kidney or heart, etc. is removed from a recently killed animal and maintained at constant temperature in suitable apparatus. The metabolic fate of the perfused compound in the organ may then be determined as for perfusion *in vivo*. The perfusion fluid may be passed through the organ either by gravity feed or by using a small pump. If perfusion fluid is recirculated, pumps must be used. The flow of fluid through the organ is usually made using peristaltic pumps (Section 10.3.6), which ensure a pulsatile flow that closely resembles the type of flow produced by the heart.

Superfusion is a simpler technique in which incubation fluid simply trickles over the surface of organs suspended in a bath. When mechanical responses of isolated tissue are being investigated, clamped organs are attached to a force transducer that records tissue movement.

Electrophysiological techniques may also be used for organs such as brain or muscle, in which case microelectrodes attached to the tissue detect voltage changes across membranes. These changes are detected on an oscilloscope or similar recording device.

Slice techniques utilise tissue slices thin enough to enable oxygen to reach the innermost layers of the slice and for adequate removal of waste products by diffusion. Slices 0.5 to 5 mm thick generally meet the above requirements, whilst allowing the proportion of disrupted cells to intact cells to remain small.

The organ under study should be removed immediately after the animal is killed in order to minimise any post-mortem changes. Cutting of the slices is achieved by using razor blades, or, for more uniform preparations, by using a microtome. The tissue slices are transferred to a vessel containing a suitable suspending medium. The metabolism of the slices may then be studied and the effects of added compounds on the metabolic processes determined. Slices need 95% (v/v) oxygen to be used in the gas phase in order to achieve aeration of the innermost cells of the slices. An obvious disadvantage of this technique, however, is that by trying to ensure adequate diffusion of oxygen to the innermost cells of the slices, one may allow toxic concentrations of oxygen to come into contact with the outer layers of cells. Figure 1.1 illustrates the preparation of rat liver slices for use in a dynamic organ culture system for the study of drug metabolism.

1.3   Physiological solutions

Physiological solutions may be classified broadly according to their use either for short-term incubations, e.g. perfusion studies, for supporting long-term growth of microbes, plant or animal tissue cultures, or for nutritional studies involving whole animals and plants. The chemical and physical properties of physiological solutions are designed to minimise artefacts and their chemical composition is sometimes based on reformulating mixtures after biochemical analysis of cells, tissues or natural body fluids. Occasionally a medium is formulated empirically. Chemically

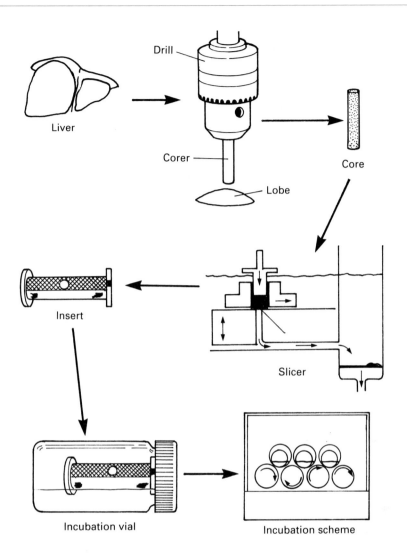

Drill

Liver

Corer

Lobe

Core

Insert

Slicer

Incubation vial

Incubation scheme

Fig. 1.1. Schematic representation of the preparation of rat liver slices in a dynamic organ culture system. *Slice preparation*: the slices are produced by pulling an immobilised core across a motor-driven oscillating blade. The thickness of the slices produced by the slicer is controlled by a combination of the distance between an adjustable plate at the bottom of the holder and the weight applied to the piston on top of the tissue core. Freshly sectioned slices, produced at a rate of one every 3–4 s, are swept away to a collecting chamber by a stream of buffer. (Reproduced, with permission, from Barr, J., Weir, A.J., Brendel, K. and Sipes, I.G. (1991) *Xenobiotica*, **3**, 331–9.)

defined solutions are preferable but not always possible because of the very stringent growth requirements of many tissue cultures. Crucial factors are to maintain the medium isotonic with respect to the tissue (i.e. possessing the same water potential) and to buffer to a physiological pH.

1.3.1   Buffer solutions

Organisms and cells can generally withstand large variations in the pH of their external environment. In contrast, cellular processes are sensitive to pH changes and take place in a medium the pH of which is carefully regulated. There may, however, be localised intracellular pH variations, e.g. at membrane surfaces. The majority of intracellular processes occur at a pH maintained near neutrality. This neutral pH is generally the one at which the various metabolic processes occur at their maximum rate. The hydrolases of lysosomes, however, have their maximum activity at a pH in the region of 5, a pH that prevails following the death of the cell. The gastric juice of mammals is quite exceptional in having a pH of approximately 1, a pH at which the enzyme pepsin, which initiates the digestion of dietary proteins in the stomach, has its maximum activity.

The control of a virtually constant pH in biological systems is achieved by the action of efficient buffering systems the chemical nature of which is such that they can resist pH changes due to the metabolic production of acids such as lactic acid and bases such as ammonium. The major buffering systems found in cellular fluids involve phosphate, bicarbonate, amino acids and proteins.

The sensitivity of biological processes to pH may be due to one of several reasons. The process may be catalysed by hydrogen ions or may involve a hydrogen ion as a reactant or product. Alternatively, a pH change may alter the distribution of a compound or ion across a membrane, possibly by altering the permeability of the membrane. Membranes, like many other biological structures and molecules, possess ionisable groups the precise state of ionisation of which influences their molecular conformation and thus their biological activity. This is particularly true of proteins and thus of enzymes. Some proteins rely on a slight change in the pH of their environment to complete their biological function. In the case of haemoglobin, whose primary function is to transport oxygen from the lungs to the tissues, a slight decrease in the pH of the tissues, due to carbon dioxide and hydrogen ion production as a result of the active respiration of the tissue, helps to facilitate the unloading of the oxygen when and where it is required. The unloading of oxygen is accompanied by the uptake of protons by the haemoglobin, thus simultaneously helping to buffer the system. *In vitro* studies of metabolic processes require the use of buffers. Deliberate changes in pH, however, may help in the analytical study of certain groups of molecules, e.g. amino acids, proteins and nucleic acids, by such techniques as electrophoresis and ion-exchange chromatography.

A buffer solution resists change in hydrogen ion concentration on the addition of acid or alkali. This resistance is called the buffer action. The magnitude of the buffer action is called the buffer capacity (β), and is measured by the amount of strong base required to alter the pH by one unit:

$$\beta = \frac{db}{d(pH)} \qquad (1.1)$$

where d(pH) is the increase in pH resulting from the addition db of base.

In practice, buffer solutions usually consist of a mixture of a weak acid or base and its salt, e.g. acetic acid and sodium acetate (on the Brönsted and Lowry nomenclature, a mixture of a weak acid and its conjugate base). In a solution of a weak acid (RCOOH) and its salt (RCOO$^-$), added hydrogen ions are neutralised by the anions of the salt, which act, therefore, as a weak base, and, conversely, added hydroxyl ions are removed by neutralisation of the acid. It is clear from this that the buffer capacity of a particular acid and its conjugate base will be a maximum when their concentrations are equal, i.e. when pH = pK_a of the acid. Buffer capacity also depends upon the total concentrations of acid and salt as well on their relative proportions – the greater the total concentration, the greater the buffer capacity. The usual concentration of acid and salt in buffer solutions is of the order of 0.05–0.20 M and generally the mixtures possess acceptable buffer capacity within the range pH = pK_a ± 1. The criteria for buffers suitable for use in biological research may be summarised as follows. They should:

  (i)   possess adequate buffer capacity in the required pH range;
  (ii) be available in a high degree of purity;
  (iii) be very water soluble and impermeable to biological membranes;
  (iv) be enzymically and hydrolytically stable;
  (v)   possess a pH that is minimally influenced by their concentration, the temperature and the ionic composition or salt effect of the medium;
  (vi) not be toxic;
  (vii) only form complexes with cations that are soluble;
  (viii) not absorb light in the visible or ultraviolet regions.

Needless to say, not all buffers that are commonly used meet all these criteria. Thus, phosphates tend to precipitate polyvalent cations and are often metabolites or inhibitors in many systems and Tris is often toxic or may have inhibitory effects. However, a number of zwitterionic buffers of the Hepes and Pipes type fulfil the requirements and are often used in tissue culture media containing sodium bicarbonate or phosphate solutions as nutrients. They suffer from the disadvantage that they interfere with Lowry protein determinations.

Physiologically, proteins comprise an important group of buffers. By virtue of their large numbers of weak acidic and basic groups in the amino acid side chains,

Table 1.1. pK_a *values of some acids and bases that are commonly used as buffer solutions*

Acid or base	pK_a (at 25°C)
Acetic acid	4.75
Barbituric acid	3.98
Carbonic acid	6.10, 10.22
Citric acid	3.10, 4.76, 5.40
Glycylglycine	3.06, 8.13
Hepes[a]	7.50
Phosphoric acid	1.96, 6.70, 12.30
Phthalic acid	2.90, 5.51
Pipes[b]	6.80
Succinic acid	4.18, 5.56
Tartaric acid	2.96, 4.16
Tris[c]	8.14

[a] Hepes, N-2-hydroxyethylpiperazine-N'-2-ethanesulphonic acid.
[b] Pipes, piperazine-N-N'-bis(2-ethanesulphonic acid).
[c] Tris, 2-amino-2-hydroxymethylpropane-1,3-diol.

proteins have a very high buffer capacity. For example, haemoglobin is the main compound responsible for the buffering capacity of blood.

Some commonly used buffers are listed in Table 1.1. To obtain buffer solutions covering an extended pH range, but derived from the same ions, mixtures of different systems may be used. Thus the McIlvaine buffers cover the pH range 2.2–8.0 and are prepared from citric acid and disodium hydrogen phosphate.

The state of ionisation of a weak electrolyte is dependent upon the prevailing pH and the numerical value of its ionisation constant. For weak acids, which ionise according to the equation:

$$RCOOH = RCOO^- + H^+$$

Acid   Conjugate base

the ionisation constant K_a, is given by the expression:

$$K_a = \frac{[RCOO^-][H^+]}{[RCOOH]} \tag{1.2}$$

In the case of weak bases such as amines, which ionise according to the equation:

the ionisation constant may be expressed in terms of a K_b value:

or more commonly in terms of the K_a value of the conjugate acid:

In such cases, the product of K_a and K_b for a weak base is equal to K_w, the ion product of water.

In practice, since K_a values are numerically very small, it is customary to use pK_a values, where $pK_a = -\log_{10}K_a$. It follows that, for weak bases, $pK_a + pK_b = 14$. The precise way in which the state of ionisation of a weak electrolyte varies with pH is given by the Henderson–Hasselbalch equation. For a weak acid, this takes the form:

$$pH = pK_a + \log_{10} \frac{[\text{conjugate base}]}{[\text{acid}]}$$

or:

$$pH = pK_a + \log_{10} \frac{[\text{ionised form}]}{[\text{unionised form}]} \qquad (1.5)$$

In the case of a weak base, expressing the equation in terms of the ionisation of the conjugate acid we obtain:

$$pH = pK_a + \log_{10} \frac{[\text{base}]}{[\text{conjugate acid}]}$$

or:

$$pH = pK_a + \log_{10} \frac{[\text{unionised form}]}{[\text{ionised form}]} \qquad (1.6)$$

It can be appreciated from these equations that weak acids will be predominantly unionised at low pH values and ionised at high pH values. The exact opposite is the case for weak bases; at low pH values the conjugate acid will predominate and at high pH values the unionised free base will exist. This sensitivity to pH of the state of ionisation of weak electrolytes is important both physiologically, as for example in the absorption by passive diffusion from the gastrointestinal tract of weak electrolytes and their excretion by the kidney, and in investigations *in vitro* employing such techniques as electrophoresis and ion-exchange chromatography.

1.3.2   Microbial cell media

The composition of the medium for incubation of microorganisms *in vitro* is determined largely by their nutritional classification as either a chemolithotroph, chemoorganotroph, photolithotroph or photoorganotroph, and whether or not they are wild-type or mutant cell lines. Typically, a minimal medium for the growth of a fully biosynthetically competent (prototrophic) chemoorganotroph would contain salts of Na^+, K^+, Ca^{2+}, Mg^{2+}, NH_4^+ or NO_3^-, Cl^-, HPO_4^{2-}, and SO_4^{2-} and a simple carbon source such as glucose. A chemoorganotroph that mutated, such that synthesis of alanine is prevented, for example, would be an alanine auxotroph and as such must be supplied with alanine in the culture medium. Different auxotrophic mutants may be exploited in the elucidation of metabolic pathways. Complex organic supplements may sometimes be added either when the nutritional requirements are ill-defined or to accelerate the growth rate of chemoorganotrophs in culture. The gelling agent, agar, may be added to solidify the medium, thus facilitating the surface growth of microorganisms. Preformulated medium is available in powdered form from commercial suppliers.

1.3.3   Higher plant cell media

The mineral composition of a nutrient solution supporting the hydroponic growth of plants (Section 1.2.1) is qualitatively similar to that used for microorganisms. However, it usually contains additional essential microelements such as Mn^{2+}, B^{3+}, Zn^{2+}, Cu^{2+} and Mo^{2+}.

A variety of different types of medium is available for the proliferation of plant cells in culture. In addition to a balanced mixture of macro- and micro-elements, carbon sources are provided because higher plant cells are not normally photosynthetically competent in culture. Nitrogen is supplied as either as NO_3^- or NH_4^+ or as an organic nitrogen supplement such as urea or glutamine. Vitamin inclusions are normally from the B group. Most non-tumour cultures require auxin-like regulators that promote cell expansion, such as naturally occurring indol-3-yl acetic acid (IAA) or synthetic compounds such as naphthalene acetic acid (NAA) or 2,4-dichlorophenoxyacetic acid (2,4-D). Cytokinins, included to stimulate cell division, may be the naturally occurring zeatin or synthetic analogues such as 6-furfurylaminopurine (kinetin). The induction and maintenance of growth depends additionally on an appropriate balance of hormones as well as on their absolute concentrations. The majority of media currently used are chemically defined but some may include complexes such as coconut milk or yeast extract. Organic buffers and antibiotics are not normally included in a medium. Conditioned medium, which has already supported the growth of cells, may be required for inducing the growth of low densities of either protoplasts or free cells (Section 1.5.3). A medium may contain either activated charcoal (0.2–3% (w/v)) or polyvinylpyrrolidone (PVP) at concentrations up to 250 mg dm^{-3} or antioxidants such as citrate, ascorbate or thiourea to prevent browning due to phenol release, oxidation and polymerisation. Growth retarding chemicals such as ICI's Paclobutrazol may be added to the medium to ameliorate physiological abnormalities associated with so-called vitrification symptoms, which sometimes arise as a result of very high relative humidity inside culture vessels. Such vitrification is reduced when the medium is solidified by agar supplements to approximately 1% (w/v) final concentration. Agar is comparatively expensive, however, and synthetic gel polymerising agents are often substituted for it. A wide variety of different types of plant tissue culture medium is available from commercial sources. Tissue culture medium has to be sterilised by either autoclaving or filtration before use. Examples of tissue culture medium are Murashige and Skoog, Gamborg B5, Nitsche, and Schenk and Hildebrandt. These are all available commercially in powder form.

1.3.4   Animal cell media

Numerous physiological salt solutions have been developed for short-term work with animal tissues, e.g. Tyrode's, Young's, Locke's, Meng's and Da Jalon's, many being derived from Ringer's solution, which was one of the first to be formulated. Extra buffering capacity is provided in Krebs–Ringer bicarbonate, the bicarbonate solution having been designed, originally, to approximate to the ionic composition of mammalian serum. Typically such salt solutions contain $NaCl$, KCl, $MgSO_4$, $CaCl_2$, $NaHCO_3$, and KH_2PO_4 in various amounts.

The control of gaseous exchange and pH are critically important for long-term maintenance of cells in culture. Medium buffered with 30 mM bicarbonate has to be maintained in air containing 5% (v/v) CO_2 to keep the pH between 7.3 and 7.5. Phenol red dye, which adopts an orange-red colour at this pH, is usually included as a check on correct pH. Medium containing zwitterionic buffers such as Hepes does not require this treatment but, unfortunately, cannot always be used alone. Calcium ion levels are critical for the growth of anchorage-dependent cells but have to be omitted for suspension cultures (Section 1.5.2). Other inclusions are typically vitamins, carbon sources, amino acids and proteins. Protein inclusions may include either plasma or serum. Serum contains important growth factors such as insulin-like growth factor (IGF), epidermal growth factor (EGF) and platelet-derived growth factor (PDG), which are essential for long-term growth of more highly differentiated excised tissues.

The use of serum-free medium (SFM) is particularly important when effects of hormones and growth factors on cell differentiation are being investigated. A range of SFM types is available commercially, containing known amounts of some of the serum components, but unfortunately such media are very expensive. Antibiotics active against a range of contaminating microorganisms may be incorporated, filter sterilised, at concentrations from 2.5 $\mu g\,cm^{-3}$ to 100 $\mu g\,cm^{-3}$ and are claimed to be effective for five to six days at 37 °C.

Most medium is supplied in liquid form and has a limited shelf-life under refrigeration. Some types of medium are commercially available in powdered form, which extends storage life, especially when the compound is kept cold.

1.3.5 Media for tissue homogenisation and separation

There is no unequivocal way of choosing the medium in which to suspend tissue during homogenisation. Some guide may be obtained by consulting the literature but ultimately the choice may depend largely on the results of trial experiments carried out on a range of media. In addition to a buffered, balanced salt solution, media usually contain a carbon source. Sucrose is commonly used to provide sufficient osmotic potential to prevent organelles swelling and bursting but it interferes with certain enzyme assays and is sometimes replaced with either mannitol or sorbitol, both of which are usually very slowly metabolised by plant and animal tissues. Many different recipes are used to try to preserve organelle and metabolic integrity, sometimes by preventing enzyme inactivation.

As well as maintaining isoosmotic conditions and an appropriate pH, critical levels of certain inorganic ions may also be important. For example Mg^{2+} helps to maintain the integrity of both nucleus and ribosomes. On the other hand, chelating agents such as EDTA or EGTA may be added to remove divalent cations such as Mg^{2+} or Ca^{2+} when it is important to inactivate membrane proteases. Inhibitors such as diisopropylfluorophosphate (for serine residues) or N-ethyl-maleimide (for sulphydryl groups) may be included in extraction media to inhibit protease activity (Section 4.5.2). Alternatively, artificial protease substrates such as bovine serum albumin (BSA) may also be added to ameliorate protease activity.

Many enzymes need to be maintained with their sulphydryl groups in a reduced state, which means adding to the extraction medium reducing agents such as 2-mercaptoethanol, dithiothreitol, reduced glutathione or cysteine at concentrations of around 30 mM. PVP either in water-soluble form or as a cross-linked polymer Polyclar may be added to a medium to remove phenolic compounds by filtration.

An organic-based aqueous medium is sometimes preferred to salt solutions in tissue extractions. For example, citrate has been used in the isolation of nuclei because of its ability to inactivate neutral deoxyribonucleases, and glycerol, ethylene glycol and ethylene glycol polymers have been used for the preparation of plastids. A non-aqueous medium may also be used in the isolation of organelles. The suspending fluid is usually a blend of a light and a heavy organic solvent such as ether–chloroform or benzene–carbon tetrachloride. The density of the medium can be varied so that the required organelles either float or sediment from the remainder of the homogenate in the subsequent centrifugation stages. Use of non-aqueous fractionation procedures has been made in the preparation of chloroplasts and haemosiderin granules from spleen. The technique does have some disadvantages in that morphological alterations may occur in some tissues and most enzymes are inactivated by organic solvents.

1.4   Cell disruption

1.4.1   Introduction

Cell disruption is the first stage in any analytical process in which the contents of cells are to be released and investigated. It is especially important in cell fractionation experiments in which the objective is to isolate, partially purify and assay subcellular organelles. Homogenisation disrupts tissue and releases intracellular compounds to form a 'brei' or homogenate. Such crude mixtures may be used directly for enzyme assay or for studies where uptake of a metabolite or a xenobiotic into intact tissue may be difficult. Homogenisation essentially involves disorganisation of a biological system and may, therefore, introduce artefacts where new morphological and metabolic states are produced *in vitro*. The choice of tissue, the physical and chemical properties of physiological solutions used and the choice of cell disruption method employed in practice are very important for successful homogenisation. The biological material should be rich in specific organelles and be very active in the particular metabolic pathways of interest. For example, fresh liver or cauliflower curd are excellent model animal and plant tissues, respectively, for crude mitochondrial preparations; hepatocytes are a good source of microsomes and thymus tissue is excellent for isolating nuclei. The specific activity of key enzymes can be used to assess the quality of the preparation.

The major use of homogenisation is as a preliminary step for the separation of cellular components in investigations into metabolic compartmentalisation. Hence the need to minimise artefacts. Unfortunately much homogenisation methodology is still empirical. In view of the wide variations between different tissues, both in fragility of different organelles and in the resistance of cells and tissues to disruption, the homogenisation of different tissue types poses separate problems that can be solved only by trial and error.

Separation involves grouping components of the homogenate into a new, ordered system having physical and chemical properties that are common to all components. Centrifugation methods are most often used for separations. The ideal isolation procedure would yield high concentrations of the desired components with their structure and function unchanged. In practice, however, cell fractionation, like homogenisation, is also an empirical procedure, and methods that appear to preserve morphology may destroy activity.

1.4.2   Methods of disrupting tissues and cells

Many enzymes are thermolabile and consequently cell disruption is best carried out at a low temperature, which is generally taken to be about 4 °C. Cell disruption is often performed inside purpose-built low temperature laboratories (cold rooms) but may also be carried out on a small scale inside vessels held within ice buckets in a laboratory.

The methods used to break cells depend largely on the fragility of the cells. For example, cell lysis of soft animal tissue can be achieved by bursting cells by osmotic shock, by exposing to alternating freeze/thaw conditions, by enzymic digestion by combinations of lipases and proteases (collagenase, trypsin, hyaluronidase), or by using organic solvents such as toluene. Plant tissue can be digested by pectinase and cellulase combinations, and microbial cells are susceptible to lysozyme treatment. More vigorous physical techniques usually have to be employed, however, and these can be grouped according to whether disruption is caused either by shearing forces between cells and solids or by liquid shear.

The pestle and mortar remains a versatile and convenient method for grinding up small amounts of material in the presence of an abrasive such as coarse or fine sand or alumina. The solid shear method is not so suitable for more delicate animal tissues or for microbial extractions. Mechanical shaking with abrasives may be achieved in Mickle shakers, which oscillate vigorously in suspension (300 to 3000 times min^{-1}) in the presence of small glass beads up to 200 μm diameter. This method is very suitable for bacteria or fungal lysis but may cause organelle damage in eukaryotes.

Liquid shear may be produced in tissue contained in blenders, which have cutting blades inclined at different angles to permit efficient mixing of the vessel contents and work in association with a specially designed stainless-steel cup which has indented walls to aid maximum suspension turbulence. The technique is good for plant and animal tissue extraction but poor for microbes.

Alternatively, homogenisers may be used. They involve the upward and downward movement of a ball-headed plunger and consist of either a hand-operated (Dounce or Tenbroeck) or power driven (Potter-Elvejham) pestle made of Pyrex glass, Teflon or leucite, and a glass tube. The tissue is forced between the walls of the fixed vessel and the moveable plunger as it is rotated. Shear forces are greatest at the surface of the pestle and least at the walls of the tube. The clearance between the pestle and the tube is manufactured precisely, usually being between 0.05 mm and 0.5 mm. The rotation rate of the pestle will also partly control the shear forces established. Gentle homogenisation is suitable for soft, delicate tissue such as polymorphonuclear leukocytes. This milder treatment has the advantage of not generating much heat. Where heat generation is a factor, the glass vessel is usually immersed between successive plungings in an ice bucket. Safety precautions need to be adhered to when homogenisers are used as it is quite easy to shatter the vessels by over vigorous manipulation. The method is suitable for plant or animal tissue that is not highly differentiated, but unsuitable for microorganisms, which are simply stirred.

More controlled solid shear operates in the Hughes Press, where a piston forces a frozen paste of cells through a narrow 0.25 mm diameter slot in a pressure chamber. This high pressure extrusion method is suitable for tough plant material, bacteria or yeasts, since pressures up to 55×10^6 Pa may be achieved. However, the operation is small scale. The most frequently used equipment for disruption of microbial cells is the French pressure cell, in which pressures of 10^8 Pa (1.5×10^4 lbf in.$^{-2}$) are commonly employed. A pressure cell consists of a stainless steel chamber that opens to the outside by means of a needle valve. The cell suspension is placed in the chamber with the needle valve in the closed position. After inversion of the chamber, the valve is opened and the piston pushed in to force out any air in the chamber. The chamber valve is then closed again and hydraulic pressure exerted on the piston. At a desired pressure, a needle valve is opened fractionally to release the pressure slightly. As the cells expand, they burst and a trickle of smashed cells is collected (Fig. 1.2). Excellent cell homogenates may be obtained using pressure cells for enzyme and similar assays. The technique is not suitable for large-scale homogenisations.

Ultrasound is thought to disrupt cell walls through amplified harmonic beats that develop shear forces in cells. The main disadvantage is that heat is generated, which may cause denaturation. Various designs of glass container are available in which the vessel is cooled in ice and side arms allow the brei to move away from the ultrasonic probe. Trial and error is used to find the best frequency for a particular cell disruption. The use of ultrasonics is usually confined to small scale operations. For safety reasons the apparatus should not be used in confined spaces and ear protection should always be used.

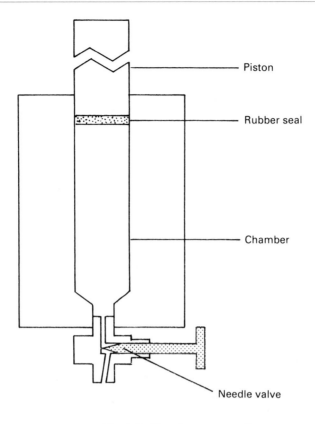

Fig. 1.2. French pressure cell.

1.5   Cell and tissue culture

1.5.1   Microbial cell culture

The term microorganism usually includes bacteria, fungi, yeasts, unicellular algae, filamentous algae and protozoa. As such organisms can be grown in pure culture axenically for experimental or industrial purposes, they are simpler to use than animal cells, plant cells or tissue cultures. A large number of biochemical and physiological studies aimed at understanding fundamental processes in the whole organism have therefore been conducted using bacteria (e.g. *Escherichia coli, Bacillus subtilis*), yeasts (e.g. *Saccharomyces cerevisiae, Candida albicans*) and algae (e.g. *Chlorella vulgaris, Chlamydomonas dysosmos*). The structural simplicity of bacteria in particular is a great advantage in studies such as self-assembly processes, in morphogenesis (e.g. the life cycle of lysogenic bacterial viruses), in DNA replication, RNA transcription and translation, and in control of gene expression, especially because their haploid nature facilitates the use of mutants. Types of mutation, their induction, selection and application are dealt with in Section 1.6.

Besides being used in biochemical and physiological studies, microbial cultures have many industrial applications as sources of, for example, alcohol, amino acids, antibiotics, coenzymes, organic acids, polysaccharides, solvents, sterols and converted sterols, sugars, surfactants and vitamins. The ability to engineer microbial cells genetically, as discussed in Chapter 3, is leading to a rapid extension of this list to include 'foreign' compounds (i.e. compounds not naturally produced by the microbe itself). These include protein products from higher organisms (e.g. insulin and human growth factor), as the DNA coding for their synthesis can be isolated and introduced (engineered) into a bacterial host, which then begins to produce the protein.

Microbial degradation of waste products, particularly those from the agricultural and food industries, is another important industrial application of microbial cells in culture, especially in the degradation of toxic wastes (e.g. pesticides) or in the bioconversion of wastes to useful end products. Biomass production as single cell protein (SCP) by the growth of microorganisms on various substrates is a further important use of microbial cultures. Microorganisms also have value in replacing animals in the preliminary toxicological testing of xenobiotics.

The composition of microbial growth media has been discussed above (Section 1.3.2). Microbial growth can be in either closed or open systems. In closed systems a finite amount of medium is supplied and growth continues until either one factor in the medium becomes limiting or a toxic by-product of growth accumulates. In open systems of so-called continuous culture, outflow from the system of both cells and medium is balanced by the addition of an equivalent volume of fresh medium.

Closed systems, which involve a single portion of growth medium, are called batch cultures. The disadvantage of batch cultures is that they usually give rise to unbalanced growth, in which a relatively heterogeneous population is present at any given point during the growth cycle. If fresh medium is added periodically, thereby increasing the total culture volume and stimulating further growth, they are called fed-batch cultures. The objective of fed-batch culture is to control important nutrients within a narrow concentration range. Open, continuous culture systems enable both nutrients, waste products and biomass to be controlled by varying the dilution rate (i.e. the rate at which culture is removed and replaced). Open systems are referred to as either chemostats, where growth rate is controlled at submaximum levels by the rate of dilution of the medium containing a single limiting nutrient, or biostats, where the cell population is maintained within narrow predetermined limits by monitoring some physiological property of the culture (e.g. concentration of outlet gas or cell population) that then acts as a signal for dilution and wash out of excess cells. Fresh medium is introduced from reservoirs to the vessel by triggering the opening of solenoid valves. The specific advantage of continuous cultures over batch cultures is that they facilitate growth under steady-state conditions in which there is tight coupling between biosynthesis and cell division. Thus all facets of metabolism are proceeding in an equilibrium set by the dilution rate employed (i.e. the supply of the limiting nutrient). It follows that if the rates of biosynthesis and cell division are constant and equal, the mean composition of cells in the culture does not vary with time and growth is said to be balanced. This is extremely useful in many biochemical studies and is in contrast to the situation in batch or fed-batch culture where cellular composition and physiological state varies throughout the growth cycle.

1.5.2   Animal cell culture

The proliferation of animal cells in culture has many applications, including their use as model systems for biochemical, physiological and pharmacological studies, and in the production of growth factors, blood factors, monoclonal antibodies, interferons, enzymes, vaccines and hormones. A brief outline of some of the specific uses of particular animal cell and tissue cultures is given in Table 1.2. The distinction between simple techniques of incubation *in vitro*, such as perfusion and cell and tissue culture, is to some extent arbitrary, because conditions chosen for cell culture may not permit cell division. However, the maintenance of viable cells for periods in excess of 24 h generally leads to them being referred to as cells in culture.

Table 1.2. *Some examples of applications of animal cell and tissue culture*

Cell type	Use
Baby hamster kidney (BHK) cells	Large-scale production of foot and mouth vaccines
WI 38 (ex human embryo lung tissue)	Production of rubella vaccine
Simian kidney cells	Production of polio vaccines
Murine lymphoblastoid cells	Production of milligram quantities of histocompatibility (H2K) and differentiation-specific (Thy-1) antigens
Myeloma cell lines	Production of monoclonal antibodies
T-helper cells	Production of lymphokines
Leukocytes	Production of α–interferon
Diploid fibroblasts	Production of β–interferon
B or T lymphocytes	γ–Interferon
Porcine kidney cells	Production of urokinase (plasminogen activator)

Table 1.3. *Strengths and weaknesses of animal cell and tissue culture*

Strengths
1. Range of cell types in culture is now very large, including genetically defined clones
2. Ready availability of media, cells and culture vessels from commercial suppliers
3. Methods for isolation of primary cells and storage of cell lines is now well documented
4. Viability and physiology of tissue is easily monitored
5. Increased control of environmental variables, e.g. temperature, pH, medium constituents
6. Ease of application of test compound, with improved tissue specificity and easier recovery of metabolites
7. Large-scale replication of experiments may be performed readily in a small amount of space and time
8. Scale of experimentation in model system is from single cell to organ
9. Scale-up offers scope for biotechnology
10. Reduction in numbers of animals killed for experimentation/bioproducts processing, e.g. cytotoxicity testing and antibody production.

Weaknesses
1. Method may be considered to be too specialised and laborious for routine use
2. Resources needed are expensive, e.g. fetal calf serum
3. Lack of specificity *in vivo*

Advances in animal cell and tissue culture applications have been profound in recent years because of the versatility of the techniques and the ready availability of media, cells and culture vessels from specialist suppliers. The major drawback is that such techniques may be expensive to operate and require specialist training. Table 1.3 lists some strengths and weaknesses of the methodology.

Excised tissue is usually heterogeneous and cell types are typically categorised according to their mesenchyme or haemopoietic embryonic origin. Examples of cells in culture include macrophages, T and B lymphocytes and polymorphonuclear leukocytes. Pioneering studies established that varying degrees of growth of tissue was possible *in vitro* in sterile medium containing salts, carbohydrates, vitamins, and amino acids and serum supplements. Animal cells may be subdivided broadly into those that remain viable only when attached to a solid substrate (e.g. primary cultures or diploid fibroblast cells) and those that will proliferate as suspensions (e.g. human tumour cells, HeLa cells and hybridomas). Newly isolated or primary cell cultures of non-tumour cells typically show anchorage dependence by growing *in vitro* only when attached to a surface, which may be either other cells (as *in vivo*) or glass, plastic, gelatin or collagen. Primary cell lines spread over the surface as monolayers but will not overlayer due to contact inhibition of adjacent cell membranes. Such cultures are said to show confluence, which severely limits cell density in culture.

In general terms, highly differentiated excised tissue is more difficult to establish *in vitro* as primary cell lines and frequently will not subculture in medium devoid of growth factors or steroid hormone supplements typically found in serum, e.g. IGF, EGF, PDGF or oestrogen.

Continuous subculture is possible only in transformed cell lines that have become naturally aneuploid or chemically mutated (e.g. with 3,4-benzpyrene, nitrosomethyl urea) or infected with a virus, e.g. simian virus 40 or Rous sarcoma virus. Transformed cells, e.g. baby hamster kidney (BHK) cells typically show reduced contact inhibition, an ability to grow in suspension, changes in shape and loss of differentiation (neoplasia), and a requirement for much less stringently defined culture medium. Interestingly, some transforming viruses may introduce viral oncogenes that code for proteins substituting for growth factors that are normally supplied in serum. Transformed cell lines frequently induce tumour formation when injected into suitable hosts. The close similarity between transformed and tumour cells is often exploited experimentally, with transformed cells being used instead of tumour cells because of their tendency to be cytologically more homogeneous.

Cells may be grown either in batch or in continuous culture. Either way, control of gaseous exchange and pH are critically important, as described in Section 1.3.4. If the vessels are sealed then medium must either be gassed with 5% or 10% CO_2 at inoculation. Vented flasks have to be kept within CO_2 incubators that automatically control temperature and CO_2 concentration.

The growth characteristics of cells and the objectives of cell culture have had profound effects on the design of culture vessels. Small-scale static culture of cells up to, say, $1 \, cm^3$ is conveniently established either in microtitre plate wells or specialised tissue culture plates having wells, of varying cubic capacity, that may be precoated with gelatin. Scintillation vials may also be usefully employed. Plastic Petri dishes and flat-sided bottles of glass or plastic may be used for slightly larger cultures of up to $50 \, cm^3$. Batches of roller bottles that rotate at 1 r.p.m. or less, on proprietary or home-made machines, provide increased surface area for up to 1 dm^3 of growth medium, especially when the inner surface is corrugated. However, cells have to be removed by treatment with trypsin. Specialised, and expensive, cell factories or capillary perfusion beds are also available, which employ either glass bead columns or hollow fibres to support cells, in which medium previously equilibrated with 5% CO_2 in air is circulated.

Suspension culture of lymphocytes or hybridoma cells, for example, is possible in small-scale fermenters of up to 10 dm^3 capacity, in which cells are magnetically stirred gently. Alternatively, small-scale air lift fermenters may be employed. Free cell suspension culture has very limited application, however, and has been surpassed in biotechnological application by the development of microcarriers, which allows for large-scale suspension culture of anchorage-dependent cells. A summary of the advantages of microcarriers may be found in the suggestions for further reading (Section 1.10).

Microcarriers consist of a cross-linked polymer, such as dextran, polyacrylamide, polystyrene or plastic, usually containing either fixed positive or fixed negative charges that facilitate attachment. Trade name examples of positively charged microcarriers are Cytodex and Superbeads; and of negatively charged microcarriers are Rapid Cell G and Biosolon. Non-charged, gelatin-coated beads are also commercially available, e.g. Gele-beads.

A major problem associated with the isolation of free cells and cell aggregates from organs is that of release of the cells from their supporting matrix without affecting the integrity of the cell membrane. The earliest approaches to this problem employed mechanical techniques such as forcing the tissue through cheese cloth or silk or shaking the tissue with glass beads in an appropriate buffer. These procedures inevitably caused considerable cell damage and resulted in a low cell yield, which raised the practical problem of determining whether or not the isolated cells were representative of those originally present in the tissue. The use of a biochemical, rather than a mechanical, approach has largely overcome these problems. In the case of hepatocyte isolation, the use of collagenase and hyaluronidase in a calcium-free medium to digest the matrix has enabled cells with a viability in excess of 95% to be isolated on a routine basis. The liver may be removed, thinly sliced and then incubated with the enzymes, or it can be cannulated *in situ*, perfused with an oxygenated, calcium-free medium containing the enzymes, and then removed and broken up with a blunt spatula. This latter technique requires greater practical expertise but appears to give a greater and more reproducible cell yield and viability. In all cases, viability is generally assessed by the ability of the cells to exclude the dye trypan blue, but other procedures based upon respiration or protein synthesis are probably better.

1.5.3 Plant cell culture

This term refers to the long-term incubation *in vitro* of parts of mosses, liverworts and vascular plants (ferns, gymnosperms and angiosperms) in or on a suitable medium under defined environmental conditions (Section 1.2.2). The technique started as a model system for investigating the potential of plant cells, incubated *in vitro*, to show totipotency, i.e. the ability of isolated cells, removed from the *in vivo* constraints of mutual chemical and physical contact, either to form any type of differentiated cell in the organism or to develop as zygotes, when provided with an appropriate artificial chemical and physical environment.

Cell and tissue experimental methods are now used very widely as the most effective model system to investigate problems across the whole range of the plant sciences. The biotechnological applications of plant cell and tissue culture methods is also increasing rapidly, particularly after scientific breakthroughs in plant molecular biology.

The technique begins with the excision and surface sterilisation of explants, which may be chosen from any part of the plant depending on the objective of the study. Thus leaves are frequently preferred for protoplast isolation (cells without walls), anthers for production of haploids, shoot meristems for proliferating shoot cultures and root tips for root cultures. The explant may be either axenic (totally sterile) or systemically infected, e.g. with a virus. Explants may be excised either in the field or from glasshouse-grown plants or seedlings incubated under aseptic conditions.

Surface sterilisation most frequently employs several minutes of exposure to a 1 to 10% (w/v) sodium hypochlorite solution (proprietary domestic bleaches will do) in which chlorine gas acts as biocide. After the set time, excess hypochlorite must be removed immediately by copious washing in sterile distilled water.

Explants are most likely to survive *in vitro* when the tissue chosen is physiologically active (i.e. non-dormant). Many types of tissue culture require cambial cells capable of meristematic activity (cell division) to be present in the explants. Callus, comprising mainly masses of undifferentiated parenchymatous cells, is often formed soon after excision. Such cells have little cytoplasm but large vacuoles and are metabolically less active than meristematic tissue. Such tissue may, however, develop localised growth centres, called meristemoids, from which caulogenesis (shoot induction) or rhizogenesis (root initiation), or both may ensue. The ability of callus to undergo such organogenesis is genetically controlled but may be encouraged *in vitro* by manipulation of cytokinin to auxin ratios in the medium. High cytokinin to auxin ratios favour shoot proliferation and low cytokinin to auxin ratios promote rooting.

In practical terms whole plantlets can be produced *in vitro* from organogenic callus by encouraging shoot proliferation on a cytokinin-rich medium and then transferring leafy shoots to auxin-rich medium for root induction. Roots may also be developed *in vivo*, provided plants are protected from desiccation. However, the biochemical explanation of this phenomenon, which has been known for nearly 50 years, is still incomplete. Organogenesis from callus is not generally recommended for plantlet multiplication *in vitro* because there is considerable evidence that it results in genetically aberrant plants being recovered.

This can be avoided by using shoot meristems as initial explants for reasons that are, as yet, ill-defined. Meristem cultures are consequently the preferred starting material for plantlet multiplication *in vitro* (micropropagation). Meristem culture can also be used to cure plants, infected with viruses, by the use of heat (thermotherapy) and chemicals (chemotherapy).

Callus may be used as the starting material for suspension culture in which a mixture of single cells and cell aggregates are incubated in liquid medium that has to be artificially aerated to prevent the cells from becoming water logged. The ease of suspension formation is largely genetically controlled but friability can be improved empirically in certain cases by reducing the calcium levels in the medium, by altering the gas mix, by changing the plant growth regulator regime, and/or by adsorbing phenols with PVP.

Plant cell suspensions may be grown as batch, fed-batch and continuous culture on a small, medium and large scale. For example, single individual cells may be grown in indented microscope slide chambers, where the clonal origin of plants is a prerequisite for the study. The use of wide-necked Erlenmeyer flasks, shaken in horizontal platform orbital shakers, is a favourite method for small-scale batch cultures. Fermenters from 1 to $50\,000$ dm^3 capacity have been designed in which plant cells are aerated mainly by sparging with air and not by agitating with metallic impeller blades, because of the low shear strength of the cell walls.

Plant cell suspensions provide excellent model systems for studies of cell division, cell expansion and differentiation, and intermediary metabolism in general, because of the ease of adding test compounds and harvesting cells and medium for analysis. The physiological properties of plant cell suspensions render these cultures more difficult to exploit than microbial cells, however, because they are much larger (greater than 10^5 μm^3), they contain more than 90% (v/v) water, the cell-doubling time is slow (one to two days usually) and normally they do not excrete metabolites into the medium but compartmentalise them within vacuoles. Synthesis of secondary products in plant cell suspensions is also linked mainly to growth. This, together with the higher cost of plant culture medium, long fermentation time (usually weeks), higher downstream processing costs and ploidy instability on prolonged subculture, significantly detracts from the usefulness of such suspensions as an industrial source of economically important drugs, food additives, perfumes, biocides, etc.

Despite these difficulties, much research is being conducted using plant cell fermenters for secondary product synthesis because such compounds are either not possible or impractical to synthesise chemically and involve complex biochemical pathways which are difficult to engineer genetically in microorganisms. Cell lines yielding higher levels of secondary product *in vitro* are obtained initially by pouring low density cell suspensions on to Petri dishes (Bergmann plating) and then screening the colonies for elevated secondary product synthesis using RIA or

ELISA techniques (Sections 2.5 and 2.6). Clones with superior yields are produced in a growth medium and are usually transferred to a different medium to stimulate secondary product synthesis. The fact that many desirable compounds are currently derived from plants found only in environmentally sensitive areas such as tropical rain forests is hastening the desire of manufacturers to use cell culture methods for synthesis.

Higher yields of secondary metabolites after cell immobilisation in plant cell cultures has been claimed (Section 4.12), although so far the method has rather limited application in higher plant systems. Biotransformation using either free or immobilised cells, in which a low value precursor is converted into higher value-added product, has also been shown to be feasible using plant cell culture systems.

Plant cell suspensions are also useful in selection experiments where cell colonies that may have resulted from mutagenic treatment are often grown on agar medium containing the selecting agent (Section 1.6.3). Ideally, suspensions comprising only single haploid cells should be used for such Bergmann plating. Single cell suspensions may be obtained in this case by sequential filtration through a series of filters graded down to 50 μm. Regenerated cells must transmit the selected trait via seed production. Certain species, e.g. carrot or oil palm, form somatic embryos in suspension, as expressions of totipotency, from which artificial 'seeds' may be produced by coating the propagules with protective resins. This offers potential for the direct liquid drilling of artificial seeds into the soil.

Single cell suspensions may also be derived from pectinase digestion of leaves or other starting material. Such pectinases are supplied under various trade names such as Macerozyme R10, Pectolyase Y23 and Rhozyme HP150 and are usually used as solutions in the concentration range 0.03% to 0.2% (w/v). Such pectinases are usually used, however, in conjunction with cellulases to prepare protoplasts. Cellulases are supplied under trade names such as Cellulase, Cellulysin, Driselase, etc. and are used normally within the 1% to 2% (w/v) concentration range.

Medium for preparing protoplasts has to be appropriately buffered and of sufficient water potential to prevent protoplasts bursting; mannitol or sorbitol up to 15% (w/v) is often used for this. Protoplasts are best prepared from thin strips of leaf by static digestion overnight in small Petri dishes in the dark at 25 °C. Protoplasts released into the incubation medium are then purified following initial filtration to clear all debris by low speed density gradient centrifugation (Section 6.3). Protoplasts, concentrated at interfaces, are removed by drawing them off with Pasteur pipettes. Strict attention to maintaining aseptic conditions is necessary if protoplasts are to be used for plating experiments – in which case antibiotics are often incorporated into the culture medium.

Protoplasts may be incubated successfully in a variety of ways, but, prior to incubation, protoplasts must be washed and counted to find an effective inoculum density for growth. Cell counting may be performed in a haemocytometer or in a particle counter, e.g. the Coulter counter (Section 1.5.5). Protoplast viability must also be assessed either using fluorescein diacetate (FDA) and ultraviolet microscopy or Evans Blue stain using visible light microscopy. FDA staining works on the principle that the dye, which is non-fluorescent, readily crosses membranes to be converted into fluorescein by endogenous esterases. Evans Blue on the other hand is excluded by living membranes. Protoplasts may be plated on agar (a 1.2% (w/v) medium) or on liquid medium overlayering agar medium, on 4.5% (w/v) agarose medium or in a hanging drop on the lid or base of a Petri dish. The inoculum density is usually about 10^4 cells cm^{-3}. Protoplast cultures are usually static. The temperature range is usually 25–30 °C, and low intensity continuous illumination may be provided up to 50 µmol m^{-2} s^{-1}. Protoplasts will not start to divide until new walls have formed, which takes about 24 h. Growth of colonies will ensue over a two-week interval and to enhance growth rate the colonies are transferred sequentially to batches of fresh medium having decreasing osmotic potential. If plantlets are required, regenerative medium has to be used, incorporating a balance of growth regulators and carbon source for the species in question.

Protoplasts have very widespread application in plant sciences, including the uptake of virus particles and organelles and cell wall regeneration studies. Probably their greatest use, however, is in somatic hybridisation and cell transformation.

Successful hybridisation depends on developing effective techniques for fusion of protoplasts, for selection of fusion hybrids, for regeneration of plantlets and for analysis of regenerated plants. Fusion methods are either chemical or physical (electrofusion). Chemical methods use incubation solutions incorporating either high concentrations, about 30% (w/v), of polyethylene glycol (PEG) of relative molecular mass 6000 or high pH calcium solutions (pH 10.4 in 1.1% (w/v) $CaCl_2.6H_2O$ in 10% (w/v) mannitol). Fusion is generally achieved in fewer than 30 min. PEG is the preferred compound when two types of mesophyll protoplast are fused, because protoplasts tend to burst at high pH. Fusion medium must be removed before protoplasts are transferred to the incubation medium.

Electrofusion is an alternative method for protoplast fusion that involves two steps. In the first, protoplasts are placed in a medium of low conductivity between two electrodes (platinum wires arranged in parallel on a microscope slide). A high frequency alternating field (0.5 to 1.5 MHz) is applied between the electrodes, which causes the protoplasts to align in a process known as dielectrophoresis. In the second step, one or more short (10–200 µs), direct current pulses (of 1–3 kV cm^{-1}) are applied, which causes pores to form in the membranes of the protoplasts and allows fusion to take place where there is close membrane contact. This technique allows a higher degree of control over the fusion process than do chemical methods and consequently is becoming more widely used.

Fusion is a random process and the result of a fusion treatment can be a mixture of unfused homokaryon parental protoplasts, fused homokaryons and fused heterokaryons. Following fusion, heterokaryons containing nuclei of both species in a common (mixed) cytoplasm regenerate a new wall. Nuclear fusion, if achieved, will result in a somatic hybrid cell that must be totipotent if hybrid whole plants are to be produced. Various techniques have been devised for identification and recovery of somatic hybrids, that are frequently based on complementation of biochemical mutants (Section 1.6). Morphologically distinct or fluorescently labelled protoplasts can also be identified under an inverted microscope and removed by micromanipulation. Fluorescence-activated cell sorting methods may also be used for separating different protoplast populations (Section 1.5.4).

Protoplasts are also often used for genetic transformation using either naked DNA or a cloning vector such as the Ti plasmid of *Agrobacterium tumefaciens* (Section 3.8). With the former, electroporation is a favoured method, in which protoplasts are incubated with cloned DNA in specially designed electrode cuvettes. A direct current of short duration is applied from a power supply that causes pore formation in the membranes of protoplasts by a process called dielectric breakdown. This allows DNA to penetrate the interior of the cell and integrate with nuclear DNA. Regeneration of such electroporated plants yields transgenic forms.

Where transformation is mediated by isolated Ti plasmids, sterile protoplast suspensions are suspended at 25 °C in a buffered plasmid preparation (concentration 10 μg cm^{-3} in the presence of 2 μg poly-L-ornithinc cm^{-3}. After incubation, washed protoplasts are plated, at a suitable inoculum density, on to an appropriate selection medium. The selection medium can take a variety of forms depending upon the extent of prior manipulation of vector DNA but is usually based on antibiotic resistance introduced with the vector, e.g. resistance to kanamycin or hygeromycin.

Haploid plant tissue and cell cultures are also important in the context of providing a source of mutants for comparative biochemical studies (Section 1.6), for protoplast fusion experiments and for improvement of agronomic traits principally by micropropagation of resistance mutants (e.g. salinity, herbicides, desiccation). Such haploid material is derived by diverting an immature gameto-phyte (pollen grain) away from its normal developmental pathway. Haploid tissue can be produced *in vitro* either by anther culture or microspore (pollen) culture. In anther culture, immature anthers are removed from donor plants, grown under very precise environmental conditions to ensure that the stage of pollen development is best for recovery of haploids. This generally corresponds either to the period immediately following meiosis or to the first pollen grain mitosis, according to the species. Aseptically removed anthers from different genera have separate but exacting nutritional requirements. Liquid overlays of agar medium are preferable for enhanced viability. Isolated microspores are much more difficult to grow *in vitro* because their nutritional requirements are so exacting. If regeneration of haploid plantlets can be achieved, homozygous diploids may be induced by treating the roots with dilute solutions of colchicine.

The range of applications of plant tissue and cell culture is truly enormous. Table 1.4 gives very limited examples and the reader is referred to Adams (1988) (see Section 1.10) for a more detailed coverage of this important experimental system.

Table 1.4. *Some examples of the applications of plant cell and tissue culture*

System	Process investigated
Protoplasts	
Nicotiana tabacum	Cell wall regeneration; clonal propagation of mutant plants; interspecific hybrids; uptake of *Rhizobium* homologous and heterologous gene expression
Solanum tuberosum	Somaclonal variation in cultivar Russet Burbank; genetic transformation using Ti plasmid; transformation by co-cultivation of protoplasts with isolated *Agrobacterium tumefaciens* cells; direct protoplast transformation
Arabidopsis thaliana	Intergenetic hybridisation, model for molecular biological investigations
Glycine max	Isolation of viable bacteroids from root nodule protoplasts, molecular model for investigating nitrogen fixation
Cell suspensions	
Acer pseudoplatanus	Control of cell division and expansion in single cells, batch and continuous cultures, and monitoring the associated biochemical changes
Daucus carota	Demonstration of somatic embryogenesis in higher plants
Nicotiana tabacum	Industrial fermentation of biomass for tobacco industry. Ubiquinone production. Selection, regeneration and sexual transmission in regenerants resistant to the herbicide picloram
Digitalis purpurea	Biotransformation of digitoxin to digoxin
Catharanthus roseus	Ajmalacine and serpentine production
Callus	
Phaseolus vulgaris	Physiology and biochemistry of differentiation
Vicea faba	Cytogenetic studies indicating chromosome instability
Elaeis guineesis	Cloning of oil palm plantlets and crop improvement
Solanum tuberosum	Somaclonal variation production
Anther and microspores	
Hordeum vulgare	Embryogenesis from cultured anthers
Brassica napus	Embryogenesis from cultured microspores

Table 1.4 contd

System	Process investigated
Embryos	
Papaver somniferum	Plant breeding; direct pollination of excised ovules to overcome prezygotic barriers to fertility
Hordeum vulgare	Production of mutants resistant to lysine analogues
Meristem	
Solanum tuberosum	Elimination of potato virus X by thermotherapy followed by plantlet regeneration
Chrysanthemum morifolium	Breeding of radiation-induced mutations
Leaf discs	
Brassica napus	Transformation, using *Agrobacterium tumefaciens*, with Glyphosate resistance genes
Lycopersicum esculentum	Transformation, using *Agrobacterium tumefaciens*, with antisense genes to inhibit fruit ripening. Organogenesis is thought to arise directly from single cells within the explant

1.5.4   Cell sorting

The isolation in bulk of individual cells or populations of particular cells from a mixed population is often essential to the cell biologist. Cells may often be separated in bulk on the basis of biological survival in an adverse environment, exploiting the capacity of a particular cell for division. Survival, and therefore selection, may be based on many factors including resistance to either an infection or a cytotoxin, due to the absence of receptors or to prior infection that has led to either an immune response or cell transformation.

Bulk separation may also be achieved by exploiting the different sedimentation characteristics of cells by differential centrifugation, density gradient centrifugation, isopycnic centrifugation and continuous flow centrifugation (Section 6.6). The different surface charge properties of cells facilitate their electrophoretic separation particularly by continuous flow electrophoresis and isoelectric focusing (Section 9.3). Other surface properties of cells, as described by Ormerod (1990), may be important in their bulk separation by affinity chromatography (Section 10.9), by agglutination involving attachment of polyvalent ligands (immunoglobulins and lectins, Section 10.9) and by phase partitioning (Section 10.6). Phase partitioning is also known as cell surface chromatography and frequently employs phase separation of cells into PEG or dextran on a multiple countercurrent distribution basis, separation being dependent on the nature of exposed polar groups of phospholipids and membrane carbohydrates of animal cells. A final method of bulk separation involves the simple approach of sequential filtration through meshes of differing pore size to separate cells on the basis of cells size, shape and deformability.

Individual cells may be isolated by a selection procedure similar to those mentioned above, particularly when resistance is based on a rare mutation. Microscopy and micromanipulation may be employed, as in somatic cell hybridisation studies.

More sophisticated cell sorting methods are now used routinely in many laboratories, using principles of flow cytometry to separate out cell subpopulations from a mixed population for function assays (e.g. selection and cloning of hybridomas, or transfected cells, or of lymphocytes). Such cell populations are morphologically similar but functionally distinct.

Cell sorting by cytometry relies on surface tension in fluids causing separate droplets to form at a distinct break-off point in a stream of liquid emerging from a nozzle. The position of the break-off point is controlled accurately by the diameter of the orifice, the stream velocity, and the frequency and amplitude of vibration applied to the nozzle. The general principle of cell sorting is illustrated in Fig. 1.3. Satellite droplets, visible under stroboscopic illumination, form in the fluid at the bottom of the main droplets and merge quickly. Droplets breaking away may be either uncharged or may carry a positive or negative charge when an appropriate voltage is applied to the stream. As the droplets pass through an electrostatic field formed between high voltage deflector plates, they separate according to charge. Droplets containing only the desired cells can be deflected by applying the voltage as a precisely timed pulse that coincides accurately with the characteristic delay between the particular cell-type passing through the laser beam and its appearance in the droplet. The degree of deflection of the stream can be controlled, in part, by controlling the voltage applied to the plates. Complex logic control circuitry is needed to make the charge pulse and the arrival of the cell at the break-off point exact. The purity of sorted cells can be checked by reanalysing a sample of sorted cells. Relatively pure (95% to 98%) subpopulations are possible using this technique.

Fluorescence-activated cell sorting (FACS) is a more sophisticated flow sorting technique that measures the fluorescence emitted by individually labelled cells in a mixed cell population in a cytofluorimeter (also termed a cytofluorograph) (Section 7.8). FACS separates cells that contain preselected fluorescence characteristics by enclosing them inside droplets that can be deflected into reservoirs under the influence of an applied electrostatic field according to the polarity and quantity of charge. Unselected cells, held in uncharged droplets, fall into another reservoir. Fluorescence may be generated from dyes specific to cell components (e.g.

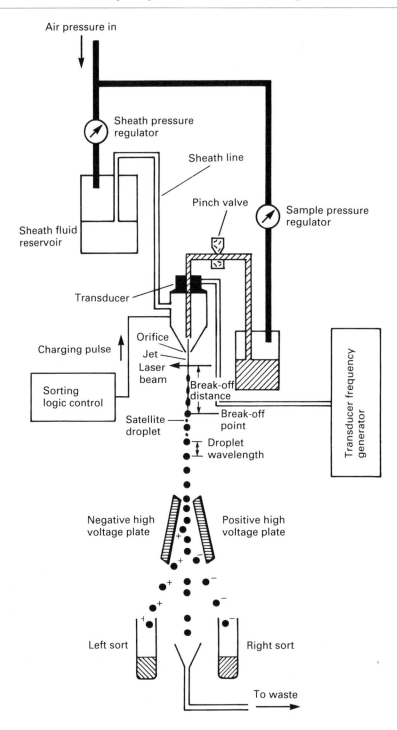

Fig. 1.3. A flow-sorting system. (Modified, with permission, from Carter, N.P. and Meyer, E.W. (1990) in *Flow Cytometry – A Practical Approach*. Ed. M.G. Ormerod. IRC Press, Oxford.)

propidium iodide for DNA) or from antibodies to surface antigens. The instrument uses an argon-laser beam to excite fluors of fluorescently-labelled cells being passed in single file through a laminar flow chamber, as above. The data from cytofluorimeters is generally displayed on a visual display unit (VDU) as histograms of fluorescence intensity, as a function of the number of cells with that intensity, in order to depict subpopulations of cells. If the fluorescence intensity, size and viability of a given cell meet the criteria preselected by the FACS operator, a droplet containing the cell is charged and recovered. The procedure is very sensitive and cytofluorimeters capable of measuring a single virus particle and less than 100 fluorophores per particle have been developed. Automation allows up to 5000 cells to be sorted per second into viable subpopulations.

A new approach to adherence methods of cell separation, known as magnetophoretic separation, involves the use of magnetic microspheres bearing conjugated ligands. So far the microspheres used have consisted of magnetite (Fe_3O_4) cores surrounded by polyacrylamide and/or agarose layers to which fluorescence labels and protein are chemically attached. Cells are allowed to mix with these microspheres in suspension and the mixture is then passed through a magnetic field. Those cells attached to the ferromagnetic beads are attracted to the poles of the magnet and washed free from the remaining cells. Application of this technique to neuroblastoma cells with expression of surface gangliosides has allowed a purification greater than 98%, with viability of cells remaining unaffected.

1.5.5 Cell counting

Biochemical measurements are often related either to growth cycle phases or to the developmental state of the organism concerned. Results may be expressed in relation to cell number, fresh or dry weight, cell volume or per unit of nitrogen. Where cell numbers are used, these must be determined by a suitable counting method. Non-motile bacteria are conveniently counted by observing colony formation of cell clones on the surface of a solid medium after inoculations of serial dilutions of the original sample.

Before cell counting can be undertaken in suspension, multicellular aggregates must be digested gently either enzymically or chemically, e.g. with a dilute solution of chromium trioxide. A known volume of cell suspension may then be introduced to a counting chamber and cell numbers determined microscopically. The haemocytometer is the most commonly used counting chamber for blood cells, microorganisms and dispersed animal and plant cells. Some counting chambers for algae, plankton and surface-aggregated animal cells in culture bottles use inverted microscopes in which the light source shines down through the counting chamber and the microscope optics are arranged to view the sample from underneath the chamber.

Counting of cells under the microscope is slow, tedious and may be inaccurate, especially at low cell densities. Electronic particle counters resolve these difficulties; many are fully automated and include facilities for estimating mean cell sizes. The Coulter particle counter operates on the following principles. A current flows between two electrodes immersed in a liquid electrolyte. One of the electrodes is inside a glass counting tube and the other is outside the tube, but immersed in the sample that is suspended in the same electrolyte. A small orifice in the counting tube ensures the flow of current. A known volume of sample, usually $0.5\,cm^3$, is drawn automatically into the tube via the orifice and any particles present increase the resistance between the two electrodes as they pass through the orifice. This results in a pulse appearing on an oscilloscope screen and a single count recording on a digital scaler. The size of each pulse is directly proportional to the size of the particle. The pulses are very small and amplification is necessary to record accurately the pulse produced. The degree of amplification depends on the size of the particles being counted and must be determined for each kind of sample. Similarly, the size of the orifice tube must be chosen with care. An orifice of 100 μm diameter is appropriate for blood cells and most bacteria. Pulse height analysis resolves cell numbers and biovolume for subpopulations. Mean cell size is determined by taking cell counts at increasing settings of the pulse height analyser and calculating the mean threshold setting when half the total population has been counted. Various standard-sized particles such as pollen grains or, more usually, latex particles are used to calibrate the instrument so that a particular mean pulse height analyser setting can be related to a particular particle volume. From a knowledge of these data the mean cell size can be determined. Multiplying this by the total cell count gives the total cell biovolume. Errors in sizing may be introduced by osmotic effects on cells caused by the suspending electrolyte, whilst errors in counting may be introduced by cell settling, by debris in suspension and by partial blocking of the orifice. The technique is very suitable for relatively simple populations but not for natural populations of, for example, phytoplankton, where the range of sizes of organisms precludes effective choice of orifice.

1.5.6   Cryopreservation

Repeated subculture of cell lines may result in genetic aberration and hence loss of biosynthetic capacity or morphogenetic potential. Various strategies are possible to alleviate these difficulties, including methods either for slowing down growth in stock cultures or for freeze-drying certain microbial cultures. Cryopreservation in liquid nitrogen is the preferred method for most eukaryotes and is based on the principle that the only reactions that occur at $-196\,°C$, the boiling point of liquid nitrogen, are ionisations due to background irradiation, which are very rare. Biochemical reactions cease at about $-130\,°C$. Various experimental protocols have been devised, often empirically, for different tissue types.

Successful cryopreservation requires that viability is maintained after the sample has been held for indefinite periods in liquid nitrogen. The rate of freezing is very important, as too rapid freezing causes large crystals of ice to form intracellularly. Such crystals, on thawing, lyse cell membranes. Slow freezing rates favour the deposition of extracellular ice at subzero temperatures, provided the internal concentration of solutes is sufficiently high to make the internal water potential lower than that externally. This process may be assisted by adding a penetrating cryoprotective agent such as glycerol to the medium supporting the specimen. When ice has formed extracellularly, water is withdrawn from the cells because the vapour pressure density of the external ice is lower than that of the predominantly aqueous cytoplasm. Water moves out until the vapour pressure densities are equilibrated. During this process, the protoplast shrinks and plant and microbial cells may become temporarily plasmolysed. Alternatively, non-penetrative cryo-protective agents such as PEG and PVP may be used, to ensure that cell dehydration proceeds smoothly during slow freezing. If the dehydration is too rapid, dena-turation of proteins may occur because of localised high salt deposits (solution effects), which may cause protein precipitation. At a critical temperature cor-responding to the eutectic point for the cytoplasm, the cell contents freeze, without ice crystals forming.

Slow cooling rates are best achieved with samples held in ampoules within commercial units, which cool at rates of 0.1 to 10 deg.C min^{-1}. Such units consist of containers enclosing a low melting point liquid alcohol, cooled by a refrigeration coil. Stepwise cooling, which may be equally effective, involves holding the specimen for prescribed periods at a given temperature in improvised units. When the temperature reaches either -50 to $-70\,°C$ for the improvised units, or $-100\,°C$ for controlled freezing, the ampoules are transferred directly to liquid nitrogen.

Thawing of samples is best achieved rapidly by removing them to a water bath held at 30 to $40\,°C$ for a few minutes, whilst applying gentle agitation. Washing to remove the cryoprotectant should not be necessary as cryoprotectants should be non-toxic. Damage to tissue during an experimental protocol may be estimated by employing light or electron microscopic techniques (Section 1.7). Viability may be assessed by monitoring respiration, by cytological staining (e.g. exclusion of Evans Blue stain from viable cells (Section 1.5.3)), or by regrowth in appropriate media.

1.5.7   Culture collections

The increasing importance of culture techniques for experimental purposes relies heavily on reference collections of plant, animal and microbial cell lines being readily available, so as to enable coincidental comparative studies to be made in different laboratories. Special characteristics of particular cell lines may include biochemical markers, karyotype analyses, nutritional mutants, drug resistance or sensitivity, tumour production, cellular inclusions (e.g. microbial contaminants, phage or other episomes) or secretory products (e.g. hormones or immunoglobulins). An example of a culture collection is the National Collection of Type Cultures, Central Public Health Laboratory, Colindale Avenue, London. Additional sources of cultures may be found in literature given in the bibliography.

1.6   Mutants in biochemistry

1.6.1   Introduction

Much of the biochemistry associated with molecular biology is concerned with defining the structural and metabolic consequences of mutations, i.e. altering the base sequences of DNA. Our knowledge of the principles of linkage and genetic mapping of the genetic code and the control of transcription and translation in prokaryotic and eukaryotic organisms is based largely on the exploitation of mutant organisms. The especial value of mutations, if expressed, is that they form the basis of a comparative study with the normal (wild-type) cells.

1.6.2   Mutant classification

In chemical terms, there are several classes of mutation, which may either occur naturally or be induced by treating cells with physical or chemical mutagens, e.g. X-rays, ultraviolet radiation, hydroxylamine, alkylating agents such as ethyl-methanesulphonate, or acridine dyes such as proflavin. The use of restriction endonucleases and genetic recombination *in vitro* (Section 3.7) facilitates directed mutagenesis, the intentional alteration of a gene at a specific location for a specific purpose. A point mutation is said to occur when a single base in DNA is changed and a double mutation when there are two such changes. Deletions and deletion substitutions refer either to removal or to exchange base-pairing in larger parts of a gene.

   Microorganisms have been used traditionally for studies in biochemical genetics because they are predominantly haploid, have short generation times, can be grown in or on defined media (making mutant isolation easier), and can be propagated as large biomasses of biochemically uniform cell populations in industrial-scale fermenters. The genetic transfer systems of microorganisms are also simpler than those of eukaryotes and are now relatively well defined both in terms of genetic mapping and in complementation tests. Largely as a result of microbial studies, mutations may be further classified according to the conditions under which the mutation is expressed. So-called non-conditional mutants display the mutant phenotype under all (natural) conditions. Examples of non-conditional mutants include auxotrophy, where mutation results in an enzyme deficiency or inactivity that may be rectified by supplements of the product of those defective enzymes, provided they can be transported into the cell. Prototrophs are non-conditional mutants in which auxotrophs gain a biosynthetic function leading to autonomy from a nutritional supplement. Regaining a wild-type phenotype in this way is called reversion and is usually associated with point mutations. A resistance mutation confers increased resistance to an antimetabolite or environmental stress. Antimicrobial (drug) resistance in bacteria, yeast or plasmids is an example of resistance mutation often of single gene origin.

A conditional mutant is one that does not always display the mutant phenotype. Suppressor-sensitive mutations (*sus*) do *not* display the mutant characteristic in the presence of the gene product from a wild-type suppressor gene. Hence the mutation expression is being suppressed. For example, a bacteriophage carrying a *sus* mutation will produce progeny when it infects a host cell expressing the suppressor gene (su^+ genotype) (the so-called permissive condition) but will not do so when the host cell lacks the suppressor gene (su^- genotype), i.e. so-called non-permissive conditions. Another good example of a conditional mutant is the temperature-sensitive mutant in which a physiologically active protein is rapidly inactivated above or below a critical temperature. The particular value of conditional mutants for biochemical studies is that they are conditionally or potentially lethal and enable investigations to be made on enzyme mutations that cannot be corrected for nutritional supplements. For example, mutations in DNA and RNA polymerases or amino acyl tRNA synthetases cannot be corrected for by simple nutritional supplements to the medium, as these enzymes are involved in the metabolism of complex macromolecules.

1.6.3   Mutant selection

The application of mutants to genetic, biochemical or physiological studies presupposes that such mutants can be selected from a mixed population. *In vitro* techniques are widely exploited for such selection procedures. Selection of biochemical auxotrophs is usually based on incorporating effective concentrations of antibiotics (e.g. penicillin or streptomycin) into minimal media of bacterial suspensions to kill dividing, wild-type cells. Auxotrophs present will be incapable of growth but remain viable and may be recovered after centrifugation and washing to remove excess penicillin. Surviving cells can then be incubated on a supplemented medium. The nature of the supplement to minimal medium will biochemically characterise the auxotroph. Replica plating from complete medium to minimal medium using a sterile velvet pad is another means of detecting auxotrophic colonies on subsequent transfer to complete medium.

A convenient way of isolating prokaryotic auxotrophs is by initial exposure of the microorganism to an effective concentration of bromodeoxyuridine (BrUdr). Prototrophs then take up the analogue and are killed when the BrUdr is degraded by exposure to light. Auxotrophs that fail to take up the analogue survive but will grow only on a medium enriched with nutrients. Specific auxotrophs can be picked up by progressively simplifying the nutrient supplements.

In eukaryotes where there is still no reliable replica plating technique, selection for auxotrophism is based currently either on laborious total isolation, used originally for *Neurospora*, in which supplements are added on a trial and error basis, or on the use of alternative scavenging pathways. In the latter case, enzyme deletion means that, when antimetabolite analogues are added to minimal medium to kill wild-type dividing cells, auxotrophs do not incorporate the compounds and survive when alternative pathways for synthesis are available. For this to occur, either chromosomal abberation of the allele must have created a stage of functional hemizygosity or the gene in question must lie on the X chromosome. Excellent examples of this principle are:

(i) the isolation of thymidine kinase (TK^-) mutants that survive because cells fail to incorporate 5-BrUdr but can synthesise dTMP from an alternative folic acid pathway;

(ii) $HRPT^-$ (hypoxanthine phosphoribosyl transferase) mutants that fail to incorporate 8-thioguanine into inosine monophosphate (IMP) but survive because IMP is again synthesised from the secondary scavenging folic acid pathway. In this pathway selective amplification of the key enzyme dehydrofolate reductase gene can be achieved by selecting for resistance to aminopterin and amethopterin (methotrexate) that are analogues of folic acid. TK^- and $HRPT^-$ strains are of critical importance in the HAT selection technique for the production of monoclonal antibodies (Section 2.2).

Eukaryotic cell lines are also critically important as starting material for selection of somatic cell hybrids following protoplast fusion experiments. A good example of this approach is the selection of a somatic hybrid from leaf protoplasts of *Nicotiana glauca* and *Nicotiana langsdorfii*. A sexual cross of these species produces a teratoma that overproduces auxins and cytokinins, which are necessary to support growth of cell lines of *N. glauca* and *N. langsdorfii*. Only the fusion hybrid colonies will grow in medium devoid of auxin and cytokinin. Scion grafts of the somatic hybrid on to either *N. glauca* or *N. langsdorfii* as stock causes a teratoma to develop.

1.7 Microscopy

1.7.1 Introduction

Biochemical analysis is frequently accompanied by light and electron microscopic examination of tissue, cell or organelle preparations to evaluate the integrity of samples and to correlate structure with function. Microscopy serves two independent functions of enlargement (magnification) and improved resolution (the rendering of two objects as separate entities). Light microscopes employ optical lenses to focus sequentially the image of objects, whereas electron microscopes use electromagnetic lenses. Light and electron microscopes may work either in a transmission or scanning mode depending on whether the light or electron beam passes through the specimen and is diffracted or whether it is deflected by the specimen surface. Polarised-light microscopes detect optically active substances in cells, e.g. particles of silica or asbestos in lung tissue or starch granules in amyloplasts. Light microscopes in phase contrast mode are often used to improve image contrast of unstained material, e.g. to test cell or organelle preparations for lysis. Changes in the phase of emergent light may be caused either by diffraction or by changes in the refractive index of material within the specimen, or even by differences in thickness of the specimen. At their point of focus the converging light rays show interference, resulting in either increases or decreases in the amplitude of the resultant wave (constructive or destructive interference, respectively), which the eye detects as differences in brightness.

Microscopes using visible light will magnify approximately 1500 times and have a resolution limit of about 0.2 μm whereas a transmission electron microscope is capable of magnifying approximately 200 000 times and has a resolution limit for biological specimens of about 1 nm. The excellent resolving power of transmission electron microscopy (TEM) is largely a function of the very short wavelength of electrons accelerated under the influence of an applied electric field. (An accelerating voltage of 100 kV produces a wavelength of 4×10^{-3} nm.) Scanning electron microscopy (SEM) uses a fine beam of electrons to scan back and forth across the metal-coated specimen surface. The secondary electrons that are generated from this surface are collected by a scintillation crystal that converts each electron impact into a flash of light. Each light flash inside the crystal is amplified by a photomultiplier and used to build up an image on a fluorescent screen. The principal application of SEM is the study of surfaces such as those of cells. The resolution limit of a scanning electron microscope is about 6 nm. Figure 1.4 illustrates the application of TEM and SEM to biological investigations.

1.7.2   Ion probe analysis

Electron microscopes may be equipped to perform X-ray spectrochemical analysis. When a specimen is irradiated by an electron beam, an electron may be displaced from an inner to outer orbital. When this happens the vacated orbital may be infilled by an electron from a higher energy orbital with the emission of an X-ray photon characteristic of the difference in energy levels of the two orbitals. The binding energy of the electrons in an orbital is related to the charge on the nucleus; hence each element produces its own characteristic emission spectrum (Section 7.3). Emitted photons are usually detected by energy-dispersive analysis, using lithium drifted solid-state detectors. Each photon emitted reacts with silicon atoms to produce an electrical pulse proportional to the energy of the photon. An electronic pulse height analyser and microcomputer are used to process the spectral data, which is normally displayed on a video monitor. X-ray spectrochemical analysis permits the measurement of ion distributions *in situ* in SEM and TEM by combining the high spatial resolution of electron microscopy with the ability to determine subcellular elemental composition. Areas as small as 100 nm^2 can be analysed under optimal conditions and detection of any element with an atomic number at or greater than 11 is possible. Measurement is made in terms of total concentration of the element rather than free ion activity and in this sense the method is comparable to flame spectrophotometry (Section 7.12), although with a higher sensitivity.

Another highly sensitive method for the analysis of elements in biological materials uses a high energy proton beam to excite characteristic X-rays. This technique is known as proton-induced X-ray emission (PIXE) and has a spatial resolution of 1 μm with normally an analytical sensitivity of 10 p.p.m., although, under optimal conditions, a sensitivity of 1 part in 10^7 can be achieved for many elements. The principle of the technique closely resembles that of X-ray spectrochemical analysis, although, in PIXE, X-rays are produced by collisions between protons rather than between electrons and the target atoms. The technology associated with electron focusing cannot be applied directly to the PIXE method as the lenses are too weak to cope with high energy protons. Consequently, strong focusing magnetic quadruple lenses have been developed. An advantage of PIXE is that high energy protons penetrate much further than do electrons, being less easily deflected by atomic collisions. Thus, the resolution of the proton beam is maintained through thicker specimens.

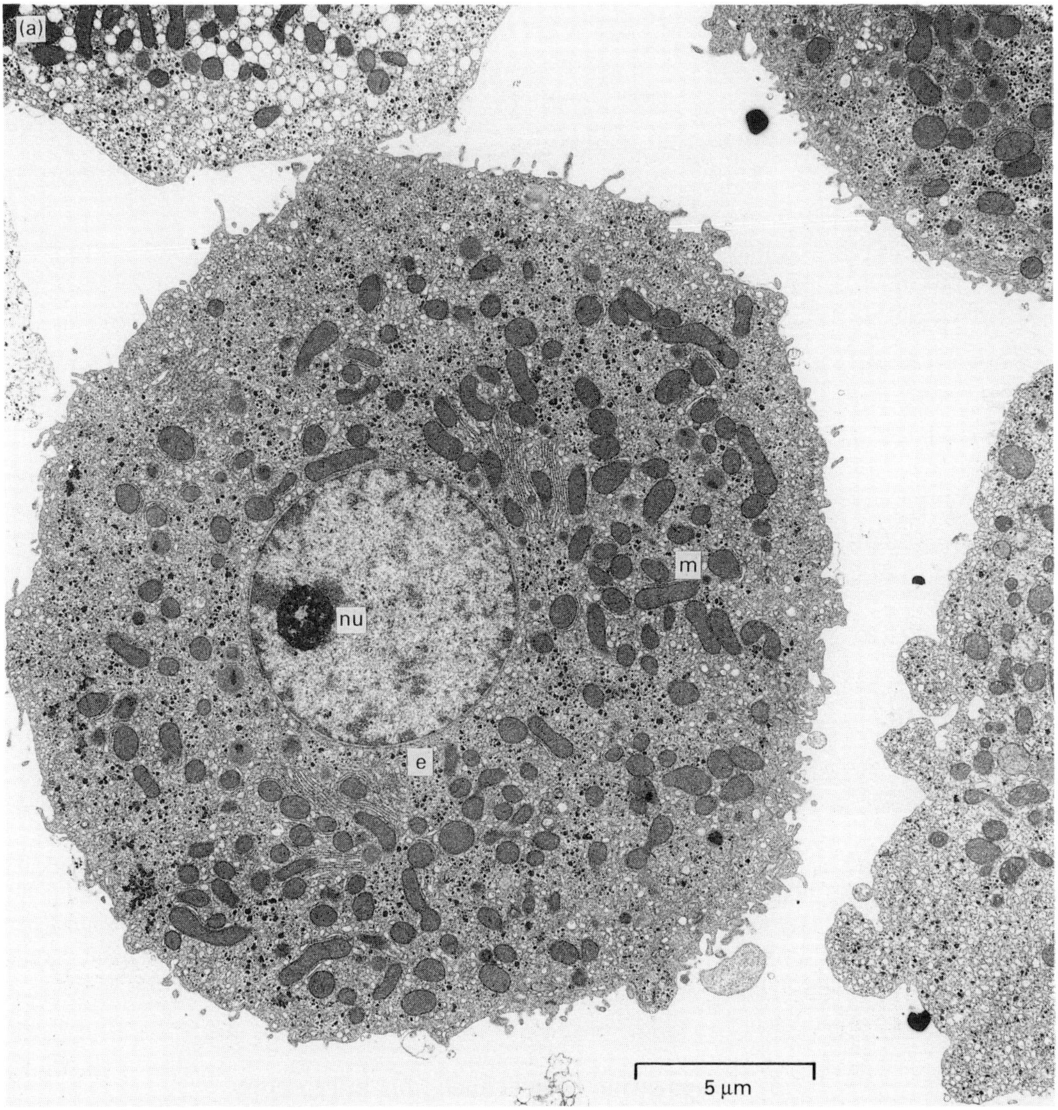

Fig. 1.4.(a) Transmission electron micrograph of a transverse section of an isolated rat hepatocyte, demonstrating intact cytoplasmic and nuclear membranes with well-defined microvilli around the cytoplasmic membrane. Cell organelles are intact, as are the nucleus (nu), mitochondria (m), smooth and rough endoplasmic reticulum (e). Cells were fixed in phosphate-buffered (0.067 M, pH 7.4) 2% (w/v) glutaraldehyde. Selected specimens were post fixed in 1% (w/v) osmium tetroxide in 0.12 M phosphate buffer (pH 7.4) and processed into epoxy resin. Ultrathin sections (60–90 nm) were stained with uranyl acetate and lead acetate. Photomicrographs were exposed during the examination of specimens on photographic film, which was then processed.

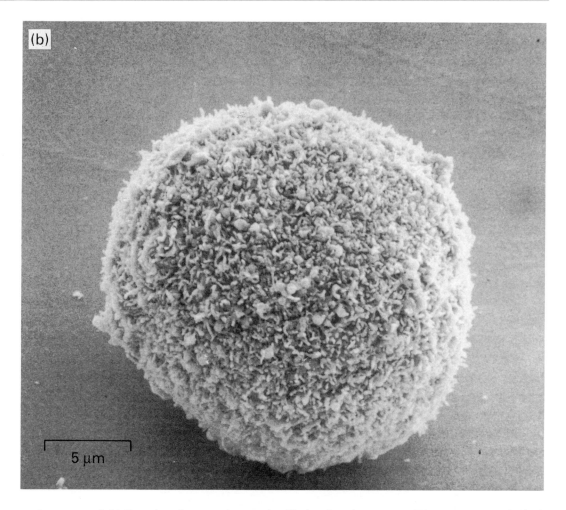

(b)

5 μm

Fig. 1.4. *contd* (b) Scanning electron micrograph of isolated rat hepatocyte. Hepatocytes are spherical with prominent microvilli covering the surface. Cells were attached to glass coverslips coated with poly-L-lysine for 2 h before they were fixed in phosphate-buffered (0.067 M, pH 7.4) 2% (w/v) glutaraldehyde for at least 4 h at room temperature. Serial dehydration in alcohol was followed by critical point drying. (Photomicrograph kindly provided by Glaxo Group Research Ltd, Ware.)

1.7.3   Preparation of specimens for microscopy

Transmission microscopy requires thin sample specimens, i.e. squashes, smears, hanging drops or very thin sections. Preservation and integrity of cellular components requires initial tissue fixation, which may be achieved either by rapid freezing or by chemical treatment to stabilise and cross-link protein and lipid components of membranes. Fixation with formaldehyde (principally used in light microscopy) and glutaraldehyde (alone or in combination with formaldehyde for electron microscopy) is the result of the formation of methylene bridges with side chain amino groups of proteins, whereas osmium tetroxide, a common fixative in electron microscopy, cross-links mainly with unsaturated fatty acid side chains (Fig. 1.5). Fixed tissue is then subjected to sequential processes of staining and sectioning.

Samples for ion probe analysis (Section 1.7.2) are prepared either by chemical fixation or, more successfully, by ultrarapid freezing methods, such as immersion in liquid propane or nitrogen slush, that serve to immobilise ions. Where frozen tissue is used, sections must be extremely thin and may require the use of an ultracryomicrotome.

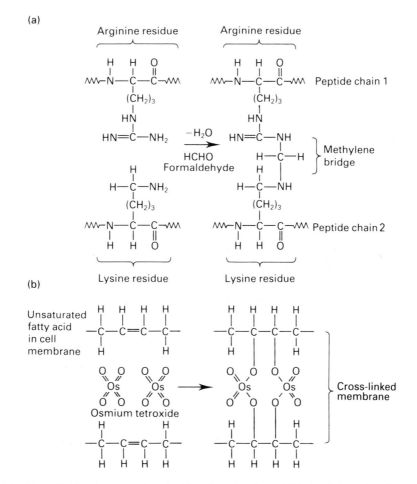

Fig. 1.5. Cross-linking by some chemical fixatives: (a) formaldehyde, (b) osmium tetroxide.

Histological stains are used to produce contrast that aids resolution.

Many stains used in light microscopy rely on anionic and cationic reactions with intracellular ampholytes, and as such their efficiency is profoundly influenced by pH. Cytoplasmic components are mostly cationic in the slightly acid pH range, inferring preferential binding to anionic (acidic) stains such as eosin. Chromatin and DNA at pH 6.0 are anionic and thus bind to cationic stains such as methylene blue. Contrast in material for TEM is improved by incorporating heavy metal salts into the specimens to induce a greater extent of electron absorption. This can be achieved by exploiting the binding of uranium to nucleic acids and proteins, and that of lead to lipids.

Fixed tissue, when not frozen, needs supporting before sections are cut for microscopic study. Embedding media such as waxes and epoxyresins (Araldite or Epon) are immiscible both with water and with alcohol, which necessitates the initial dehydration and equilibration of fixed tissues by passing them through solutions containing increasing concentrations of ethanol followed by transfer to xylene or propylene oxide, before infiltration with an appropriate embedding medium in its liquid phase. Sections 10 μm thick when cut from tissue frozen at − 20 °C in a refrigerated microtome, called a cryostat, are good for rapid examination by light microscopy but suffer damage when thawed. Wax sections, 5 μm thick, are floated on warm water, placed on glass slides, dried and subsequently dewaxed and stained for light-microscope examination. Ultra-thin sections for TEM (less than 100 nm) are cut with ultramicrotomes, using either fractured glass or diamond as the cutting edge from an area of block face trimmed to about $0.1 \, mm^2$. Section ribbons are floated on to water and mounted on fine copper grids for staining and examination. Cell organelles are examined in the electron microscope in isolated, intact form rather than in thin sections.

In negative staining, a heavy-metal stain, often phosphotungstic acid, is allowed to dry in a puddle around the surfaces of isolated cell particles supported on a thick carbon or plastic film. The stain deposits into surface crevices in the specimen during drying and produces a ghost image in which the specimen appears light against a dark background, often outlining details very clearly. The contrast and surface details of isolated specimens are also frequently increased by the technique of shadowing. A thin layer of a heavy metal such as platinum is deposited on the surface of the specimen from one side. The effect in the electron microscope is as if a strong light is directed toward the specimen from one side, placing surface depression in deep shadow.

Antibodies, prepared against specific cellular proteins, that have absorbed colloidal gold are frequently used for staining in TEM because the gold is electron dense. Several antigens can be located simultaneously if different antibodies are labelled with gold particles of different sizes (Section 2.10.4).

Freeze-fracture techniques, in which frozen samples are cleaved with a knife along fracture planes in membranes, utilise this shadowing technique to produce replicas of broken cellular material that show the membrane surface structure on, and within, cell organelles.

Microscopical images using light microscopy and electron microscopy are best recorded using photomicrography, which requires considerable experience of microscopes, film and processing to reproduce high quality images that are accurate and repeatable. Further details of photomicrographic methods may be found in the suggestions for further reading (Section 1.10).

1.7.4   Cytochemistry

Cytochemical techniques may be used to identify specific chemical components, especially enzymes, in cells, by direct microscopic observations of tissue sections *in situ*. The technique is based on the colorimetric detection of the components or of their metabolic products. The products of most enzyme reactions are soluble and must be made insoluble before they can be seen. The formation of insoluble electron-dense precipitates suitable for electron microscopic studies involves such so-called capture mechanisms. Table 1.5 gives some specific examples of cytochemical techniques for enzyme location. Such techniques have been very useful in locating the enzymes associated with oxidative phosphorylation, β-oxidation, fatty acid synthesis etc.

Immunocytochemical microscopy exploits the capacity of cell constituents to act as antigens and to bind specifically to antibodies produced against them. The immunoglobulin antibody is isolated and purified chromatographically from the blood of the inoculated animal (say a rabbit) and this is used to produce a second antibody in another animal (usually a goat). Methods for visualisation of the label commonly include fluorescence, enzymes, and gold with silver enhancement. A fluorescent dye, such as fluorescein or rhodamine isocyanate, is attached to the goat anti-rabbit antibody, which may be observed to detect conjugation between rabbit antibodies and the host antigens as a characteristic fluorescence under an ultraviolet light microscope fitted with a suitable filter. This indirect method has the advantage that goat anti-rabbit antibody may be used with different rabbit antibodies. The method is also quick and can reveal more than one antigen on the same section, but has the disadvantage that little structural detail is revealed. Further details of immunological methods are found in Chapter 2.

Fluorescent analogue cytochemistry permits a study of the spatial organisation of cellular components in living cells by covalent labelling of functional molecules or organelles with fluorescent probes and reincorporation of these fluorescent analogues into cells. For example, the molecular organisation of the cytoskeletal protein actin, labelled with *S*-iodoacetamidofluorescein and injected into cells by hypoosmotic shock treatment, may be followed using fluorescence microscopy. Fluorescent analogues and fluorescent antibodies may both be used in cell sorting methods (Sections 1.5.4, 2.7.3).

Table 1.5. *Some examples of the cytochemical assay of enzymes by light microscopy (LM) and electron microscopy (EM)*

Enzyme	Cellular function	Localisation methods
Acyl transferases	Catalyse transfer of acyl group through CoA	Localisation depends on production of free sulphydryl group and can involve: (i) production of electron-dense ferrocyanide precipitate, or (ii) incubation in the presence of cadmium/lanthanum ions to form insoluble mercaptides
Cytochrome oxidase	Terminal cytochrome in electron transfer chain involved in the transfer of electrons from flavoproteins to O_2	'Nadi' reaction. Naphthol plus aromatic diamine mixed in the presence of cytochrome *c* gives coloured indophenol. Modification for EM involves the formation of an indoaniline product that yields a coordination polymer
DNase and RNase	Release nucleotides from DNA and RNA	Precipitation with lead, plus an extra initial step, e.g. hydrolysis of nucleotide by phosphatase
Esterases	Hydrolyse a range of carboxylic acid esters	For EM, thioacetic acid is used as substrate. If incubation medium includes lead nitrate, an electron-dense lead sulphide is produced. Alternatively, for LM and EM, azo dye methods can be used
Sulphatases	Catalyse the hydrolytic cleavage of sulphate esters	For LM, naphthol sulphates substituted as substrates and liberated naphthols coupled to diazonium salts give an insoluble dye product. (Azo dye method.) For EM, heavy-metal trapping agents used, e.g. lead and barium

Enzymic antibody labels may also be exploited when conjugated with suitable cross-linking reagents. For example, horseradish peroxidase may be bound by a bifunctional disulphide agent such as *N*-succinimidyl 3-(2-pyridyldithio)propionate (SDPD), which forms disulphide bonds with lysine residues. Antigen–antibody reactions are detected by applying suitable chromogenic reagents. Gold particles attaching to specific cell antigen must be visualised by silver precipitation in light microscopy but are very sensitive.

1.7.5 Micrometry, image analysis and video microscopy

These techniques permit measurements of length, area, depth, shape, surface density and other topographical features that may then be correlated with biochemical parameters. Micrometry, in simple terms, may be carried out using the eyepiece of a compound microscope fitted with an eyepiece graticule (calibrated against a stage micrometer) to measure, for example, the degree of opening of stomata in epidermal strips to a range of different incubation conditions. So-called quantitative light microscopy may be used to define three-dimensional images in two-dimensional form as in morphometry or, alternatively, similar methods may take a two-dimensional image and use it to derive a three-dimensional structure, as in stereology. These methods usually include a range of eyepiece graticules with either simple set ruling or short or wavy lines that permit scoring of the number of points or intersections of reference objects of interest when the slide is moved randomly with a mechanical stage. Bias may be avoided in electron microscopy by the taking of random micrographs. An especially important measurement based on these principles is the ratio of surface area to volume, which is frequently particularly important in physiological investigations.

The advent of microcomputers has greatly extended manipulation of numerical data for microscopic investigations. For example, digitiser pads allow operators electronically to trace with a pen the outline of images either projected directly on to the pad, or on photographs attached to the pad. The movements of the pen are then passed electronically to a digital computer programmed for morphometry. Modern image analysers use closed-circuit television cameras to project the image on to a photocathode tube to produce a digitised picture of say 512×512 elements (pixels). Usually, each pixel can take the form of 256 shades of grey from 0 (black) to 255 (white). Digitised images are then stored in the computer memory or on disc for later processing and analysis.

Video microscopy can improve microscopic images in three ways. These comprise: video enhancement microscopy (VEM), which increases low contrast images electronically; video intensification microscopy (VIM), which amplifies low light images, e.g. from bioluminescence or a fluorescence; and digital image processing. The basic equipment needed for video microscopy is shown in Fig. 1.6.

Video microscopy has great advantages for biochemical investigations because it allows not only for image enhancement and improved resolution of low contrast images of living material, but also for quantitative measurement of specific molecules to be followed over a specific time. For example, the use of video microscopy has revealed structures that include micelles, actively transcribing rDNA genes and microtubule gliding, involving ATPase activity, all of which were invisible by conventional light microscopy. Further details of this

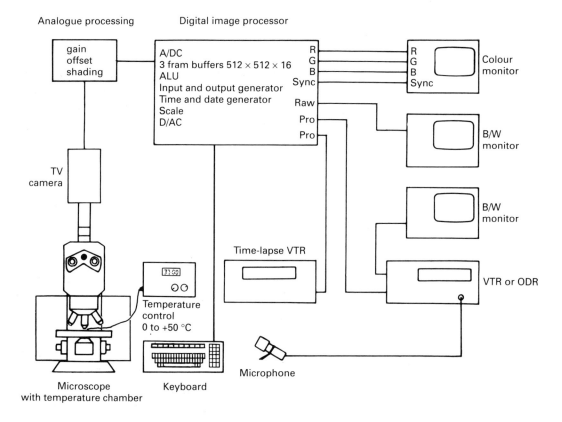

Fig. 1.6. A possible schematic layout for video microscopy. At least one black-and-white (B/W) monitor is required. The colour monitor and time-lapse recorder are optional. Pro, processed images; Sync, synchroniser; Raw, raw image; VTR, video taperecorder; ODR, optical disc recorder. (Reproduced, with permission, from Weiss, D.G. and Galfe, G. (1992) in *Image Analysis in Biology*. Ed. D.-P. Häder. CRC Press, Boca Raton, FL.)

specialised but important technique is given in the suggestions for further reading (Section 1.10).

From a practical viewpoint there is an important distinction to be made in the requirements for video microscopy, depending on its projected use, either in short-term recording (including relay for class use) or long term for possible inclusion in production videos or research presentations. For the latter, additional facilities, including, at the microscope to camera interface, a steady movement pointer, a cartridge for pre-shooting effects and a clapper board for synchronisation of sound and vision. This additional equipment should be considered at the time of initial purchase and not as an addendum, as the design of such equipment changes rapidly.

1.8 Elucidating metabolic pathways

Certain generalities may be applied to the way in which metabolic pathways are elucidated and, once the latter are known, to determine whether or not a particular organelle, cell, organ or organism is exploiting a specific pathway. These approaches are mainly, though not exclusively, applied to preparations *in vitro* and thus may also be used, along with microscopy, to evaluate the integrity of organelles, etc. Experiments may be performed in which all intermediates and enzymes involved in a pathway are first identified and quantified by chromatographic, spectroscopic or other means. When tissue extracts are supplemented with these intermediates, the rate of processes in the pathway would be expected to accelerate in a predictable manner due to faster turnover. Changes in metabolic rate may also be achieved by altering environmental conditions, e.g. glucose breakdown via the Embden–Meyerhof pathway (EMP) in the facultative anaerobe *Saccharomyces cerevisiae* is accelerated by switching from aerobic to anaerobic conditions. Such metabolic perturbations are closely related to changes in the kinetic activities of key regulatory enzymes as they respond to the binding of different positive or negative effector ligands that induce conformational (allosteric) changes in the enzymes (Section 4.7). The identification and characterisation of rate-limiting enzymes is thus an important component in the elucidation of metabolic pathways, particularly with respect to how the pathways are regulated. In general terms, there must be a positive correlation between the rate of the overall physiological process and the kinetic and regulatory properties of key enzymes. In particular, the specific activity of key enzymes involved in a metabolic pathway would be expected to be high. Addition of enzyme inhibitors will cause intermediates to accumulate, thereby aiding in their identification, e.g. iodoacetate inhibition of glyceraldehyde-3-phosphate in the EMP, or citrate synthetase inhibition by monofluoroacetic acid in the tricarboxylic acid cycle.

Isotopic tracers are a very powerful tool for identifying the metabolic fate of precursors and for following metabolic turnover (Section 5.6). Time course studies and pulse-chase techniques are the two main approaches employed in metabolic studies in which isotopically labelled compounds are identified and recovered. Often different elements in compounds may be labelled and this aids in the discrimination of the biosynthetic origins of pathways, e.g. amino acid biosynthesis. The great advantages of radiolabelling are its sensitivity and its specificity.

Preliminary evidence for identifying whether or not a particular pathway is operating may be inferred from yield (biomass production) studies on micro-organisms growing on a particular substrate, or by measurements of gaseous exchange processes. The value of these approaches relies on prior knowledge of stoichiometry and of the pathways for dissimilation.

Manometry may be used to measure either uptake or evolution of both CO_2 and O_2 whereas O_2 and CO_2 electrodes (Section 11.6) simply monitor changes in O_2 and CO_2 levels, respectively, albeit with greater sensitivity. Manometry also has the distinctive feature of allowing the simultaneous determination of O_2 and CO_2 exchange and has the advantage that the magnitude of the exchange is independent of the partial pressure of the gas at the beginning of the experiment. Manometric studies are carried out in a small flask attached to some form of manometer that measures changes in the amount of gas in the flask. In all types of manometer, the flasks, immersed in a water bath with a temperature control of ± 0.5 deg.C, are shaken mechanically at rates of 100 to 120 oscillations min^{-1} to ensure that respiratory gas exchange is not limited by diffusion of gas into the liquid phase. The total volume of liquid in the flask should generally not exceed $4 \, cm^3$ because of gas diffusion limitations. The two principal types of manometer are the Warburg constant volume manometer, illustrated in Fig. 1.7, and the Gilson constant pressure manometer.

The biological applications of manometry are extensive. Respiratory quotients (RQ), defined as the relationship between the volume of CO_2 produced and the volume of O_2 consumed during respiration, i.e.:

$$RQ = \frac{CO_2 \text{ evolved}}{O_2 \text{ absorbed}} \tag{1.7}$$

may give an indication of the nature of the endogenous substrate being metabolised. The complete oxidation of a simple carbohydrate gives an RQ of 1, whereas that of an average fat gives a value of approximately 0.7 and that of a protein gives 0.8. Deviations from these values are sometimes obtained in practice, as some CO_2 may be incorporated into cellular material, lowering the volume of the CO_2 evolved. The rate at which an organism or tissue consumes O_2 or evolves CO_2 is expressed by a metabolic quotient Q_x, where x is the gas being measured. Thus Q_{O_2} is defined as the volume of O_2 taken up per milligram dry weight of biological material per hour. In some cases it may be expressed in terms of milligram of nitrogen (generally determined by the Kjeldahl method, Section 4.4) or milligram of protein or DNA. In such cases it is expressed as follows:

$$Q_{O_2}(N) = mm^3 \, O_2 \, (mg \text{ tissue N})^{-1} \, h^{-1}$$

Metabolic quotients may also be determined in atmospheres of different or varying gas composition, e.g. pure N_2, in which case an additional suffix is added to the quotient, thus:

$$Q_{CO_2}^{N_2}(N) = mm^3 \, CO_2 \, (mg \text{ tissue N})^{-1} \, h^{-1} \text{ in an atmosphere of } N_2 \text{ gas.}$$

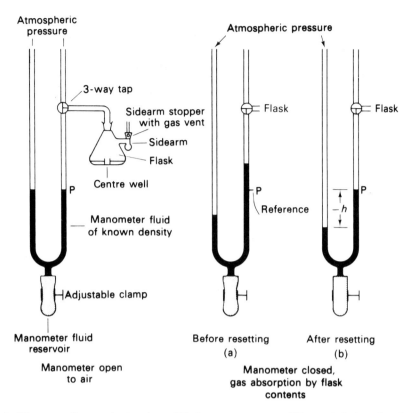

Fig. 1.7. Diagram of a constant volume Warburg manometer. CO_2 gas is being absorbed in the experimental flask and the resulting decrease in pressure forces the fluid level in the right-hand limb to rise and that in the left-hand limb to fall. At regular time intervals the meniscus of the fluid in the left-hand limb is returned to the reference point P by withdrawing fluid into the reservoir using the adjustable clamp, thereby measuring the resultant decrease in pressure at constant volume as $-h$ mm. The change h is related to the quantity X of gas evolved at s.t.p. by the equation:

$$X = h \left[\frac{V_g \dfrac{273}{T} + V_1 \alpha}{P_o} \right]$$

where V_g is the volume of the gas space in the flask, including that of the capillary from the flask to the reference mark P (mm³); V_1 is the volume of liquid in the flask (mm³); α is the solubility of the gas in the liquid in the flask; T is the temperature of the water bath in K; P_o is the standard pressure expressed in millimetres of manometric fluid. Since, for a given experimental flask being used to study a particular reaction under defined conditions, all values within the brackets in the equation are constant, $X = kh$, where k is the flask constant. Flask constants for particular manometers are usually supplied by manufacturers or can be obtained by a suitable form of calibration.

If there is a possibility of confusion, the quotients may be indicated as positive or negative, indicating the production or removal, respectively, of metabolite. Manometric techniques have been applied in studies on tissue slices and homogenates, with attendant problems of homogeneity of gas supply and artefacts, in studies of respiratory control, and the effect of inhibitors on mitochondrial respiration (although the technique is less sensitive than the O_2 electrode). When used for photosynthetic studies a control determination in darkness should be performed and the partial pressure of one gas kept constant during the experiment. This may be achieved by maintaining a constant partial pressure of CO_2 by using carbonate–bicarbonate buffers or by removing all the O_2 by chemical means. This has limited value, however, since it is known that the rate of CO_2 exchange is dependent on the O_2 content of the atmosphere.

1.9   Quality control

1.9.1   The work environment

Most biochemical work is performed in purpose-built laboratory suites the ergonomic design of which is critically important for efficient work and management.

Biochemical laboratories are most effectively designed with large centrally sited areas for housing either heavy equipment for general application (e.g. high speed centrifuges) or stores, wash up and supportive administrative facilities. Specialised laboratories for weighing, spectrophotometry or various forms of chromatography may also be found in larger departments. Many laboratories are now designed on a flexible modular basis for possible change of purpose. All laboratories are potentially dangerous places to work and thus eating, drinking, smoking and horseplay of any kind are forbidden.

Health and safety for all laboratory workers, especially laboratory managers, has become a key issue in most of Europe and the USA in recent years, with government legislation enforcing strict codes of practice. In the UK, for example, detailed COSHH forms (Control of Substances Hazardous to Health) must be completed either personally or on behalf of laboratory workers detailing precisely the levels of exposure to potentially hazardous chemicals. The extent of knowledge of the toxicity of potentially hazardous substances is available through a variety of databases. Many toxicity data are incomplete, however, so great care should be taken in cases of uncertainty. Hazard warning labels are frequently provided by suppliers of laboratory equipment and chemicals. The amount of personal protection necessary will vary according to the nature of the experiment. Safety spectacles and lightweight disposable gloves should be worn at all times, along with buttoned-up laboratory coats. High-necked buttoned gowns must be worn for microbiological work. Heavy-duty leather gloves should be worn either for ultralow temperature work or for handling highly corrosive liquids. The main physical dangers in laboratories comes from fire and smoke, from explosion of volatile solvents and gases, from vacuum implosion of glass vessels, from sources of ionising radiation, from lasers and ultraviolet light and from ultrasonic vibration. Potential biological hazards arise also from: laboratory animals, which may cause allergic reactions or transmit certain diseases to humans; body fluids (especially as hepatitis B virus may be passed from infected serum); pathogenic animal and cell tissue cultures or *all* microorganisms, including genetically engineered micro-organisms. Work with either tissue cultures or microorganisms should, therefore, take place inside safety cabinets in which a sterile air flow should move vertically away from the operator. The acquisition of skilled manipulative aseptic techniques is essential for all experiments using such experimental systems.

Despite even the most stringent attention to good safety practice accidents do happen and it is very important to have available qualified practising first-aiders to deal with emergencies. Key telephone numbers for emergency services should be clearly and conveniently posted.

The disposal of unused chemicals, radioactive isotopes and infected material should be subject to clear management guidelines, and detailed records of such disposals should be kept. Corrosive chemicals such as acids and alkalis must be neutralised before being washed away down the sink. All organic solvents must be stored in metal drums, with chlorinated solvents stored separately, before under-going incineration at high temperature. Radioactive waste, in particular, is subject to stringent disposal regulations that are legally enforced. All spent or contaminated tissue or cell or microbial cultures must be sterilised by autoclaving before disposal.

When viewed holistically, laboratory work of any kind needs an elaborate management system to plan, control and coordinate activities, which means, as far as safety is concerned, that either line managers or a specific safety officer has special responsibility for safety issues. The working environment as a whole must comply with ordered bureaucratic principles, which in practical terms inevitably means lists of rules, regulations and procedures. This is most conveniently com-municated to laboratory workers, many of whom have different job specifications, through a handbook. This should fully describe lines of accountability and responsibility, with flow diagrams showing operational procedures for all depart-mental activities from capital acquisitions to safety practices.

1.9.2   Workmanship and statistical analysis

In today's enterprise culture the desire for ever-increasing efficiency and productivity is the key objective for 'competitive advantage' in both profit- and non-profit-making organisations that employ biochemists. The notion of total quality man-agement, with its emphasis on 'getting it right first time', includes setting and monitoring high standards of workmanship from its employees. Academic and

professional qualifications are particularly important in the employability of biochemists. On-the-job (in-house) training for employees is becoming the norm for
practising biochemists. In the context of the world of work, good personal
transferrable skills (PTS) are just as important as technical knowledge and skills
for ensuring the efficient working of biochemical laboratories at all levels, from
leaders of research units or heads of quality control sections to students on
supervised work experience. PTS skills include learning, good verbal com-
munication, writing, self-management and coping, thinking and group skills,
together with project management and information retrieval methods.

Researching biochemists in particular need to become familiar with a range of
specialist library facilities and services, and to develop convivial personal relationships with the subject librarian supporting the researcher's individual research area.
Good information retrieval skills are absolutely essential for workers to become
knowledgeable about new developments impinging on fields of interest. Training
should be expected to be subject specific and focus on modern methods of retrieval
(abstracts/indices, CD-ROMS, on-line databases, etc.), on types of literature
(methods, reviews, patents, statistics, accounts, current awareness, conferences,
theses, etc.). For studies nearing the frontiers of knowledge, search strategies might
be usefully deployed using Bolean/proximity operators, synonyms, natural language
versus controlled language, etc. The overall objective is to develop a critical
approach to acquiring relevant information from the plethora of data available
and to develop an awareness of the advantages and limitations of different types
of information tool. Setting aside dedicated library time is frequently the most
difficult of a researcher's tasks and here the 'Pareto effect' or 80/20 management
rule certainly should apply, i.e. 80% of results come from doing 20% of tasks. In
practical terms, this most often means being highly selective in the use of time and
effort.

All scientifically based work is by definition international. The need to harmonise
scientific nomenclature internationally is reflected in the adoption of SI units
(Système International d'Unités) for most, but unfortunately not all, biochemical
work. Numerous updating reference texts are available that transform older
nomenclature to SI units and specify universal constants for key equations.

Long experience of many undergraduates' difficulties in distinguishing units of
mass from concentration, in transforming units of concentration and in doing
simple proportional calculations has prompted the inclusion of Table 1.6 as an
aide memoire to the relationship between moles (mol) and molarity.

Such calculations are important because many physiological properties of cells
depend on molar concentration, e.g. water potential (Ψ), freezing point, boiling
point, etc. Occasionally concentration is more conveniently expressed as weight
per unit volume (w/v), e.g. when the relative molecular mass is unknown.

Table 1.6. *Interconversion of mol, mmol and µmol in different volumes to give different final concentrations (molarity)*

M	mM	µM
1 mol dm^{-3}	1 mmol dm^{-3}	1 µmol dm^{-3}
1 mmol cm^{-3}	1 µmol cm^{-3}	1 nmol cm^{-3}
1 µmol nm^{-3}	1 nmol mm^{-3}	1 pmol mm^{-3}

The basis of much scientific work in biochemistry is the experimental approach popularly ascribed to Sir Francis Bacon (1561–1626). All experiments are designed to test a hypothesis, in which experimenters should alter *one* experimental factor at a time (the variable) in a defined way to test the response of the 'system' to the factor being applied. Appropriate controls should always be included to serve as a standard against which the effect of the variable is measured. A series of experiments may be necessary to complete a project in which the objectives of this experimentation are more ambitious.

Designing project work demands considerable theoretical and technical knowledge and managerial experience. The five-P maxim that 'proper preparation prevents poor performance' is never more true. Key decisions to be made in designing an investigation would comprise a clear statement of aims, the approach to be adopted, techniques, apparatus, instruments and procedures to be used, what methods will be used to analyse the results, the nature of the controls and the degree of replication necessary to apply appropriate statistical analysis.

Statistical tests are frequently necessary because of the limited number of replicated measurements or observations that can be made in any investigation. A sample is taken from which probability estimates are made either of the accuracy or conversely of the error of the results, error representing the difference between the estimated value and the true value. Accuracy must not be confused with precision, which is simply a measurement made in very small units. The precision required of instruments must reflect the objectives of the experiment and the accuracy anticipated. Similarly, recording results very precisely to several places of decimals is frequently meaningless if the methods used are not very accurate. The simplest type of deviation is how far a particular value (x_i) of a sample differs from the arithmetic mean ($\bar{x}$), i.e. sample deviation $= (x_i - \bar{x})$. The *mean deviation* ($\bar{d}$) is then calculated from $\bar{d} = \Sigma |x_i - \bar{x}|/n$, where n measurements are made and the modulus bars (| |) indicate that all deviations are taken to be as positive. Results using this approach may be expressed therefore as ($\bar{x} \pm \bar{d}$). High accuracy is estimated if $\bar{d}$ is low in relation to $\bar{x}$. This is most conveniently expressed as percentage deviation, which is $(\bar{x}/\bar{d}) \times 100\%$.

Standard deviation (σ) is a more useful expression than mean deviation because it takes into account sample deviation being either positive or negative with respect to $\bar{x}$. This is achieved by squaring both $\bar{d}$ and $\Sigma(x_i - \bar{x})$. Thus:

$$\bar{d}^2 = \sum(x_i - \bar{x})^2/n = \text{variance}/n$$

$$\sigma = \sqrt{\bar{d}^2} = \sqrt{\sum(x_i - \bar{x})^2/n}$$

Expressed in words, this becomes: standard deviation is the square root of the arithmetic average of the square of deviations around the mean.

In statistical terms, slightly more accurate values of σ are obtained from the expression:

$$\sigma = \frac{\sqrt{(x_i - \bar{x})^2}}{n - 1} \tag{1.8}$$

where $n - 1$ represents the number of so-called degrees of freedom, i.e. there are $n - 1$ alternative choices for n samples. Hence σ is the square root of variance divided by the number of degrees of freedom. Standard deviation measurements relate to Gaussian (normal) continuous frequency distributions, shown graphically in Fig. 1.8.

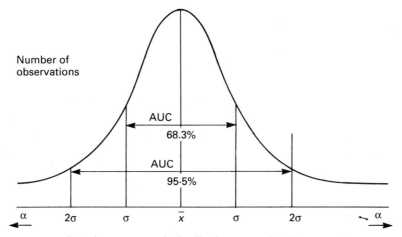

Fig. 1.8. Gaussian, or normal, distribution curve. AUC, area under curve.

In this bell-shaped curve the area under the curve between any two values of x_i represents probability. A simple application of this curve to biochemical analysis can be related to the following example. Suppose the protein content of a plasma membrane preparation was estimated with $\bar{x}$ of 5.5 $\mu g\,cm^{-3}$ and σ of $\pm 1.2\ \mu g\,cm^{-3}$. This would mean that there would be an approximate probability of 17% that an additional assay would show a protein content either at or less than 4.3 $\mu g\,cm^{-3}$ or at 6.70 $\mu g\,cm^{-3}$. Similarly, there would be an approximate probability of 2.3% of additional values being either at or less than 3.1 $\mu g\,cm^{-3}$ or at or greater than 7.9 $\mu g\,cm^{-3}$.

If individual experimental values differ from $\bar{x}$ by greater than 2σ of the mean, the probability of that measurement being a true member of that set of values is less than 2.3%, which suggests the result has not occurred as a chance variation but by mistake. This value should therefore be discarded for analysis.

A very useful way of finding out the probability of any result being obtained once the mean and standard deviation are known is to transpose the normal distribution into a table that converts any value of x_i into proportional values of σ to derive the normal deviate, z, which is the number of standard deviations between x_i and $\bar{x}$. Thus $z = (x_i - \bar{x}/\sigma)$. Table 1.7 shows values for z and probability values.

In the example just given, suppose we wish to know the probability of a protein content being 3.20 $\mu g\,cm^{-3}$ or less. The calculation is as follows:

An initial sketch diagram shows that in relation to the normal distribution curve we are dealing with a tail problem (Fig. 1.9):

$$z = \frac{(x_i - \bar{x})}{\sigma}$$

$$= \frac{2.30}{1.20}$$

$$= 1.92$$

From Table 1.7 the area under the curve (i.e. the probability) $= 0.4726$. Because we are dealing with a tail problem, probability $= (0.50 - 0.4726) = $ approx. $0.03 = 3\%$. This value would just be acceptable because it lies within two standard deviations of the mean, as explained above. Body calculations can be similarly worked out.

If σ is low in relation to $\bar{x}$, i.e. there is a sharp peak around the mean, the set of measurements is said to show a low coefficient of variation. This is normally expressed in percentage terms where $c = [(\sigma/\bar{x}) \times 100)]\%$. This expression is useful in cases where alternative methods are being tried out, since the procedure with the lowest coefficient of variation is the more effective.

The standard deviation around the mean might be expected to decrease after

Table 1.7. *Showing relationship between normal standard deviate (z) and probability (area under curve) for normal distributions*

		$z \rightarrow$									
		.01	.02	.03	.04	.05	.06	.07	.08	.08	.09
	.0	.0000	.0040	.0080	.0120	.0160	.0199	.0239	.0279	.0319	.0359
	.1	.0398	.0038	.0478	.0517	.0557	.0596	.0636	.0675	.0714	.0753
	.2	.0793	.0832	.0871	.0910	.0948	.0987	.1026	.1064	.1103	.1141
	.3	.1179	.1217	.1255	.1293	.1331	.1368	.1406	.1443	.1480	.1517
	.4	.1554	.1591	.1628	.1664	.1700	.1736	.1772	.1808	.1844	.1879
	.5	.1915	.1950	.1985	.2019	.2054	.2088	.2123	.2157	.2190	.2224
	.6	.2257	.2291	.2324	.2357	.2389	.2422	.2454	.2486	.2518	.2549
	.7	.2580	.2612	.2642	.2673	.2704	.2734	.2764	.2794	.2823	.2852
	.8	.2881	.2910	.2939	.2967	.2995	.3023	.3051	.3078	.3106	.3133
	.9	.3159	.3186	.3212	.3238	.3264	.3289	.3315	.3340	.3365	.3389
	1.0	.3413	.3438	.3461	.3485	.3508	.3531	.3554	.3577	.3599	.3621
	1.1	.3643	.3665	.3686	.3708	.3729	.3749	.3770	.3790	.3810	.3830
	1.2	.3849	.3869	.3888	.3907	.3925	.3944	.3962	.3980	.3997	.4015
	1.3	.4032	.4049	.4066	.4082	.4099	.4115	.4131	.4147	.4162	.4177
	1.4	.4192	.4207	.4222	.4236	.4251	.4265	.4279	.4292	.4306	.4319
z	1.5	.4332	.4345	.4357	.4370	.4382	.4394	.4406	.4418	.4429	.4441
↓	1.6	.4452	.4463	.4474	.4484	.4495	.4505	.4515	.4525	.4535	.4545
	1.7	.4554	.4564	.4573	.4582	.4591	.4599	.4608	.4616	.4625	.4633
	1.8	.4641	.4649	.4656	.4664	.4671	.4678	.4686	.4693	.4699	.4706
	1.9	.4713	.4719	.4726	.4732	.4738	.4744	.4750	.4756	.4761	.4767
	2.0	.4772	.4778	.4783	.4788	.4793	.4798	.4803	.4808	.4812	.4817
	2.1	.4821	.4826	.4830	.4834	.4838	.4842	.4846	.4850	.4854	.4857
	2.2	.4861	.4864	.4868	.4871	.4875	.4878	.4881	.4884	.4887	.4890
	2.3	.4893	.4896	.4898	.4901	.4904	.4906	.4909	.4911	.4913	.4916
	2.4	.4918	.4920	.4922	.4925	.4927	.4929	.4931	.4932	.4934	.4936
	2.5	.4938	.4940	.4941	.4943	.4945	.4946	.4948	.4949	.4951	.4952
	2.6	.4953	.4955	.4956	.4957	.4959	.4960	.4961	.4962	.4963	.4964
	2.7	.4965	.4966	.4967	.4968	.4969	.4970	.4971	.4972	.4973	.4974
	2.8	.4974	.4975	.4976	.4977	.4977	.4978	.4979	.4979	.4980	.4981
	2.9	.4981	.4982	.4982	.4983	.4984	.4984	.4985	.4985	.4986	.4986
	3.0	.4986	.4987	.4987	.4988	.4988	.4989	.4989	.4989	.4990	.4990

Example: for $z = 2.25$, area under curve (probability) = 0.4875.

Taken, with permission, from Anderson, D.R., Sweeney, D.J. and Williams, T.A. (1993) *Statistics for Business and Economics*, 5th Edn, West Publishing Company, St Paul, MI.

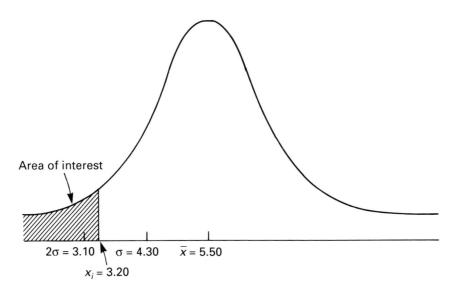

Fig. 1.9. Distribution curve showing tailing.

the number of replicates in an experiment is increased. This relationship is expressed by the term standard deviation of the mean or standard error of the mean, $\sigma_{\bar{x}}$, where

$$\sigma_{\bar{x}} = \frac{\sigma}{\sqrt{N}} \qquad (1.9)$$

and N is the number of measurements.

There is, of course, much more to statistics than standard deviation and standard error measurements and experienced biochemists should know how to apply paired and unpaired t tests, analysis of variance (ANOVA, one and two way), correlation and regression analysis and the use of non-parametric statistics, including the chi-squared test used to validate hypotheses. Further details of such methods can be found in the suggestions for further reading (Section 1.10).

The successful application of statistical methods is critically dependent, however, on the accuracy of the primary data on which statistics are based. Good standards of workmanship must be achieved by laboratory workers and much vigilance is therefore required by supervisors, because in biochemistry, as in other fields, workers are more likely to blame their tools than themselves. Elementary mistakes, such as blowing out run out pipettes, transposing numbers on weight values and improper calibration of equipment, should be avoided. It is particularly important that primary experimental data are recorded honestly and that protocol can be

Table 1.8. *Showing common errors in students' presentations of experiments*

1. All sentences have verbs, which should be approximately in the middle of the sentence and not at the end
2. Number each table or figure
3. Legends to figures usually go *below* the diagram, whereas legends to tables go *above* the table
4. Try to avoid the use of common grammatical errors, e.g. unrelated participles, split infinitives, commencing a sentence with a conjunction, ending a sentence with a preposition, etc.
5. Do not issue instructions in the text
6. Reports should be written in the past tense, as objective descriptions
7. Do not change tense in the middle of a sentence
8. Avoid the use of the first person
9. Avoid the use of: actual, fact, actual fact, from the results it can be seen, the results show
10. Please leave wide margins for comments by the practical assessor
11. Do not use 'x' or '+' as a symbol on figures
12. Do not abbreviate words except where appropriate, e.g. ATP
13. ATP, NAD, etc. have no dots between the letters
14. Do not repeat information already given in the schedule
15. Use chemical formulae where appropriate
16. The discussion should always include a criticism of methodology and give suggestions for further work
17. If statistical analysis is appropriate, use it
18. Ensure that your writing is legible

replicated by other workers. Results should, preferably, be recorded either on proformas or in a log book and certainly not on scruffy bits of paper! Raw data should be transferred to files or databases for storage and recovery. Back-up computer discs or hard copies of results should always be kept in case of computer virus malpractice or similar catastrophe. Experiments should be written up concisely as soon as possible after completion and to a standard format, including an Introduction (however brief), Materials and Methods (which should include changes to routine protocol), Results (including statistical treatment, if appropriate) and brief Discussion. A short list of common criticisms of undergraduate practical reports is included in Table 1.8 for reference purposes.

Many commercial laboratories are now having to work to Good Laboratory Practice (GLP) regulations, which are concerned with the organisational process and the conditions under which laboratory studies are planned, performed, monitored, recorded and repeated, and by so doing, ensuring as far as possible the quality and integrity of laboratory data. GLP regulations were first published by the USA's Food and Drug Administration (FDA) in 1976. At that time the FDA identified significant problems in the manner in which non-clinical studies were being performed in the USA. Deficiencies were found during inspections of major pharmaceutical firms, private contract testing facilities and government laboratories. The result of these experiences heralded the birth of GLP. The purpose of GLP is to establish systems and procedures that ensure the quality and integrity of data generated during the conduct of a non-clinical laboratory study. Essential elements of GLP cover personnel, management, study directors, quality assurance, laboratory areas and operation, equipment, standard operating procedures (SOPs), protocol, conduct of study, final reports, archives and regulatory compliance. Further details of GLP and associated topics of good clinical practice (GCP) and good manufacturing practice (GMP) may be found in specialist literature cited in the next section.

1.10   Suggestions for further reading

ADAMS, R.L.P (1988) *Laboratory Techniques in Biochemistry and Molecular Biology*, volume 8. Ed. R.H. Burdon and P.H. van Knippenburg, Elsevier Science Publishers, B.V., Amsterdam. (A specialist text on animal cell and tissue culture.)

ANON (1990) *Substances Hazardous to Health.* Available from Croner Publications, Croner House, London Road, Kingston-upon-Thames, Surrey, KT2 6SR. (As well as a large section on chemical hazards it also has much information on new legislation, sources of information, biological hazards, assessment and control.)

CARSON, P.A. and DENT, N.J. (Ed.) (1990) *Good Laboratory and Clinical Practices – Techniques for Quality Assurance for Professionals.* Heinemann Newnes, Oxford. (A comprehensive and up-to-date specialist text on GLP and GMP.)

GUIRDHAM, M. (1990) *Interpersonal Skills at Work.* Prentice Hall Int. (UK) Ltd, London. (A systematic coverage of most aspects of face-to-face interaction in the workplace. The text includes practical exercises, role plays and performance analysis questionnaires to aid monitoring progress.)

LACEY, A.J. (Ed.) (1989) *Light Microscopy in Biology – A Practical Approach.* IRL Press, Oxford. (Recommended for more specialist practitioners of light microscopy.)

LINDSEY, K. and JONES, M.G.K. (1989) *Plant Biotechnology in Agriculture.* Open University Press, Milton Keynes. (Good coverage of applications and practical limitations of the rapidly expanding area of biotechnology in the plant sciences.)

ORMEROD, M.G. (Ed.) (1990) *Flow Cytometry – A Practical Approach.* IRL Press, Oxford. (Another specialist text for the more advanced practitioner.)

POLLARD, J.W. and WALKER J.M. (Ed.) (1990) *Methods in Molecular Biology* vol. 6. *Plant Cell and Tissue Culture.* Humana Press, Clifton, NJ. (A good basic text on methodology.)

Sigma-Aldrich Material Safety Data Sheets on CD-ROM (IBM PC version or Apple Macintosh version). (Mainly for laboratory workers needing rapid updating on COSSH regulations.)

SOKAL, R.R. and ROTILF, F.J. (1987) *Introduction to Biostatistics*, 2nd ed. W.H. Freeman and Co., New York. (A useful introduction to biometrics and statistics.)

STANBURY, P.F., WHITAKER, A. and HALL, S.J. (1994) *Principles of Fermentation Technology*. Pergamon Press, Oxford. (An excellent introduction to fermentation processes.)

WEDGEWOOD, M. (1989) *Tackling Biology Projects*. Macmillan Education Ltd., Basingstoke. (Most of what you need to know about organising project work.)

Immunochemical techniques

2.1 General principles

2.1.1 Introduction

Immunology is the study of the immune responses, which are the processes by which an animal's body defends itself against invasion by foreign organisms. Immune responses can be divided into two general types: (a) the humoral (antibody-mediated) response and (b) the cell-mediated response. Both types of immune response involve cells of the lymphoreticular system. Immunochemical techniques employ the antibodies involved in humoral immunity.

When an antigen, which may be a foreign organism or compound, enters the tissues of an animal, lymphocytes are stimulated to divide and differentiate; in the case of the humoral response, this results in the eventual production of plasma cells. Plasma cells secrete specific antibody molecules containing specific binding sites capable of binding tightly but non-covalently to the original antigen, and hence causing its precipitation, neutralisation or death, via phagocytosis or complement-mediated cell lysis, depending on whether the antigen is a macromolecule, toxin or microorganism. Antibodies belong to a group of proteins known as the immunoglobulins (Ig), which are divided into five classes – IgG, IgM, IgA, IgD and IgE – some properties of which are listed in Table 2.1.

Immunochemical techniques utilise mainly IgG, which constitutes 80% of the serum immunoglobulin. All immunoglobulins have structures based on four polypeptide chains comprising two identical heavy chains and two identical light chains.

Table 2.1. *Physico-chemical and biological properties of immunoglobulin classes*

Immunoglobulin class	IgG	IgM	IgA	IgD	IgE
Heavy-chain symbol	γ	μ	α	δ	ε
Heavy-chain M_r	50 000	70 000	55 000	65 000	65 000
Molecular composition	$\gamma_2\kappa_2$	$(\mu_2\kappa_2)_5$; J	$a_2\lambda_2, a_2\kappa_2$	$\delta_2\kappa_2$	$\varepsilon_2\kappa_2$
	$\gamma_2\lambda_2$	$(\mu_2\lambda_2)_5$; J	$(a_2\lambda_2)_2Sc^a$; J	$\delta_2\lambda_2$	$\varepsilon_2\lambda_2$
			$(a_2\kappa_2)_2Sc^a$; J		
Physical state	Monomer	Pentamer	Monomer, dimer	Monomer	Monomer
Relative molecular mass (M_r)	150 000	900 000	160 000– 380 000^a	180 000	200 000
Valency	2	5–10	2, 4^a	2	2
Carbohydrate content (%)	3	12	7.5–10	12	12
Mercaptoethanol sensitivity	–	+	–	–	–
Heat stability (56 °C, 4 h)	+	+	+	?	–
Complement fixation	+	+ +	–	–	–
Normal serum levels (mg cm^{-3})	8–16	0.6–2	1–4	0.001	0.0003
Half-life (days)	23	5	6	3	2

^{a}Dimer in external secretions carries the secretory component, Sc.

The immunoglobulin classes listed in Table 2.1 are defined by the nature of their heavy chains: γ, μ, α, δ, and ε. Only two types of light chain are known, κ and λ, and these are common to all classes of immunoglobulin. The proteolytic enzyme papain cleaves IgG antibody molecules into three fragments with similar molecular weights: two identical Fab (antigen binding) fragments and one Fc (crystallisable) fragment. The latter is capable of binding and activating the complement system. This information is summarised in Fig. 2.1, which also indicates that papain actually cleaves the IgG molecules in the so-called hinge region. This hinge region allows limited flexibility in the spatial positioning of the Fab and Fc fragments in the intact IgG molecule. It should be noted that different IgG subclasses contain 2–13 interchain disulphide bridges in the hinge region, and that both the heavy and light chains of all immunoglobulins contain a number of intrachain disulphide links, which results in a compact globular structure.

Many of the N-terminal 100 to 110 amino acid residues of both the light and the heavy chains vary in different IgG molecules; these therefore comprise the variable region. The other amino acid residues of the light and heavy chains are virtually constant within an IgG subclass (constant region).

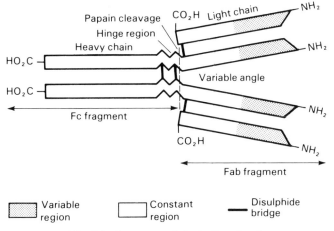

Fig. 2.1. Structure of the IgG molecule.

Variations in the composition of a few of the amino acid residues (the hypervariable residues) in the N-terminal portion (variable region) of each chain of an immunoglobulin result in a unique topography of the antigen binding site on each immunoglobulin and account for the uniqueness of each specific antibody. IgG, which is a monomer containing only two identical light and two identical heavy chains, has an antigen valency of two, i.e. it has two identical antigen binding sites. The variable regions of one light and one heavy chain interact to form a globular domain (variable domain). The constant region of one light chain and the appropriate constant region of one heavy chain form a globular constant domain, and the remaining constant region of the heavy chains form two further constant domains. Thus each Fc and Fab fragment contains two globular domains, and a single IgG molecule contains two variable domains and four constant domains, with limited flexibility between each compact domain.

The specificity of the immune reaction is well known to all those who find themselves immune to a disease caused by one microorganism but fall ill to infection with a different microorganism. It is even more apparent to those renal transplant patients who eventually reject a kidney from another person who might be a very close relative who apparently had an antigenic make-up almost identical with that of the recipient. The analytical biochemist, however, is more interested in the *in vitro* use of immunochemical techniques rather than the detailed *in vivo* interactions involved in immunology. It should be stressed that it is not necessary to be a student of immunology to use and understand most immunochemical techniques.

The major attraction of immunochemical techniques is the extreme specificity conferred on them by biological recognition at the molecular level, even in the presence of high levels of contaminating molecules; for example, at the submolecular level monospecific antisera can differentiate between macromolecules containing the D and L forms of a single amino acid. The multivalency of most antigens and antibodies enables them to interact to form a precipitate. Labelling antigens and antibodies with radioisotopes enables the biochemist to combine the specificity of biological activity with the sensitivity of liquid scintillation counting techniques (Section 5.2.2). Thus immunochemical techniques extend the ability of the biochemist to detect and quantify specific molecular structures, even in the presence of closely related material in, for example, sera, microorganism culture filtrates, tissue extracts, fractions from gradient centrifugation or the effluent of chromatography columns.

2.1.2   Definitions

Immunology is a subject that tends to have a vocabulary of its own and it is advisable to begin with a series of definitions of important terms used throughout this chapter.

Adjuvant A substance, inoculated with antigen that increases the rate of biosynthesis of antibody to that antigen.

Affinity The intrinsic binding power of a single antibody combining site with a single antigen binding site, which may be expressed in terms of a single binding constant.

Antibody An immunoglobulin capable of specific combination with the antigen that caused its production in a susceptible animal.

Antibody valency The number of antigen binding sites on an antibody molecule.

Antigen Any foreign substance that elicits an immune response (e.g. the production of specific antibody molecules) when introduced into the tissues of a susceptible animal and which is capable of combining with the specific antibody molecules produced. Antigens are usually of high relative molecular mass and are commonly either proteins or polysaccharides.

Antigen combining site or paratope The site on the antibody that combines specifically with its corresponding antigenic determinant. It usually involves only a few amino acid residues that line a shallow cleft in the antibody molecule but are not usually directly linked to each other.

Antigen valency The number of antibody binding sites on the antigen.

Antigenic determinant or epitope A small site on an antigen to which a complementary antibody molecule may be specifically bound through its combining site. This is usually one to six monosaccharide residues or amino acid residues on the surface of the antigen, not necessarily covalently linked to each other.

Antigenicity The potential of an antigen to stimulate an immune response in a particular host.

Antiserum A serum containing antibodies against a specific antigen or antigen mixture; for example anti-ovalbumin serum, or anti-sheep-erythrocyte serum.

Autoantibody An antibody to one of the constituent molecules of the individual producing the antibody, e.g. anti-DNA or anti-thyroid antibodies. Such antibodies do not arise naturally in healthy individuals and are indicative usually of a pathological condition.

Avidin A protein that specifically binds biotin with very high affinity.

Avidity The net combining power of an antibody molecule with its antigen.

B lymphocyte A cell that, on exposure to a particular epitope, is stimulated to divide and differentiate to form a clone of plasma cells synthesising an immunoglobulin capable of combining specifically with that epitope.

Biotin A water-soluble vitamin.

Clone A family of cells of genetically identical constitution derived asexually from a single cell by repeated division.

Complement A group of serum proteins essential for antibody-mediated immune lysis.

Conjugation The covalent joining together of molecules. For example, haptens (q.v.) may be conjugated to proteins and enzymes conjugated to antibodies.

Epitope See antigenic determinant.

Equivalence The ratio of antigen and antibody giving complete precipitation of both molecules, leaving neither unbound in solution.

Hapten A substance that can combine with a specific antibody but lacks antigenicity, i.e. it cannot initiate an immune response unless bound to an antigenic carrier molecule such as bovine serum albumin. Haptens are usually compounds with relative molecular mass of less than 1000, e.g. simple sugars, amino acids, small oligopeptides, phospholipids, triglycerides, drugs.

HAT medium A growth medium containing hypoxanthine, aminopterin and thymidine, used for selecting HPGRT (q.v.) deficient cell lines.

HPGRT The enzyme hypoxanthine-guanine phosphoribosyl transferase, involved in nucleic acid synthesis.

Hybridoma The name given to cell lines created by fusing B lymphocytes with myeloma cells. Hybridomas produce monoclonal antibodies.

Immunoglobulin A member of the largest family of proteins produced by animals. See Antibody.

J chain (joining chain) A polypeptide involved in the polymerisation of IgA and IgM antibody molecules.

Lymphocyte A cell that, on exposure to a specific antigenic determinant, is capable of triggering an immune response.

Macrophage A general name for a large number of morphologically dissimilar, relatively long lived, phagocytic cells, which may be sessile or motile, and which play an important role in the immune response by processing antigens and presenting them to B and T lymphocytes.

Monoclonal antibody Immunoglobulin derived from a single clone and therefore homogeneous.

Myeloma cell A malignant B lymphocyte or plasma cell. Such cells are immortal.

Phagocytosis The ability of a cell to engulf particles such as bacteria.

Plasma cell The terminal cell of the differentiation of a B lymphocyte on exposure to its specific epitope.

Plasmacytoma Another word for myeloma.

Polyclonal antiserum An antiserum, possibly to a single antigen, containing a number of antibodies to that antigen from different plasma cell clones. (All antisera are in fact polyclonal.)

Protein A Protein derived from *Staphylococcus aureus* that binds to the Fc fragment of IgG.

Streptavidin An avidin-like protein isolated from streptococci.

T cells or T lymphocytes Lymphocytes that have undergone antigen-independent differentiation in the thymus.

Valency See Antibody valency and Antigen valency.

2.1.3   The precipitin reaction

The basic principle involved in most immunochemical techniques is that a specific antigen will combine with its specific antibody to give an antigen–antibody complex, as illustrated in Fig. 2.3. Since antigens and antibodies are both multivalent, the antigen–antibody complex is usually insoluble and may be seen with the naked eye (Section 2.3.1).

2.1.4   Preparative uses of the antigen–antibody interaction

Most immunochemical techniques are used in an analytical capacity. However, a specific antiserum can be used in a preparative mode to isolate the appropriate antigen from a heterogeneous solution by mixing and centrifugation to remove the antigen–antibody precipitate.

If the protein–antibody complex is not completely insoluble, it might be precipitated by the addition of a second antibody capable of precipitating the first antibody or by addition of Protein A from *Staphyloccocus aureus*, which has the property of cross-linking IgG molecules. An alternative procedure, if the aim of the experiment is to isolate and purify a particular protein, is to use the specific antiserum to prepare an immunoadsorbent that may then be used to isolate the protein, either by affinity chromatography (Section 10.9) or as a batch procedure.

Advantages of immunoadsorbent procedures include the production of the antigen in purer form (no other serum components present), far less chance of proteolysis by serum proteases, and more rapid separation of antigen and antibody in the final stages of purification. Advantages of precipitation methods are that less antiserum is required and higher yields of antigen are often obtained.

Radiolabelled, detergent-solubilised membrane antigens have been prepared and purified by the use of monoclonal antibodies and immunoadsorbants. However, mouse and rat monoclonal antibodies bind at best only weakly to staphylococcal Protein A attached to Sepharose, which is an efficient immunoaffinity isolating agent for human and rabbit antibodies. Biotinylated monoclonal antibodies and streptavidin-agarose (Section 2.8) form a superior method for isolating immune complexes formed with monoclonal antibodies that have only a weak affinity for Sepharose A.

The hardest part of using antisera to purify protein antigens is the final dissociation of the antigen–antibody complex. The dissociation conditions must be commensurate with the minimum denaturation of the desired protein, since dissociation usually requires exposure of the complex either to a very low pH or to high concentrations of guanidinium chloride or urea. The correct conditions must be identified by trial and error. It is usual to try to dissociate the complex using the least destructive system from a series such as: 10% (w/v) dioxane, pH 7.2; 25% (w/v) ethylene glycol, pH 6.5; 3 M potassium thiocyanate, pH 6.4; 0.2 M glycine/HCl, pH 2.8; 1 M propanoic acid, pH 2.4. One advantage of the column chromatography method is that the eluted antigen can be collected into another buffer (e.g. 0.3 M borate, pH 8.2) to reduce the exposure time to the denaturing pH.

A powerful and novel use of antibodies, particularly monoclonal antibodies (Section 2.2.2), as preparative agents involves binding a mouse antibody specific for a particular cell type to the surface of polystyrene micro beads containing a core of superparamagnetic ferric oxide. Mixing this preparation with a hetero-geneous single-cell suspension allows the specific cells to bind to the bead, which may then be separated and purified from all other cell types and contaminants using a magnet (Section 1.5.4). For most purposes the purified cells may be used attached to the beads, which retain no magnetism. Alternatively the purified cells may be released by the addition of a goat polyclonal antiserum to mouse Fab regions, marketed under the trade name Detachabead. Mouse IgM antibodies will physically adsorb tightly to the surface of the beads. Mouse IgG antibodies are best attached by first chemically binding sheep anti-mouse IgG to the beads with p-toluenesulphonyl chloride.

2.2   Production of antibodies

2.2.1   Production of antisera (polyclonal antibodies)

Most of the antibodies used in immunochemistry are raised or induced by injection of a solution or suspension of the appropriate antigen into a rabbit. After a suitable period of time, 5–50 cm^3 of blood is obtained from the immunised rabbit by making an incision in the posterior marginal vein of the ear. The blood is allowed to clot at 37 °C for 1 h. The clot is detached from the sides of its container to allow it to retract, and left at 4 °C to contract and exude 2–25 cm^3 of serum. The serum is separated from the clot and free cells by centrifugation. Proteases and complement are inactivated by incubating the serum at 56 °C for 45 min. The serum is usually stored in small portions at -20 °C. A control serum is usually obtained from the same rabbit prior to immunisation. Sheep, goats and horses are used for large-scale antiserum production.

Inoculating a rabbit with a single injection of a saline solution of a strongly antigenic compound results in the production of specific antibodies that are detectable in its serum after about 10 days, reach a maximum after 15–20 days, and then decline over a period of weeks. This is the primary humoral immune response and results mainly in the production of IgM antibodies. A secondary humoral response may be induced by a further injection of the same antigen at any period after the primary response. The secondary response is more rapid and increased antibody levels may be detected after three days, with the maximum level of antibody occurring only 10 days after the second injection. A level of IgM antibody similar to that in the primary response is found but, in addition, 3–10 times that amount of IgG antibody may also be detected. Further injections of the same antigen, at say fortnightly intervals, result in a hyperimmunised rabbit whose serum contains vastly increased levels of specific IgG antibody (1–5 mg specific IgG cm^{-3}).

When weakly antigenic compounds are used, there are two approaches for obtaining antisera of a reasonable titre; that is, antisera containing readily detectable levels of specific antibodies. Firstly, the period of time that the immune system is exposed to the antigen may be extended, either by repeated inoculation or by establishing within the rabbit depots of antigen that slowly release the antigen over a period of weeks. The latter may be achieved by intramuscular, subcutaneous or intradermal inoculation of particulate or precipitated antigen. Secondly, the antigenicity of the compound may be enhanced by the use of adjuvants. Aluminium potassium sulphate is a simple adjuvant used to precipitate soluble protein antigens such as tetanus or diphtheria toxin prior to their inoculation into most animal species. This technique results in the slow release of the antigen, which mimics a prolonged series of injections and which causes the antigen to be more readily trapped or phagocytosed by the macrophages involved in the immune response.

The most commonly used adjuvant used in non-human species is Freund's complete adjuvant, which contains a pharmaceutical grade white mineral oil such as liquid paraffin, an emulsifier such as mannide monooleate, and heat-killed *Mycobacterium tuberculosis*. Often 1 cm^3 of an emulsion of equal parts of Freund's complete adjuvant and the antigen solution is divided into three portions and inoculated into three sites within the rabbit. For example, one portion of the emulsion might be inoculated subcutaneously and the other two inoculated into the hindleg muscles, instead of a single large injection. This extends the number of lymph nodes, and hence a larger number of potentially active B lymphocytes, that the antigen in the emulsion droplets can reach via the lymphatic system. Antigen entering the bloodstream in emulsion droplets may activate lymphocytes in the spleen and bone marrow. A depot of antigen in emulsion tends to persist at each inoculation site and the killed tubercle bacilli stimulate invasion of the site by reticuloendothelial cells. This results in the formation of a granuloma, which further enhances the overall immune response to the antigen. When antisera are raised for use in immunochemical studies, the final inoculations before bleeding usually contain the antigen in an emulsion with Freund's incomplete adjuvant, which lacks the tubercle bacilli and hence tends to stimulate only the humoral immune response.

Muramyl dipeptide (*N*-acetylmuramyl-L-alanyl-D-isoglutamine) is the simplest adjuvant-active component of water-soluble peptidoglycan isolated from *M. tuberculosis*. It results in an enhanced immune response to Freund's complete adjuvant and can be used in aqueous solution rather than an oil–water emulsion. Glucosaminylmuramyl peptides are novel, synthetic analogues of bacterial cell wall glycopeptides and can be used as modulators of the immune response.

Non-antigenic, micrometre-sized β-glucan particles provide an alternative to Freund's adjuvant that yields a comparable response, without causing an inflammatory reaction or granuloma formation. These particles act as sustained-release delivery vehicles targeted to macrophages by virtue of their binding to specific β-glucan receptors and hence activating a cascade of both specific and non-specific immune responses, resulting in the production of sera with high antibody titres.

Antibodies are readily raised to particulate antigens, such as erythrocytes or killed bacteria, in isotonic saline following intravenous inoculation of the antigens into an appropriate mammal. Before haptens such as drugs and non-peptide hormones can be used to raise antisera they must be coupled to antigenic structures such as proteins, polysaccharides or sheep erythrocytes by methods similar to those used to prepare affinity chromatography material (Section 10.9). When the antiserum has been raised, addition of excess carrier macromolecules or erythrocytes to the antiserum precipitates antibodies to the carrier molecules or erythrocytes, leaving only antibodies to the hapten in solution. After centrifugation, the antiserum, now monospecific for the hapten, may be decanted.

It is most important to understand that even a small hapten of relative molecular mass as low as 200 can induce a mammal to produce up to six different immunoglobulins, each differing in amino acid composition and capable of combining with different parts of the surface of the hapten. Therefore it is not surprising that large antigens, such as proteins, induce the production of a multitude of different immunoglobulins, each reacting with a slightly different epitope on the surface of the antigen. This in turn can lead to slight cross-reactions between an antiserum and antigens sharing only a minimum amount of common structural feature. The different immunoglobulins are produced as a result of the interaction between the epitopes in the antigen surface and different clones of B lymphocytes each carrying a specific receptor for that epitope – hence the name polyclonal antiserum. The relative amount of each different immunoglobulin in such a polyclonal serum depends on the route of inoculation of antigen, the species and strain inoculated, as well as the adjuvants used and frequency of inoculation. Thus it is not surprising that the standardisation of so-called antigen-specific polyclonal antisera is very difficult because most animals contain many different clones of B lymphocytes each capable of reacting with a different epitope on the antigen, which may or may not be induced to produce IgG.

In practice it is impossible to obtain a homogeneous population of a single molecular species of immunoglobulin molecules (i.e. a monoclonal antibody) from polyclonal antiserum, despite intensive efforts by immunochemists for many decades. However, modern cell biological techniques, including fusion of B lymphocytes with immortal cell cultures to produce hybridomas, have facilitated the production of monoclonal antibodies.

2.2.2   The production of monoclonal antibodies

The IgG fraction of a normal healthy human individual's serum has a large range of electrophoretic mobilities, reflecting the $10^5 - 10^6$ different IgG molecules present. However, the IgG fraction of the serum of unhealthy individuals suffering from myelomatous disease is much simpler and is dominated by a single type of IgG molecule. Most of the circulating lymphocytes of such individuals are myeloma cells; that is, they comprise a single clone of neoplastic B lymphocytes. Myeloma cells, unlike normal B lymphocytes, continue to divide rapidly and secrete a single species of immunoglobulin molecule in the absence of any stimulating antigen. Although normal B lymphocytes do not survive for more than a few hours *in vitro*, myelomas, like other neoplastic cells, are virtually immortal in cell culture.

Normal mammalian cells have two different ways of synthesising DNA: a *de novo* pathway in which nucleic acids are built up from purine and pyrimidine bases, deoxyribose and inorganic phosphate; and a 'salvage pathway' that converts similar nucleotides into the correct nucleic acid. The *de novo* pathway is completely inhibited by the metabolic poison aminopterin. Cells containing the enzyme HGPRT (Section 2.1.2) can utilise the salvage pathway and survive in the presence of aminopterin. However, as a result of the extra demand placed on the salvage pathway, extra hypoxanthine and thymidine must be added to the so-called HAT medium (hypoxanthine, aminopterin, thymidine). Thus the HAT medium will selectively allow cells containing HGPRT to grow, whilst preventing the growth of HGPRT$^-$ cells. Mouse myeloma cells have been selected that lack HGPRT (HGPRT$^-$ clones) and hence cannot survive in HAT medium. Furthermore, HGPRT$^-$ myeloma cells have been selected that secrete little or even no unwanted immunoglobulin.

The currently standard type of methodology used to produce monoclonal antibodies is illustrated in Fig. 2.2a. The spleen from a mouse immunised with the appropriate antigen is used as a source of a suspension of sensitised B lymphocytes. Sensitised B cells and HGPRT myeloma cells grown *in vitro* are fused in the presence of 30% to 50% (w/v) polyethyleneglycol (PEG) for only a few minutes, as PEG is toxic to cells. The fusion mixture is seeded into culture vessels containing HAT medium. Lymphocytes and fused lymphocytes soon die, as they cannot be cultured *in vitro*. Myeloma cells and fused myeloma cells are poisoned by the aminopterin in the HAT medium, because they lack HGPRT. Only hybridomas, i.e. fusion products of B lymphocytes and HGPRT$^-$ myeloma cells, can survive, because the myeloma cell confers on the hybrid the ability to grow *in vitro* and the B lymphocytes contribute the HGPRT essential for growth in the HAT medium. After 10–14 days, hybridomas will be the only surviving cells. However, most of these will be producing either unwanted immunoglobulins or no immunoglobulin. Unfortunately, non-immunoglobulin-producing hybridomas have a tendency to outgrow the immunoglobulin-producing hybridomas, so that it is essential over the next 7–14 days to test continually individual cultures for required immunoglobulin production and to reject those that are non-productive. Also cultures that are making useful immunoglobulin should be subcultured and portions stored frozen in liquid nitrogen (Section 1.5.6) in order to prevent contamination of later cultures with microorganisms that might destroy months of work.

Finally, single clones are obtained in one of two ways. Cultures may be diluted to limiting dilution and distributed so that each well of a microtitre plate contains only one hybridoma cell. In order for such clones to grow, it is usually necessary to supplement the medium regularly with feeder layers of normal cells such as macrophages. When the clones have grown sufficiently they may be tested for

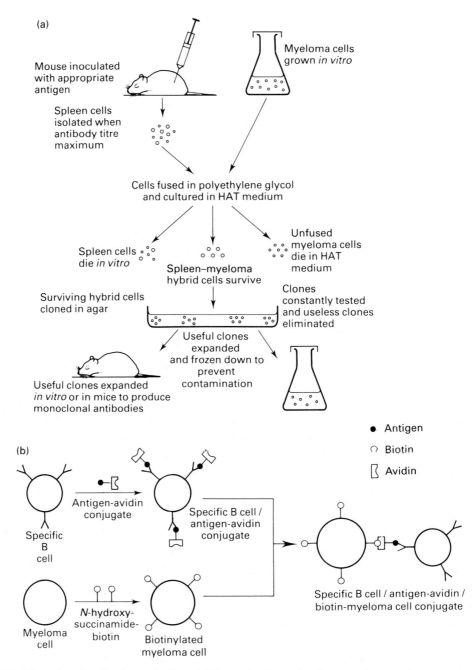

Fig. 2.2. Monoclonal antibody production. (a) General scheme for the procedure most commonly used; (b) scheme for ensuring a high proportion of specific hybridomas by antigen-specific adhesion of the B cells of interest to the myeloma cells.

specific immunoglobulin production using sophisticated immunochemical techniques such as radioimmunoassay or enzyme-linked immunoassays (Sections 2.5 and 2.6, respectively). Alternatively, hybrids may be diluted and seeded into sloppy nutrient agar so that individual growing clones may be observed and picked out with a Pasteur pipette. These clones can then be grown on microtitre plates and tested for specific immunoglobulin production. Portions of each useful clone may be further characterised and stored frozen in liquid nitrogen. It is usual to reclone at least once to ensure that the immunoglobulin produced is a true monoclonal antibody. Usable quantities of monoclonal antibody can then be obtained either by large-scale culture of cells *in vitro*, which is expensive and labour intensive, or more easily by growing the hybridoma cells as an ascites (single-cell) tumour in the peritoneal cavity of a suitable mouse. A disadvantage of the latter procedure is that the monoclonal antibody becomes contaminated with the other proteins found in mouse ascitic fluid, but this is rarely important.

A major disadvantage of the usual methods of monoclonal antibody production is that they rely on the random fusion of a heterogeneous population of B lymphocytes with myeloma cells, resulting in much time being spent in growing and testing hybrids that secrete a vast variety of immunoglobulins. Alternative procedures endeavour to ensure that only B lymphocytes capable of producing immunoglobulins complementary to the antigen of interest lie close enough to myeloma cells for fusion to occur, as illustrated in Fig. 2.2b. This may be engineered by chemically coupling biotin to the surface of the myeloma cells and avidin to the antigen of interest. When the avidin–antigen conjugate is mixed with lymphocytes from the spleen of a mouse immunised with the antigen, it binds specifically with those B lymphocytes that carry receptors for that antigen. When this suspension is mixed with the biotin-conjugated myeloma cell suspension, the avidin and biotin combine and ensure that the myeloma and antigen-specific B lymphocytes are held together by a biotin–avidin–antigen bridge. An intense electric field across the bulk cell suspension (4 kv cm^{-1} for 5 s at 30 °C) then causes fusion of cells that are in close proximity; that is, the myeloma cells and those B lymphocytes capable of making the appropriate immunoglobulins. The appropriate hybrids may then be selected in HAT medium and cloned as described above.

A major advantage of monoclonal antibodies is the extreme confidence with which claims can be made about results obtained with them. However, because of the time, labour and thus cost of producing, and therefore the expense of buying, monoclonal antibodies, polyclonal sera are used whenever there is no distinct advantage to be gained by using monoclonal antibodies. Furthermore, there are occasions on which the extreme specificity of monoclonal antibodies that recognise only single epitopes is a disadvantage and polyclonal sera are of more use.

Despite the labour and expense of their production, monoclonal antibodies have already been prepared against a vast range of antigens, including serum proteins, enzymes, cell surface receptors, hormones, drugs, tumour-specific antigens, viruses, and differentiation antigens. The latter are proving invaluable in the subclassification and isolation by affinity chromatography (Section 10.9) of cell populations, e.g. functional subpopulations of T lymphocytes. The extreme specificity of monoclonal antibodies means that, in general, a clonal product will identify one target molecule. Thus, monoclonal antibodies raise the exciting possibility of identifying and purifying previously unknown, or at least uncharacterised, molecules, such as the T lymphocyte receptor for antigen. This is likely to become the area of research in which monoclonal antibodies have their biggest impact.

Unfortunately, the affinity of monoclonal antibodies for antigen is often less than that of antisera raised against the same antigen. This can result in a loss of sensitivity, which may be offset by utilising mixtures of two or more different monoclonal antibodies to the same antigen, i.e. ultrasensitive, cooperative immunoassays. For example, the sensitivity of a double-antibody radioimmunoassay with two monoclonal antibodies to human chorionic gonadotrophin has been increased 100–1000 times over that possible with the individual monoclonal antibodies. The increased sensitivity is due mainly to the formation of a circular complex consisting of each type of monoclonal antibody and two antigen molecules. In addition, specificity can be greatly affected in cooperative immunoassays, but whether this will be increased or decreased cannot be predicted and can be determined only by trial and error.

Monoclonal antibodies do have limitations in that they are monoclonal with respect to their constant regions as well as their variable regions, and hence monoclonal antibody has a limited number of the potential effector functions of antibodies. For example, not all monoclonal antibodies fix complement, bind staphylococcal Protein A or carry out antibody-dependent cell-mediated cytotoxicity. Additionally, whilst it is reasonably straightforward to generate mouse monoclonal antibodies of virtually any specificity, it is much more difficult to generate human monoclonal antibodies that would be preferential for some applications.

Transfectomas (transfected cell lines) provide one approach to overcoming some of the limitations inherent in monoclonal antibodies. Genetically engineered antibody genes can be expressed after transfection into lymphoid cells. Chimaeric human–mouse antibodies, which deal in part with the species limitations of monoclonal antibodies, have been produced to haptens, tumour antigens, and the CD4 antigen of helper T cells. Potentially any binding specificity can be engineered into an antibody by the use of synthetic oligonucleotides and site-directed mutagenesis. Furthermore, antibodies with tailor-made effector functions can be produced, as well as antibodies with constant regions that are never found in nature, including non-immunoglobulin proteins, such as enzymes.

Mouse hybridomas yield 10–$100\,\mu g\ cm^{-3}$ in cell culture. Both scale-up of cell culture and use of ascites culture can yield gram and even kilogram quantities, but both are costly. The bacterium *Escherichia coli* and the yeast *Saccharomyces cerevisiae* are both capable of expressing, assembling and secreting Fab fragments and possibly whole antibody molecules. These two well-characterised systems are powerful new tools for the cost-effective manufacture of antibody fragments for a wide variety of uses.

2.3   The precipitin reaction in free solution

2.3.1   Principles

If increasing amounts of an antigen (e.g. human serum albumin) are mixed with portions of suitable antibody solution (e.g. rabbit anti-human serum albumin) a precipitate forms in some of the tubes. After equilibration, the antigen–antibody precipitate may be isolated from the supernatant by centrifugation and may be estimated with a suitable protein assay (Section 4.4). The presence of excess antigen or antibody in each supernatant sample may be checked by looking for the formation of an antigen–antibody precipitate when further antibody or antigen solution is added to a portion of the supernatant. The results of an experiment in which increasing amounts of an antigen are added to a fixed amount of antibody are illustrated in Fig. 2.3.

As expected, initially the amount of antigen–antibody precipitate increases with increasing amounts of antigen added. However, a sharp plateau indicating that all of the antibody has been precipitated is not obtained and furthermore at higher concentrations of antigen the precipitate apparently dissolves. This is due to the solubility of antigen–antibody complexes containing a single antigen molecule, even if antibody molecules are bound to every antigenic determinant. The resulting curve of Fig. 2.3 may be divided into three zones:

   (i) zone of antibody excess, where the addition of further antigen leads to a substantial increase in the amount of precipitate;
   (ii) zone of equivalence, when the maximum amount of antigen–antibody precipitate is formed;
   (iii) zone of antigen excess, when the precipitate dissolves. Thus valid data can be obtained only for the precipitin reaction in the zone of equivalence.

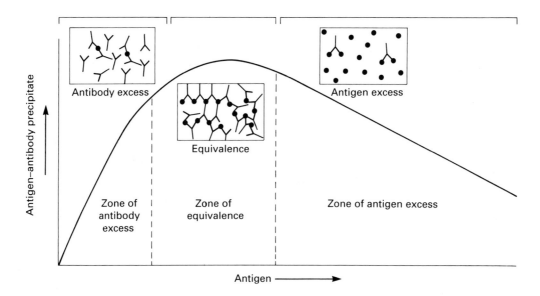

Fig. 2.3. Typical precipitin curve for a protein antigen (•) reacting with rabbit antiserum.

It should be noted that monoclonal antibody preparations can be used to precipitate only antigens that contain more than one copy of the epitope recognised by the monoclonal antibody.

2.3.2   Qualitative analysis of antigen

It is possible to detect the presence of an antigen by adding a range of concentrations of the suspected antigen to a fixed amount of monospecific antiserum (containing antibodies to one antigen only) and looking for precipitation. In this way it is possible to ascertain the presence of any one particular antigen (e.g. specific protein or carbohydrate) in the presence of many related compounds, e.g. during the isolation of any one component of serum or a cell extract. A major disadvantage of this procedure is that false negative data may be obtained if the ratios of antigen to antibody in the experiment do not fall within those pertaining to the zone of equivalence.

2.3.3   Quantitative analysis

Provided that the zone of equivalence can be determined, the precipitin reaction provides a very accurate method for the quantitative assay of antigens. Usually the total reaction volume is not more than $100\,mm^3$, and precipitation is allowed to take place overnight at $4\,^\circ$C. The antigen–antibody complex is washed and its amount determined by any sensitive protein or nitrogen assay (Section 4.4), by liquid scintillation counting if the antibody is labelled with a negatron emitter (Section 5.1.3), or by γ-counting if the antibody is labelled with a γ-emitting isotope, e.g. [131]I. It should be noted that all of these methods require the construction of a calibration curve from standards.

The precipitin reaction in free solution is an important historical technique that is generally superceded by more simple methods such as single radial immunodiffusion (Section 2.4.2). It is, however, particularly useful during the quantitative assay of a protein synthesised *in vitro* under the direction of an isolated or chemically synthesised messenger RNA (mRNA). For example, the rate of synthesis of haemoglobin in the presence of mRNA extracted from chick erythrocytes, *E. coli* ribosomes and cofactors, and a radiolabelled amino acid, may be determined by measuring the rate of incorporation of radioactivity into the synthesised protein that is precipitable by an anti-haemoglobin serum. Homologous protein (chick haemoglobin) is mixed into the system as a carrier, followed by the equivalent amount of specific antibody, e.g. rabbit anti-chick haemoglobin. This coprecipitates virtually all of the protein synthesised *in vitro* that may be assayed by liquid scintillation counting after separation by centrifugation.

2.3.4   Hapten inhibition test

The specificity of the antigen–antibody reaction can be used to elucidate the chemical nature of the antigenic determinants in a macromolecule. For example, an insight into the sequence of monosaccharides in a bacterial heteropolysaccharide may be gained by mixing portions of the polysaccharide with monospecific antisera to known small oligosaccharides and from the precipitation of polysaccharide–antibody complex with the specific anti-oligosaccharide antiserum, noting which antigenic determinants are present. Unfortunately, this requires the laborious production of an antiserum to each oligosaccharide which, since they are haptens and hence not antigenic per se, also requires their prior coupling to a carrier protein. The hapten inhibition test saves most of this effort and is illustrated in Fig. 2.4.

Mixing an antigen that contains repeat determinants (epitopes) with antisera containing antibodies specific to those determinants results in a visible precipitate (Fig. 2.4a). Mixing the antiserum with a hapten that shares a determinant with the antigen also leads to a reaction, but no precipitate results. That a reaction has occurred can be demonstrated by adding the original antigen to the mixture of hapten and antiserum, since no free antibody remains to precipitate the antigen (Fig. 2.4b). Therefore, lack of precipitation at this stage indicates that the hapten

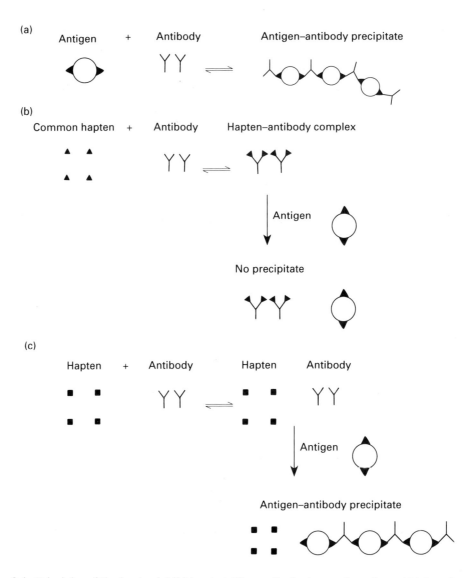

Fig. 2.4. Principles of the hapten inhibition test. The antibody detects the epitope (▲) found on the antigen.

and antigen share a common determinant. Mixing the specific antiserum with another hapten that does not share a common determinant with the antigen does not lead to complex formation. Addition of antigen to this system results in a precipitate, indicating that this hapten does not share a determinant with the antigen (Fig. 2.4c). Thus, a whole range of small molecules may be screened rapidly to elucidate the chemical nature of a complex antigen.

The hapten inhibition test provided early evidence that the maximum size of most polysaccharide and protein antigenic determinants was of the order of five to six monosaccharide or amino acid residues, respectively. The test has been used to show that N-terminal *N*-acetylgalactosamine and galactose residues constitute the major difference between the blood group A and B determinants, respectively, and it is still used to investigate blood and tissue antigens.

The major contribution of this test has been to increase our knowledge of bacterial heteropolysaccharide structure. This knowledge, in conjunction with that obtained using mono-determinant-specific antibodies, has resulted in an extensive understanding of the complexities of cross-antigenic reactions between individual strains of bacteria, which allows typing of the strains of clinical importance such as the Salmonellae. Antibodies to determinants shared by two strains of bacteria may be removed from a polyclonal antiserum raised against one strain by the addition of excess bacterial cells of the other strain. A mono-determinant-specific antiserum may be obtained eventually by repeating this absorption process with a number of related strains.

The major drawback in using the precipitin reaction in free solution is the requirement that the ratio of antigen to antibody must be in the region of equivalence. This makes such techniques lengthy to perform and may require relatively large quantities of expensive antisera or antigens.

2.4   The precipitin reaction in gels: immunodiffusion (ID)

2.4.1   Principles

When molecules such as soluble antigens diffuse from a homogeneous solution into an agar gel, the concentration falls from a maximum at the solution/gel interface to zero at the leading edge of the region penetrated. Thus the system rapidly adjusts to give a complete antigen concentration gradient. Somewhere along this concentration gradient will be an antigen concentration that will give equivalence with almost any given concentration of antibody. This fact has been used since 1905 and developed into a range of highly sophisticated techniques.

2.4.2   Single (simple) immunodiffusion

This technique usually involves the diffusion of antigen from a solution into a gel containing antibody.

In two dimensions: single radial immunodiffusion (SRID)

In this (Mancini) technique, a range of antigen concentrations is placed in wells cut in agar containing the corresponding antiserum on a microscope slide. As the antigen diffuses out radially from the wells a ring of precipitation forms and appears to move outwards, eventually becoming stationary at equivalence (Fig. 2.5a). The precipitate ring diameter at equivalence is a function of the antigen

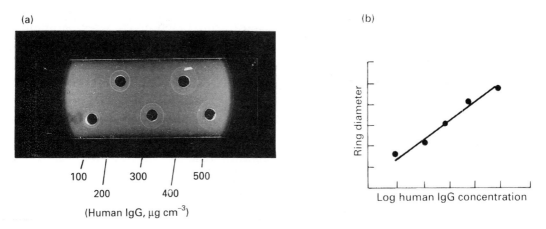

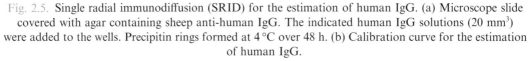

Fig. 2.5. Single radial immunodiffusion (SRID) for the estimation of human IgG. (a) Microscope slide covered with agar containing sheep anti-human IgG. The indicated human IgG solutions (20 mm^3) were added to the wells. Precipitin rings formed at 4 °C over 48 h. (b) Calibration curve for the estimation of human IgG.

concentration. By plotting ring diameter or ring area at equivalence against antigen concentration, a calibration curve may be obtained for determining the concentration of antigen in unknown solutions (Fig. 2.5b).

This technique is commonly used for estimating the concentration of various plasma proteins, e.g. IgG and IgM in patients suspected of suffering from agamma-globulinaemia and multiple myeloma, respectively.

2.4.3   Double immunodiffusion in two dimensions

Ouchterlony technique

This technique, which involves the diffusion of both antigen and antibody (towards each other), is probably the most widely used immunochemical technique. It may, for example, be used for detecting which sera, or chromatographic or cell fractions, contain a particular antigen and whether or not two antigens are identical, different or share antigenic determinants.

Originally 5–10 mm diameter wells were cut with a cork borer and removed from 1–2 mm thick layers of agar in a Petri dish. A typical pattern used for comparing different antigen preparations is shown in Fig. 2.6a. The Petri dishes were stored in a humid chamber. The wells were observed and refilled with the appropriate solutions daily. Nowadays, patterns are cut in agar layers on microscope slides with stainless steel cutters (Fig. 2.6b). This allows much smaller quantities of

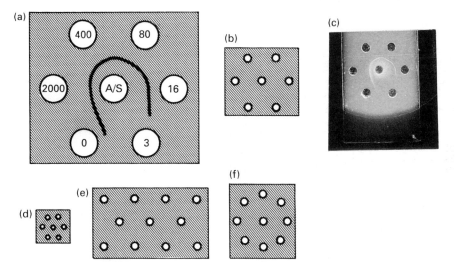

Fig. 2.6.   Double immunodiffusion in two dimensions (drawn to same scale). (a) Macroscale as used in Petri dishes, showing the predicted effect of different antigen concentrations ($\mu g\,cm^{-3}$) on the position of precipitin lines relative to the wells. The central well contains antiserum. (b) Commonest pattern used on microscope slide scale. (c) Photograph of the commonest pattern used on microscope slides, showing the results of the experiment described in (a) of the effect of antigen, human serum albumin (HSA) concentration on the position of precipitin lines relative to wells. Anti-human serum albumin was in the central well. (d) Commonest pattern using only one drop of agar and a plastic template; (e) and (f) Other common patterns.

antigens and antisera to be used and gives results in a few, rather than many, days.

Fig. 2.6c is a photograph of the result of an Ouchterlony immunodiffusion experiment. Because the antiserum used in the centre well contained antibodies to most of the components of human serum, the presence of a single precipitation line between this well and the peripheral wells indicates that these wells contained an apparently pure human serum protein, certainly not grossly contaminated with any other human serum protein. However, it should be noted that these results tell us nothing about any contaminant to which the centre well does not contain antibodies!

Two-dimensional double diffusion through a very thin layer of agar may be carried out by sticking a small plastic template containing holes cut in a conventional Ouchterlony pattern to a microscope slide with only one drop of agar, which serves as both the adhesive and the diffusion medium. The results are obtained even more rapidly (hours rather than days) and relatively dilute antigen solutions may be used, because the diffusion distance between wells is small and the ratio of well volume to agar volume is large (Fig. 2.6d).

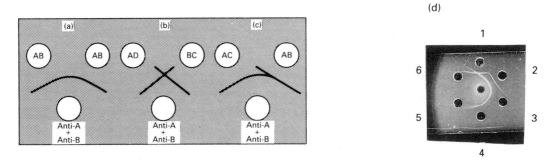

Fig. 2.7. The basic patterns of precipitation formed during double diffusion in two dimensions (Ouchterlony method). (a) Reaction of identity, lines fuse; (b) reaction of non-identity, lines cross; (c) reaction of partial identity, lines spur. A, B, C, D represent antigenic determinants, and the antiserum used in each example is anti A + anti B.
(d) Photograph of an Ouchterlony immunodiffusion slide showing reactions of both identity and non-identity. Solutions used to keep the wells filled were as follows: (1) human IgG (400 $\mu g\,cm^{-3}$), (2) human IgG (400 $\mu g\,cm^{-3}$) + human serum albumin (1000 $\mu g\,cm^{-3}$), (3) human serum albumin (1000 $\mu g\,cm^{-3}$), (4) human serum albumin (100 $\mu g\,cm^{-3}$), (5) human serum albumin (10 $\mu g\,cm^{-3}$), (6) human serum albumin (1 $\mu g\,cm^{-3}$). Centre well, rabbit anti-human serum.

A reaction of identity occurs between an antibody and antigens containing identical antigenic determinants, and results in smoothly fused precipitation lines (Fig. 2.7a). A reaction of non-identity occurs when the antiserum contains antibodies to both antigens but the two antigens do not share a common determinant. The two lines are formed independently with different antibody molecules and cross without interaction (Fig. 2.7b). A reaction of partial identity occurs when two antigens have at least one common determinant, but where the antisera contains antibodies to a determinant in one antigen that is absent from the other (Fig. 2.7c). Figure 2.7d clearly shows (i) a reaction of identity between the human IgG in wells 1 and 2, (ii) reactions of identity between the human serum albumin in wells 2, 3 and 4; (iii) a reaction of non-identity between the human IgG in well 1 and the human serum albumin in well 2.

The relative position of a precipitation line yields an estimate of antigen concentration, i.e. the more concentrated the antigen solutions the further from the antigen well the precipitation line is formed, as illustrated in Fig. 2.6a and c. The shape of a precipitation line can give a rough estimate of the relative molecular mass of a globular protein antigen. The major antibody is usually IgG, with a relative molecular mass of 150 000. Globular proteins with relative molecular masses significantly less than 150 000 will diffuse through agar more rapidly and give an arc of precipitation as shown in Fig. 2.8a. Those with molecular weights significantly greater than 150 000 will diffuse more slowly and give an arc in the opposite direction (Fig. 2.8c). Antigens with relative molecular masses similar to that of IgG will give a straight precipitation line (Fig. 2.8b).

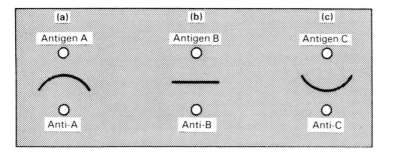

Fig. 2.8. The effect of relative molecular mass/M_r of an antigen on the shape of the precipitin lines formed in the Ouchterlony method. (a) M_r of antigen A $\ll M_r$ of IgG (e.g. human serum albumin, 68 000); (b) M_r of antigen B $= M_r$ of IgG (e.g. IgA); (c) M_r of antigen C $\gg M_r$ of IgG (e.g. keyhole limpet haemocyanin 3×10^5).

2.4.4 Immunoelectrophoresis (IE)

Qualitative immunoelectrophoresis

This technique combines the specificity of immunoprecipitin reactions with the separation of molecules by electrophoresis in a molecular sieving medium. Usually the analysis is carried out in an agarose gel containing barbitone buffer on a microscope slide. A suitable pattern, like that shown in Fig 2.9a, is cut with a gel punch and 1–5 mm^3 solutions containing 1–100 µg of antigen are added to the wells. The slides are connected by thick wet filter paper wicks to the electrode wells, and a direct electric current of about 8 mA per slide is passed for 1–2 h, giving a voltage drop of 4–8 V cm^{-1}. Charged molecules will have been separated electrophoretically, as indicated in Fig. 2.9b, but of course will not usually be visible. Immediately after the voltage supply has been disconnected, the troughs are filled with appropriate antisera and incubated overnight at room temperature in a humid chamber. The antigens diffuse radially and the antibodies diffuse laterally, as shown in Fig. 2.9c, resulting in the antigen–antibody precipitation arcs shown in Fig. 2.9d. Despite the use of agarose, which acquires a smaller charge than agar, the electroosmotic flow of water during electrophoresis (Section 9.1) moves all of the antigens towards the cathode. This results in the apparent cathodic

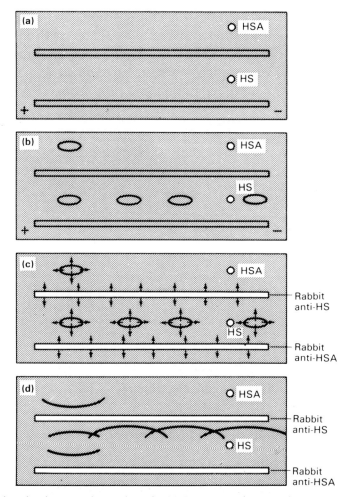

Fig. 2.9. Stages in microimmunoelectrophoresis. (a) Pattern cut in agar. Agar removed from wells and wells filled with antigens, human serum albumin (HSA) and human serum (HS), respectively. (b) After electrophoresis the proteins will have migrated as indicated but will not be visible. Agar removed from troughs. (c) On addition of antisera to troughs, diffusion of antigens and antibodies occurs as shown. (d) The final precipitin lines expected for the systems described.

migration of γ-globulins, including IgG antibodies, although they migrate very little on electrophoresis on uncharged support media. This technique may be used to investigate the purity of, or to detect, particular antigens in sera, culture filtrates, tissue or cell extracts, or fractions from any preparative procedure.

Cross-over immunoelectrophoresis

As illustrated in Fig. 2.9d, most proteins show anodic migration at pH 8.0 but γ-globulins are exceptional in apparently migrating towards the cathode, due to electroosmosis (Section 9.1). Cross-over electrophoresis, as illustrated in Fig. 2.10, takes advantage of this by moving IgG antibodies (γ-globulins) and antigens towards each other and resulting in the formation of precipitation lines. This technique is therefore more rapid (15–20 min) than the Ouchterlony method, which may take days to produce a clear result. In addition it is more sensitive because all of the molecules migrate towards each other rather than diffusing radially. The technique is particularly useful in forensic science for establishing the species of origin of body fluids such as blood, semen and saliva. Up to 12 samples may be tested on one agar-covered microscope slide, saving time, valuable reagents and samples.

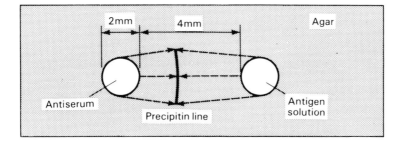

Fig. 2.10. The principles of cross-over electrophoresis.

Quantitative immunoelectrophoresis

Laurell's 'rocket' electrophoresis is related to single radial immunodiffusion in the same way that cross-over electrophoresis is related to the Ouchterlony technique. As in single radial immunodiffusion, the antigen sample is placed in wells cut in agar containing specific antiserum to the antigen to be assayed. When a direct electric current is applied, most antigens migrate towards the anode and the IgG antibodies migrate towards the cathode. Initially, soluble antigen–antibody complexes are formed in the presence of excess antigen. When all of the antigen has migrated into the gel, equivalence is reached and antigen–antibody complexes precipitate, as shown in Fig. 2.11.

The area under the rocket shape is directly proportional to the antigen concentration. When the precipitation arcs have become stationary (1–10 h) a plot of rocket height against concentration will be linear. Thus, by the use of standards and the preparation of a calibration curve, the concentration of antigen solutions

Fig. 2.11. A typical rocket immunoelectrophoresis. The gel has been stained with Coomassie Brilliant Blue to enhance the immunoprecipitates. The gel contained anti-bovine serum albumin (BSA) and 1 mm³ of different concentrations of BSA were loaded into individual wells. (Courtesy of J. Walker, University of Hertfordshire.)

may be determined. Obviously, unlike single radial immunodiffusion, the technique cannot be used to determine the concentration of IgG because of its cathodic migration!

Two-dimensional immunoelectrophoresis

This technique combines electrophoretic separation in a molecular sieving medium with the specificity and speed of rocket electrophoresis. Initially, antigens are separated by agar electrophoresis in one dimension (Fig. 2.9b). A suitable slice of the gel is transferred on to a square glass plate and a layer of agar containing a suitable antiserum is solidified against it over the rest of the plate. After rocket electrophoresis in the second dimension, precipitin arcs like those shown in Fig. 2.12, are formed. By comparison of the area under the arcs with those of standard systems, a semi-quantitative estimate of the amount of individual antigens present may be made.

Fig. 2.12. A typical two-dimensional immunoelectrophoresis of 0.5 mm³ of a human serum sample. The gel has been stained with Coomassie Brilliant Blue to enhance the immunoprecipitates. The large peak on the right-hand side is human serum albumin, the major serum protein.

2.4.5 Visualisation and recording of precipitin lines in gels

Precipitin lines in gels should always be developed in a humid atmosphere and wells should not be allowed to dry out. Usually the precipitin lines can be seen clearly with the naked eye, particularly if the gel is illuminated from the side and viewed against a dark background (Fig. 2.13). If staining is desired, excess proteins are first removed from the gels by washing for 12–24 h in several changes of phosphate-buffered saline. Precipitin arcs may be stained before or after drying

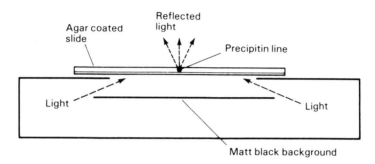

Fig. 2.13. Principle of the dark background viewer for studying precipitin lines.

with most protein stains, for example Coomassie Brilliant Blue. Gels are dried overnight by covering with a good-quality filter paper. The filter paper is removed after slight dampening, leaving a permanent record that is easily stored in a slide box. Alternatively, permanent records may be made, without drying, by drawing or photography. Polaroid cameras enable a series of records to be made and stored as precipitation progresses.

2.5   Radioimmunoassay (RIA)

2.5.1   Principles

Radioimmunoassay is one of the most important techniques in the clinical and biochemical fields for the quantitative analysis of hormones, steroids and drugs. It combines the specificity of the immune reaction with the sensitivity of radioisotope techniques. Alternative names used for RIA include saturation analysis, displacement analysis and competitive radioassay.

   The technique is based on the competition between unlabelled antigen and a finite amount of the corresponding radiolabelled antigen for a limited number of antibody binding sites in a fixed amount of antiserum. At equilibrium in the presence of an antigen excess there will be both free antigen and antigen bound to the antibody. Under standard conditions the amount of labelled antigen bound to antibody will decrease as the amount of unlabelled antigen in the sample increases, for example:

$$4Ag^* + 4Ab \rightleftharpoons 4Ag^*Ab$$

$$4Ag + 4Ag^* + 4Ab \rightleftharpoons 2Ag^*Ab + 2AgAb + 2Ag^* + 2Ag$$

$$12Ag + 4Ag^* + 4Ab \rightleftharpoons Ag^*Ab + 3AgAb + 3Ag^* + 9Ag$$

where Ab, Ag, Ag* and AgAb represent one equivalent of each of antibody, unlabelled antigen, labelled antigen and antigen–antibody complex, respectively.

Known amounts of unlabelled antigen and a fixed amount of antibody and labelled antigen may be used to measure the amount of labelled antigen bound as a function of the total antigen added, and a calibration curve can be constructed (Fig. 2.14). This enables an unknown amount of antigen to be determined.

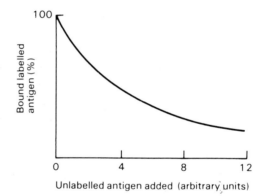

Fig. 2.14. A typical radioimmunoassay calibration curve.

2.5.2   Practical aspects

Pure antigen, labelled antigen and a suitable antiserum are all required for RIA. Pure antigen is required for standard samples, production of labelled antigen and the production of a specific antiserum. Labelled antigen must be produced with minimal alteration of the immunoreactivity of the molecule. It should also have a high specific activity (Section 5.1.6) in order to give an assay with maximum sensitivity. Usually the phenolic group of a tyrosine residue in protein antigens is labelled with ^{125}I by either the chloramine T or the less drastic lactoperoxidase method, as illustrated in Fig. 2.15. The labelled protein is immediately separated from unreacted iodide, and any denatured protein removed by gel filtration. Non-protein haptens are usually labelled with tritium. The availability of a suitable antiserum is probably the most important single factor for satisfactory RIA. Therefore care must be taken to ensure the specificity and avidity of the antiserum produced. It is usual to work with an antiserum dilution that binds 30% to 60% of the labelled antigen.

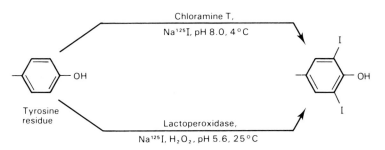

Fig. 2.15. Methods of labelling the tyrosine residues of protein antigens with ^{125}I.

Antibody-bound and free antigen are separated in order that the radioactivity in one or both fractions may be measured. This enables the proportion of labelled antigen bound to be estimated and hence the amount of unlabelled antigen in the system to be determined from the calibration curve. Separation methods based on the removal of free antigen include ion-exchange chromatography and adsorption on to charcoal or silica. After centrifugation, the antibody-bound antigen is left in the supernatant for radioassay. Separation methods based on the removal of antibody-bound antigen include:

(i) the double-antibody technique, in which an antiserum capable of reacting specifically with the first antibody to form a separable precipitate is used;

(ii) the non-specific precipitation of antibody-bound antigen with salt or organic solvent, e.g. saturated ammonium sulphate solution, dioxane or ethanol.

Alternatively, if the antibodies used in RIA are initially attached to Sephadex beads or the walls of test tubes it is possible to centrifuge, decant and wash away the unbound antigen.

Whenever RIA is being carried out it is essential that a number of standards are assayed under identical conditions, because environmental conditions are almost never identical. Standards and samples should always be prepared at least in duplicate. Many RIAs are carried out at room temperature, but to prevent proteolysis and microbial growth 4 °C is used if incubation times exceed 6 h.

Finally, γ-emitting radioisotopes such as ^{131}I are assayed in a γ-counter, and negatron-emitting isotopes such as ^{3}H are measured in a liquid scintillation counter (Section 5.2.2).

The major advantages of RIA include the following (see also Table 2.2):

(i) It can be used to assay any compound that is immunogenic, available in a pure form, and can be radiolabelled.

(ii) It has high sensitivity – some compounds may be detected at a level of $pg\,cm^{-3}$.

(iii) It shares the high specificity of all immunoassays.

(iv) Its precision is comparable to that of other physico-chemical techniques and far better than that of bioassays.

(v) It can be automated so that a minimum of manual handling and data processing is necessary, which allows a large number of samples to be processed at minimal cost.

The major disadvantages of RIA include the following:

(i) The relatively high cost of equipment and reagents. Gamma-scintillation counters are expensive to buy and maintain, and radioiodine is not a particularly cheap reagent.

(ii) The shelf-life of reagents – the half-lives of ^{125}I and ^{131}I are 60 days and 8 days, respectively, necessitating relatively frequent labelling of antisera.

(iii) The radiological hazards of using radioiodine, particularly during the labelling of antisera, which must be repeated fairly regularly. Staff should have regular thyroid scans and be rested if the level of radio-activity increases significantly.

(iv) The duration of assays (usually days rather than hours) results in extensive usage of highly skilled technical staff.

2.5.3   Immunoradiometric assay (IRMA)

IRMA utilises purified radiolabelled antibody to the antigen to be estimated and is often more sensitive than RIA. IRMA is a direct binding assay in which the radiolabelled antibody reagent is in excess and the amount bound to the antigen is actually measured. RIA, however, is a competitive displacement assay in which the radioactive antigen reagent, which is limiting, is measured. Increased specificity may be obtained by removing the antigen to be assayed from solution with antibody against one subunit or epitope and using radiolabelled antibody to another subunit or epitope. For example, thyroid-stimulating hormone (TSH) may be assayed by allowing it to bind to an antibody complementary to the α-subunit of TSH attached to Sephadex and then reacting it with radiolabelled antibody against the TSH β-subunit (Fig. 2.16). As follicle-stimulating hormone (FSH) and luteinising hormone (LH) share a common α-subunit with TSH, but have antigenically distinct β-subunits, FSH and LH can be assayed in the same samples as TSH, provided that radiolabelled anti-β-subunit antisera are available, with a minimum of extra effort.

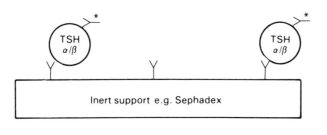

Y   Anti-TSH α-subunit

Y*   Radiolabelled anti-TSH β-subunit

Fig. 2.16. The immunoradiometric assay of thyroid-stimulating hormone.

2.6   Enzyme-linked immunosorbent assay (ELISA)

2.6.1   Principles

Enzyme immunoassays combine the specificity of antibodies with the sensitivity of simple spectrophotometric enzyme assays by using antibodies or antigens coupled to an easily assayed enzyme that also possesses a high turnover number. ELISA is replacing RIA, despite the latter already being established, extensively automated and sometimes more sensitive. This is because ELISA is relatively cheap to operate, lacks the radiological hazards of RIA and is suitable for use in small laboratories lacking γ-radioactive counting facilities, such as some hospital diagnostic laboratories and the laboratories of the developing nations. A comparison of the two techniques, with that of fluorescence immunoassay, is shown in Table 2.2.

Table 2.2. *A comparison of enzyme immunoassay (ELISA), radioimmunoassay (RIA), and fluorescence immunoassay (FIA)*

	ELISA	RIA	FIA
Specificity	+	+	+
Sensitivity	+	+	+
Reproducibility	+	+	+
Cost	+	−	±
Reagent shelf life	+	−	+
Result reading			
Objective	+	+	+
Automated	±	+	±
Subjective	+	−	−
Safety	+	−	+
Usefulness			
Large central laboratories	+	+	+
Small diagnostic laboratories	+	−	±

+, favourable; − unfavourable.

ELISA may be used for assaying antigens by either a competitive or a double-antibody method and for assaying a specific antibody by an indirect method. All these methods require the preparation of a calibration curve during the assay.

The competitive method is summarised in Fig. 2.17. A mixture of a known amount of enzyme-labelled antigen and an unknown amount of unlabelled antigen is allowed to react with a specific antibody attached to a solid phase (Section 2.6.2). After the complex has been washed with buffer, the enzyme substrate is added and the enzyme activity measured. The difference between this value and that of a sample lacking unlabelled antigen is a measure of the concentration of unlabelled antigen. A major disadvantage of this method is that each antigen may require a different method to couple it to the enzyme; this is not so for the double-antibody method.

The double-antibody method for ELISA is summarised in Fig. 2.18. The unknown antigen solution is reacted with specific antibody attached to a solid phase, washed and treated with enzyme-labelled antibody (directed against a different epitope, if monoclonal antibodies are being used). After a further wash the enzyme substrate is added. The amount of enzyme activity measured under standard conditions is directly proportional to the amount of antigen present (cf. IRMA, Section 2.5.3). An advantage of this method is that only one procedure is required to couple the enzyme to all antibody preparations.

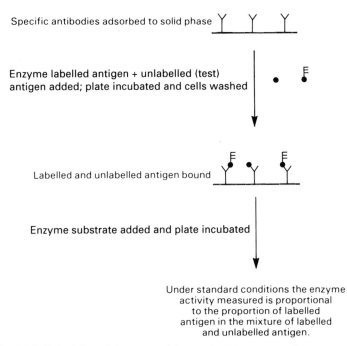

Specific antibodies adsorbed to solid phase

Enzyme labelled antigen + unlabelled (test) antigen added; plate incubated and cells washed

Labelled and unlabelled antigen bound

Enzyme substrate added and plate incubated

Under standard conditions the enzyme activity measured is proportional to the proportion of labelled antigen in the mixture of labelled and unlabelled antigen.

Fig. 2.17. Principles of the competitive method for enzyme immunoassay.

An indirect ELISA method may be used to measure antibody levels, as summarised in Fig. 2.19 (cf. indirect immunofluorescence, Fig. 2.22b). The putative antiserum is reacted with specific antigen attached to a solid phase. Any specific antibody molecules bind to the antigen and all other material is washed away. Exposure of the complex to enzyme-labelled anti-immunoglobulin antibody results in binding to any specific antibody molecules adsorbed from the original serum. The complex is washed and the substrate for the enzyme added, resulting in activity proportional to the amount of specific antibody in the original serum.

The sensitivity of any type of ELISA may be greatly enhanced by means of enzyme amplification. In its simplest form the primary enzyme product is used to trigger a secondary enzyme system that can generate a large quantity of coloured product (Section 4.9). Since the product of the first enzyme need not be measurable but acts catalytically only on the second system, enzymes not currently used for ELISA may become important in these systems, e.g. aldolase or glucose-6-phosphatase. Enzyme amplification of a double-antibody assay in which alkaline phosphatase, the primary enzyme, degrades the substrate $NADP^+$ to NAD^+ can be achieved by using alcohol dehydrogenase as the amplifying enzyme and a tetrazolium dye as a redox acceptor. When the alcohol dehydrogenase is triggered, the tetrazolium dye is reduced to a coloured formazan product, which can be

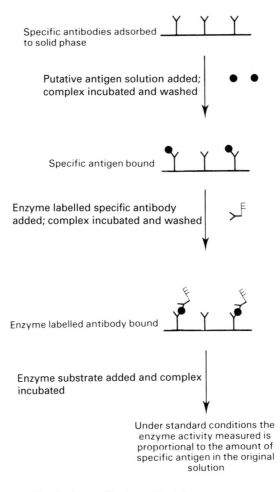

Specific antibodies adsorbed
to solid phase

Putative antigen solution added;
complex incubated and washed

Specific antigen bound

Enzyme labelled specific antibody
added; complex incubated and washed

Enzyme labelled antibody bound

Enzyme substrate added and complex
incubated

Under standard conditions the
enzyme activity measured is
proportional to the amount of
specific antigen in the original
solution

Fig. 2.18. The double-antibody method for enzyme immunoassay.

assayed spectrophotometrically, e.g. iodonitrotetrazolium violet may be reduced to a red formazan and the yellow thiazolyl blue may be reduced to a blue formazan. Under appropriate conditions 500 molecules of formazan are produced per minute by such a redox amplifier, for each molecule of NAD^+ produced by the primary enzyme system. Enzyme amplification can be carried out as (a) a one-step process, in which both enzymes and substrates are reacting at the same time; or (b) a two-step amplification in which the first enzyme is inhibited before or during the addition of the second enzyme and substrate.

Enzyme amplification immunoassays have already been developed for hormones, pathogenic viruses and bacteria, and tumour markers. A possible significant advantage of enzyme amplification immunoassays over the technological demands

Specific antigen attached to solid phase

Serum containing putative antibodies
added; plate incubated and cells washed

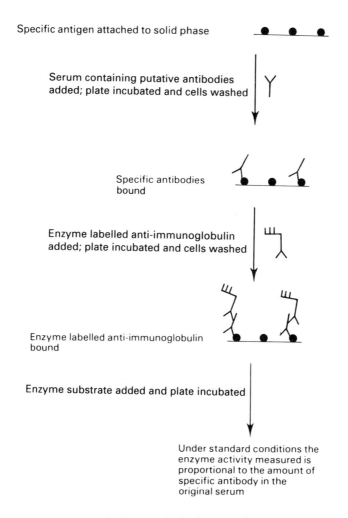

Specific antibodies
bound

Enzyme labelled anti-immunoglobulin
added; plate incubated and cells washed

Enzyme labelled anti-immunoglobulin
bound

Enzyme substrate added and plate incubated

Under standard conditions the
enzyme activity measured is
proportional to the amount of
specific antibody in the
original serum

Fig. 2.19. The indirect method of enzyme immunoassay.

of RIA (Section 2.6) and FIA (Section 2.8) is that they may be assayed with a simple spectrophotometer or even by visual inspection.

2.6.2   Practical aspects

Solid phases used in ELISA include cross-linked dextran or polyacrylamide beads, filter paper (cellulose) discs or polypropylene tubes, and disposable polystyrene microtitre plates are particularly convenient for large numbers of samples. The appropriate antigen or antibody may be attached to the solid phase by passive adsorption or covalent coupling with cyanogen bromide (Section 10.9.2). Test samples of antigen or antibody are usually diluted with the same buffer containing a wetting agent, such as Triton X-100, used for washing the antigen-bound solid phase to prevent non-specific interactions (hydrophobic bonding) that would lead to erroneous results.

The conjugate used must contain a highly reactive antibody coupled to an enzyme with a high turnover number. Alkaline phosphatase and horseradish peroxidase are commonly used enzymes. Glutaraldehyde is a popular coupling agent but requires reactive amino groups to be part of the enzyme as well as the antibody. Periodate oxidation of the extensive carbohydrate group of horseradish peroxidase produces aldehyde groups that will combine with the amino groups of immunoglobulins to produce Schiff base conjugates that can be stabilised by reduction with sodium borohydride.

Ideally enzyme substrates should be stable, safe and inexpensive. Colourless substrates that are converted to a coloured product by the enzyme are popular; for example p-nitrophenylphosphate (pNPP) is converted to the yellow p-nitrophenol by alkaline phosphatase. Substrates used with peroxidase include: 2,2'-azo-bis(3-ethylbenzthiazoline-6-sulphonic acid), o-phenylenediamine (OPD) and 3,3'5,5'-tetramethylbenzidine base (TMB), which yield green, orange and blue colours, respectively. It should be noted that these substrates are considered hazardous materials; some are simply irritants but others are classified as potential carcinogens. Reference positive and negative samples must be included in each series of tests to ensure accurate and reproducible results.

The dot immunobinding assay

Nitrocellulose paper is widely used as a support for solid phase immunoassays, particularly for proteins blotted from polyacrylamide gels (Section 9.3.9). Antigens may be applied directly to nitrocellulose filter papers as a dot to give a simple and reliable assay with advantages over conventional immunoassays. Sometimes they are more sensitive than conventional ELISAs because the dot permits the colour reaction to be viewed against a white background; they are very sparing in their use of antigen (0.05 mm^3 yields a 0.3 mm dot) and the dot immunobinding assay may be used with a roster of different antigens simultaneously to detect and characterise a range of antibodies of interest.

The antigen is applied as a discrete spot and any remaining binding sites are blocked using a solution of non-cross-reacting protein. The dot is then incubated with the putative antibody containing solution and washed. Antibody binding is detected with a second species-specific antibody conjugated to horseradish per-oxidase. The assay can also be used to detect other ligands, e.g. lectins, proteins, synthetic polypeptides, nucleic acids, glycoproteins, glycolipids, membrane fractions, viruses, yeasts, protozoans and aldehyde-fixed cells.

2.6.3 Applications

Whilst ELISA can be used for the assay of virtually any antigen, hapten or antibody, it is used predominantly in clinical biochemistry laboratories to measure, for example, IgG and IgE, oncofetal proteins, haematological factors, immune complexes, and hormones such as insulin, oestrogens and human chorionic gonadotrophin. Examples of its use in the study of infectious diseases include the detection of bacterial toxins, *Candida albicans*, rotaviruses, herpes simplex viruses and the hepatitis B surface antigen.

ELISA has also been used extensively for the assay of antibodies in infectious diseases including: anti-viral antibodies, e.g. to Epstein–Barr virus and rubella virus; anti-bacterial antibodies, e.g. to *Brucella*, *Rickettsia* and *Salmonella* spp.; anti-fungal antibodies, e.g. to *Aspergillus* and *Candida* spp.; anti-parasite antibodies, e.g. to *Plasmodium*, *Schistosoma* and *Trypanosoma* spp.; and autoantibodies e.g. anti-DNA and anti-thyroglobulin.

2.7 Fluorescence immunoassay (FIA)

Fluorescence immunoassays combine the specificity of antibodies with the sensitivity of fluorimetric assays (Section 7.8), usually by using antibodies coupled to a fluorescent chromophore.

2.7.1 Homogeneous substrate-labelled fluorescent immunoassay (SLFIA)

SLFIA uses the principles of competitive binding reactions to measure the concentration of compounds, such as drugs of low relative molecular mass, in body fluids. The compound is covalently linked to a fluorogenic enzyme substrate (e.g. galactosylumbelliferone) to produce a non-fluorescent reagent which can be hydrolysed by an appropriate enzyme (e.g. β-galactosidase) to produce a fluorescent product. It is essential that the binding of antibody to the non-fluorescent reagent prevents the enzymic hydrolysis of the reagent. Thus, the enzyme can produce fluorescence from unbound reagent but not from antibody-bound reagent. In the presence of the original compound and of limited amounts of reagent and specific antibody the enzyme will produce fluorescence as a function of the concentration of the compound (c.f. RIA, Section 2.5). Unlike in RIA, it is not necessary to separate the free and antibody-bound forms of the reagent because only the free reagent can be enzymically hydrolysed to produce fluorescence. A calibration curve of percentage fluorescence against concentration may be constructed by incubating known amounts of compound with appropriate fixed amounts of reagent, adding enzyme and measuring the fluorescence produced. The antiserum and the enzyme can be mixed and added together because the affinity of the antibody for the reagent is usually very much greater than the affinity of the enzyme for the reagent.

As an example, under appropriate conditions SLFIA produces a linear relationship between the antibiotic gentamicin and fluorescence over the range 0–24 ng gentamicin, with a mid-range coefficient of variation of less than 3%. This means that human sera being monitored for drug level can be diluted 10 times and hence it is unnecessary to carry out controls for any native fluorescence of the sera. Other compounds routinely assayed by SLFIA include tobramycin, kanamycin, theophylene, phenobarbitone, IgG and IgM.

2.7.2 Delayed enhanced lanthanide fluorescence immunoassay (DELFIA)

This technique combines the sensitivity, reproducibility and accuracy normally associated with RIA with the speed and versatility usually associated with enzyme immunoassays (see Table 2.2). Usually it involves the use of antibody labelled with a chelate of the lanthanide metal europium. The weak fluorescence of the europium chelate is intensified 10-fold on addition of an enhancing solution, which releases the europium and rechelates it into micelles that protect it from the fluorescence-quenching effects of water molecules. The lanthanide chelates have long fluorescent decay times (10–1000 μs) after excitation. This allows time-resolved fluorescent immunoassays to measure the lanthanide fluorescence after interference from scattered light and the short-lived (1–20 ns) natural fluorescence of biological materials has disappeared, as illustrated in Fig. 2.20a. Usually the fluorescence emitted by the lanthanide chelate is measured 400–500 μs after it has been irradiated for 1 μs at 340 nm wavelength with a xenon lamp. Other advantages of using lanthanide chelates include the following:

(i) A very large Stokes' shift, e.g. from 340 nm to 613 nm, as illustrated in Fig. 2.20b, and a very narrow emission peak with a half-intensity bandwidth of about 10 nm (Fig. 2.20b); together these allow great spectral selectivity in the assay.

(ii) A broad excitation region with a half-intensity bandwidth of about 50 nm (Fig. 2.20b), resulting in great sensitivity of the assay and great stability and biological inertness (in contrast to radioisotopes and enzymes, respectively).

DELFIA is usually carried out by means of a solid phase, two-site, direct

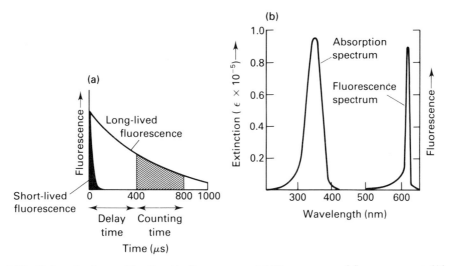

Fig. 2.20.   Delayed enhanced lanthanide fluorescence. (a) Time course of fluorescence at 613 nm wavelength after excitation by a 1 μs flash at 340 nm; (b) absorption and fluorescence spectra of a europium chelate.

sandwich technique analagous to the double-antibody ELISA illustrated in Fig. 2.18 but using excess europium-labelled second antibody in place of the enzyme-labelled antibody and enhancer solution in place of the enzyme substrate. Under standard conditions the intensity of fluorescence produced is directly proportional to the amount of specific antigen in the original solution. As little as 10^{-16} mol europium per sample can be measured and the precision of DELFIA is better than 5% up to 10^{-12} mol europium. Commercial kits utilising microtitre plates are already available for DELFIA of hepatitis B antigen, digoxin, thyroxine, testosterone, progesterone, oestradiol, cortisol, human follicle-stimulating hormone and human luteinising hormone. Results are usually available within 1–5 h.

2.7.3   Flow cytofluorimetry and fluorescence-activated cell sorting (FACS)

The cytofluorograph or flow cytofluorimeter measures the fluorescence emitted by individual cells in a mixed cell population (Section 1.5.4). Monoclonal antibodies have vastly increased the effectiveness of flow cytofluorimetry and FACS. In particular the availability of a vast array of monoclonal antibodies to leukocyte surface antigens is greatly facilitating studies on the different lineages of leukocytes and their numerous roles in the immune response.

2.7.4   Bioluminescent immunoassays

Bioluminescent immunoassays, comparable in sensitivity to RIA, have been developed by covalently linking antigens and firefly luciferase (Section 4.8.4). A competitive binding assay has been used for haptens. Use of glucose-6-phosphate dehydrogenase promises even greater sensitivity.

2.8   Avidin–biotin-mediated immunoassays

Avidin is a $66\,000\ M_r$ glycoprotein present in egg white. It binds with extremely high avidity to the small molecule biotin (vitamin H). The dissociation constant (K_d) for the avidin–biotin complex is 10^{-15} M, as compared to a K_d of 10^{-8}–10^{-10} M for antigen–antibody interactions and a K_d of 10^{-4}–10^{-5} M for lectin–carbohydrate interactions. Biotin is easily linked to proteins, retains its high affinity for avidin and barely affects the biological activities of the protein, e.g. antibodies.

Whenever improvements in immunodiagnostic assays have been required, the suitability of avidin–biotin complex technology has been rapidly demonstrated.

A major application of avidin–biotin technology has been the enhancement of signal and/or the speed of an immunoassay. Signal enhancement is due to the fact that many biotin residues can be introduced chemically into an antibody molecule. Thus, multiple avidin-containing probes (radiolabels, enzymes or fluors) can be incorporated into the final step of the detection system. Also, the four biotin-binding sites on the avidin molecule allow further signal enhancement by enabling a biotinylated detection probe to be used either sequentially or complexed with avidin in a predetermined ratio. An alternative approach is to use the avidin–biotin complex in the capture system by means of biotinylated antibody to Protein A immobilised to an avidin–containing matrix.

Streptavidin, isolated from *Streptomyces avidinii*, which has an affinity for biotin comparable to that of avidin, is tending to supercede the use of the highly positively charged avidin, because it is less basic and has no carbohydrate residues, thus limiting non-specific reactions with acidic groups and lectins.

In immunoassays, biotinylated antibodies react with antigen, and then avidin linked to a reporter substance (radiolabel, enzyme or fluor) is added. In ELISAs, there are two protocols, described below, for linking enzyme to the antibody.

(i) Labelled avidin–biotin technique. Biotin-labelled antibody is reacted with immobilised antigen and, after washing, enzyme-labelled avidin is added. After the mixture has undergone further incubation and washing, substrate is added and the

enzyme-associated antigen is assayed spectrophotometrically, as illustrated in Fig. 2.21a.

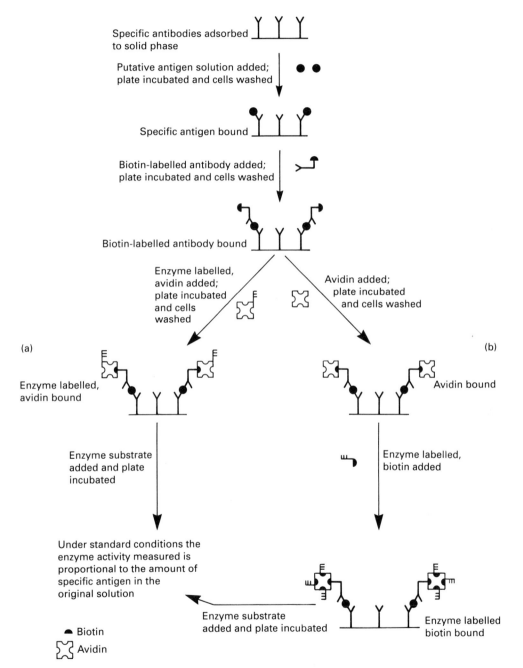

Fig. 2.21. Assay of antigens by (a) the labelled avidin–biotin system and (b) the bridged avidin–biotin system.

(ii) Bridged avidin–biotin technique. Biotin-labelled antibody is reacted with immobilised antigen. After excess labelled antibody has been washed off, avidin is added. After incubation and washing have been carried out, biotin-labelled enzyme is added. After a further round of incubation and washing, substrate is added and the enzyme-associated antigen measured spectrophotometrically, as illustrated in Fig. 2.21b. This technique relies on the fact that avidin possesses four active sites, not all of which react with biotin residues associated with the biotinylated antibody–antigen complex. The remaining free active sites can bind to the added biotin-labelled enzyme.

2.9   Particle-counting immunoassays (PACIA)

PACIA relies on the principles that the number of free particles coated with antigen will decrease during agglutination and that the angle at which a beam of light is scattered is a function of particle size.

Cell agglutination has been used to detect cell surface antigens for decades. The concentrations of antibodies relative to those of cell surface antigens can be determined by using doubling dilutions of antiserum, carefully standardised conditions and noting the end point of cell agglutination. The technique has been extended to other antibodies by binding the corresponding antigen to erythrocytes. The relative concentration of an antigen solution can be determined by agglutination inhibition assays, using doubling dilutions of the antigen. However, because the erythrocytes used are labile they cannot be stored indefinitely, and these techniques are laborious and difficult to automate. Polystyrene beads (usually 0.8 µm minimum diameter and called latex) are stable and can be coated with antibodies and used to assay the corresponding antigen by agglutination (latex fixation test). Antibodies are either physically adsorbed to plain latex or covalently coupled with carbodiimide to carboxylated latex.

Similar techniques can be used to assay antibodies and immune complexes. The optical size counter is set to count only non-agglutinated particles and ignore those with diameters smaller than 0.6 µm and greater than 1.2 µm. Alternatively a system that measures electrical resistance can be used for counting non-agglutinated particles. Both of these techniques of counting give a sensitive and precise evaluation of the extent of the agglutination reaction by a simple mixing of antigen and antibody-coated latex, followed by dilution to give a count of 3000–4000 particles s^{-1}.

Advantages of PACIA include: the ease of attaching proteins to the latex; the relative stability of the latex suspension; the sensitivity of the assay, 0.1–1 ng antigen cm^{-3}; and ready automation of the technique.

The limitations of PACIA are its susceptibility to non-specific agglutination and agglutination inhibition, but these phenomena can be minimised by the use of Fab fragments of antibody in place of whole Ig molecules, high ionic strength media and control serum. Antibodies and haptens can be assayed by PACIA using agglutination inhibition.

2.10   Labelled antibody techniques for detecting antigens: immunohistochemistry

In these techniques an antibody is used to link a cellular antigen specifically to a stain that can be more readily seen with the microscope.

2.10.1   Immunofluorescence (IF)

Immunofluorescence provides sensitive assays for the detection of antigens in frozen or fixed tissue sections or in viable cells. The location of specific antigens in a tissue or cell preparation may be studied by staining with specific antibody conjugated to a fluorescent chromophore and illuminating with ultraviolet light. Fluoroscein emits green light and rhodamine emits orange light, whereas the weak natural fluorescence of some biological materials occurs in the blue region of the spectrum. Immunofluorescence assays may be performed in a number of different ways, illustrated in Fig. 2.22. The more complicated methods are more sensitive as they amplify the amount of fluorescent antibody bound. Samples must be very carefully washed between the incubations with each reagent, and both positive and negative controls should be included. Immunofluorescence is particularly useful for detecting autoantibodies to thyroglobulin, DNA, mitochondria, nuclei and nucleoli, immune complexes and cell surface receptors such as the immunoglobulin receptors on lymphocytes. Monoclonal antibodies have vastly increased the range and effectiveness of immunofluorescent measurements, e.g. identification of helper and suppressor subpopulations of T-lymphocytes.

2.10.2   Immunoenzyme probes

Enzyme-labelled antibody is used in light microscopy for localizing antigens (c.f. ELISA, Section 2.6). For enzyme immunohistochemistry, an insoluble end product is required. 5-Bromo-4-chloro-3-indolylphosphate/nitro blue tetrazolium (BCIP/ NBT) and naphthol AS-TR phosphate/Fast Red RC are used with alkaline phosphatase to give purple and red precipitates, respectively. 3,3'-Diaminobenzidine (DAB) and 3-amino-9-ethylcarbazole (AEC) may be used with peroxidase to yield brown and red precipitates, respectively.

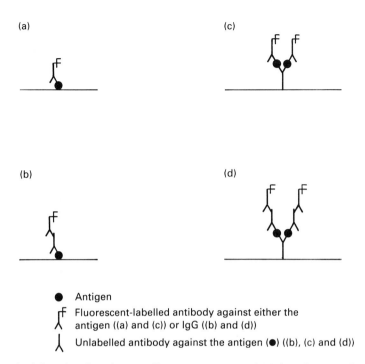

(a)

(b)

(c)

(d)

●   Antigen

ℸ   Fluorescent-labelled antibody against either the
Λ   antigen ((a) and (c)) or IgG ((b) and (d))

Λ   Unlabelled antibody against the antigen (●) ((b), (c) and (d))

Fig. 2.22. The principles of various immunofluorescence assays. (a) Direct immunofluorescence test or
single-layer technique, (b) indirect immunofluorescence test or double-layer technique, (c) simple
sandwich technique for the detection of specific antibodies, and (d) multiple sandwich technique.

2.10.3   Immunoferritin probes

Ferritin-labelled antibody may be used for locating antigens by electron microscopy.

2.10.4   Immunogold probes

Colloidal gold may be used directly and indirectly for the location of antigens with
light, transmission and scanning electron microscopy. The preparation of gold
probes is a simple two-step procedure involving first the generation of a colloidal
gold sol, containing particles of a narrow range of sizes from 5 to 40 nm. Then
the primary or species-specific antibody, or Protein A, is adsorbed on to the surface
of the gold particles, giving rise to immunogold and Protein A–gold probes,
respectively, which are stable for long periods. The basic staining technique usually
involves incubating cryostat or fixed sections of specimen first with primary
antibody and then with the gold probe; obviously washing off excess reagent after
each incubation. The labelling is highly specific and lends itself to quantitative
studies, merely by permitting the counting of the number of probes present in a

sample. The use of antibodies to different antigens and probes labelled with different sizes of colloid gold facilitates the determination of the distribution of more than one antigen in a sample. For light microscopy, the location of the gold probe is developed by a silver precipitation reaction; so-called immunogold–silver staining. Immunogold technologies have probably become the most popular cytochemical marking system in biological research.

2.10.5   Immunoradioisotope probes

Radioisotope-labelled antibody may be used for detecting and locating antigens by autoradiography (c.f. RIA, Section 2.5).

2.11   Suggestions for further reading

BENJAMINI, E. and LESKOWITZ, S. (1991) *Immunology: A Short Course*, 2nd edn. Wiley-Liss, New York. (An excellent introduction to immunology.)

CATTY, D. (Ed.) (1988 and 1989) *Antibodies: A Practical Approach*, volumes I and II. IRL Press at Oxford University Press, London. (Useful discursive texts about many immunochemical procedures.)

COLEMAN, R.M., LOMBARD, M.F. and SICARD, R.E. (1992) *Fundamental Immunology*, 2nd edn. W.C. Brown, Dubeque, IA. (An excellent textbook on immunology.)

COLOWICK, S.P. and KAPLAN, N.O. (Eds.) (1980–90). Immunochemical techniques in *Methods in Enzymology*, volumes 70, 73, 74, 84, 92, 93, 108, 121, 178, 184. Academic Press, London. (These volumes contain comprehensive and exhaustive accounts of all the techniques discussed in this chapter.)

DAVEY, B. (1989) *Immunology: A Foundation Text*. J. Wiley and Sons, Chichester. (Possibly the best value for money as an introductory text to immunology.)

GOLUB, E.S. (1987) *Immunology: A Synthesis*. Sinauer Association, New York. (An excellent textbook for the biochemically inclined.)

HUDSON, L. and HAY, F.C. (1989) *Practical Immunology*, 3rd edn. Blackwell Scientific Publications, Oxford. (An excellent introduction to the practice of many immunological and some immunochemical techniques.)

LANE, H.D. (Ed.) (1988) *Antibodies: A Laboratory Manual*. Cold Spring Harbor Laboratory Press, Cold Spring Harbor, NY. (An up-to-date manual of immunological methods.)

ROIT, S.M. (1991) *Essential Immunology*, 7th edn. Blackwell Scientific Publications, Oxford. (An excellent textbook on immunology.)

WEIR, D.M. (Ed.) (1986) *Handbook of Experimental Immunology* (4 volumes), 4th edn. Blackwell Scientific Publications, Oxford. (The standard reference work on all immunological and immunochemical techniques, containing comprehensive details on each.)

3

Molecular biology techniques

3.1 Introduction

The shapes and functions of cells are determined by their proteins. Enzymes catalyse the reactions which synthesise membranes, cell walls and pigments, and they are needed to extract energy from substrates. Proteins in membranes are responsible for the transport of molecules from one cellular compartment to another, and between the inside and outside of the cell. Synthesis of proteins is itself catalysed by proteins, but this process is ultimately directed by DNA, which carries all the information needed to specify the structure of every protein the cell can make. The realisation that DNA lies behind all the cell's activities led to the development of molecular biology, which aims to explain biological processes in terms of the structures and interactions between nucleic acids and proteins. Although it is a relatively young discipline, it has already transformed our understanding of the way in which cells store and express their genetic information, and has had an enormous impact on many fields of study, such as immunology, medicine, plant breeding, microbiology and forensic science.

Molecular biology has also led directly to the immensely powerful, and potentially profitable, techniques of genetic engineering. A great deal of effort (and money) is being directed into the genetic manipulation, or genetic engineering, of microorganisms, to make them produce a range of valuable polypeptides such as insulin, blood clotting factor VIII, growth hormone, or interferons, which they would not normally produce and which are expensive to prepare by conventional biochemical means. Since it is easy and cheap to grow microorganisms on a large scale, they are very attractive as potential sources of these polypeptides. Genetic manipulation of plants and livestock should also permit the introduction of beneficial characteristics, such as resistance to herbicides or diseases, that could not readily be obtained by conventional breeding.

3.2   Structure of nucleic acids

3.2.1   Components and primary structure of nucleic acids

In order to appreciate the ways in which nucleic acids can be analysed and manipulated, it is essential to have a basic understanding of their structures and functions. Such information can be found in any general biochemistry textbook but the most important features will be summarised here.

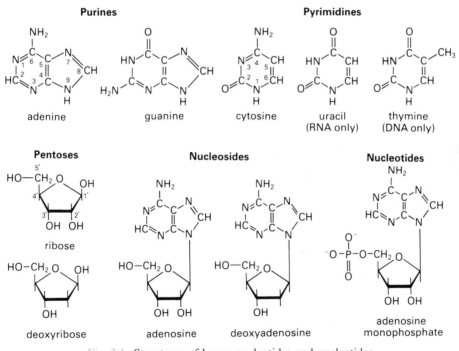

Fig. 3.1. Structures of bases, nucleotides and nucleotides.

In spite of their ultimate complexity, ribonucleic acid (RNA) and deoxyribonucleic acid (DNA) are made up of relatively few components. Both contain a pentose sugar (ribose in RNA, 2'-deoxyribose in DNA) to which is attached a purine or pyrimidine base, forming a nucleoside. The pentose sugar carbon atoms are numbered, as shown in Fig. 3.1, using a prime (′) to indicate that the carbon is part of the sugar rather than of the purine or pyrimidine base, and it can be seen that the base is attached to the 1′ position of the pentose. A nucleotide, or nucleoside phosphate, is formed by the attachment of a phosphate group to the 5′ position of a nucleoside by an ester linkage. Such nucleotides can be joined together by the formation of a second ester bond by reaction between the phosphate of one nucleotide and the 3′ hydroxyl group of another, thus generating a 5′ to 3′ phosphodiester bond between adjacent sugars; this process can be repeated indefinitely to give long polynucleotide molecules. Each polynucleotide chain will have a free phosphate group at one of its ends, and a free 3′ hydroxyl group at its other end; thus the molecule has polarity, and we can refer to its 3′ and 5′ ends (Fig. 3.2).

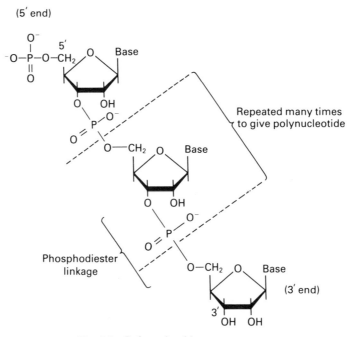

Fig. 3.2. Polynucleotide structure.

The purine bases adenine and guanine are found in both RNA and DNA, as is the pyrimidine cytosine. The other pyrimidines are each restricted to one type of nucleic acid: uracil occurs exclusively in RNA, whereas thymine is limited to DNA. Thus, we can distinguish between RNA and DNA on the basis of the presence of ribose and uracil in RNA, and deoxyribose and thymine in DNA. However, it is the sequence of bases along a molecule that distinguishes one RNA (or DNA) type from another. It is conventional to write a nucleic acid sequence starting at the 5′ end of the molecule, using single capital letters to represent each of the bases, e.g. CGGATCT. Note that there is usually no point in including the sugar or phosphate groups because these are identical throughout the length of the molecule. Terminal phosphate groups can, when necessary, be indicated by use of a 'p'; thus 5′ pCGGATCT 3′ indicates the presence of a phosphate on the 5′ end of the molecule.

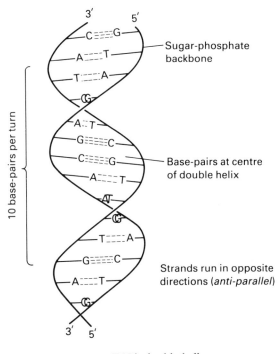

Fig. 3.3. **DNA double helix.**

3.2.2   Secondary structure of nucleic acids

DNA is not usually found in a single-stranded form, but as a double-stranded molecule in the shape of a double helix, in which the bases of the two strands lie in the centre of the molecule, with the sugar-phosphate backbones on the outside (Fig. 3.3). A crucial feature of this structure is that it depends on the sequence of bases in one strand being complementary to that in the other. As shown in Fig. 3.4, thymine can form hydrogen bonds with adenine, and cytosine with guanine, in such a way that the distance between the 1′ carbon atoms of their respective deoxyriboses is the same. Thus, only if adenine is always base-paired with thymine, and cytosine with guanine, will a stable double helix structure result, in which the

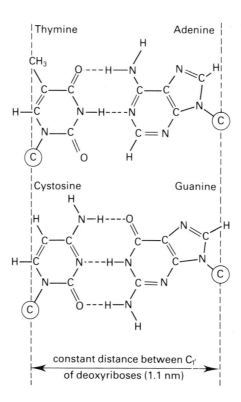

Fig. 3.4. Base-pairing in DNA. © represents carbon at 1' position of deoxyribose.

backbones of the two strands are, on average, a constant distance apart. The base sequence does, in fact, cause significant local variations in DNA shape and these variations are vital for specific interactions between the DNA and various proteins.

If the sequence of one strand is known, that of the other strand can be deduced. The strands are designated as plus and minus, depending on which is copied during transcription (Section 3.3.3). Another important feature of DNA is that the two strands are anti-parallel, i.e. they run in opposite directions. For example:

$$5' \quad C\;G\;G\;T\;A\;A\;C\;T \quad 3'$$
$$3' \quad G\;C\;C\;A\;T\;T\;G\;A \quad 5'$$

It should be noted that the two strands of DNA are held together only by the weak forces of hydrogen bonding between complementary bases, and by hydrophobic interactions between adjacent, stacked base-pairs. Little energy is needed to separate a few base-pairs, and so, at any instant, a few short stretches of DNA will be opened up to the single-stranded conformation. However, such stretches immediately pair up again at room temperature, so the molecule as a whole remains predominantly double stranded. If, however, a DNA solution is heated to about 90 °C there will be enough kinetic energy to denature the DNA completely, causing it to separate into single strands. This denaturation can be followed spectrophotometrically by monitoring the absorbance of light at 260 nm wavelength. The stacked bases of double-stranded DNA are less able to absorb light than are the less constrained bases of single-stranded molecules, and so the absorbance of DNA at 260 nm increases as the DNA becomes denatured, a phenomenon known as the hyperchromic effect.

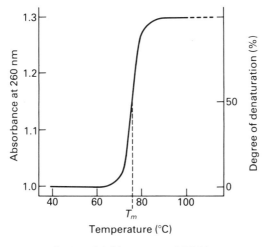

Fig. 3.5. Melting curve of DNA.

If absorbance at 260 nm is plotted against the temperature of a DNA solution (Fig. 3.5), it is seen that little denaturation occurs below about 70°C, but further increases in temperature result in a marked increase in the extent of denaturation. Eventually a temperature is reached at which the sample is totally denatured, or melted. The temperature at which the DNA is 50% melted is called the melting temperature or T_m, and will depend on the nature of the DNA. If several different samples of DNA are melted, it is found that T_m is highest for those DNAs that contain the highest proportion of cytosine and guanine, and T_m can be used to estimate the percentage (C + G) content in a DNA sample. This relationship between T_m and (C + G) content arises because cytosine and guanine form three hydrogen bonds when base-pairing, whereas thymine and adenine form only two (Fig. 3.4); consequently, more energy is needed to separate C·G pairs.

If melted DNA is cooled it is possible for the separated strands to reassociate, a process known as renaturation. However, a stable double-stranded molecule will be formed only if the complementary strands collide in such a way that their bases are paired precisely, and this is an unlikely event if the DNA is very long and complex (i.e. if it contains a large number of different genes). Measurements of the rate of renaturation can give information about the complexity of a DNA preparation (Section 3.5.2).

Although RNA almost always exists as a single strand, it often contains sequences within the same strand that are complementary to each other, and therefore can base-pair if brought together by suitable folding of the molecule. This is most obvious in the case of transfer RNA, which has four pairs of complementary sequences within its 70 to 80 nucleotide length (Fig. 3.6); consequently the single strand folds up to give a clover leaf secondary structure. As with DNA, cytosine pairs with guanine, but in RNA adenine pairs with uracil.

Strands of RNA and DNA will associate with each other, if their sequences are complementary, to give double-stranded, hybrid molecules (uracil of RNA base-pairs with adenine of DNA). Similarly, strands of radioactively labelled RNA or DNA, when added to a denatured DNA preparation, will act as probes for DNA molecules to which they are complementary. This hybridisation of complementary strands of nucleic acids is very useful for pulling a specific piece of DNA out of a complex mixture (Section 3.9).

3.3   Functions of nucleic acids

3.3.1   Classes of RNA

The genetic information of cells and most viruses is stored in the form of DNA. This information is used to direct the synthesis of RNA molecules, which fall into three classes.

(i) Messenger RNA (mRNA) contains sequences of ribonucleotides which code for the amino acid sequences of proteins. A single mRNA molecule codes for a single polypeptide chain in eukaryotes, but may code for several polypeptides in prokaryotes.

(ii) Ribosomal RNA (rRNA) forms part of the structure of ribosomes, which are the sites of protein synthesis. Each ribosome contains only three or four different rRNA molecules, complexed with a total of between 55 and 75 proteins.

(iii) Transfer RNA (tRNA) molecules carry amino acids to the ribosomes, and interact with the mRNA in such a way that their amino acids are joined together in the order specified by the mRNA. There is at least one type of tRNA for each amino acid.

Each block of DNA that codes for a single RNA or protein is called a gene,

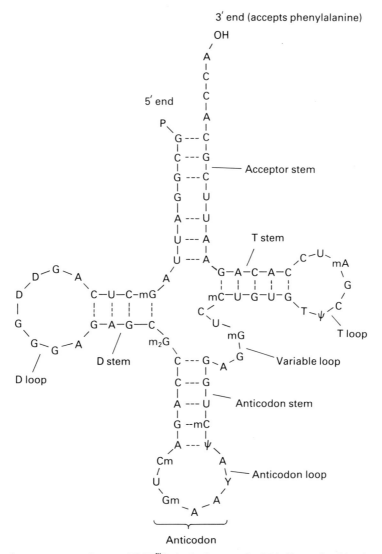

Fig. 3.6. Secondary structure of yeast tRNA^Phe. A single strand of 76 ribonucleotides forms four double-stranded 'stem' regions by base-pairing between complementary sequences. The anticodon will base-pair with UUU or UUC (both are codons for phenylalanine), and phenylalanine is attached to the 3' end by a specific aminoacyl tRNA synthetase. Several 'unusual' bases are present: D, dihydrouridine; T, ribothymidine; Ψ, pseudouridine; Y, very highly modified, unlike any 'normal' base. mX indicates methylation of base X (m₂X shows dimethylation); Xm indicates methylation of ribose, on the 2' position.

and the entire set of genes in a cell, organelle or virus forms its genome. Cells and organelles may contain more than one copy of their genome.

3.3.2 DNA replication

Chromosomal DNA must be replicated at a rate that will at least keep up with the rate of cell division. Replication begins at a sequence called the origin of replication, and involves the separation of the two DNA strands over a short length, and the binding of enzymes, including DNA and RNA polymerases (Fig. 3.7). In prokaryotes, RNA polymerase synthesises a short, complementary RNA chain on each exposed strand, using the DNA as a template. Then DNA polymerase III (pol III) also uses the DNA as a template for the synthesis of a DNA strand, using the short RNA as a primer. Synthesis of the DNA strand occurs only in a 5′ to 3′ direction, but, since the two strands of DNA are antiparallel, only one can be synthesised in a continuous fashion. The other is synthesised in relatively short stretches, still in a 5′ to 3′ direction, using an RNA primer for each stretch. These RNA primers are then removed by DNA pol I, acting as a 5′ to 3′ exonuclease, the gaps are filled by the same enzyme acting as a polymerase, and the separate fragments are joined together by DNA ligase to give a continuous strand of DNA. The replication of eukaryotic DNA is less well characterised, and is certainly more complex than that of prokaryotes; however, both processes involve 5′ to 3′ synthesis of new DNA strands.

The net result of the replication is that the original DNA is replaced by two molecules, each containing one 'old' and one 'new' strand; the process is therefore known as semi-conservative replication.

3.3.3 Transcription

At any instant only a fraction of all the genes in a genome are active. These genes, which are being expressed, undergo the process of transcription, in which an RNA molecule complementary to one of the gene's DNA strands is synthesised.

Most prokaryotic genes are made up of three regions (Fig. 3.8). At the centre is the sequence that will be copied in the form of RNA, called the transcription unit. To the 5′ side (upstream) of the strand that will be copied (the plus (+) strand) lies a region called the promoter, and downstream from the transcription unit is the terminator region. Transcription begins when DNA-dependent RNA polymerase binds to the promoter region and moves along the DNA to the transcription unit. At the start of the transcription unit the polymerase begins to synthesise an RNA molecule complementary to the minus (−) strand of the DNA, moving along this strand in a 3′ to 5′ direction, and synthesising RNA in a 5′ to 3′ direction, using nucleoside triphosphates. The RNA will therefore have the same sequence as the plus strand of DNA, apart from the substitution of uracil for

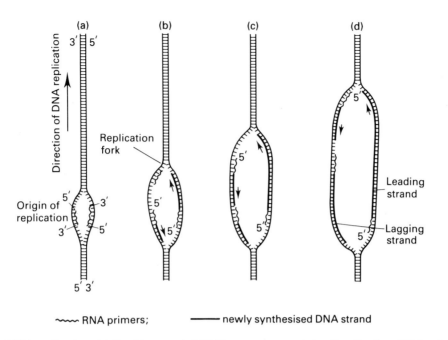

~~~~ RNA primers;   ──── newly synthesised DNA strand

Fig. 3.7. DNA replication. (a) Double-stranded DNA separates at origin of replication. RNA polymerase synthesises short RNA primer strands complementary to both DNA strands. (b) DNA polymerase III synthesises new DNA strands in a 5′ to 3′ direction, complementary to the exposed, old DNA strands, and continuing from the 3′ end of each RNA primer. Consequently DNA synthesis is in the same direction as DNA replication for one strand (the leading strand) and in the opposite direction for the other (the lagging strand). RNA primer synthesis occurs repeatedly to allow the synthesis of fragments of lagging strand. (c) As the replication fork moves away from the origin of replication, pol III continues the synthesis of the leading strand, and synthesises DNA between RNA primers of the lagging strand. (d) DNA polymerase I removes RNA primers from the lagging strand, and fills the resulting gaps with DNA. DNA ligase then joins the resulting fragments, producing a continuous DNA strand.

thymine. When the terminator region is reached, transcription is stopped, and the RNA molecule is released.

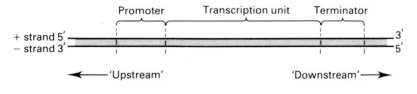

Fig. 3.8. Gene structure.

Both tRNA and rRNA are produced as precursors, which are then trimmed to their correct sizes by ribonucleases. The mRNA of prokaryotes can be used without any modification to direct protein synthesis, but post-transcriptional processing is needed in eukaryotes. In this processing a CAP sequence is added to the 5′ end of the RNA, and about 150 to 200 adenosine residues are added to the 3′ end, forming a poly(A) tail. The majority of eukaryotic genes contain lengths of non-coding DNA, called introns, which interrupt their coding regions (exons) (Fig. 3.9). Transcription of these genes results in the production of heterogeneous nuclear RNA (hnRNA), which must have its introns excised, leaving the exons spliced together to form mRNA.

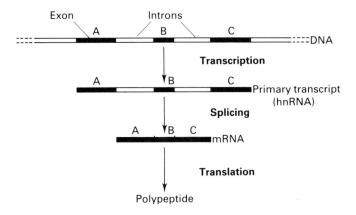

Fig. 3.9. Splicing of RNA to remove introns. Regions A, B and C, the exons, are sequences that, when joined together, code for the polypeptide product. The primary transcript still contains introns, and is called heterogeneous nuclear RNA (hnRNA).

3.3.4   Translation

Each mRNA codes for the primary amino acid sequence of a protein, using a triplet of nucleotides to represent each of the amino acids. Each triplet is known as a codon, and, as there are 64 possible triplet codons but only 20 different amino acids (plus 'start' and 'stop' codons), most amino acids are coded for by more than one codon. The genetic code has been worked out and is found to be universal for all chromosomal and chloroplast DNAs or RNAs so far examined, but a few differences have been found in the codes used by mitochondria.

The mRNAs are 'read', and proteins assembled, on the ribosomes, which are structures formed of a complex of rRNAs and proteins. Each ribosome consists of a large and a small subunit, which associate during the process of protein synthesis, or translation. The ribosomes of prokaryotes and of organelles have a sedimentation coefficient (Section 6.2) of 70 S, whilst those of the eukaryotic cytoplasm are 80 S. Transfer RNA molecules are also needed for translation. Each of these can be covalently linked to a specific amino acid, forming an aminoacyl tRNA, and each has a triplet of bases exposed that is complementary to the codon for that amino acid. This exposed triplet is known as the anti-codon, and allows the tRNA to act as an 'adapter' molecule, bringing together a codon and its corresponding amino acid.

After binding to a specific sequence at the 5′ end of the mRNA, known as a Shine–Dalgarno sequence (after its discoverers) in prokaryotes, the ribosome moves towards the 3′ end, allowing an aminoacyl tRNA molecule to base-pair with each successive codon, thereby carrying in amino acids in the correct order for protein synthesis. The ribosome forms peptide bonds between these amino acids as it moves along the mRNA, and releases a completed polypeptide chain when it reaches a termination codon (UAA, UGA or UAG) (Fig. 3.10). The first codon after the ribosome binding site, the initiation codon, is always AUG, which acts as a start signal.

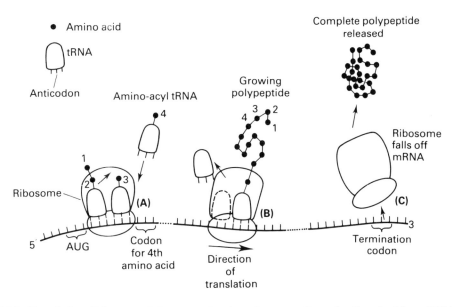

Fig. 3.10. Translation. Ribosome A has moved only a short way from the 5′ end of the mRNA, and has built up a dipeptide (on one tRNA), which is about to be transferred on to the third amino acid (still attached to tRNA). Ribosome B has moved much further along the mRNA, and has built up an oligopeptide that has just been transferred on to the most recent aminoacyl tRNA. The resulting free tRNA leaves the ribosome and will receive another amino acid. The ribosome moves towards the 3′ end of the mRNA by a distance of three nucleotides, so that the next codon can be aligned with its corresponding aminoacyl tRNA on the ribosome. Ribosome C has reached a termination codon, has released the completed polypeptide, and has fallen off the mRNA.

Since the mRNA is read in triplets, an error of one or two nucleotides in positioning of the ribosome will result in the synthesis of an incorrect polypeptide. Thus is it essential for the correct reading frame to be used during translation. This is ensured in prokaryotes by base-pairing between the Shine–Dalgarno sequence and a complementary sequence of one of the ribosome's rRNAs, thus establishing the correct starting point for movement of the ribosome along the mRNA.

3.4   Isolation of nucleic acids

3.4.1   DNA

Before nucleic acids can be cut or otherwise manipulated, they must be isolated and purified to some extent, the degree of purity being dependent on the intended use of the preparation.

DNA is very easily damaged by shear forces; even rapid stirring of a solution can break high molecular weight DNA into much shorter fragments. The other main threat comes from digestion by deoxyribonucleases (DNases), which are found in most cells and may also be present in dust that could contaminate laboratory glassware. Consequently, DNA is recovered from cells by the gentlest possible method of cell rupture, in the presence of EDTA to chelate the magnesium ions needed for DNase activity. Ideally, cell walls, if present, should be digested enzymatically (e.g. lysozyme treatment of bacteria), and the cell membrane should be solubilised using detergent. If physical disruption is necessary, it should be kept to a minimum, and should involve cutting or squashing of cells, rather than the use of shear forces. Cell disruption (and most subsequent steps) should be performed at $4\,°C$, using glassware and solutions that have been autoclaved, to destroy DNase activity.

After release of nucleic acids from the cells, RNA can be removed by treatment with ribonuclease (RNase) that has been heat treated to inactivate any DNase contaminants; RNase is relatively stable to heat as a result of its disulphide bonds, which ensure rapid renaturation of the molecule on cooling. The other major contaminant, protein, is removed by shaking the solution gently with water-saturated phenol, or with a phenol/chloroform mixture, either of which will denature proteins but not nucleic acids. Centrifugation of the emulsion formed by this mixing produces a lower, organic phase, separated from the upper, aqueous phase by an interface of denatured protein. The aqueous solution is recovered and deproteinised repeatedly, until no more material is seen at the interface. Finally, the deproteinised DNA preparation is mixed with two volumes of absolute ethanol, and the DNA allowed to precipitate out of solution in a freezer. After centrifugation, the DNA pellet is redissolved in a buffer containing a little EDTA for protection against DNases, and this solution can be stored at $4\,°C$ for at least a month. DNA solutions can be stored frozen, but repeated freezing and thawing tends to damage long molecules by shearing; preparations in frequent use are therefore normally stored at 4 °C.

The procedure described above is suitable for total cellular DNA. If the DNA from a specific organelle or viral particle is needed, it is best to isolate the organelle or virus before trying to extract its DNA, since the recovery of a particular type of DNA from a mixture is usually rather difficult. The isolation of plasmids (Section 3.8.1) is described later.

3.4.2 RNA

The methods used for RNA isolation are very similar to those described above for DNA; however, RNA molecules are relatively short, and therefore less easily damaged by shearing, so cell disruption can be rather more violent. RNA is also very vulnerable to digestion by RNases, which are present on fingers, so gloves should be worn by the experimenter, and a strong detergent should be included in the isolation medium to denature any RNases immediately. Subsequent deproteinisation should be particularly rigorous, since RNA is often tightly associated with proteins. DNase treatment can be used to remove DNA, and RNA can be precipitated by ethanol.

Frequently one wishes to isolate mRNA, for translation *in vitro* or for the synthesis of a cDNA probe (Section 3.10.1). Since almost all mRNA molecules encoded by chromosomal DNA have lengths of adenosine units at their 3' ends, forming poly(A) tails, they can be separated from a mixture of RNA molecules by affinity chromatography on oligo(dT)-cellulose columns (Section 10.9). At high salt concentrations, the poly(A) tails will bind to the complementary oligo(dT) units of the affinity column, and so mRNA will be retained; all other RNA molecules can be washed through the column by a further solution with a high concentration of salt. Finally, the bound mRNA can be eluted using a low concentration of salt.

3.5 Physical analysis of DNA

3.5.1 Electrophoresis

Electrophoresis in agarose or polyacrylamide gels is the most usual way to separate DNA molecules according to size. The technique can be used analytically or preparatively, and can be qualitative or quantitative. Electrophoresis is discussed in Chapter 9, so no practical details will be given here. The easiest and most widely applicable method is electrophoresis in horizontal agarose gels, followed by staining with ethidium bromide. This dye binds to DNA by insertion between stacked base-pairs (intercalation), and it exhibits a strong orange-red fluorescence when illuminated with ultraviolet light (Section 9.4.1). Very often electrophoresis is used to check that restriction (Section 3.7.1) or ligation (Section 3.7.2) reactions have gone to completion, or to assess the purity and intactness of a DNA preparation. For such checks 'mini-gels' are particularly convenient, since they need little sample material and give results quickly. Agarose gels (Section 9.4.1) can be used to separate molecules larger than about 200 base-pairs (bp), but polyacrylamide gels (Section 9.4.2) must be used for shorter molecules.

By calibration against DNA molecules of known sizes, such as restriction fragments (Section 3.7.1), the lengths of molecules in a sample can be measured. This can be used for such purposes as determining the sizes of insert in cloning vectors (Section 3.8), or to map the positions of restriction sites on a length of DNA. Restriction mapping involves the size analysis of restriction fragments produced by several restriction enzymes (Section 3.7.1) individually and in combination. The principle of this mapping is illustrated in Fig. 3.11, in which the restriction sites of two enzymes, A and B, are being mapped. Cleavage with A gives fragments 2 and 7 kilobases (kb) from a 9 kb molecule; hence we can position the single A site 2 kb from one end. Similarly, B gives fragments of 3 and 6 kb length, so it has a single site 3 kb from one end; but it is not possible at this stage to say if it is near to A's site, or at the opposite end of the DNA. This can be resolved by a double digestion. If the resultant fragments are 2, 3 and 4 kb in length, then A and B cut at opposite ends of the molecule; if they are 1, 2, and 6 kb, the sites are near each other. Not surprisingly, the mapping of real molecules is rarely as simple as this, and computer analysis of the restriction fragment lengths is usually needed to construct a map.

When electrophoresis is used preparatively, the piece of gel containing the desired DNA fragment is cut out, and the DNA is recovered from it in various ways, including crushing with a glass rod in a small volume of buffer, or by electroelution. In this method the piece of gel is sealed in a length of dialysis tubing containing buffer, and is then placed between two electrodes in a tank containing more buffer. Passage of an electric current between the electrodes causes DNA to migrate out of the gel piece, but it remains trapped within the dialysis tubing, and can therefore be recovered easily.

It is, of course, necessary to discover which band on a gel contains the particular piece of DNA that is needed. This can be achieved by transferring the DNA from the intact gel on to a piece of nitrocellulose paper placed in contact with it, using denaturing conditions, so that the DNA becomes bound to the paper in exactly the same pattern as that originally on the gel. This transfer, named a Southern blot after its inventor, can be performed electrophoretically, or by drawing large

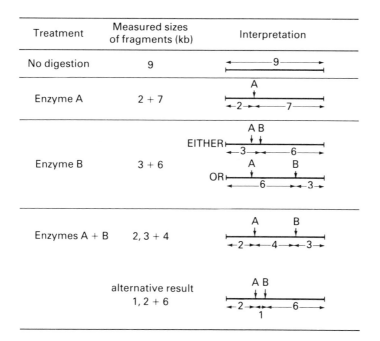

Treatment	Measured sizes of fragments (kb)	Interpretation
No digestion	9	
Enzyme A	2 + 7	
Enzyme B	3 + 6	
Enzymes A + B	2, 3 + 4	
alternative result	1, 2 + 6	

Fig. 3.11. Restriction mapping of DNA. Note that each experimental result and its interpretation should be considered in sequence, thus building up an increasingly unambiguous map.

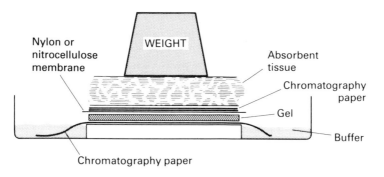

Fig. 3.12. Southern blot apparatus.

volumes of buffer through both gel and paper, thus transferring DNA from one to the other (Fig. 3.12). The point of this operation is that the paper can now be treated with a radioactive DNA molecule, e.g. a cDNA (Section 3.10.1) acting as a probe, in the same way as in colony hybridisation (Section 3.10.3), to discover which bands of DNA contain sequences complementary to the probe. The same process can be used to transfer RNA from gels on to nitrocellulose paper, for identification of specific sequences by hybridisation, and it is then known as

Northern blotting. When the transfer procedure is used with proteins separated on polyacrylamide gels it is termed Western blotting or protein blotting.

3.5.2   Renaturation kinetics

When preparations of double-stranded DNA are denatured by heat or alkali, and then allowed to renature, measurement of the rate of renaturation can give valuable information about the complexity of the DNA, i.e. how much information it contains (measured in base-pairs). The complexity of a molecule may be much less than its total length if some sequences are repetitive, but complexity will equal total length if all sequences are unique, appearing only once in the genome. In practice, the DNA is first cut randomly into fragments about 1 kb in length (Section 3.10.2), and is then completely denatured by heating above its T_m (Section 3.2.2). Renaturation at a temperature about 10 deg.C below the T_m is monitored either by decrease in absorbance at 260 nm wavelength (the hypochromic effect) or by passing samples at intervals through a column of hydroxylapatite (Section 10.5.2), which will adsorb only double-stranded DNA, and measuring how much of the sample is bound. The degree of renaturation after a given time will depend on C_o, the concentration (in nucleotides per unit volume) of double-stranded DNA prior to denaturation, and t, the duration of the renaturation.

For a given C_o, it should be evident that a preparation of bacteriophage λ DNA (genome size 49 kb) will contain many more copies of the same sequence per unit volume than does a preparation of human DNA (haploid genome size 3×10^6 kb), and will therefore renature far more rapidly because there will be more molecules complementary to each other per unit volume in the case of λ DNA and therefore more chance of two complementary strands colliding with each other. In order to compare the rates of renaturation of different DNA samples, it is usual to measure C_o and the time taken for renaturation to proceed half way to completion, $t_{\frac{1}{2}}$, and to multiply these values together to give a $C_o t_{\frac{1}{2}}$ value. The larger $C_o t_{\frac{1}{2}}$, the greater is the complexity of the DNA; hence λ DNA has a far lower $C_o t_{\frac{1}{2}}$ than does human DNA.

In fact, the human genome does not renature in a uniform fashion. If the extent of renaturation is plotted against log $C_o t$ (this is known as a Cot curve), it is seen that part of the DNA renatures quite rapidly, whereas the remainder is very slow to renature (Fig. 3.13). This must mean that some sequences have a higher concentration than others; in other words, part of the genome consists of repetitive sequences. These repetitive sequences can be separated from the single-copy DNA by passing the renaturing sample through a hydroxylapatite column early in the renaturation process, at a time that gives a low value of $C_o t$. At this stage only the rapidly renaturing sequences will be double stranded, and they will therefore be the only ones able to bind to the column.

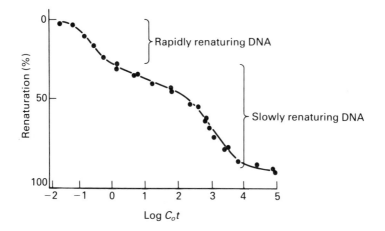

Fig. 3.13. Cot curve of human DNA. DNA was allowed to renature at 60°C after being completely dissociated by heat. Samples were taken at intervals and passed through a hydroxylapatite column to determine the percentage of double-stranded DNA present. This percentage was plotted against log $C_o t$ (original concentration of DNA×time of sampling).

3.6   Outline of genetic manipulation

The molecular biologist has a wide range of techniques available for the investigation and alteration of gene structure and function, but the key steps are those of cutting and joining lengths of DNA in a precise way, using restriction endonucleases and ligases.

By such cutting and joining, it is possible to divide a complex genome into a large number of small fragments, each about the size of a single gene, and to insert each of those fragments into a carrier, or vector DNA molecule, which can then be replicated indefinitely within microbial cells. In this way genes can be cloned to provide sufficient material for detailed analysis, or for insertion into the genome of a cell that is to be genetically engineered. The principal steps involved in gene cloning are illustrated in Fig. 3.14. Each step, and its key enzymes, is discussed in more detail below.

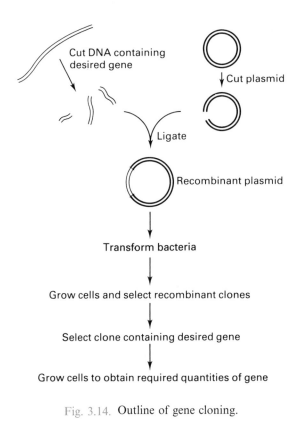

Cut DNA containing
desired gene

Cut plasmid

Ligate

Recombinant plasmid

Transform bacteria

Grow cells and select recombinant clones

Select clone containing desired gene

Grow cells to obtain required quantities of gene

Fig. 3.14. Outline of gene cloning.

3.7 Enzymes used in genetic manipulation

3.7.1 Restriction endonucleases

It was in 1970 that the first enzyme was isolated that would recognise a specific
sequence of DNA and cut the molecule within that sequence. Such enzymes are
known as restriction endonucleases, and they are used by bacteria as a defence
mechanism against foreign DNA, e.g. viral DNA, since they recognise and digest
such invading molecules. Bacterial DNA is protected from digestion by its own
enzymes as a result of methylation of bases within vulnerable sequences. A large
number of restriction enzymes have been isolated, and the class that is most useful
to the molecular biologist is known as type II. These enzymes recognise a specific
DNA sequence, usually four, five or six nucleotides in length, and cleave the DNA
within this restriction site. Clearly a tetranucleotide sequence will occur more
frequently in a given molecule than a hexanucleotide sequence, so more fragments

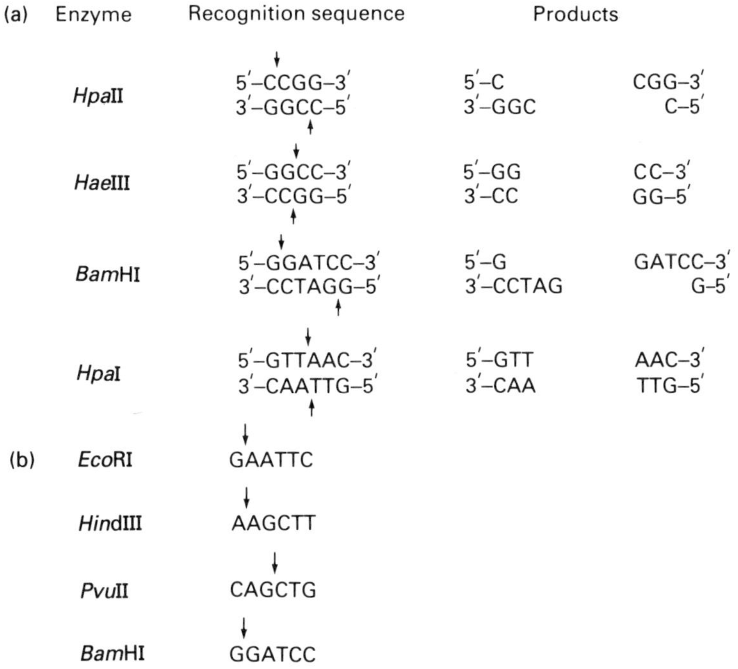

(a) Enzyme | Recognition sequence | Products

*Hpa*II
$$\overset{\downarrow}{\text{5}'-\text{CCGG}-\text{3}'}$$
$$\text{3}'-\text{GGCC}-\text{5}'$$
5′–C   CGG–3′
3′–GGC   C–5′

*Hae*III
$$\overset{\downarrow}{\text{5}'-\text{GGCC}-\text{3}'}$$
$$\underset{\uparrow}{\text{3}'-\text{CCGG}-\text{5}'}$$
5′–GG   CC–3′
3′–CC   GG–5′

*Bam*HI
$$\overset{\downarrow}{\text{5}'-\text{GGATCC}-\text{3}'}$$
$$\underset{\uparrow}{\text{3}'-\text{CCTAGG}-\text{5}'}$$
5′–G   GATCC–3′
3′–CCTAG   G–5′

*Hpa*I
$$\overset{\downarrow}{\text{5}'-\text{GTTAAC}-\text{3}'}$$
$$\underset{\uparrow}{\text{3}'-\text{CAATTG}-\text{5}'}$$
5′–GTT   AAC–3′
3′–CAA   TTG–5′

(b) *Eco*RI
$$\overset{\downarrow}{\text{GAATTC}}$$

*Hind*III
$$\overset{\downarrow}{\text{AAGCTT}}$$

*Pvu*II
$$\overset{\downarrow}{\text{CAGCTG}}$$

*Bam*HI
$$\overset{\downarrow}{\text{GGATCC}}$$

Fig. 3.15. Recognition sequences of some restriction enzymes showing (a) full descriptions and (b) conventional representations. Arrows indicate positions of cleavage. Note that all the information in (a) can be derived from knowledge of a single strand of the DNA, whereas in (b) only one strand is shown, drawn 5′ to 3′; this is the conventional way of representing restriction sites.

will be generated by an enzyme that recognises a tetranucleotide sequence.

Some enzymes cut straight across the DNA to give blunt ends, whereas others make staggered single-strand cuts, producing short, single-stranded projections at each end of the cleaved DNA (Fig. 3.15). Since the restriction sites are symmetrical, so that both strands have the same sequence when read in the 5′ to 3′ direction, such staggered cuts will generate identical single-stranded projections on either side of the cut. These ends are not only identical, but complementary, and will base-pair with each other; they are therefore known as cohesive or sticky ends. It is most important to realise that, because of the specificity of restriction enzymes, every copy of a given DNA molecule will give the same set of fragments when cleaved with a particular enzyme, and different DNA molecules will, in general, give different sets of fragments when treated with the same enzyme. Well over 1000 type II enzymes, recognising more than 160 different restriction sites, have been characterised, and the list is growing steadily.

3.7.2   Ligases

Although cutting DNA precisely is very useful for DNA analysis, its full potential is revealed only when the fragments produced are joined together to give a new structure, known as recombinant DNA. This joining, or ligation, is achieved by the use of a DNA ligase enzyme, the most common being that isolated from the bacterial virus known as T4 phage.

   If two different DNA preparations are treated with the same restriction enzyme to give fragments with sticky ends, these ends will be identical in both preparations. Thus, when the two sets of fragments are mixed, base-pairing between sticky ends will result in the coming together of fragments that were derived from different molecules. There will, of course, also be pairing of fragments derived from the same molecule. All these pairings are transient, owing to the weakness of hydrogen bonding between the few bases in the sticky ends, but they can be stabilised by use of DNA ligase, which forms a covalent bond between the 5′ phosphate group at the end of one strand and the 3′ hydroxyl group of the adjacent strand (Fig. 3.16). This reaction, which is driven by ATP, is often carried out at 10 °C to lower the kinetic energy of molecules, and so reduce the chances of base-paired sticky ends parting before they have been stabilised by ligation. However, long reaction times are needed to compensate for the low activity of DNA ligase in the cold.

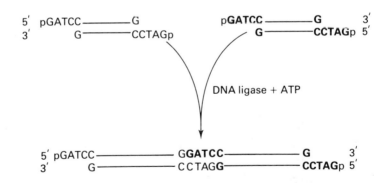

Fragments produced by cleavage with *Bam*HI

Fig. 3.16. Ligation of molecules with cohesive ends. Complementary cohesive ends base-pair, forming a temporary link between two DNA fragments. This association of fragments is stabilised by the formation of 3′ to 5′ phosphodiester linkages between cohesive ends, a reaction catalysed by DNA ligase.

   Since ligation reconstructs the site of cleavage, recombinant molecules produced by ligation of sticky ends can be cleaved again at the 'joins', using the same restriction enzyme that was used to generate the fragments initially. Consequently a fragment can be inserted into a vector DNA, and recovered again after cloning of the recombinant molecule.

Lengths of blunt-ended DNA can be ligated, but, since there is no base-pairing to hold fragments together temporarily, concentrations of DNA and ligase must be high. However, blunt-ended ligation is a useful way of joining together DNA fragments that have not been produced by the same restriction enzyme, and

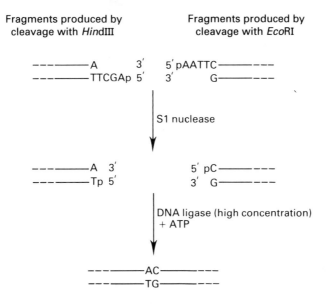

Fig. 3.17. Production and ligation of blunt ends. Different blunt-ended fragments cannot be held together temporarily by base-pairing between cohesive ends, and so both DNA and ligase must be used at high concentrations to increase the chances of two DNA fragments occupying the ligase active site simultaneously.

therefore probably have incompatible sticky ends. These ends are removed prior to ligation, using the enzyme S1 nuclease, which digests single-stranded DNA (Fig. 3.17). In such cases a restriction site will not be regenerated, and this may prevent recovery of a fragment after cloning. For this reason molecules called linkers are frequently used for joining DNA. Linkers are short, double-stranded oligo-nucleotides, with blunt ends, containing at least one restriction site within their sequence (Fig. 3.18). These linkers can be joined to one preparation of DNA by blunt-ended ligation, and then sticky ends can be created by cleavage of the linkers with a suitable restriction enzyme. The linker is chosen so that the sticky end it produces is identical with that on the other DNA preparation; consequently, the two can then be joined by ligation of their sticky ends. Some very versatile linkers are available that contain restriction sites for several different enzymes within a sequence of only eight to ten nucleotides (Fig. 3.19).

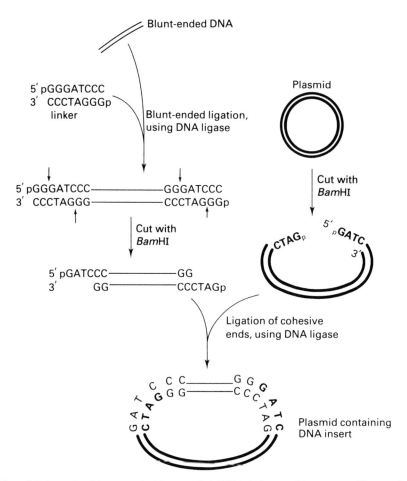

Fig. 3.18. Use of linkers. In this example blunt-ended DNA is inserted into a specific restriction site on a plasmid, after ligation to a linker containing the same restriction site.

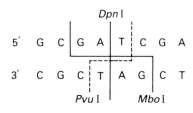

Fig. 3.19. A versatile linker. Cleavage sites of three different restriction enzymes are marked. Recognition sequences are: *Dpn*I, GA↓TC; *Mbo*I, ↓GATC; *Pvu*I, CGAT↓CG. Thus *Dpn*I generates blunt-ended fragments, whereas *Mbo*I and *Pvu*I produce fragments with cohesive ends.

Using a technique called homopolymer tailing, sticky ends can be built up on blunt-ended molecules (Fig. 3.20). For example, one preparation of DNA could be treated with the enzyme terminal transferase in the presence of dATP, resulting in the addition of a poly(dA) chain to the 3′ end of each strand. The other preparation would then be given 3′ tails of poly(T), using the same enzyme with TTP. On mixing, there would be base-pairing between complementary sticky ends, which could then be ligated. An attractive feature of this method is that ligation will not occur between fragments from the same preparation.

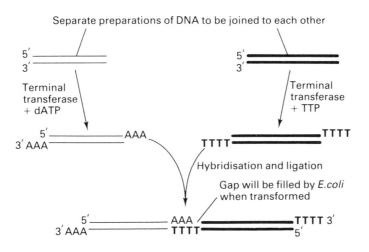

Fig. 3.20. Homopolymer tailing. One DNA preparation is given poly(dA) tails at its 3′ ends, the other receives poly(T) tails. These tails will hybridise with each other when the two preparations are mixed, and ligation can be used to stabilise this association. If the tails are long enough, hybridisation will be so stable that ligation is not needed.

3.8   Cloning vectors

3.8.1   Plasmids

By cloning, we can produce unlimited amounts of any particular fragment of DNA. In principle, the DNA is introduced into a suitable host cell, most usually a bacterium such as *Escherichia coli*, where it is replicated as the cell grows and divides. However, replication will occur only if the DNA contains a sequence that is recognised by the cell as an origin of replication. Most DNA samples do not contain such sequences and therefore the DNA to be cloned has to be attached to a carrier, or vector DNA that does contain an origin of replication. Many bacteria contain such a piece of DNA, called a plasmid, which is a relatively small, circular, extrachromosomal molecule, carrying genes for such properties as antibiotic resistance, conjugation, metabolism of 'unusual' substrates, etc. Some of these plasmids are replicated at a high rate by bacteria, and so they are excellent potential vectors. Starting from a selection of natural plasmids, artificial plasmids have been

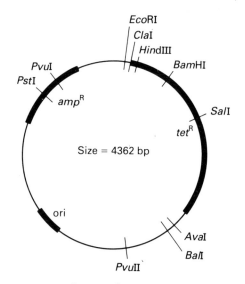

Fig. 3.21. The plasmid pBR322: amp^R and tet^R are genes for resistance to ampicillin (amp) and tetracycline (tet), respectively; ori is the origin of DNA replication. Some unique sites of cleavage by a selection of restriction endonucleases are indicated (e.g. *EcoRI, ClaI*).

constructed as vectors, by a complex series of cutting and joining reactions.

One of the most widely used plasmids, pBR322, illustrates the desirable features that have been incorporated into these vectors (Fig. 3.21):

(i) The plasmid is much smaller than a natural plasmid, since this makes it more resistant to damage by shearing, and increases the efficiency of uptake by bacteria during transformation (see below).

(ii) A bacterial origin of DNA replication ensures that the plasmid, will be replicated by the host cell. Some replication origins display stringent regulation of replication, in which rounds of replication are initiated at the same frequency as cell division. Most plasmids, including pBR322, have a relaxed origin of replication, whose activity is not tightly linked to cell division, and so plasmid replication will be initiated far more frequently than chromosomal replication. Hence a large number of plasmid molecules will be produced per cell.

(iii) Two genes coding for resistance to antibiotics have been included. One of these allows the selection of cells that contain plasmid: if cells are plated on medium containing an appropriate antibiotic, only those that contain plasmid will grow to form colonies. The other resistance gene can be used, as described below, for detection of those plasmids that contain inserted DNA.

(iv) There are single restriction sites for a number of enzymes, scattered around the plasmid, that can be used to open the circle at a specific point prior to insertion of a piece of DNA to be cloned. The variety of sites not only makes it easier to find a restriction enzyme that is suitable for both vector and inserted DNA, but, since some of the sites are placed within an antibiotic resistance gene, the presence of an insert can be detected by loss of resistance to that antibiotic (insertional inactivation).

The value of pBR322 can be illustrated by considering its use for cloning a fragment of DNA. It is assumed that this fragment has been produced by cleavage of a larger molecule with the restriction enzyme *Bam*HI, followed by separation of the different fragments using gel electrophoresis, and recovery of the desired fragment from the gel by crushing or electroelution. The plasmid would also be cut at a single site, using *Bam*HI, and both samples would then be deproteinised to inactivate the restriction enzyme. *Bam*HI cleaves to give sticky ends, and so it is easy to obtain ligation between plasmid and fragment, using T4 DNA ligase. The products of this ligation will include plasmid containing a single fragment of the DNA as an insert, but there will also be unwanted products, such as plasmid that has recircularised without an insert, dimers of plasmid, fragments joined to each other, and plasmid with an insert composed of more than one fragment (Fig. 3.22). Most of these unwanted molecules can be eliminated during subsequent steps.

The ligated DNA must now be used to transform *E. coli*. Bacteria do not normally take up DNA from their surroundings, but can be induced to do so by prior treatment with Ca^{2+} in the cold; they are then said to be competent, since DNA added to the suspension of competent cells will be taken up during a mild heat shock. Small, circular molecules are taken up most efficiently, whereas long, linear molecules will not enter the bacteria.

After a brief incubation to allow expression of the antibiotic resistance genes the cells are plated on to medium containing the antibiotic ampicillin. Colonies that grow on these plates must be derived from cells that contain plasmid, since this carries the gene for resistance to ampicillin. It is not, at this stage, possible to distinguish between those colonies containing plasmids with inserts and those that simply contain recircularised plasmids. To do this, the colonies are replica plated, using a sterile velvet pad, on to plates containing the antibiotic tetracycline in their medium (Fig. 3.23). Since the *Bam*HI site lies within the tetracycline resistance gene, this gene will be inactivated by the presence of insert, but will be intact in

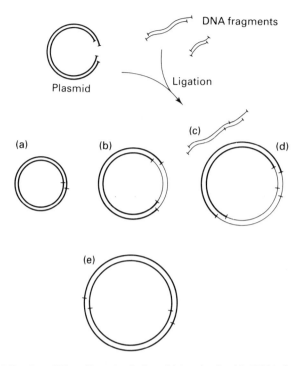

Fig. 3.22. Products of ligation. When linearised plasmid is mixed with DNA fragments in the presence of DNA ligase, there will be a mixture of products consisting of (a) recircularised plasmid, (b) recircularised plasmid containing a single DNA insert, (c) fragments of DNA joined together, (d) recircularised plasmid containing an insert made up of more than one fragment, and (e) two plasmids joined together and recircularised to give a dimer.

those plasmids which have merely recircularised. Thus colonies that grow on ampicillin but not on tetracycline must contain plasmids with inserts. Because replica plating gives an identical pattern of colonies on both sets of plates, it is easy to recognise the colonies with inserts, and to recover them from the ampicillin plate for further growth. This illustrates the importance of a second gene for antibiotic resistance in a vector.

Although recircularised plasmid can be selected against, its presence decreases the yield of recombinant plasmid containing inserts. If the cut plasmid is treated with alkaline phosphatase prior to ligation, recircularisation will be prevented, since the enzyme removes the 5′ phosphoryl groups that are essential for ligation. Links can still be made between the 5′ phosphate group of the insert and the 3′ hydroxyl group of the plasmid, so only recombinant plasmids and chains of linked DNA fragments will be formed. It does not matter that only one strand of the recombinant DNA is ligated, since the nick will be repaired by bacteria transformed with these molecules.

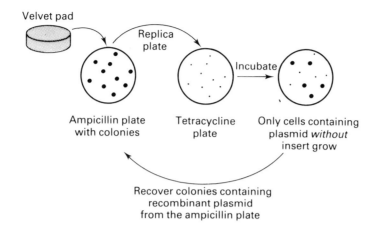

Fig. 3.23. Replica plating to detect recombinant plasmids. A sterile velvet pad is pressed on to the surface of an agar plate, picking up some cells from each colony growing on that plate. The pad is then pressed on to a fresh agar plate, thus inoculating it with cells in a pattern identical with that of the original colonies. Clones of cells that fail to grow on the second plate (e.g. owing to the loss of antibiotic resistance) can be recovered from their corresponding colonies on the first plate.

Both prior to the insertion of foreign DNA and after cloning, plasmid must be isolated from the bacteria. It is possible to obtain a DNA preparation highly enriched for plasmid by very gentle cell lysis, using lysozyme and then detergent, followed by clearing of the lysate by centrifugation. Centrifugation sediments the high molecular weight DNA (predominantly chromosomal) and cell debris, leaving the small plasmid molecules and RNA in the supernatant. Undamaged plasmid is particularly compact, since it is supercoiled as a consequence of having slightly too few turns of the double helix per unit length. Such supercoiling can easily be demonstrated by attempting to unwind a piece of string that is clamped at one end. Other methods rely on the virtually irreversible denaturation of linear DNA by heat or alkali, and the use of centrifugation to remove denatured DNA from the easily renatured, circular plasmid.

Further purification of the plasmid can be achieved by caesium chloride density gradient ultracentrifugation (Section 6.7) of the nucleic acid preparation in the presence of ethidium bromide. Ethidium bromide causes unwinding of DNA as it binds to it, simultaneously producing a decrease in its buoyant density. Since the supercoiled plasmid DNA can unwind to only a very limited extent, it will not bind as much dye as will the linear and open circle forms of DNA; hence, supercoiled plasmid will have a higher density than the other types of DNA in the presence of saturating levels of ethidium bromide. Because of this density difference, plasmid DNA can be separated from other DNA by isopycnic ultracentrifugation (Section 6.6.2).

3.8.2 Viral DNA

The cloning of single genes is usually best carried out using plasmids because the insert will rarely be larger than about 5 kb. As will be shown later, there are several reasons for cloning much larger pieces of DNA, particularly when gene libraries (Section 3.10.2) of higher-eukaryote genomes are being constructed. Large inserts increase plasmid size to the point at which efficient transformation cannot occur, and so another way must be found to move recombinant DNA into the bacterial cells.

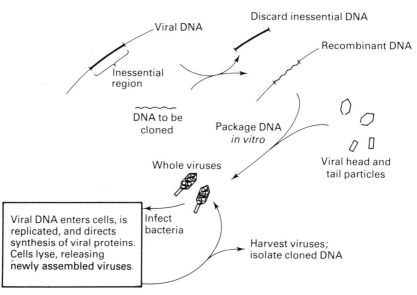

Fig. 3.24. Cloning DNA in bacteriophage λ. Inessential DNA is cut out of the DNA using restriction enzymes, and is separated from essential DNA fragments by electrophoresis. Essential DNA and DNA to be cloned are ligated and packaged in empty head particles, and tails are added to give infective viruses. See Fig. 3.25 for details of packaging.

One way of doing this is by using the DNA of bacteriophages (bacterial viruses) as a vector because these large molecules are injected into bacteria by the viral particles. A commonly used vector is that of bacteriophage λ, which is 49 kb in length. For the cloning of long DNA fragments, up to about 20 kb, much of the inessential λ DNA is removed and replaced by the insert. The recombinant DNA is then packaged within viral particles *in vitro*, and these are allowed to infect bacterial cells that have been plated out on agar (Fig. 3.24). Since the DNA is injected into the cells, a very high efficiency of transformation can be obtained. Once inside the cells, the recombinant viral DNA is replicated. All the genes needed for normal lytic growth are still present in the DNA, and so multiplication of the virus takes place by cycles of cell lysis and infection of surrounding cells, giving rise to plaques of lysed cells on a background, or lawn, of bacterial cells. Cloned DNA can be recovered from the viruses in these plaques.

3.8.3   Cosmids

Even longer fragments of DNA must be cloned for the analysis of highly complex genomes, and, in order to understand how this can be achieved, it is necessary to

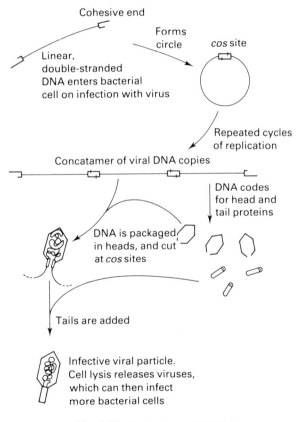

Fig. 3.25. Packaging of DNA.

know in outline how viral DNA is packaged. Viral DNA is injected into the cell as a linear molecule, at each end of which are cohesive ends, complementary to each other, 12 bases in length. Once inside the cell, these ends base pair and become permanently joined by ligation to form a region known as the *cos* site, giving a circular DNA molecule (Fig. 3.25). Replication of the DNA by a rolling circle mechanism gives rise to a concatamer, which is a long molecule made up of many copies of the viral DNA linked end to end through the *cos* sites. DNA is packaged by looping regions between *cos* sites into the precursor of the viral head;

when a head is full, the *cos* sites should be at the mouth of the head, where they will be cleaved to generate a linear molecule with cohesive ends. Subsequently, tail proteins are added to give infective particles.

Thus the only requirement for a length of DNA to be packaged into viral heads is that it should contain *cos* sites spaced the correct distance apart; in practice this spacing can range from 37 to 52 kb. Consequently vectors called cosmids have been constructed; they contain a *cos* site plus essential features of a plasmid, such as the plasmid origin of replication, a gene for drug resistance, and several unique restriction sites for insertion of DNA to be cloned (Fig. 3.26). When a cosmid preparation is linearised by restriction, and ligated to DNA for cloning, the

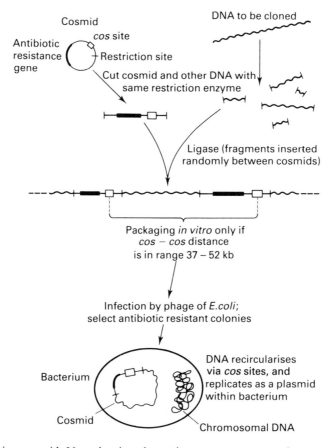

Fig. 3.26. Cloning in a cosmid. Note that in eukaryotic systems upper-case letters are used to denote sites (i.e. *COS*).

products will include concatamers of alternating cosmid and insert. Such DNA can be packaged *in vitro* if bacteriophage head precursors, tails and packaging proteins are provided, so long as the *cos* sites are spaced from 37 to 52 kb apart.

Since the cosmid is very small, inserts about 40 kb in length will be most readily packaged. Once inside the cell, the DNA recircularises through its *cos* site, and from then onwards behaves exactly like a plasmid.

3.8.4 Vectors used in eukaryotes

Plasmids are also used for cloning DNA in eukaryotic cells, but they need a eukaryotic origin of replication and marker genes that will be expressed by eukaryotic cells. At present the two most important applications of plasmids to eukaryotic cells are for cloning in yeast and in plants.

Although yeast has a natural plasmid, called the 2μ circle, this is too large for use in cloning. Plasmids have been created by genetic manipulation, sometimes using replication origins from the 2μ circle, and usually incorporating a gene that will complement a defective gene in the host yeast cell. If, for example, a strain of yeast is used that has a defective gene for the biosynthesis of an amino acid, an active copy of that gene on a yeast plasmid can be used as a selectable marker for the presence of that plasmid. Yeast, like bacteria, can be grown rapidly, and it is therefore well suited for use in cloning. Moreover, it is not pathogenic, and can carry out such post-translational modifications of the polypeptide as glycosylation and limited proteolysis. Such modifications are sometimes required for the activation or export from the cell of a polypeptide, and so yeast is particularly attractive for the expression of cloned genes on an industrial scale.

The bacterium *Agrobacterium tumefaciens* infects plants that have been damaged near soil level, and this infection is often followed by the formation of plant tumours in the vicinity of the infected region. It is now known that *A. tumefaciens* contains a plasmid called Ti, part of which is transferred into the nuclei of plant cells that are infected by the bacterium. Once in the nucleus, this DNA is maintained by integrating with the chromosomal DNA. The integrated DNA carries genes for the synthesis of opines (which are metabolised by the bacteria but not by the plants) and for tumour induction (hence 'Ti'). DNA inserted into the correct region of the Ti plasmid will be transferred to infected plant cells, and in this way it has been possible to clone and express foreign genes in plants. This is an essential prerequisite for the genetic engineering of crops.

In cases where protoplasts can be generated from cells, if necessary by digestion of the cell wall, transformation may be achieved by electroporation. In this process the protoplasts are subjected to pulses of a high voltage gradient, causing many of them to take up DNA from the surrounding solution. This technique has proved to be useful with cells from a range of animal, plant and microbial sources. An even more dramatic transformation procedure involves firing microscopically small tungsten pellets, or microprojectiles, coated with DNA, into plant cells in intact tissues. An explosive charge is used to propel the microprojectiles and the cells appear to reseal themselves after being entered. This is a particularly promising technique for use with plants whose protoplasts will not regenerate whole plants.

3.9   Sequencing of DNA

The advent of methods for the sequencing of DNA has revolutionised our understanding of gene structure, and it is now routine to sequence any newly isolated DNA fragment of interest. There are two methods in use, the dideoxy, or chain termination method, of F. Sanger and the chemical cleavage method of A. Maxam and W. Gilbert. Both methods are based on the high resolution electrophoresis of four sets of radioactive oligonucleotides produced from the DNA to be sequenced, but they differ in the procedures used to generate the oligonucleotides.

3.9.1   Sanger sequencing

For one version of the Sanger method, single-stranded DNA is required. This is readily prepared by cloning the DNA in bacteriophage M13. M13 bacteriophage DNA has been constructed so that it is well suited to DNA sequencing. The circular DNA molecule contains a bacterial origin of replication and other genes required for the production of viruses by infected bacterial cells. The DNA also includes part of the gene for β-galactosidase. This incomplete gene will complement a defective β-galactosidase gene in a suitable host cell, resulting in the appearance of β-galactosidase activity. Hence, cells containing M13 DNA can be detected by their ability to hydrolyse the artificial substrate X-gal (5-bromo-4-chloro-3-indolyl-β-D-galactoside) to yield a blue product. Such cells can be observed as blue plaques on a lawn of bacteria when plated out. Most importantly, within the partial β-galactosidase gene of the M13 DNA is a short sequence of DNA containing sites that can be cleaved by a variety of restriction enzymes. Foreign DNA can therefore easily be inserted into this polylinker, provided that both insert and polylinker have been cleaved with the same (or compatible) restriction enzyme. The insertion of foreign DNA in this way will disrupt the partial β-galactosidase gene, thus preventing host cells from producing an active β-galactosidase. These cells will give rise to colourless plaques and it is such cells that are of use for sequencing.

Restriction and ligation of the M13 DNA is performed using the double-stranded, replicative form (RF) of the molecule and cells are transformed as already described for plasmids. Once within the bacterial cell the M13 DNA is replicated and directs the synthesis of viral coat proteins. When the cell contains about 200 RF molecules there is a switch to the production of many copies of the plus strand

and this single-stranded DNA is then packaged into the viral coat. The viral particles so formed are extruded from the cell and are then able to infect other cells. Immediately after infection, the single-stranded DNA is used to generate an RF molecule and so initiate another cycle of virus production.

The attraction of M13 is that it provides a convenient way to generate large quantities of single-stranded DNA, all containing a piece of 'foreign' DNA to be sequenced, and all being copies of the plus strand. Since the viral particles are extruded into the growth medium they can be easily separated from the bacterial cells by differential centrifugation; simple deproteinisation of the virus then yields pure, single-stranded M13 DNA.

For Sanger sequencing this DNA is allowed to hybridise with an oligonucleotide,

Fragment to be sequenced, cloned in M13 phage

3′ – – – AG – – – CT**GCTCGCAT** – – – 5′
     TC – – – GA
     ⎵‿⎵
      Primer

| DNA polymerase
| 4 dNTPs (radioactive)
| ddGTP
↓

Synthesis of complementary second strands:

5′ TC – – – GAC**ddG** 3′

5′ TC – – – GA**CGA**dd**G** 3′

5′ TC – – – GA**CGAGC**dd**G** 3′

Denature to give single strands

Run on sequencing gel alongside products of
ddCTP, ddATP and ddTTP reactions

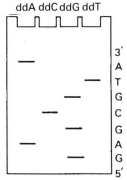

ddA ddC ddG ddT   Read sequence of *second strand*
from autoradiograph

3′
A
T
G
C
G
A
G
5′

Fig. 3.27. Sanger sequencing of DNA.

about 15 nucleotides long, which is complementary to a sequence in the M13 DNA just to the 3′ side of its inserted DNA (Fig. 3.27). The oligonucleotide will then act as a primer for synthesis of a second strand of DNA, catalysed by DNA polymerase. Since the new strand is synthesised from its 5′ end, virtually the first DNA to be made will be complementary to the inserted DNA, which we wish to sequence. One of the deoxyribonucleoside triphosphates (dNTPs) that must be provided for DNA synthesis is radioactively labelled with ^{32}P or ^{35}S, and so the newly synthesised strand will be labelled. If a 2′,3′-dideoxynucleoside triphosphate (ddNTP) is also present in the reaction mixture, there is a chance that it will be added to the growing DNA chain in place of the normal dNTP, since it is identical with its corresponding dNTP, apart from the absence of a 3′-hydroxyl group. Once this has happened, the chain growth terminates, since a 5′ to 3′ phosphodiester linkage cannot be formed without a 3′ hydroxyl. Since the incorporation of ddNTP rather than dNTP is a random event, the reaction will produce new molecules varying widely in length, but all terminating at the same type of base. If the cloned DNA is split into four portions and a different ddNTP used for each, four sets of molecules are generated, each terminating at a different type of base, but all having a common 5′ end (the primer).

The four samples so produced are then denatured and loaded next to each other on a polyacrylamide gel for electrophoresis. Electrophoresis is performed at about 70°C in the presence of urea, to prevent renaturation of the DNA, because even partial renaturation would seriously alter the rates of migration of DNA fragments. Very thin, long gels are used for maximum resolution over a wide range of fragment lengths (Section 9.4.2). After electrophoresis, the positions of radioactive DNA bands on the gel are determined by auto-radiography (Section 5.2.3). Bearing in mind that every band in the track from the dideoxyadenosine triphosphate sample must contain molecules that terminate with adenine, and that those in the ddGTP terminate with guanine, etc., it is possible to read the sequence of the newly synthesised strand from the autoradiograph, provided that the gel can resolve differences in length equal to a single nucleotide (Figs. 3.27 and 3.28). Under ideal conditions, sequences up to about 300 bases in length can be read from one gel.

Sanger sequencing may also be carried out using double-stranded DNA, thus avoiding the need to clone in M13. Usually the DNA to be sequenced is cloned in a pUC plasmid (e.g. pUC18). These plasmids have a gene for resistance to ampicillin, a bacterial origin of replication, and a partial β-galactosidase gene containing a multiple insertion site (just as in M13 DNA). DNA can be cloned and manipulated by standard plasmid procedures and clones containing recombinant DNA can be identified by insertional inactivation of the partial β-gal-actosidase gene. Sequencing is as described for M13, using the same primer, but

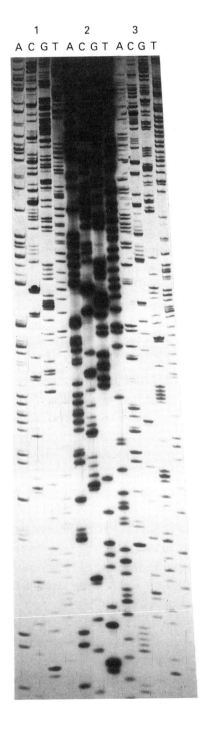

Direction of electrophoretic movement

Fig. 3.28. Autoradiograph of a DNA
sequencing gel. Samples were prepared
using the Sanger, dideoxy method of DNA
sequencing. Each set of four samples was
loaded into adjacent tracks, indicated by A,
C, G and T, depending on the identity of
the dideoxyribonucleotide used for that
sample. Two sets of samples were labelled
with ^{35}S (1 and 3) and one was labelled with
^{32}P (2). It is evident that ^{32}P generates
darker but more diffuse bands than does ^{35}S,
making the bands near the bottom of the
autoradiograph easy to see. However, the
broad bands produced by ^{32}P cannot be
resolved near the top of the autoradiograph,
making it impossible to read a sequence
from this region. The much sharper bands
produced by ^{35}S allow sequences to be read
with confidence along most of the
autoradiograph and so a longer sequence
of DNA can be obtained from a single gel.

the double-stranded DNA must be denatured prior to annealing with primer.

3.9.2   Maxam and Gilbert sequencing

The chemical cleavage method of Maxam and Gilbert usually starts with the enzymic addition of a radioactive label to *either* the 3′ *or* the 5′ ends of a double-stranded DNA preparation (Fig. 3.29). The strands are then separated by electrophoresis under denaturing conditions, and analysed separately. DNA labelled at one end is split into four portions and each is treated with chemicals that will act on a specific base (or, in some cases, either of two bases) by methylation or removal of the base. Conditions are chosen so that, on average, each molecule is modified at only one position along its length; every base in the DNA strand has an equal chance of being modified. After the modification reactions, the separate samples are cleaved by piperidine, which breaks phosphodiester bonds exclusively at the 5′ side of nucleotides whose base has been modified. The result is similar to that produced by the Sanger method, since each sample now contains radioactive molecules of various lengths, all with one end in common (the labelled end), and with the other end cut at the same type of base. Analysis of the reaction products by electrophoresis is as described for the Sanger method.

3.9.3   Sequencing strategies

Because the Sanger method produces oligonucleotides that are radioactively labelled throughout their lengths, rather than only at one end, the molecules can be made a lot more radioactive, and therefore easier to detect; less DNA therefore is needed for sequencing. Once M13 or pUC cloning has been set up in a laboratory, it provides a very convenient and rapid way to obtain DNA for sequencing. For these reasons, dideoxy sequencing of DNA is probably the most commonly used sequencing method, though the chemical procedure is still used by some laboratories.

It should be remembered that a lot of effort is involved in the isolation of a specific gene, and this, rather than sequencing, is likely to be the rate-limiting step in the process of finding a gene's sequence. Sequencing technology has now reached such a level of sophistication that it is quite common for a large stretch of DNA to be sequenced and then that sequence is searched by computer for open reading frames, i.e. sequences beginning with a start codon (ATG) and continuing with a significant number of 'coding' triplets before a stop codon is reached. Such sequences are potentially able to code for a polypeptide and their identities can sometimes be deduced by comparison with other sequences in the vast databases now available.

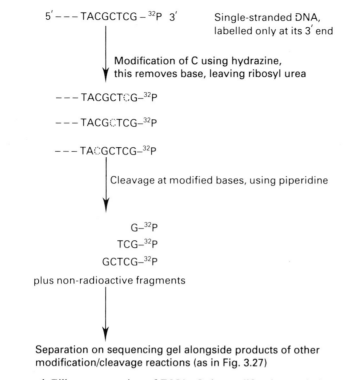

5′ – – – TACGCTCG – ³²P 3′   Single-stranded DNA,
labelled only at its 3′ end

Modification of C using hydrazine,
this removes base, leaving ribosyl urea

– – – TACGCT⊙G–³²P

– – – TACG⊙TCG–³²P

– – – TA⊙GCTCG–³²P

Cleavage at modified bases, using piperidine

G–³²P
TCG–³²P
GCTCG–³²P

plus non-radioactive fragments

Separation on sequencing gel alongside products of other
modification/cleavage reactions (as in Fig. 3.27)

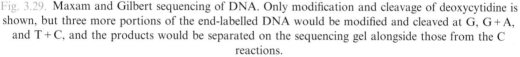

Fig. 3.29. Maxam and Gilbert sequencing of DNA. Only modification and cleavage of deoxycytidine is shown, but three more portions of the end-labelled DNA would be modified and cleaved at G, G + A, and T + C, and the products would be separated on the sequencing gel alongside those from the C reactions.

The largest project of this type is the Human Genome Project, which aims to generate the sequence of an entire human genome over a period of about 15 years. This has been made possible by the development of automated sequencing techniques. One automatic method involves the use of dideoxyribonucleotides labelled with coloured or fluorescent groups. After a standard Sanger reaction all four samples are run on the same track of an electrophoresis gel and an optical sensor further down the gel detects the passage of each fragment of DNA and identifies whether it has been terminated by ddA, ddC, ddG or ddT from its colour or wavelength of fluorescence. The output from the sensor can be fed directly into a computer for storage and further processing.

3.10   Isolation of specific nucleic acid sequences

3.10.1   Complementary DNA

The most difficult part of genetic manipulation is not the cloning of DNA, but isolation of the particular piece of DNA to be cloned. If the aim is to clone a gene, then it is enormously helpful to have as much information as possible about the gene product. Usually this product is a protein, and ideally antibodies to the protein should be available for detection or precipitation of the protein or its precursors (see below); but even knowledge of its relative molecular mass can be of use.

Frequently an attempt is made to isolate the mRNA transcribed from a desired gene. If this codes for a major protein of the cell, it should form a major fraction of the total mRNA, as in the case of the B cells of the pancreas, which contain high levels of proinsulin mRNA. It is sometimes possible to precipitate polysomes, which are translating the mRNA, by using antibodies to their proteins; mRNA can then be dissociated from the precipitated ribosomes. More usually, the mRNA needed is only a minor component of the total cellular mRNA. In such cases, total mRNA is fractionated by size, using sucrose density gradient centrifugation (Section 6.6.2). Then each fraction is used to direct the synthesis of proteins, using an *in vitro* translation system derived from lysates of rabbit reticulocytes or from wheat germ extracts. Immunoprecipitation (Section 2.3) or polyacrylamide gel electrophoresis (Section 9.3.1) can then be used to detect the 'target' protein amongst the many other products.

When a fraction containing the desired mRNA has been identified, it is used to direct the synthesis of DNA molecules complementary to all of the mRNAs in that fraction. This cDNA (complementary DNA) is made using the enzyme reverse transcriptase, as shown in Fig. 3.30. Reverse transcriptase will synthesise a DNA strand complementary to an mRNA template, using a mixture of the four deoxyribonucleoside triphosphates, provided that a short length of primer is base-paired with the 3′ end of the RNA. Since mRNA has a poly(A) tail at its 3′ end, a short oligo(dT) molecule will act as the primer for reverse transcriptase. After synthesis of the first DNA strand, a poly(dC) tail is added to its 3′ end, using terminal transferase and dCTP. This will also, incidentally, put a poly(dC) tail on the poly(A) of mRNA. Alkaline hydrolysis is then used to remove the RNA strand, leaving single-stranded DNA that can be used, like the mRNA, to direct the synthesis of a complementary DNA strand. The second-strand synthesis requires an oligo(dG) primer, base-paired with the poly(dC) tail, and it can be catalysed by the Klenow fragment (so named after its discoverer) of DNA pol I. This is prepared by cleavage of DNA polymerase with subtilisin, giving a large fragment

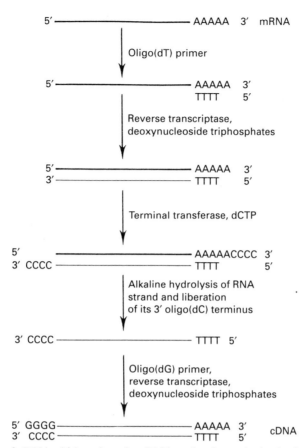

Fig. 3.30. Synthesis of cDNA. Although each mRNA molecule results in the formation of only one cDNA molecule, the cDNA can be cloned, using a plasmid vector, to give an unlimited number of copies.

that has no 5′ to 3′ exonuclease activity, but which still acts as a 5′ to 3′ polymerase. Surprisingly, since the template is now DNA, the reaction can also be catalysed by reverse transcriptase. The final product is double-stranded DNA, one of whose strands is complementary to the mRNA.

The mixture of cDNA molecules can now be inserted into plasmids and used to transform bacteria, which are then grown to give colonies. Since the cloned cDNAs lack promoters, and will therefore not be expressed in a plasmid such as pBR322, the wanted sequence cannot be detected using antibodies to its corresponding protein. Consequently, a rather devious method is used to pick out the sequence.

Plasmid is extracted from part of each colony, and each preparation is then denatured and immobilised on a nitrocellulose filter (Fig. 3.31). The filters are soaked in total cellular mRNA, under stringent conditions (temperature only a few degrees below T_m) in which hybridisation will occur only between

Colonies from cDNA library, growing on agar plate

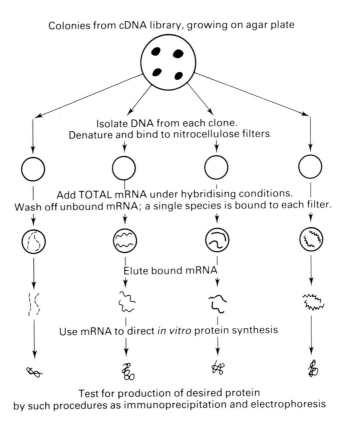

Isolate DNA from each clone.
Denature and bind to nitrocellulose filters.

Add TOTAL mRNA under hybridising conditions.
Wash off unbound mRNA; a single species is bound to each filter.

Elute bound mRNA

Use mRNA to direct *in vitro* protein synthesis

Test for production of desired protein
by such procedures as immunoprecipitation and electrophoresis

Fig. 3.31. Hybrid release translation.

complementary strands of nucleic acid. Hence, each filter will bind just one species of mRNA because it has only one type of cDNA immobilised on it. Unbound mRNA is washed off the filters, and then the bound mRNA is eluted and used to direct *in vitro* translation. By immunoprecipitation or electrophoresis of the translation products, the mRNA coding for a particular protein can be detected, and the clone containing its corresponding cDNA isolated. This technique is known as hybrid release translation. In a related method called hybrid arrested translation a positive result is indicated by the *absence* of a particular translation product when total mRNA is hybridised with excess cDNA. This is a consequence of the fact that mRNA cannot be translated when it is hybridised to another molecule.

In some cases the cDNA is all that needs to be cloned; as, for example, when the aim is to persuade bacteria to synthesise a 'foreign' protein. The cDNA sequence can be inserted into an expression vector (Section 3.11) and should then result in production of the desired protein. It should be noted that the cDNA will often not be identical with the original gene, if the gene is eukaryotic, because the majority of such genes contain introns (Section 3.3.3), which interrupt their coding regions (exons). During maturation of mRNA in eukaryotes, the introns are excised from the molecule, leaving only the exons spliced together; it is this spliced molecule that is used to make cDNA. Genes also have regions flanking them that are of importance in the regulation of their expression, and are not transcribed as part of the mRNA. When a complete gene must be isolated, cDNA can be used as a probe to search through a gene library for the desired gene.

3.10.2   Gene libraries

Gene libraries are constructed by isolating the complete genomic DNA from a cell, and cutting it almost randomly into fragments of the desired average length. This can be achieved by partial restriction with an enzyme that recognises tetra-nucleotide sequences. Complete restriction with such an enzyme would produce a large number of very short fragments (Section 3.7.1), but, if the enzyme is allowed to cleave only a few of its potential restriction sites before the reaction is stopped, each DNA molecule will be cut almost randomly into relatively large fragments. Average fragment size will depend on the concentrations of DNA and restriction enzyme, and on the conditions and length of incubation.

The mixture of fragments is ligated with a vector, and cloned. If enough clones are produced there will be a very high chance that any particular gene will be present in at least one of the clones. Such a collection of clones is known as a gene library. To keep the number of clones to a manageable size, fragments about 10 kb in length are needed for prokaryotic libraries, but the length must be increased to about 40 kb for mammalian libraries.

A cDNA library can be created similarly by cloning cDNA prepared from the total mRNA of a tissue. Such libraries are particularly useful for the study of tissue-specific gene expression.

3.10.3   Colony hybridisation

The technique of colony hybridisation is used to pull a particular gene out of a gene library (Fig. 3.32). A large number of clones is grown up to form colonies on one or more plates, and these are then replica plated on to nitrocellulose or nylon membranes placed on solid agar medium. Nutrients diffuse through the membranes and allow colonies to grow on them. The colonies are then lysed, and liberated DNA is denatured and bound to the membranes, so that the pattern of colonies is replaced by an identical pattern of bound DNA. If the membranes are incubated with denatured, radioactive cDNA under hybridising conditions, the cDNA will bind only to cloned fragments containing at least part of its corresponding gene.

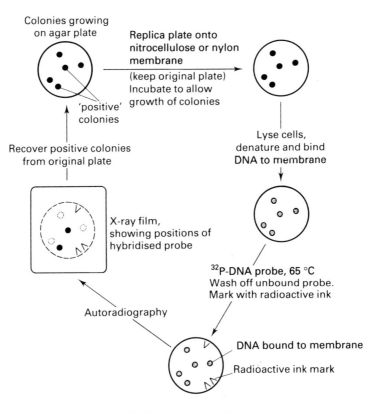

Fig. 3.32. Colony hybridisation.

Such binding can be detected by autoradiography of the washed membranes (Section 5.2.3). It is worth noting that the introns of a gene do not interfere with its hybridisation to cDNA, since there is a strong interaction between the exons and the cDNA. By comparison of the autoradiographs with the original plates of colonies, those which contain the desired gene (or part of it) can be identified and used for gene isolation.

A similar procedure is used to identify desired genes cloned in viral DNA. In this case it is the DNA contained in the viral particles found in each plaque that is immobilised on a membrane and then probed.

3.10.4   Nick translation

Probes must be labelled before use and radioactive labelling of cDNA is easily carried out by nick translation, in which low concentrations of DNase I are used to make occasional single-strand nicks in the DNA. DNA polymerase then fills in the nicks, using an appropriate deoxynucleoside triphosphate (dNTP), at the same time making a new nick to the 3′ side of the previous one (Fig 3.33). In this way the nick is translated along the DNA. If radioactive dNTPs are added to the reaction mixture, they will be used to fill the nicks, and so the DNA can be labelled to a very high specific radioactivity.

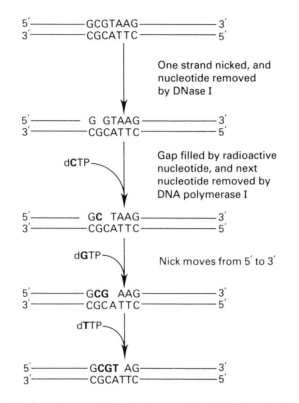

Fig. 3.33. Nick translation. If one (or more) of the deoxynucleoside triphosphates supplied is radioactive, the DNA will become progressively more highly radioactive as nick translation proceeds.

An alternative to nick translation is the method of random primer labelling. The DNA to be labelled is first denatured and then placed under renaturing conditions in the presence of a mixture of many different sequences of hexanucleotides. These hexanucleotides will bind to the DNA sample wherever they encounter a complementary sequence and so the DNA will rapidly acquire an approximately random sprinkling of hexanucleotides annealed to it. Now each of these hexanucleotides can act as a primer for the synthesis of a fresh strand of DNA catalysed by DNA polymerase (see Section 3.3.2). In fact the Klenow fragment of DNA polymerase (see Section 3.10.1) is used for random primer labelling because of its lack of 5′ to 3′ exonuclease activity. Thus when the Klenow enzyme is supplied to the annealed DNA sample in the presence of deoxyribonucleotides, including at least one which is labelled, many short stretches of labelled DNA will be generated that are ideal for use as probes.

3.10.5   Polymerase chain reaction (PCR)

An extension of the principles behind random primer labelling and Sanger sequencing has resulted in one of the most powerful techniques in molecular biology: the polymerase chain reaction (PCR). The main feature of PCR is that, if we know the sequences of DNA flanking an 'unknown' region of a DNA molecule, we can, selectively, copy the unknown DNA repeatedly to generate large quantities for further analysis. Fig. 3.34 illustrates how this is achieved.

The DNA is denatured by heating and then allowed to anneal with an excess of two oligonucleotide sequences, each complementary to a stretch of DNA to the 3' side of the unknown DNA, one oligonucleotide for each of the two DNA strands. These annealed oligonucleotides can now act as primers for the synthesis of DNA, catalysed by DNA polymerase. As with Sanger sequencing (Section 3.9.1), almost the first DNA to be synthesised will be complementary to the 'unknown' DNA.

DNA synthesis is allowed to proceed until the new strands have been extended along and beyond the unknown DNA. The sample is then heated to denature the DNA and then cooled to allow the two primers (remember they are in considerable excess) to anneal to their complementary sequences. It is important to note that, since the new strands extend beyond the unknown DNA, they will contain a region near their 3' ends that is complementary to primer. Thus, if another round of DNA synthesis is allowed to take place not only will the original strands be used as templates but so also will the new strands. Most interestingly, the products obtained from the new strands will have a precise length, delimited exactly by the two regions complementary to the primers. As the system is taken through successive rounds of denaturation, annealing and DNA synthesis, all the new strands will act as templates and so there will be an exponential increase in the amount of DNA produced. The net effect is selectively to amplify the unknown DNA and the primer regions flanking it.

The temperature needed to denature DNA would also denature normal DNA polymerase. Fortunately it is possible to avoid having to add fresh polymerase at the start of each synthetic step in the cycle by using *Taq* DNA polymerase, a thermostable enzyme isolated from the thermophilic bacterium *Thermus aquaticus*. This enzyme survives prolonged exposure to temperatures as high as 95 °C and so will still be active after the denaturation steps. This allows the use of a PCR machine, which automatically takes the solutions through controlled cycles of temperature changes for denaturation, annealing and synthesis, each complete cycle taking about 4 min.

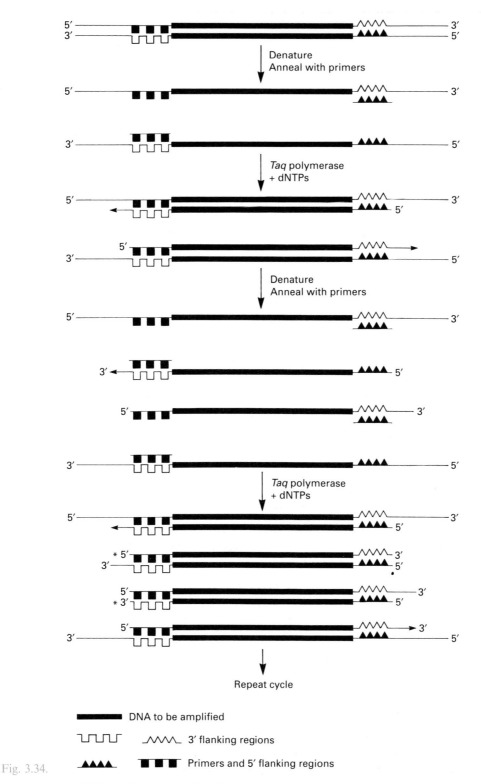

5′ ▬▬▬▬ DNA to be amplified

⊔⊔⊔⊔ ⌒⌒⌒⌒ 3′ flanking regions

▲▲▲▲ ■■■ Primers and 5′ flanking regions

dNTPs   Deoxyribonucleoside triphosphates

Fig. 3.34. The polymerase chain reaction. The figure shows two cycles of DNA amplification, producing four molecules of double-stranded DNA from the single starting molecule. Note that, by the end of the second cycle, half the newly synthesized strands (marked by asterisks) are delimited by the primers; in further cycles virtually all the products will be the same length.

The beauty of PCR is that it does not require any prior knowledge of the sequence of the 'unknown' DNA and the enormous amplification factors that can be obtained (20 cycles give an amplification of about 10^6) mean that only traces of DNA are needed initially. Radioactive material can be readily prepared by including a radioactive deoxyribonucleotide in the reaction mixture and so the method is well suited to the production of probes.

It is not only the DNA but also the applications of PCR that are increasing exponentially. Forensic scientists are exploiting its high amplification factor to produce sufficient DNA for analysis from samples as small as single hair roots. DNA inserted into a vector can be labelled and amplified ready for use as a probe in one process, rather than needing to be cloned, isolated from the host, excised from the vector and labelled in separate steps. By including far more of one primer than the other a single strand can be preferentially amplified and then used as the template for Sanger sequencing without any need for cloning. It is already evident that PCR is making some of the 'standard' techniques of molecular biology redundant and no molecular biology laboratory will be complete without its PCR machine.

Even the best technique has some limitations and PCR is no exception. Its huge amplification makes the system very vulnerable to contamination: even a trace of foreign DNA, such as may be contained in dust particles, will be amplified to significant levels and may give misleading results. Hence cleanliness is paramount when PCR is carried out. The other limitation results from the need to supply primers that are complementary to sequences on either side of the DNA to be amplified. This is not a problem when the DNA is inserted in a vector but can cause difficulties in the case of, for example, genomic DNA. Depending on the reasons for needing to amplify the DNA, there are many approaches to solving this problem, including the use of highly repetitive sequences as targets for the primers.

3.10.6   Oligonucleotide probes

If the amino acid sequence of a protein is known, there is no need to prepare a cDNA probe for its gene. From our knowledge of the genetic code, we can predict the DNA sequences that would code for the protein, and can then synthesise appropriate oligonucleotide sequences chemically. Since most amino acids are coded for by more than one codon, there will be more than one possible nucleotide sequence that could code for a given polypeptide (Fig. 3.35). The longer the polypeptide, the greater the number of possible oligonucleotides that must be synthesised. Fortunately, there is no need to synthesise a sequence longer than about 20 bases because this should hybridise efficiently with any complementary

Polypeptide:		Phe	Met	Pro	Trp	His	
Corresponding nucleotide sequences:	5′	T TTC	ATC	T CCC A G	TGG	T CAC	3′

Fig. 3.35. Oligonucleotide probes. Note that only methionine and tryptophan have unique codons. It is impossible to predict which of the indicated codons for phenylalanine, proline and histidine will actually be present in the gene to be probed, and so all possible combinations must be synthesised (16 in the example above).

sequences and should be specific for one gene. Ideally, a section of the protein should be chosen that contains as many tryptophan and methionine residues as possible, since these have unique codons, and there will therefore be fewer possible base sequences that could code for that part of the protein. The synthetic oligonucleotides can then be used as probes in a colony hybridisation, as described for cDNA.

3.11   Expression of genes

One of the main purposes of genetic engineering is to obtain the expression within bacterial cells of a foreign gene that codes for some valuable polypeptide. However, for a gene to be expressed in a bacterial cell, it must have particular sequences of bases forming a promoter upstream from the coding region, to which the RNA polymerase will bind prior to transcription of the gene (Section 3.3.3). It must also contain a Shine–Dalgarno sequence, placed just before the coding region, which is transcribed and then acts as a ribosome binding site at the start of translation (Section 3.3.4). Unless a cloned gene contains both of these sequences, it will not be expressed in its bacterial host cell. If the gene has been produced via cDNA from a eukaryotic cell, then it will certainly not have such sequences.

Consequently, expression vectors have been developed which contain promoter and Shine–Dalgarno sequences sited just before one or more restriction sites for the insertion of foreign DNA. These regulatory sequences, such as that from the *lac* operon of *E. coli*, are usually derived from genes which, when induced, are strongly expressed in bacteria. Since the mRNA produced from the gene is read as triplet codons, the inserted sequence must be placed so that its reading frame is in phase with the regulatory sequence. This can be ensured by the use of three vectors that differ only in the number of bases between promoter and insertion site, the second and third vectors being, respectively, one and two bases longer than the first. If an insert is cloned in all three vectors it is bound to be in the correct reading frame in one of them. The resulting clones can be screened, using antibodies, enzyme assays, etc., for the production of a functional foreign protein; approximately a third of the colonies should be positive.

It is not only possible but usually essential to use cDNA instead of a eukaryotic gene to direct the production of a functional protein by bacteria. This is because bacteria are not capable of processing RNA to remove introns, and so any foreign genes must be pre-processed as cDNA if they contain introns. A further problem arises if the protein must be glycosylated, by the addition of oligosaccharides at specific sites, in order to become functional. Again, this is something that bacteria are not equipped to carry out. Yeast cells can perform such post-translational modifications, producing a glycosylation pattern that is sometimes adequate, though not identical with that which would be produced in animal cells. Although yeast genes often contain introns, which are spliced out at the RNA stage, yeast is not able to process the introns of other eukaryotes, and so it is still necessary to use cDNA when expression of foreign genes in yeast is needed.

3.12   Safety of cloning procedures

As soon as DNA cloning became possible, fears were expressed about the dangers of the process. An obvious concern was that the gene for some toxic polypeptide could be cloned in a bacterium, which could then escape from the laboratory and either infect humans itself or transfer the cloned DNA to other organisms that were pathogenic, and which might express the cloned gene in humans. Even more worrying was the possibility that a piece of DNA that was in some way able to cause cancer might be similarly transferred to humans. Similar fears concerned the inadvertent transfer of harmful genes into animals and crops. Consequently, very strict guidelines were drawn up that imposed a ban on the cloning of viral or tumour DNA and demanded high levels of physical containment in most other cases.

Fortunately, it was not long before safe host cells and plasmids were developed, with features which made it highly unlikely that the cells could survive outside the laboratory, and which prevented any transfer of DNA into other organisms. Experience has shown also how difficult it can be to obtain expression of cloned DNA, and so the chances of expression of a rogue gene occurring spontaneously are very low. Thus it is now widely accepted that, provided containment of microorganisms is at a level in keeping with their pathogenicity and that of any genes being deliberately cloned, genetic manipulation should not present any new hazard. There is still debate about the release of genetically engineered organisms into the environment, for such purposes as the microbial degradation of pollutants, and such programmes are subject to very tight regulation.

3.13   Applications of molecular biology

In terms of pure research, all the techniques of molecular biology, from cloning to sequencing, have revolutionised our understanding of gene structure and function. Promoter regions, control sites, ribosome binding sites, introns, etc. have been sequenced, and secondary structures have been postulated that could help to explain how these regions act. Oncogenes (genes implicated in the causation of cancer) have been discovered through the application of molecular biology, and this has changed the pattern of research into the mechanisms and prevention of cancer.

When a DNA sample is restricted with a given enzyme it will be cut into a precise mixture of fragments that will form a reproducible pattern of bands if subjected to agarose gel electrophoresis. By comparing the patterns produced after restriction of small molecules such as plasmid or viral DNA it is often possible to distinguish between the types of DNA and to identify the sources of the DNA. These differences in pattern are known as restriction fragment length polymorphisms (RFLPs) and result from differences in positions of restriction sites between DNA species. In other words, the RFLP reflects differences in the sequences of the DNA samples.

More complex DNAs, such as human genomic DNA, give so many fragments after restriction that they cannot be resolved by electrophoresis and so RFLPs cannot be examined by using band patterns after staining with ethidium bromide. However, by making only a small subset of the restriction fragments visible one can generate a much simpler picture in which a distinct pattern of bands is visible. This is achieved by probing a Southern blot of the restricted and electrophoresed DNA with a probe that hybridises to only a few regions of the genomic DNA. Such probes can be based on regions of DNA that are repeated in various parts of the genome or several different probes can be used, each hybridising to only one part of the genome. By the use of a suitable combination of restriction enzyme and probe, the pattern of radioactive bands produced from Southern blots will be unique to a particular species, strain or even individual.

Forensic scientists are now using RFLP analysis for matching unambiguously samples of blood, semen, hair roots or other tissue. Such DNA profiles are becoming a common feature in law courts and generally provide much firmer evidence, in terms of statistics, than that given by conventional blood groupings. RFLP analysis is also being used to aid plant breeding and by taxonomists to help them work out the evolutionary relationships between organisms.

Over 500 inherited diseases, such as sickle-cell anaemia and β-thalassaemias, are known to be caused by a mutation in a single gene. More than 40 of these diseases can be detected early in the development of a fetus by biochemical assays carried out on cells sampled by amniocentesis. This process involves the insertion of a needle through the abdominal wall into the amniotic cavity, and withdrawal of some of the amniotic fluid. Suspended in the fluid are fetal cells, which can be isolated and cultured in the laboratory to provide sufficient material for diagnostic tests. Techniques such as chorionic sampling, which involves the collection of small samples of chorionic villus tissue from the placenta, are being developed for even earlier sampling of fetal cells. A positive result allows an informed decision to be taken over allowing the pregnancy to continue.

Unfortunately such tests are applicable only to those cases where the gene is normally expressed in the sampled cells. However, since inherited disorders result from mutations in the DNA, it is sometimes possible to detect the mutation by its effect on the pattern of bands seen in RFLP analysis. Using specific probes it can sometimes be shown that the presence or absence of a particular band is closely linked to the presence of a defective gene. In those cases where a mutant gene has been sequenced it is possible to discriminate between the normal and mutant forms by their relative abilities to hybridise to an oligonucleotide probe that has been constructed to be an exact match for only the normal form of the gene. Under conditions of high stringency the probe will not bind to mutant DNA, thus revealing the presence of a defective gene.

Molecular biology has probably received most publicity in connection with genetic engineering. Already it has been possible to clone the genes for such polypeptides as human insulin, growth hormone, interferons, tumour necrosis factor, blood clotting factor VIII, and viral coat proteins (for vaccines) in bacteria in such a way that they are expressed, and the polypeptides can be recovered from the cell cultures. Genes have been altered *in vitro* to produce slightly altered enzymes with increased stability or different reaction kinetics; this is likely to be of great importance for the production of enzymes for use on an industrial scale.

In the long term it is anticipated that genetic engineering will be used by plant and animal breeders to introduce genes for disease resistance, improved yield, etc., into crops and livestock. A major problem here is that we can rarely attribute such characteristics to a single gene, and, until our understanding of the biochemistry and physiology of plants and animals is more profound, we will not really know which genes to 'engineer'.

3.14 Suggestions for further reading

BROWN, T.A. (1990) *Gene Cloning*. Chapman & Hall, London. (A good introductory text.)

OLD, R.W. and PRIMROSE, S.B. (1989) *Principles of Gene Manipulation*, 4th edn. Blackwell, Oxford. (An advanced textbook, including explanations of the latest techniques, that also illustrates the applications of the technology.)

SAMBROOK, J., FRITSCH, E.F. and MANIATIS, T. (1989) *Molecular Cloning: A Laboratory Manual*, 2nd edn. Cold Spring Harbor Laboratory Press, Cold Spring Harbor, NY. (An invaluable source of practical details and recipes, now in three huge volumes; found on the shelves of most molecular biology laboratories.)

STRYER, L. (1988) *Biochemistry*, 3rd edn. Freeman, New York. (An excellent general biochemistry textbook, superbly illustrated.)

WALKER J.M. (Ed.) (1984, 1988) *Methods in Molecular Biology*, volumes 1, 2 and 4. Humana, Clifton, NJ. (An extensive collection of procedures for use in protein and nucleic acid biochemistry.)

4

Protein and enzyme techniques

▲ ● ■ ▲ ● ■ ▲ ● ■ ▲ ● ■ ▲ ● ■ ▲ ● ■ ▲ ● ■

4.1   Ionic properties of amino acids and proteins

Twenty amino acids varying in size, shape, charge and chemical reactivity are found in proteins and each has at least one codon in the genetic code (Section 3.3.4). Nineteen of the amino acids are α-amino acids (i.e. the amino and carboxyl groups are attached to the carbon atom that is adjacent to the carboxyl group) with the general formula $RCH(NH_2)COOH$, where R is an aliphatic, aromatic or heterocyclic group. The only exception to this general formula is proline, which is an imino acid in which the $-NH_2$ group is incorporated into a five-membered ring. With the exception of the simplest amino acid glycine (R = H), all the amino acids found in proteins contain one asymmetric carbon atom and hence are optically active and have been found to have the L configuration.

For convenience, each amino acid found in proteins is designated by either a three-letter abbreviation, generally based on the first three letters of their name, or a one-letter symbol, some of which are the first letter of the name. Details are given in Table 4.1.

Since they possess both an amino group and a carboxyl group, amino acids are ionised at all pH values, i.e. a neutral species represented by the general formula does not exist in solution irrespective of the pH. This can be seen as follows:

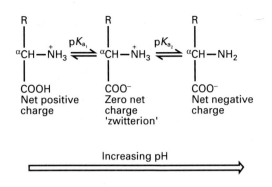

Table 4.1. *Abbreviations for amino acids*

Amino acid	Three-letter symbol	One-letter symbol
Alanine	Ala	A
Arginine	Arg	R
Asparagine	Asn	N
Aspartic acid	Asp	D
Asparagine or aspartic acid	Asx	B
Cysteine	Cys	C
Glutamine	Gln	Q
Glutamic acid	Glu	E
Glutamine or glutamic acid	Glx	Z
Glycine	Gly	G
Histidine	His	H
Isoleucine	Ile	I
Leucine	Leu	L
Lysine	Lys	K
Methionine	Met	M
Phenylalanine	Phe	F
Proline	Pro	P
Serine	Ser	S
Threonine	Thr	T
Tryptophan	Trp	W
Tyrosine	Tyr	Y
Valine	Val	V

Thus, at low pH values an amino acid exists as a cation and at high pH values as an anion. At a particular intermediate pH the amino acid carries no net charge, although it is still ionised, and is called a zwitterion. It has been shown that, in the crystalline state and in solution in water, amino acids exist predominantly as this zwitterionic form. This confers upon them physical properties characteristic of ionic compounds, i.e. high melting point and boiling point, water solubility and low solubility in organic solvents such as ether and chloroform. The pH at which the zwitterion predominates in aqueous solution is referred to as the isoionic point, because it is the pH at which the number of negative charges on the molecule produced by ionisation of the carboxyl group is equal to the number of positive charges acquired by proton acceptance by the amino group. In the case of amino acids this is equal to the isoelectric point (pI), since the molecule carries no net charge and is therefore electrophoretically immobile. The numerical value of this pH for a given amino acid is related to its acid strength (pK$_a$ values) by the equation:

$$pI = \frac{pK_{a_1} + pK_{a_2}}{2} \tag{4.1}$$

where pK$_{a_1}$ and pK$_{a_2}$ are equal to the negative logarithm of the acid dissociation constants, K_{a_1} and K_{a_2} (Section 1.3.1).

In the case of glycine, pK$_{a_1}$ and pK$_{a_2}$ are 2.3 and 9.6, respectively, so that the isoionic point is 6.0. At pH values below this, the cation and zwitterion will coexist in equilibrium in a ratio determined by the Henderson–Hasselbalch equation (Section 1.3.1), whereas at higher pH values the zwitterion and anion will coexist in equilibrium.

For acidic amino acids such as aspartic acid, the ionisation pattern is different owing to the presence of a second carboxyl group:

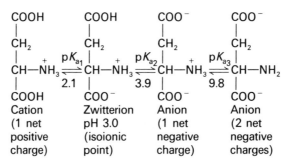

In this case, the zwitterion will predominate in aqueous solution at a pH determined by pK_{a_1} and pK_{a_2}, and the isoelectric point is the mean of pK_{a_1} and pK_{a_2}.

In the case of lysine, which is a basic amino acid, the ionisation pattern is different again and its isoionic point is the mean of pK_{a_2} and pK_{a_3}:

$$
\begin{array}{llll}
\overset{+}{N}H_3 & \overset{+}{N}H_3 & \overset{+}{N}H_3 & NH_2 \\
| & | & | & | \\
(CH_2)_4 & (CH_2)_4 & (CH_2)_4 & (CH_2)_4 \\
| \quad \overset{+}{} \quad pK_{a_1} & | \quad pK_{a_2} & | \quad \overset{+}{} \quad pK_{a_3} & | \\
CH-\overset{+}{N}H_3 \rightleftharpoons CH-NH_2 \rightleftharpoons CH-\overset{+}{N}H_3 \rightleftharpoons CH-NH_2 \\
| \qquad 2.2 \quad | \qquad 9.0 \quad | \qquad 10.5 \quad | \\
COOH & COO^- & COO^- & COO^-
\end{array}
$$

Cation (2 net positive charge) Cation (1 net positive charge) Zwitterion pH 9.8 (isoionic point) Anion (1 net negative charge)

As an alternative to possessing a second amino or carboxyl group, an amino acid side chain may contain in the R of the general formula a quite different chemical group that is also capable of ionising at a characteristic pH. Such groups include a phenolic group (tyrosine), guanidino group (arginine), imidazolyl group (histidine) and sulphydryl group (cysteine) (Table 4.2). It is clear that the state of ionisation of the main groups of amino acids (acidic, basic, neutral) will be grossly different at a particular pH. Moreover, even within a given group there will be minor differences due to the precise nature of the R group. These differences are exploited in the electrophoretic and ion-exchange chromatographic separation of mixtures of amino acids such as those present in a protein hydrolysate (Section 4.3.2).

Proteins are formed by the condensation of the α-amino group of one amino acid with the α-carboxyl of the adjacent amino acid (Section 4.2). With the exception of the two terminal amino acids, therefore, the α-amino and carboxyl groups are all involved in peptide bonds and are no longer ionisable in the protein. Amino, carboxyl, imidazolyl, guanidino, phenolic and sulphydryl groups in the side chains are, however, free to ionise and of course there may be many of these. Proteins fold in such a manner that the majority of these ionisable groups are on the outside of the molecule, where they can interact with the surrounding aqueous medium. Some of these groups are located within the structure and may be involved in electrostatic attractions that help to stabilise the three-dimensional structure of the protein molecule. The relative numbers of positive and negative groups in a protein molecule influence aspects of its physical behaviour, such as solubility and electrophoretic mobility.

The isoionic point of a protein and its isoelectric point, unlike that of an amino acid, are generally not identical. This is because, by definition, the isoionic point

Table 4.2. *Ionisable groups found in proteins*

Amino acid group	pH-dependent ionisation	Approx. pK_a
N-terminal α-amino	$-\overset{+}{N}H_3 \rightleftharpoons NH_2 + H^+$	8.0
C-terminal α-carboxyl	$-COOH \rightleftharpoons COO^- + H^+$	3.0
Asp-β-carboxyl	$-CH_2COOH \rightleftharpoons CH_2COO^- + H^+$	3.9
Glu-γ-carboxyl	$-(CH_2)_2COOH \rightleftharpoons (CH_2)_2COO^- + H^+$	4.1
His-imidazolyl		6.0
Cys-sulphydryl	$-CH_2SH \rightleftharpoons -CH_2S^- + H^+$	8.4
Tyr-phenolic		10.1
Lys-ε-amino	$-(CH_2)_4\overset{+}{N}H_3 \rightleftharpoons -(CH_2)_4NH_2 + H^+$	10.3
Arg-guanidino		12.5

is the pH at which the protein molecule possesses an equal number of positive and negative groups formed by the association of basic groups with protons and dissociation of acidic groups, respectively. In contrast, the isoelectric point is the pH at which the protein is electrophoretically immobile. In order to determine electrophoretic mobility experimentally, the protein must be dissolved in a buffered medium containing anions and cations, of low relative molecular mass, that are capable of binding to the multi-ionised protein. Hence the observed balance of charges at the isoelectric point could be due in part to there being more bound mobile anions (or cations) than bound cations (anions) at this pH. This could mask an imbalance of charges on the actual protein.

In practice, protein molecules are always studied in buffered solutions, hence it is the isoelectric point that is important. It is the pH at which, for example, the protein has minimum solubility, since it is the point at which there is the greatest opportunity for attraction between oppositely charged groups of neighbouring molecules and consequent aggregation and easy precipitation.

4.2   Protein structure

Proteins are formed by condensing the α-amino group of one amino acid or the imino group of proline with the α-carboxyl group of another, with the concomitant

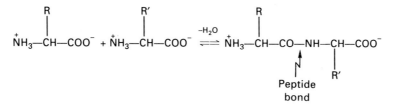

Peptide
bond

loss of a molecule of water and the formation of a peptide bond.

The progressive condensation of many molecules of amino acids gives rise to an unbranched polypeptide chain. By convention, the N-terminal amino acid is taken as the beginning of the chain and the C-terminal amino acid the end of the chain (proteins are biosynthesised in this direction). Polypeptide chains contain between 20 and 2000 amino acid residues and hence have a relative molecular mass ranging between about 2000 and 200 000. Most proteins have a relative molecular mass (Section 4.3.1) in the range 20 000 to 100 000. The distinction between a large peptide and a small protein is not clear. Generally peptides contain fewer than 50 amino acid residues and proteins more than 50. Insulin, which is a peptide hormone with 53 amino acid residues, is on the border between the two. Ribonuclease, which is a small protein, has 103 amino acid residues.

The primary structure of a protein defines the sequence of the amino acid residues and is dictated by the base sequence of the corresponding gene(s). Indirectly, the primary structure also defines the amino acid composition (which of the possible 20 amino acids are actually present) and content (the relative proportions of the amino acids present).

The peptide bonds linking the individual amino acid residues in a protein are both rigid and planar, with no opportunity for rotation about the carbon–nitrogen bond, as it has considerable double bond character due to the delocalisation of the lone pair of electrons on the nitrogen atom; this, coupled with the tetrahedral geometry around each α-carbon atom, profoundly influences the three-dimensional arrangement which the polypeptide chain adopts.

Secondary structure defines the localised folding of a polypeptide chain due to hydrogen bonding. It includes structures such as the α-helix and β-pleated sheet. Certain of the 20 amino acids found in proteins, including proline, isoleucine, tryptophan and asparagine, disrupt α-helical structures. Some proteins have up to 70% secondary structure but others have none.

Tertiary structure defines the overall folding of a polypeptide chain. It is stabilised by electrostatic attractions between oppositely charged ionic groups (-$\overset{+}{N}H_3$, COO^-),

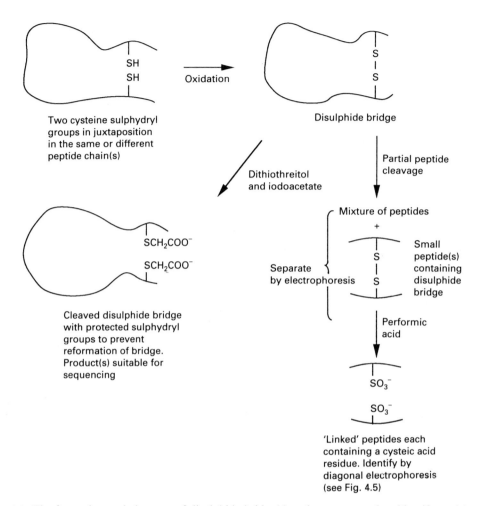

Fig. 4.1. The formation and cleavage of disulphide bridge(s) and strategy used to identify position of bridge(s).

by weak van der Waals' forces, by hydrogen bonding, hydrophobic interactions and, in some proteins, by disulphide (-S–S-) bridges formed by the oxidation of spatially adjacent sulphydryl groups (-SH) of cysteine residues (Fig. 4.1). The three-dimensional folding of polypeptide chains is such that the interior consists predominantly of non-polar, hydrophobic amino acid residues such as valine, leucine and phenylalanine. The polar, ionised, hydrophilic residues are found on the outside of the molecule, where they are compatible with the aqueous environment. However, some proteins also have hydrophobic residues on their outside and the presence of these residues is important in the processes of ammonium sulphate fractionation (Section 4.5.3) and hydrophobic interaction chromatography (Sections 4.5.4 and 10.5.3).

Quaternary structure is restricted to oligomeric proteins, which consist of the association of two or more polypeptide chains held together by electrostatic attractions, hydrogen bonding, van der Waals' forces and occasionally disulphide bridges. Thus, disulphide bridges may exist within a given polypeptide chain (intra-chain) or linking different chains (inter-chain). An individual polypeptide chain in an oligomeric protein is referred to as a subunit. The subunits in a protein may be identical or different: e.g. haemoglobin consists of two α- and two β-chains, and lactate dehydrogenase of four (virtually) identical chains.

Traditionally, proteins are classified into two groups – globular and fibrous. The former are approximately spherical in shape, are generally water soluble and may contain a mixture of α-helix, β-pleated sheet and random structures. Globular proteins include the enzymes, transport proteins, hormones and immunoglobulins. Fibrous proteins are structural proteins, generally insoluble in water, consisting of long cable-like structures built entirely of either helical or sheet arrangements. Examples include hair keratin, silk fibroin and collagen. The native state of a protein is its biologically active form.

The process of protein denaturation results in the loss of biological activity, decreased aqueous solubility and increased susceptibility to proteolytic degradation. It can be brought about by heat and by treatment with reagents such as acids and alkalis, detergents, organic solvents and heavy-metal cations such as mercury and lead. It is associated with the loss of organised (tertiary) three-dimensional structure and exposure to the aqueous environment of numerous hydrophobic groups previously located within the folded structure.

In enzymes, the specific three-dimensional folding of the polypeptide chain(s) results in the juxtaposition of certain amino acid residues that constitute the active site or catalytic site. Oligomeric enzymes may possess several such sites. Many enzymes also possess one or more regulatory site(s). X-ray crystallographic studies have revealed that the active site is often located in a cleft that is lined with hydrophobic amino acid residues but which contains some polar residues. The binding of the substrate at the catalytic site and the subsequent conversion of substrate to product involves different amino acid residues.

Some oligomeric enzymes exist in multiple forms called isoenzymes or isozymes. Their existence relies on the presence of two genes that give similar but not identical subunits. One of the best-known examples of isoenzymes is lactate dehydrogenase, which reversibly interconverts pyruvate and lactate. It is a tetramer and exists in five forms (LDH1–5) corresponding to the five permutations of arranging the two types of subunits (H and M), which differ only in a single amino acid substitution, into a tetramer:

H_4	LDH1
H_3M	LDH2
H_2M_2	LDH3
HM_3	LDH4
M_4	LDH5

Each isoenzyme promotes the same reaction but has different kinetic constants (K_m, V_{max}), thermal stability and electrophoretic mobility. The tissue distribution of isoenzymes within an organism is frequently different, e.g. in humans LDH1 is the dominant isoenzyme in heart muscle but LDH5 is the most abundant form in liver and muscle. These differences are exploited in diagnostic enzymology to identify specific organ damage, e.g. following myocardial infarction, and thereby to aid clinical diagnosis and prognosis.

4.3   Investigation of protein structure

4.3.1   Relative molecular mass

Information about the relative molecular mass, M_r, of a protein is important in both structural and purification studies. Three techniques are used routinely to investigate it and in the majority of cases the experimental aim is simply to obtain a rough estimate of size.

SDS polyacrylamide gel electrophoresis (SDS-PAGE)

Under defined conditions, the relative electrophoretic mobilities of proteins in the presence of sodium dodecylsulphate (SDS) is related to their relative molecular mass. The method is dependent upon the fact that proteins bind the same amount of SDS per gram in spite of the fact that the binding occurs predominantly to hydrophobic regions and that proteins differ in their hydrophobicities. Unreduced proteins (disulphide bridges and tertiary structure intact) bind much less SDS than do reduced proteins, and the extent of SDS binding decreases with increasing ionic concentration. SDS-PAGE separates SDS–protein complexes on the basis of their relative molecular mass, provided the charge:mass ratios and the shape of the complexes are the same.

    The experimental procedure for SDS-PAGE is discussed in Section 9.3.1. The slab gel technique is most commonly used for relative molecular mass determinations. It requires the availability of a number of reference proteins of known M_r and is suitable for determinations in the range 3000 to 300 000. SDS-PAGE is simple and requires very little material. When coupled with the silver staining method (Section 4.4) for detection, it can be used at the submicrogram level and in practice is the most commonly used method for protein relative molecular mass determination.

Exclusion chromatography

The elution volume of a protein from an exclusion chromatography column packed with a gel having an appropriate fractionation range is determined largely by its size such that there is a logarithmic relationship between protein relative molecular mass and elution volume (Section 10.8.3). By calibrating the column with a range of proteins of known M_r, the relative molecular mass of a test protein can be calculated. The experimental procedure is relatively simple and can be carried out on impure samples.

Ultracentrifugation

The sedimentation rate in a high centrifugal field of a protein in solution can be used to compute its relative molecular mass by the application of the underlying principles discussed in Section 6.10. The low speed sedimentation equilibrium method is particularly useful for proteins with an M_r below 1 000 000. The high speed equilibrium method is suitable for high relative molecular mass proteins but is unsuitable for proteins with M_r values below 50 000. Both of these techniques give information about the purity of the protein sample, since in both cases the sedimentation constant of a pure protein is characterised by a sharp symmetrical moving boundary. The disadvantage of both techniques is that they require milligram quantities of protein.

4.3.2 Amino acid composition

The determination of which of the 20 possible amino acids are present in a particular protein and in what relative amounts is achieved by hydrolysing the protein to its component amino acids and identifying and quantifying them chromatographically. Hydrolysis is achieved by heating the protein with 6 M hydrochloric acid for 24 h at 110 °C *in vacuo*. Unfortunately, the hydrolysis procedure destroys or chemically modifies all or some of the asparagine, glutamine and tryptophan residues. Asparagine and glutamine are converted to their corresponding acids (Asp and Glu) and are quantified with them. Tryptophan is best

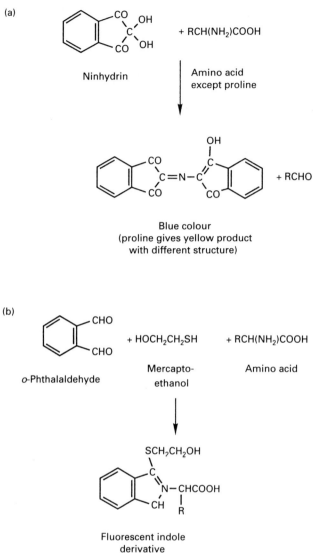

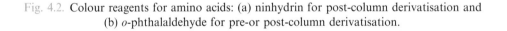

Fig. 4.2. Colour reagents for amino acids: (a) ninhydrin for post-column derivatisation and (b) *o*-phthalaldehyde for pre-or post-column derivatisation.

determined spectrophotometrically on the unhydrolysed protein.

The amino acids in the protein hydrolysate may be separated by ion-exchange chromatography on a sulphonated polystyrene column (Section 10.7.2) and quantified by post-column derivatisation to a coloured compound (Fig. 4.2). The apparatus dedicated to the analysis of amino acids in mixtures by this technique is referred to as an amino acid analyser. In the original procedure, ninhydrin was used as the colour reagent as it is sensitive down to about 50–100 pmol of the amino acid. *o*-Phthalaldehyde and fluorescamine, which give fluorescent products, enable as little as 10 pmol of an amino acid to be detected by fluorimetry. Reversed-phase high performance liquid chromatography (HPLC) of amino acids has become attractive and has generally superceded the original ion-exchange method for the quantification of amino acids in a protein hydrolysate. It is best achieved by pre-column derivatisation either by *o*-phthalaldehyde (Fig. 10.11, p. 503) or by phenylisothiocyanate to yield phenylthiohydantoin (PTH) derivatives (Section 4.3.3 and Fig. 4.3).

4.3.3   Amino acid sequence

Knowledge of the amino acid sequence of a protein is fundamental to the understanding of protein structure – function relationships. In recent years, protein sequence determination has assumed even greater importance, since knowledge of a protein sequence enables an oligonucleotide probe to be synthesised for use in recovering the gene for the protein from a gene library (Section 3.10.6).

Routine protein sequencing methods are based on the Edman degradation of polypeptides, in which the N-terminal amino acid residue is specifically removed, leaving a polypeptide one residue shorter. Variations arise in the method of identifying the removed amino acid or the newly exposed N-terminal residue. By repeated cycles of Edman degradation and identification of product, the polypeptide can be sequenced.

In the Edman reaction (Fig. 4.3) the polypeptide is treated with phenyl-isothiocyanate (PITC), which reacts with the N-terminal amino acid residue to form a phenylthiocarbamyl (PTC) derivative of the polypeptide. Anhydrous trifluoroacetic acid is then used to cleave the molecule, giving the 2-anilino-5-thiazolinone derivative of the N-terminal residue and also the polypeptide shortened by one residue. The thiazolinone derivative is separated from the polypeptide and converted into the more stable 3-phenyl-2-thiohydantoin (PTH) derivative, which is then identified by HPLC. By repeating this cycle, the polypeptide can be sequenced from its N-terminal end by either manual or automated techniques.

The alternative dansyl–Edman procedure (Fig. 4.4) is highly sensitive, allowing as little as 1 nmol of polypeptide to be sequenced, and it is therefore well suited to manual determination of sequences. It uses cycles of the Edman reaction to remove N-terminal amino acid residues sequentially, but, instead of identifying the released PTH derivatives, it identifies the newly exposed N-terminal amino acid residues. This is achieved by adding a dansyl group to the N terminus of a very small portion of the polypeptide after each cycle of the Edman reaction, followed by cleavage with hydrochloric acid to release a dansyl amino acid plus free amino

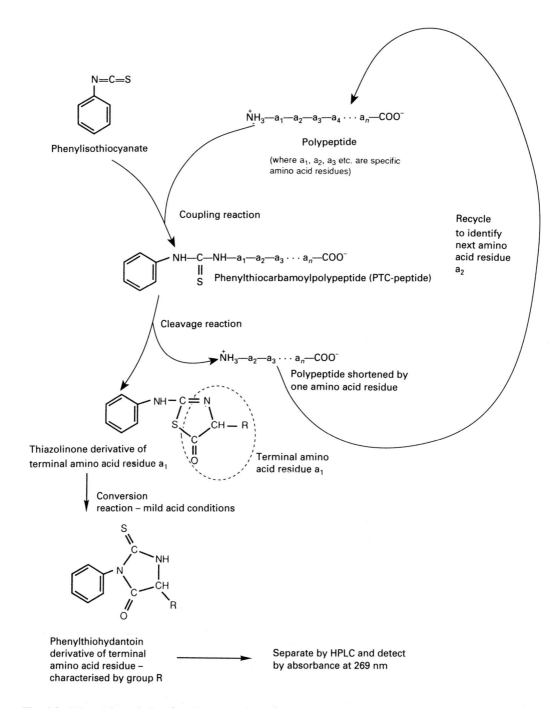

Fig. 4.3. Edman degradation for the sequencing of amino acid residues (a_1 to a_n) in a polypeptide. Each cycle releases the N-terminal amino acid residue as a phenylthiohydantoin derivative, which is then identified chromatographically against reference compounds.

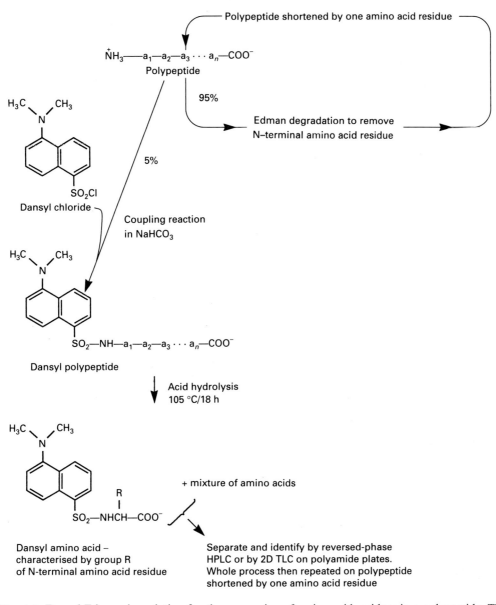

Fig. 4.4. Dansyl-Edman degradation for the sequencing of amino acid residues in a polypeptide. The high sensitivity for the detection of the dansyl amino acid more than compensates for the removal of some of the polypeptide after each Edman degradation cycle.

acids. The dansyl derivative can be identified by HPLC or by two-dimensional thin-layer chromatography on polyamide plates (Section 10.11.2). As little as 10 pmol of protein can be sequenced in this way.

The amino acid sequence of polypeptides with a molecular weight up to 10 000 can also be determined by fast atom bombardment (FAB) mass spectrometry (Section 8.6.2). The technique has a practical advantage over the Edman degradation technique in that it can cope simultaneously with a mixture of up to five peptides. The peptide(s) are dissolved in aqueous acetic acid containing glycerol, inserted into a mass spectrometer and bombarded with a beam of atoms of an inert gas such as argon or xenon, which cause ionisation and fragmentation of the polypeptide. The resulting fragmentation pattern allows the sequence of amino acids in each peptide to be deduced. The technique is faster than the Edman sequencing technique but is complementary to it in terms of its underlying principle. Its current applicability is limited by the availability of the expensive instrumentation.

Manual and automated sequencing of amino acids in polypeptides can be carried out routinely, but after approximately 50 cycles the amino acid sequence ceases to be unambiguous. This is because beyond this size the cumulative effects of incomplete reactions and side reactions of the Edman degradation reaction makes impossible the unambiguous identification of the released amino acid and hence the amino acid sequence. As proteins are larger than this upper limit, they must first be cleaved to smaller polypeptide fragments. The sequence of each fragment is then determined and the amino acid sequence of the original protein deduced by exploiting the principle of overlapping peptides.

This procedure uses the specific cleavage of peptide bonds by either chemical reagents or enzymes. Examples are given in Table 4.3. The principle is best illustrated by considering as an example the undecapeptide A. Cleavage by trypsin, by staphylococcal protease and by cyanogen bromide produces peptides whose sequences overlap:

Peptide A	Leu-Glu-Ser-Met-Arg-Ala-Glu-Val-Lys-His-Phe		
Trypsin digest	Leu-Glu-Ser-Met-Arg	Ala-Glu-Val-Lys	His-Phe
	B	C	D
Staphylococcal	Leu-Glu	Ser-Met-Arg-Ala-Glu	Val-Lys-His-Phe
protease digest	E	F	G
Cyanogen	Leu-Glu-Ser-Met	Arg-Ala-Glu-Val-Lys-His-Phe	
bromide digest	H	I	

If peptides B, C and D, E, F and G, and H and I are separated by ion-exchange or reversed-phase chromatography and the sequence of each peptide determined, the sequence of the amino acids in peptide A can be deduced. In fact it can be deduced by involving only two different types of digest. The technique can be applied only to single-chain peptides lacking disulphide bonds. If intra- or interchain disulphide bridges are present in the original protein they must be reduced with dithiothreitol or 2-mercaptoethanol and the exposed sulphydryl groups alkyl-

Table 4.3. *Specific cleavage of polypeptides*

Reagent	Specificity
Enzymatic cleavage	
Chymotrypsin	C-terminal side of hydrophobic amino acid residues, e.g. Phe, Try, Tyr, Leu
Endoproteinase Arg-C	C-terminal side of arginine
Endoproteinase Asp-N	Peptide bonds N-terminal to aspartate or cysteine residues
Trypsin	C-terminal side of arginine and lysine residues but Arg-Pro and Lys-Pro poorly cleaved
Staphylococcus aureus protease	C-terminal side of glutamate residues and some aspartate residues
Thermolysin	N-terminal side of hydrophobic amino acid residues excluding Trp
Chemical cleavage	
BNPS skatole	
N-Bromosuccinimide	C-terminal side of tryptophan residues
o-Iodosobenzoate	
Cyanogen bromide	C-terminal side of methionine residues
Hydroxylamine	Asparagine–glycine bonds
2-Nitro-5-thiocyanobenzoate	N-terminal side of cysteine residues

ated, e.g. with iodoacetate, to prevent the disulphide bridges from reforming (Fig. 4.1). If this results in the release of two or more distinct chains, they must then be separated and studied independently. Oligomeric proteins containing subunits not linked by disulphide bridges must first be dissociated with agents such as urea or guanidine hydrochloride and the subunits studied individually.

The identification of the position of disulphide bridges in the original protein presents a logistical problem if there are more than two cysteine residues in the protein, since several combinations of cross-linking are possible. For example, in a protein containing four cysteine residues, A, B, C and D, which combine to give two disulphide bridges, the pairing of the cysteine residues could be one of three:

$$A\text{-}B \quad \text{and} \quad C\text{-}D$$
$$\text{or} \quad A\text{-}C \quad \text{and} \quad B\text{-}D$$
$$\text{or} \quad A\text{-}D \quad \text{and} \quad B\text{-}C$$

In a protein such as ribonuclease, which has four disulphide bridges within a single polypeptide chain, there are 105 possible combinations. The problem can be resolved by cleaving the protein into peptides under conditions that retain the disulphide bridges. The resulting peptides are then separated by electrophoresis on thin-layer cellulose plates and the plates exposed to performic acid, which oxidises disulphide bridges thereby converting the sulphur atom of each cysteine residue to a sulphonic acid group $-SO_3^-$ and yielding cysteic acid (Fig. 4.1). The electrophoretograph is then subjected to electrophoresis at right angles to the original direction (two-dimensional electrophoresis) but under identical conditions. Peptides originally lacking a disulphide bridge have the same electrophoretic mobility as in the first direction and will become located on a single diagonal line (diagonal electrophoresis). In contrast the cysteic acid residues in peptides released by the performic acid oxidation result in the peptides having an entirely different mobility and they will be located off the diagonal, but the related pairs are linked by their

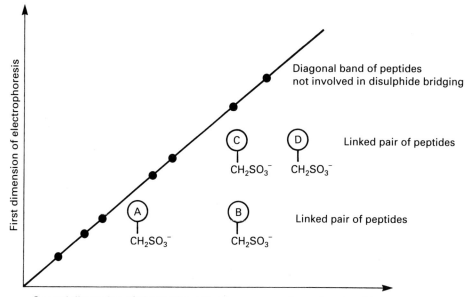

Fig. 4.5. Diagrammatic representation of diagonal electrophoresis in the study of disulphide bridges. The protein is cleaved with disulphide bridges intact and electrophoresed in the first dimension. The sheet is then exposed to performic acid and re-electrophoresed in the second dimension. Linked peptides identify the position of disulphide bridges. The identification of A, B, C and D is achieved by their removal from the electrophoretograph and subsequent amino acid sequencing.

common mobility in the first dimension (Fig. 4.5). These linked pairs of peptides may be separately eluted off the electrophoretograph and their structure determined. With a knowledge of these structures, the position of each peptide in the polypeptide sequence(s) and hence the pairing of cysteine residues can be deduced.

4.3.4 Protein conformation

The three-dimensional structure adopted by a protein is determined by its primary structure, which in turn is determined genetically by its gene base sequence. Evidence has accumulated to confirm the view that the specific three-dimensional structure of a protein is the most favoured structure thermodynamically because of the optimum interaction between amino acid residues.

The absolute way of studying protein structure is X-ray crystallography. For such studies the protein has to be purified and crystallised. The classic work of J. Kendrew, M. Perutz and D. Phillips on myoglobin, haemoglobin and lysozyme, respectively, are testimony to the power and value of such studies. However, even though considerable progress has been made on the technique, crystallographic studies remain relatively uncommon and specialised. Less detailed information can be obtained using alternative techniques.

Gel electrophoresis, particularly urea gradient electrophoresis, has been used to study transitions between native and unfolded conformations. Fluorescence spectroscopy can be used to study conformational changes in proteins by exploiting changes in the environment of tryptophan residues and the presence of FAD prosthetic groups (Section 7.8.3). Circular dichroism (CD) (Section 7.9.3) has been used to calculate the content of secondary structure of proteins, and immunological techniques and ultracentrifugation (Section 6.10.3) techniques can give information about specific aspects of conformation and of conformational changes induced by ligand binding to the protein (Section 7.9.3). Site-directed mutagenesis of a gene and the general techniques of molecular biology enable any amino acid in a protein to be changed to any other or to be deleted or an additional amino acid to be inserted. The study of the impact of this modification on the structure, stability and property of the modified protein is providing a better understanding of the factors that determine protein structure and function. Computer programs now available commercially allow predictions of secondary and tertiary structure to be made from a knowledge of primary structure. To date, their success has been limited, but no doubt as our understanding of protein structure increases the program predictions will become more accurate and valuable.

4.4 Total protein estimation

Many techniques are available for the determination of the protein content of an impure preparation. More sophisticated, less routine, methods include infrared spectrophotometry, fluorimetry, polarography, and refractometry. Several of the more common methods are not absolute methods but relative methods requiring the use of a standard. Bovine serum albumin (BSA) is almost universally used for this purpose because of its low cost, high purity and ready availability. However, because colour development in spectrophotometric assays, for example, is dependent upon the amino acid composition of the protein, and the presence of prosthetic groups (especially carbohydrate) also influence colorimetric assays, a purified sample of the protein being assayed or a closely related protein is a preferred standard.

Ultraviolet absorption

The aromatic amino acid residues tyrosine and tryptophan in a protein exhibit an absorption maximum at a wavelength of 280 nm. Since the proportions of these aromatic amino acids in proteins vary, so too do extinction coefficients but $E_{280}^{1\mathrm{mg\,cm}^{-3}}$ for most proteins lie in the range 0.4–1.5. The method is relatively sensitive, being able to measure protein concentrations as low as 10 μg cm^{-3}. However, it is subject to interference by the presence of other compounds that absorb at 280 nm. Nucleic acids fall into this category but if the absorbances (A) at 280 and 260 nm wavelengths are measured it is possible to apply a correction factor:

$$\text{Protein} \quad (\mathrm{mg\,cm}^{-3}) \ = \ 1.55\,A_{280} \ - 0.76\,A_{260}$$

The great advantage of this protein assay is that it is non-destructive and can be measured continuously, for example in chromatographic column effluents.

The sensitivity of the ultraviolet absorption assay of proteins can be increased significantly by exploiting the absorption maxima of proteins at wavelengths between 190 and 220 nm. Such measurements require more sophisticated spectrophotometers than are necessary for measurements at 260–280 nm. The additional sensitivity of the 190–205 nm wavelength method makes it particularly useful in protein purification studies aimed at producing a probe for gene structure studies, because very low concentrations of protein are encountered routinely.

Lowry (Folin–Ciocalteau) method

The phenolic group of tyrosine residues in a protein will produce a blue-purple colour, with maximum absorptions in the region of 660 nm wavelength, with Folin–Ciocalteau reagent, which consists of sodium tungstate, molybdate and phosphate. The method is sensitive down to about 10 μg cm^{-3} and is probably the most widely used protein assay despite its being only a relative method, subject to interference from Tris, zwitterionic buffers such as Pipes and Hepes and from EDTA, and the incubation time being critical for a reproducible assay. The reaction is also dependent on pH and a working range of pH 10.0–10.5 is essential. The basis of the assay is thought to be the production of the cuprous ion, which reduces the Folin–Ciocalteau reagent.

Recently, a number of alternative reagents to detect the cuprous ion have been introduced and claimed to offer a more convenient and reproducible assay. One such reagent is bicinchoninic acid (BCA). This gives a purple-blue colour with an absorbance maximum at 562 nm wavelength. The method has a sensitivity similar to that of the Lowry method and in its microassay form can measure as little as $0.5 \, \mu g \, cm^{-3}$.

Biuret method

Biuret reagent consists of alkaline copper sulphate solution containing sodium potassium tartrate. The cupric ions form a coordination complex with four -NH groups present in peptide bonds giving an absorption maximum at 540–560 nm wavelength. Since the method is based on peptide bonds, it is absolute and very reproducible. Its main disadvantage is its lack of sensitivity, being unsuitable for the assay of proteins at concentrations much less than $1 \, mg \, cm^{-3}$.

Coomassie Brilliant Blue Dye (Bradford) method

Coomassie Brilliant Blue is one of a number of dyes that complex with proteins to give an absorption maximum in the region of 595 nm wavelength. The practical advantages of the method are that the reagent is simple to prepare and that the colour develops rapidly and is stable. Although it is sensitive down to 20 μg protein cm^{-3} it is only a relative method, as the amount of dye binding appears to vary with the content of basic amino acid residues in the protein. This makes the choice of a standard difficult. In addition, many proteins will not dissolve properly in the acidic reaction medium.

Silver-binding method

This highly sensitive method relies on the ability of proteins to bind silver. The method is capable of measuring protein concentrations as low as 200 ng cm^{-3} but problems can arise due to the absorption of protein on the surface of the cuvette. This can be overcome by the addition of a detergent such as Tween 20. The method is sensitive to the amino acid composition of the protein and to the presence of EDTA, SDS (in excess of 0.01%), mercaptoethanol and dithiothreitol. These contaminants, which are commonly present, must be removed by a preliminary gel filtration step.

Kjeldahl analysis

This is a general chemical method for determining the nitrogen content of any compound. The sample is digested by boiling with concentrated sulphuric acid in the presence of sodium sulphate (to raise the boiling point) and a copper and/or selenium catalyst. The digestion converts all the organic nitrogen to ammonia, which is trapped as ammonium sulphate. Completion of the digestion stage is generally recognised by the formation of a clear solution. The ammonia is released by the addition of excess sodium hydroxide and removed by steam distillation in a Markham still. It is collected in boric acid and titrated with standard hydrochloric acid using methyl red–methylene blue as indicator. It is possible to carry out the analysis automatically in an autokjeldahl apparatus. Alternatively, a selective ammonium ion electrode (Section 11.3) may be used to determine directly the content of ammonium ion in the digest. Although Kjeldahl analysis is a precise and reproducible method for the determination of nitrogen, the determination of the protein content of the original sample is complicated by the variation of the nitrogen content of individual proteins and by the presence of nitrogen in contaminants such as DNA. In practice, the nitrogen content of proteins is generally assumed to be 16% by weight.

Turbidimetric methods

Strong organic acids such as trichloroacetic and sulphosalicylic cause proteins to precipitate. The amount of precipitate formed is measured by its light-scattering intensity (turbidimetry) (Section 7.10). The method requires precalibration and relies upon the conditions being controlled so that precipitation is complete but flocculation does not occur.

4.5   Protein purification

4.5.1   Background factors

The purification of protein is an essential first step in the study of its physical and biological properties and is one of the most common procedures in practical biochemistry. To achieve purification the protein has to be released from its biological matrix and removed selectively from other proteins by an appropriate fractionation procedure. The purification strategy adopted in this fractionation process is dependent upon three factors.

(i) Quality: The objectives of the studies on the purified protein dictate the acceptable limits of purity that need to be achieved. Thus, activity studies do not necessarily require a 100% pure preparation, whereas primary structure studies generally do. Activity studies, in contrast, require the preservation of function and hence the minimisation of denaturation and proteolysis, whereas structural studies can be carried out on a denatured enzyme.

(ii) Quantity: the quantity of the purified protein needed to achieve the study objectives will influence the choice of methods used for its purification. Some techniques have a high capacity, i.e. they can cope with large volumes and high protein concentrations, whereas others have a limited capacity. High capacity techniques such as precipitation with ammonium sulphate and adsorption chromatography are used in the early stages of protein purification, and low capacity techniques such as exclusion chromatography towards the end of the purification process.

(iii) Cost: the cost of materials such as ion-exchangers and stationary phases for affinity chromatography may become limiting factors especially for large-scale preparations. In contrast, they often have a high resolution and can drastically reduce the number of individual steps necessary to achieve protein purity.

Proteins differ in their sensitivity to denaturation during the extraction and purification processes, in particular they differ in their sensitivity to elevated (>40 °C) temperatures, to the presence of detergents and heavy metals and to extremes of pH. These differences are commonly exploited in the purification strategy for a particular protein and techniques based upon them commonly feature in the early stages of the purification process. The solubility of native proteins is influenced by pH (solubility being a minimum at the isoelectric point), by the addition of salts such as ammonium sulphate and by the presence of organic solvents, such as acetone and butanol, at low temperatures. Small differences in these properties of enzymes are also exploited in fractionation and purification procedures. Thiol-containing compounds such as mercaptoethanol, glutathione (reduced form) and dithiothreitol are commonly added to crude enzyme preparations to prevent the oxidation of sulphydryl groups, which may proceed rapidly after disruption of the cell and precede denaturation.

All proteins are susceptible to proteolysis by the action of endopeptidases and exopeptidases, which are present in all cells especially in lysosomes, but the magnitude of this susceptibility varies from one protein to another. Whereas denatured proteins are all equally susceptible to proteolysis, native proteins are not. Precautions therefore have to be incorporated into the purification strategy to minimise proteolytic degradation. The protease activity of a particular protein preparation may be assessed by the use of a proteolytically sensitive protein or peptide that on hydrolysis releases either a coloured dye or a radionuclide. Examples include azocasein, which releases small, dye-labelled peptides, or the B-chain of insulin radiolabelled with ^{125}I. A wide range of protease inhibitors is available for use in enzyme purification procedures. Each inhibitor is specific for a particular type of protease, e.g. serine proteases, cysteine proteases, aspartic proteases and metalloproteases. Common examples of inhibitors include: diisopropylphosphofluoridate (DFP), phenylmethanesulphonylfluoride (PMSF) and tosylphenylalanylchloromethylketone (TPCK) (all serine protease inhibitors); iodoacetate and cystatin (cysteine protease inhibitor); pepstatin (aspartic protease inhibitor); EDTA and 1,10-phenanthroline (metalloprotease inhibitors); amastatin and bestatin (exopeptidase inhibitors). The protease inhibitors may be used in protein purification procedures in one of two ways. The first is to maintain an effective concentration of the mixture of inhibitors in the protein preparation at all times, in which case cognisance must be taken of the fact that some purification procedures, including fractionation with ammonium sulphate and exclusion chromatography, inevitably separate the protein from the inhibitor. The alternative approach is to remove selectively the proteases from the initial crude protein preparation. This is best achieved by affinity chromatography using the inhibitor(s) as the immobilised ligand.

During the course of protein purification, dilute solutions of the protein are commonly produced. Unfortunately, dilute protein solutions, especially of purified proteins, are often unstable due either to the dissociation of subunits or to the adsorption of protein on to the surface of the container vessel. Adsorption problems can be minimised by use of siliconised containers and by the inclusion of low concentrations of detergents such as Triton X-100 or Tween 20. Dissociation problems can often be overcome by the addition of glycerol (10% to 40% (v/v)), glucose or sucrose or occasionally BSA. Storage of samples at 0–4 °C is also helpful. Long-term storage may also be facilitated by the use of high ammonium sulphate concentrations (this explains why commercial preparations of enzymes are often supplied in ammonium sulphate solution).

4.5.2 Protein extraction

The cellular location of the protein to be purified determines the procedures that can be used to obtain an extract containing the protein in soluble form. Extracellular proteins, such as those in a fermentation broth or in serum, may be simply removed from insoluble material by centrifugation or filtration. Intracellular enzymes and those found in the organelles of eukaryotes may be released into an extract by the general cell disruption techniques discussed in Section 1.4.2. It may be advantageous to the purification of organelle proteins to isolate the intact organelle (Section 6.9) before disrupting the organelle to release the soluble protein. Membrane-bound proteins present a greater problem for isolation because normally they cannot be released by simple cell disruption procedures. Such proteins are associated with phospholipids and the protein–lipid complex must be dissociated before extraction is possible. Extrinsic membrane proteins, i.e. those that are bound only to the surface of the membrane, can generally be released either by raising the ionic concentration with 1 M NaCl or by freezing and thawing. Intrinsic membrane proteins, i.e. those embedded in the membrane, are best released by treatment with either a detergent such as SDS or an organic solvent such as butanol. In all cases, precautions must be taken to minimise protein denaturation and proteolysis.

The crude protein extract is cleared of insoluble material by centrifugation or filtration across a semi-permeable membrane. At this stage it is quite common to introduce a concentration stage to facilitate subsequent fractionation procedures. Concentration can be achieved by the addition of any dry gel designed for exclusion chromatography, e.g. Sephadex G25, which has a low fractionation range (Section 10.8.2), by lyophilisation (i.e. removal of the water *in vacuo*), or by ultrafiltration (Section 4.11.3). The use of dry gels or ultrafiltration to achieve concentration has the added advantage that, concomitantly, contaminating salts such as ammonium sulphate are also removed.

Contaminating nucleic acid can be removed next from the protein extract by the addition of the strongly basic protein protamine or the long-chain cationic polymer polyethyleneimine. These form an insoluble complex with the nucleic acid, which can then be removed by centrifugation.

4.5.3 Preliminary fractionation procedures

There are several fractionation procedures available that enable the concentration of a particular enzyme in a complex, heterogeneous mixture to be relatively cheaply, quickly and simply increased at the expense of contaminating proteins. It is important to appreciate, however, that there is no universal order in which these procedures should be employed. An ideal sequence of steps for one enzyme may be unsuccessful for another and could even lead to its denaturation. This is a direct consequence of small but vital differences in protein structure and hence stability. In practice, therefore, the development of a successful protocol involves a considerable amount of trial and error and the use of a number of small-scale pilot experiments to assess the potential of each proposed stage. This, in turn, relies upon the availability of a sensitive and specific assay for both the enzyme being purified and total protein. For reasons already discussed, high capacity procedures generally feature early in a protocol and low capacity procedures late in a protocol.

Each fractionation divides the total protein in the mixture into a series of fractions each of which is then assayed for total protein and for units of activity of the enzyme being purified. A successful fractionation is recognised by a fraction with a high specific activity and hence fold purification and high yield where:

$$\text{Yield} = \frac{\text{units of enzyme in fraction}}{\text{units of enzyme in original preparation}}$$

$$\text{Fold purification} = \frac{\text{specific activity of fraction}}{\text{original specific activity}}$$

The amount of enzyme present in a particular fraction is expressed conventionally not in terms of units of mass or moles but in terms of units based upon the rate of the reaction that the enzyme promotes. The international unit (U) of an enzyme is defined as the amount of enzyme that will convert 1 µmol of substrate to product in 1 min under defined conditions (generally 25 or 30 °C and the optimum pH). The SI unit of enzyme activity is defined as the amount of enzyme that will convert 1 mol of substrate to product in 1 s. It has units of katal (kat) such that 1 kat = 6×10^{7} U and 1 U = 1.7×10^{-8} kat. For some enzymes, especially those where the substrate is a macromolecule of unknown relative molecular mass (amylase, pepsin, RNase, DNase), it is not possible to define either of these units. In such cases arbitrary units are used generally that are based upon some observable change in a chemical or physical property of the substrate.

The purity of an enzyme in a particular fraction is expressed by its specific activity, which relates its total activity to the total amount of protein present in the preparation:

$$\text{Specific activity} = \frac{\text{total units of enzyme in fraction}}{\text{total amount of protein in fraction}}$$

The measurement of units of an enzyme relies on an appreciation of certain basic kinetic concepts and upon the availability of a suitable analytical procedure. These are discussed in Sections 4.7 and 4.8.

Denaturation fractionation exploits differences in the heat sensitivity of proteins. The temperature at which the protein being purified is denatured is determined by a small-scale experiment. Once this temperature is known, it is possible to remove more thermolabile contaminating proteins by heating the mixture to a temperature 5 to 10 deg.C below this critical temperature for a period of 15 to 30 min. The denatured, unwanted protein is then removed by centrifugation. The presence of the substrate, product or a competitive inhibitor of an enzyme often stabilises it and allows an even higher heat denaturation temperature to be employed. In a similar way, proteins differ in the ease with which they are denatured by extremes of pH (<3 and >10). The sensitivity of the protein under investigation to extreme pH is determined by a small-scale trial. The whole protein extract is then adjusted to a pH not less than 1 pH unit within that at which the test protein is precipitated. More sensitive proteins will precipitate and are removed by centrifugation.

Salt fractionation is carried out by the stepwise addition of a suitable salt. In practice, ammonium sulphate is most commonly used because it is highly water soluble, can be obtained in a high degree of purity, is cheap and has no deleterious effect on the structure of proteins. After each addition of salt, care must be taken to ensure complete dissolution of the salt and the production of a homogeneous solution. The precipitated protein is centrifuged off, redissolved in fresh buffer and assayed for total protein and protein activity. All stages are generally carried out at 0 to 10 °C to minimise denaturation. A given protein is normally precipitated (salted out) over a small range of ammonium sulphate concentrations. This reflects the fact that protein–protein aggregation for a given protein suddenly becomes predominant over protein–water and protein–salt interactions. This is because the addition of the salt removes the layer of water molecules surrounding hydrophobic groups on the protein surface, which allows the hydrophobic groups to cause protein aggregation and hence precipitation. The presence of hydrophobic groups on the protein is therefore fundamental to the salting out process. Proteins lacking such groups cannot be salted out of solution. The results of a successful ammonium

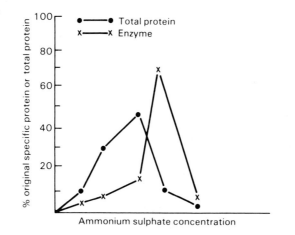

Fig. 4.6. Ammonium sulphate fractionation of an enzyme preparation.

sulphate fractionation are shown in Fig. 4.6. To calculate the amount (*g*) of ammonium sulphate to add to 1 cm^3 of solution at 20 °C to achieve a particular concentration, the following formula is used:

$$g = \frac{533(S_2 - S_1)}{100 - 0.3S_1}$$

where S_1 is the starting concentration and S_2 is the final concentration. The formula takes into account the change in volume caused by the addition of the ammonium sulphate.

Organic solvent fractionation is based upon differences in the solubility of proteins in aqueous solutions of organic solvents such as ethanol, acetone and butanol. The organic solvent lowers the dielectric constant of the medium thereby increasing the attraction between charged protein molecules and decreasing their interaction with water. The protein solubility therefore decreases. Organic solvent fractionation is in a sense the reverse of the process involved in salt fractionation. In the former process, hydrophobic groups in the protein are increasingly protected by the molecules of organic solvent, and ionic groups become dominant, whereas in the latter process hydrophobic groups are progressively exposed. Since the aqueous organic solvents may simultaneously cause denaturation, this form of fractionation must be carried out at between -10 and $-20\,^\circ$C and each fraction must be carefully redissolved in fresh buffer before it is allowed to reach room temperature. Nevertheless, even precautions such as these cannot avoid some denaturation so that, unlike salt fractionation, the technique does result in some loss of protein activity. The experimental procedure for organic solvent fractionation is similar to that for salt fractionation. To calculate the volume V (in cm^3) of organic solvent to add to 1 cm^3 of protein extract, the following formula is used:

$$V = \frac{1000(P_2 - P_1)}{100 - P_2}$$

where P_1 is the starting percentage of organic solvent and P_2 is the desired percentage of organic solvent.

One of the practical advantages of organic solvent fractionation over ammonium sulphate fractionation is that, owing to the change of density of the solution caused by the addition of organic solvent, the protein precipitate settles out more readily.

Organic polymer fractionation resembles organic solvent fractionation in its mechanism of action but requires lower concentrations to cause protein precipitation. The most commonly used polymer is polyethyleneglycol (PEG) with a relative molecular mass in the range 6000 to 20 000.

Isoelectric precipitation fractionation is based upon proteins having their minimum solubility at their isoelectric point. At this pH there is an equal number of positive and negative charges on the protein molecule and intermolecular repulsions are therefore minimised and intermolecular attractions maximised, resulting in the formation of insoluble aggregates. The principle can be exploited either to remove unwanted protein, by adjusting the pH of the protein extract so as to cause the precipitation of these proteins but not that of the test protein, or to remove the test protein, by adjusting the pH of the extract to its pI. In practice, the former alternative is preferable, since some denaturation of the precipitated protein inevitably occurs.

4.5.4 Chromatographic and electrophoretic techniques

The preliminary fractionation procedures discussed in the previous section generally result in a relatively modest increase in the specific activity of the test protein. To achieve a greater fold purification and ultimately to achieve absolute purification of the test protein, techniques with greater resolving power need to be employed. Most commonly these are based on the various forms of chromatography discussed in Chapter 10, but preparative electrophoretic techniques are also available.

The characteristics of the various forms of chromatography used in protein purification are summarised in Table 4.4. Hydrophobic interaction chromatography and covalent chromatography have been developed specifically to facilitate protein purification. One of the fundamental points of any purification protocol is the scale of the study. Traditional protein purifications aimed at obtaining sufficient quantities of the purified protein for the study of its primary, secondary and tertiary structure as well as biological properties involved many milligrams of the protein. More recent 'molecular biology' protein studies are aimed simply at obtaining sufficient protein to get a partial protein sequence in order to construct a gene probe. The detailed studies of the protein in terms of control are then carried out on the isolated gene. This type of study requires very small amounts of protein (microgram quantities). The isolation of nerve factors, growth factors and other regulatory proteins is also carried out on a very small scale. The protocols for these two contrasting-scale studies can be quite different, since, generally speaking, cost is not a limiting factor for small-scale work and so, for example, affinity chromatography can feature earlier in the protocol.

There is no universal prescription for the order in which the various techniques should be applied to the purification of a particular protein. Hydrophobic inter-action chromatography and ion-exchange chromatography are commonly used soon after the early crude fractionation stages because both have a high capacity and are relatively cheap. Hydrophobic interaction chromatography is facilitated by the presence of the ammonium sulphate, as this exposes the hydrophobic regions on the surface of the protein (Section 10.5.3). It is advantageous to design a protocol that progressively exploits different properties of the test protein. The

Table 4.4. *Summary of chromatographic techniques commonly used in protein purification*

Technique	Property exploited	Capacity	Resolution	Practical points	Further details
Hydrophobic interaction	Hydrophobicity	High	Medium	Can cope with high ionic strength samples, e.g. ammonium sulphate precipitates. Fractions are of varying pH and/or ionic strength. Medium yield. Commonly used in early stages of purification protocol. Unpredictable	Section 10.5.3
Ion-exchange	Charge	High	Medium	Sample ionic strength must be low. Fractions are of varying pH and/or ionic strength. Medium yield. Commonly used in early stages of purification protocol	Section 10.7
Affinity	Biological function	Medium (cost limited)	High	Limited by availability of immobilised ligand. Elution may denature protein. Yield medium–low. Commonly used towards end of purification protocol	Section 10.9
Dye affinity	Structure and hydrophobicity	Medium	High	Necessary to carry out initial screening of a wide range of dye–ligand supports	Section 10.9.7
Chromato-focusing	Charge and pI	High–medium	High–medium	Sample ionic strength must be low. Fractions contaminated with ampholytes	Section 10.7.3
Covalent	Thiol groups	Medium–low	High	Specific for thiol-containing proteins. Limited by high cost and long (3 h) regeneration time	Section 10.9.8
Metal chelate	Imidazole, thiol, tryptophan groups	Medium–low	High	Relatively few examples in literature. Expensive	Section 10.9.6
Exclusion	Molecular size	Medium	Low	Commonly used as a final stage of purification. Can give information about protein molecular weight. Good for desalting protein samples	Section 10.8

sequence should also be adjusted so that the product from one technique provides a suitable starting material for the next. Careful thought also has to be given to the balance between yield and purification. For example, if four consecutive stages each gave a yield of 75%, the overall yield would be 31%. If the concomitant fold purification was high, such a yield would generally be acceptable, but, if not, the protocol should be revised.

Continuous flow electrophoresis (Section 9.3.7) is particularly suitable for the preparation of very large quantities of protein, for example in the isolation of specific proteins from blood and the products of gene-cloning studies in microorganisms such as *Escherichia coli* or yeast. Other forms of electrophoresis, whilst being capable of separating complex mixtures of proteins, suffer from the disadvantages of being relatively slow and of causing the denaturation of the protein. Miniaturised electrophoresis systems are becoming available that overcome the first of these disadvantages.

Protein purification studies aimed at the production of a gene probe do not require the preservation of biological activity, so the purification process can be carried out under denaturing conditions using so-called fast protein liquid chromatography (FPLC) (Section 10.4.8). Increasingly commonly, particularly for membrane proteins, reversed-phase FPLC is used as a single-step process in such studies. Typically a silica-based packing with C_8, C_{18}, phenyl or cyano groups and a mean pore diameter of less than 30 μm is used in conjunction with a mobile phase based upon aqueous acetonitrile or isopropanol and an ultraviolet-detector measuring at 210–220 nm wavelength. The presence of compounds such as trifluoroacetic acid, heptauorobutyric acid, triethylammonium phosphate and pyridinium formate help to solubilise the proteins in the mobile phase and enhance resolution. Most proteins studied by this form of chromatography are denatured and/or dissociated by the conditions, but for sequence studies this is acceptable. If the retention of biological activity is important, high efficiency FPLC based upon hydrophobic interaction chromatography or ion-exchange chromatography is preferable.

4.5.5   Monitoring the purification proces

As outlined above (Section 4.5.3), the progress of a protein purification protocol is monitored by the determination of the specific activity and fold purification of the fractions. This in turn requires the availability of a sensitive assay for the protein and for total protein. In studies using very small amounts of protein this may be difficult and it may be necessary to resort to specialised, sensitive techniques. The two most commonly used are based on gel electrophoresis (Section 9.2) and on the exploitation of the high affinity of antibodies for specific proteins coupled with the technique of Western blotting (Section 9.3.9).

PAGE (Section 4.3.1) under denaturing conditions using SDS is probably the most commonly used technique for the routine study of protein samples (Section 9.3.1). It is simple, reproducible and capable of high resolution. The technique may be used to resolve complex mixtures. After electrophoresis, the gels are fixed by treatment with trichloroacetic acid and/or sulphosalicylic acid, stained with a dye such as Coomassie Brilliant Blue, thereby allowing a quantitative measurement of the content of each protein to be made by means of scanning densitometry. Caution has to be exercised in the interpretation of the densitometry data as proteins do not stain equally.

An alternative approach to the quantitative evaluation of the developed SDS-PAGE gel is to apply the technique of Western blotting in which the protein is transferred to a nitrocellulose sheet electrophoretically (electroblotting) (Section 9.3.9). The transferred protein can then be assayed either by a dye-binding procedure or by exploiting its high affinity and specificity for an antibody, especially a monoclonal antibody, using techniques such as immunoradiometric assay (Section 2.5.3), enzyme-linked immunosorbent assay (Section 2.6) or fluorescence immunoassay (Section 2.7). The transferred proteins on the blot can be excised and placed directly on to a protein sequencer. An alternative approach to the blotting technique is immunoelectrophoresis (Section 2.4.4).

Confirmation of the absolute purity of a given protein is difficult to obtain, since any contaminating proteins present in minor amounts may have properties very similar to that of the major protein. A single band on SDS-PAGE or a sharp moving boundary on ultracentrifugation, although encouraging, do not confirm homogeneity and such studies need to be repeated under a range of experimental conditions before one can be confident in the purity of the sample. Constant chemical composition (amino acid content, elemental analysis), solubility, and spectroscopic properties on subsequent attempted further purifications are good indicators of purity, but one of the least ambiguous indicators is the ability to obtain the protein in crystalline form and that can be a long and difficult process.

4.6   Enzyme nomenclature

The majority of cellular processes are catalysed by enzymes that have the ability to promote a specific reaction under the chemically mild conditions that prevail in living organisms. All enzymes are globular proteins whose catalytic properties rely on the three-dimensional arrangement of their polypeptide chain(s). The catalytic properties of an enzyme may also be dependent upon the presence of non-peptide cofactors or coenzymes, which may be tightly bound to the polypeptide chain as a prosthetic group. Examples of cofactors include NAD^+, $NADP^+$, FMN, FAD, more complex organic structures such as haem groups (haemoproteins) and oligosaccharides (glycoproteins), and simple metal ions such as Mg^{2+}, Fe^{2+} and Zn^{2+}. An enzyme lacking its cofactor is termed an apoenzyme, and the active enzyme with its cofactor the holoenzyme.

By international convention, an enzyme is put into one of six classes on the basis of the chemical reaction it catalyses. Each class is subdivided according to the nature of the chemical group, coenzymes and other groups involved in the reaction. In accordance with the Enzyme Commission (EC) rules, each enzyme can be assigned a unique four figure code and an unambiguous systematic name based upon the reaction catalysed. The six classes are:

- Group 1: oxidoreductases, which transfer hydrogen atoms, oxygen atoms or electrons from one substrate to another.
- Group 2: transferases, which transfer chemical groups between substrates.
- Group 3: hydrolases, which catalyse hydrolytic reactions.
- Group 4: lyases, which cleave substrates by reactions other than hydrolysis.
- Group 5: isomerases, which interconvert isomers by intramolecular rearrangements.
- Group 6: ligases (synthetases), which catalyse covalent bond formation, with the concomitant breakdown of a nucleoside triphosphate.

The comprehensive study of an enzyme involves the investigation of its molecular structure (primary, secondary, tertiary, quaternary, cofactors), protein properties (isoelectric point, electrophoretic mobility, pH and temperature stability, spectroscopic properties), enzymic properties (specificity, reversibility, kinetics), thermodynamics (activation energy, free energies and entropies), active site (number, molecular nature, mechanism) and biological properties (cellular location, isoenzymes, metabolic relevance of the reaction promoted). To undertake such a study, the enzyme has to be isolated in pure form and studied *in vitro*. The study of purified enzymes is fundamental to biochemistry because it generates data that allow the biochemist to understand and exploit the cellular situation *in vivo*. This exploitation extends to the design of selective inhibitors that can be used as drugs or biocides, to the industrial use of enzymes to promote specific chemical conversions, and to the use of enzymes in biosensors that are used to assay specific substrates.

4.7   Steady-state enzyme kinetics

4.7.1   Initial rates

When an enzyme is mixed with an excess of substrate there is an initial short period of time (a few hundred microseconds) during which intermediates leading

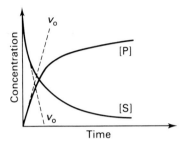

Fig. 4.7. Calculation of initial rate (v_o) from the time-dependent change in the concentration of substrate [S] and product [P] of an enzyme-catalysed reaction.

to the formation of the product gradually build up (Fig. 4.14). This so-called pre-steady state requires special techniques for study and these are discussed in Section 4.10. After this initial state, reaction rates and the concentration of intermediates change relatively slowly with time and so-called steady-state kinetics exist. Measurement of the progress of the reaction during this phase gives the relationships shown in Fig. 4.7. Tangents drawn through the origin to the curves of substrate concentration and product concentration versus time allows the initial rate, v_o, to be calculated. For simplicity, initial rates are sometimes determined on the basis of a single measurement of the amount of substrate consumed or product produced rather than by the tangent method. This approach is valid over only the short period of time when the reaction is proceeding effectively at a constant rate. This linear section comprises at the most the first 10% of the total possible change and clearly the error is smaller the quicker the rate is measured. In such cases, the initial rate is proportional either to the reciprocal of the time to produce a fixed change (fixed change assays) or to the amount of substrate reacted in a given time (fixed time assays). The potential problem with fixed-time assays is illustrated in Fig. 4.8, which represents the effect of enzyme concentration on the progress of the reaction in the presence of a constant initial substrate concentration (Fig. 4.8a). Measurement of the rate of the reaction at time t_o (by the tangent method) to give the true initial rate or two fixed times, t_1 and t_2, gives the relationship between initial rate and enzyme concentration shown in Fig. 4.8b. Only the tangent method gives the correct linear relationship. Since the determination of initial rate means that the changes in the concentration of substrate or product are relatively small, it is inherently more accurate to measure the increase in product concentration because the relative increase in its concentration is significantly larger than the corresponding decrease in substrate concentration.

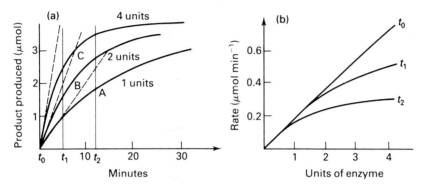

Fig. 4.8. Importance of measuring initial rate in the assay of an enzyme. (a) Time-dependent variation in the concentration of product in the presence of 1, 2 and 4 units of enzyme; (b) variation of reaction rate with enzyme concentration using true initial rate (t_o) and two fixed times (t_1 and t_2).

Measurement of the initial rate of an enzyme-catalysed reaction is fundamental to a complete understanding of the mechanism by which the enzyme works, as well as to the estimation of the activity of an enzyme in a biological sample. Its numerical value is influenced by many factors, including substrate and enzyme concentration, pH, temperature and the presence of activators or inhibitors. Each of these variables will now be discussed.

4.7.2 Variation of initial rate with substrate concentration

For the majority of enzymes, the initial rate of reaction varies hyperbolically with substrate concentration (Fig. 4.9a). At low substrate concentrations, approximately first-order kinetics are observed (i.e. the initial rate is proportional to the substrate concentration), but at high substrate concentrations saturation (zero-order) kinetics exist and the initial rate is independent of substrate concentration. The mathematical relationship between initial rate and substrate concentration, is expressed by the Michaelis–Menten equation:

$$v_o = \frac{V_{max}[S]}{K_m + [S]} \tag{4.2}$$

where [S] is the substrate concentration, V_{max} is the limiting value of v_o, and K_m is the Michaelis constant.

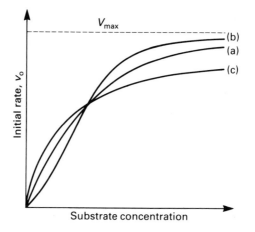

Fig. 4.9. The effect of substrate concentration on the initial rate of an enzyme-catalysed reaction: (a) when substrate binding is non-cooperative; (b) when substrate binding is positively cooperative; and (c) when substrate binding is negatively cooperative. (Note that negative homotropic cooperativity gives a curve that is neither a rectangular hyperbola nor sigmoidal.)

It can be seen from equation 4.2, that when $v_o = \frac{1}{2} V_{max}$, $K_m = [S]$. Thus K_m is numerically equal to the substrate concentration at which the initial rate is one-half of the maximum rate. K_m, which therefore has units of molarity, is independent of enzyme concentration and is a characteristic of the system. It has values usually in the range 10^{-2} to 10^{-5} M and is important because its value, in conjunction with the value of V_{max} and the relative amounts of enzymes, influences the importance of competing metabolic pathways in a cell. In interpreting K_m and V_{max} values in these terms, however, it must be appreciated that the conditions used to determine their value *in vitro* may not reflect the situation *in vivo*. In practice, K_m values are important because they enable the concentration of substrate required to saturate all of the active sites of an enzyme to be calculated. When $[S] \geq K_m$, equation 4.2 reduces to $v_o \approx V_{max}$, but a simple calculation reveals that when $[S] = 10\,K_m$, v_o is only 90% V_{max} and that when $[S] = 100\,V_m$, $v_o = 99\%\ V_{max}$. Appreciation of this relationship is vital in enzyme assays. V_{max} is related to the turnover number (k_{cat}) of the enzyme, which expresses the maximum number of moles of substrate that 1 mol of enzyme can convert to product in 1 s. Values of k_{cat} range from 1 to $10^6\,\text{s}^{-1}$.

The ratio of k_{cat}/K_m is known as the specificity constant and is a measure of the relative rate of reaction at low ($[S] \ll K_m$) substrate concentrations. Its value is limited to the maximum rate of diffusion of substrate to the enzyme molecule, which is of the order of 10^8–10^9 M^{-1} s^{-1}. Enzymes such as carbonic anhydrase and triosephosphate isomerase have specificity constants of this order, indicating that they have evolved to maximum kinetic efficiency.

A number of oligomeric enzymes that contain multiple substrate binding sites do not display simple Michaelis–Menten kinetics, but give a sigmoidal relationship between initial rate and substrate concentration (Fig. 4.9b). Such a curve is indicative of an allosteric enzyme, which is one whose molecular conformation changes as a result of the binding of a substrate molecule, a process referred to as a homotropic effect. This conformational change may result in either increased (positive cooperativity) or decreased (negative cooperativity) activity towards further substrate molecules. Changes in activity towards the substrate may also be induced by molecules other than the substrate. Compounds that induce such changes are referred to as heterotropic effectors. They are commonly key metabolic intermediates such as ATP, ADP, AMP and P_i, which bind to an allosteric (regulatory) site so that the structure of the active site is modified. Heterotropic activators increase the reactivity of the enzyme, making the curve less sigmoidal and moving it to the left, whilst heterotropic inhibitors cause a decrease in activity, making the curve more sigmoidal and moving it to the right. The operation of cooperative effects may be confirmed by a Hill plot, which is based on the equation:

$$\log \frac{v_o}{V_{max} - v_o} = h \log[S] + \log K \tag{4.3}$$

where h is the Hill constant or coefficient, and K is an overall binding constant related to the individual binding constants for n sites by the expression $K = (K_{a_1} \times K_{a_2} \times \ldots \times K_{a_n})^{1/n}$.

The Hill constant, which is equal to the slope of the plot, is a measure of the cooperativity such that: if $h = 1$, binding is non-cooperative and normal Michaelis–Menten kinetics exist; if $h > 1$, binding is positively cooperative; and, if $h < 1$, binding is negatively cooperative. At very low substrate concentrations that are insufficient to fill more than one site and at high concentrations at which most of the binding sites are occupied, the slopes of Hill plots tend to a value of 1. The Hill coefficient is therefore taken from the linear central portion of the plot. One of the problems with Hill plots is the difficulty of estimating V_{max} accurately. This is best obtained from a Lineweaver-Burk (equation 4.7) or Eadie–Hofstee (equation 4.9) plot.

It is sometimes argued that h is numerically equal to the number of binding sites, n. This is an oversimplification and very often h is not an integer. For example, h for the binding of oxygen to haemoglobin, for which the number of binding sites is known to be four, is 2.6. In practice, h can be taken to be a minimum estimate of the number of interacting binding sites as well as a measure of their cooperativity.

The Michaelis constant K_m is not used with allosteric enzymes. Instead, the term $S_{0.5}$, which is the substrate concentration required to produce 50% saturation of the enzyme, is used. It is important to appreciate that sigmoidal kinetics do not confirm the operation of allosteric effects because sigmoidicity may be the consequence of the enzyme preparation containing more than one enzyme capable of acting on the substrate. It is easy to establish the presence of more than one enzyme as there will be a discrepancy between the amount of substrate consumed and the expected amount of product produced. It is equally important to appreciate that not all enzymes subject to allosteric control display sigmoidal kinetics. Some monomeric enzymes, for example wheat hexokinase, have been shown to be subject to such control but to display simple Michaelis–Menten kinetics.

Enzyme-catalysed reactions proceed via the formation of an enzyme substrate complex (ES) in which the substrate (S) is non-covalently bound to the active site of the enzyme (E). The formation of this complex is rapid and reversible and is characterised by the dissociation constant, K_s, of the complex:

$$E + S \underset{k_{-1}}{\overset{k_{+1}}{\rightleftharpoons}} ES$$

$$K_s = \frac{[E][S]}{[ES]} = \frac{k_{-1}}{k_{+1}} \tag{4.4}$$

where k_{+1} and k_{-1} are first-order rate constants for the forward and reverse reactions.

The conversion of the bound substrate to product is a slower and rate-determining step. In the simplest situation, where the product is formed in a single step and at a rate such that the equilibrium concentration of ES is maintained, it can be shown that the observed K_m is equal to K_s, i.e. the Michaelis–Menten equation becomes:

$$v_0 = \frac{V_{max}[S]}{K_s + [S]} \tag{4.5}$$

However, in the majority of cases the conversion of ES to EP (enzyme–product complex) is such that, although the concentration of ES is essentially constant, it is not the equilibrium concentration. In these cases, Briggs–Haldane kinetics prevail and it can be shown that:

$$K_m = \frac{k_{+2} + k_{-1}}{k_{+1}} = K_s + \frac{k_{+2}}{k_{+1}} \tag{4.6}$$

where k_{+2} is the first-order rate constant for the conversion of ES to EP. Thus, in these cases, K_m is numerically larger than K_s. Detailed studies using the approaches outlined in Section 4.10 have revealed that the conversion of ES to EP frequently proceeds through a number of intermediates, some of which involve covalent bond formation. An example is the action of chymotrypsin on some amides in which an acylated form (EA) of the enzyme is an additional intermediate and two products, P_1 and P_2 are produced sequentially:

$$E + S \rightleftharpoons ES \xrightarrow{k_{+2}} EA \xrightarrow{k_{+3}} E + P_2$$
$$\downarrow$$
$$P_1$$

In such circumstances it can be shown that:

$$K_m = K_s \left(\frac{k_{+3}}{k_{+2} + k_{+3}} \right) \tag{4.7}$$

so that K_m is numerically smaller than K_s. It is obvious therefore that care must be taken in the interpretation of the significance of K_m relative to K_s. Only when the complete reaction mechanism is known can the relationship between K_m and K_s be fully appreciated.

Whilst the Michaelis–Menten equation can be used to calculate K_m and V_{max}, its use is subject to error due to the difficulty of experimentally measuring initial rates at high substrate concentrations. Linear transformations of the Michaelis–Menten equation are therefore preferred. The most popular of these is the Lineweaver–Burk equation:

$$\frac{1}{v_0} = \frac{K_m}{V_{max}} \times \frac{1}{[S]} + \frac{1}{V_{max}} \tag{4.8}$$

A plot of $1/v_0$ against $1/[S]$ (Fig. 4.10) gives a straight line of slope K_m/V_{max}, with an intercept on the $1/v_0$ axis of $1/V_{max}$ and an intercept on the $1/[S]$ axis of $-1/K_m$. This double reciprocal plot identifies substrate inhibition (Section 4.7.6) by an upward curve at high substrate concentrations (low $1/[S]$ values) (Fig. 4.11b). Positive cooperativity is characterised by a concave upwards curve and negative cooperativity by a concave downwards curve (Fig. 4.11c). Unless very careful thought is given to the planning of the series of substrate concentrations, the Lineweaver–Burk equation tends to give an unequal distribution of points and greatest emphasis to the points at low substrate concentrations. Alternative plots (Fig. 4.9) are based on the Hanes equation:

$$\frac{[S]}{v_0} = \frac{K_m}{V_{max}} + \frac{[S]}{V_{max}} \tag{4.9}$$

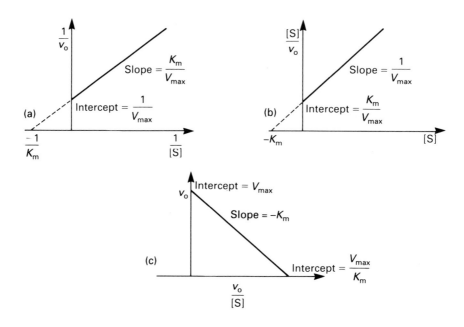

Fig. 4.10. Linear plots for the calculation of K_m and V_{max}: (a) Lineweaver–Burk, (b) Hanes, and (c) Eadie–Hofstee.

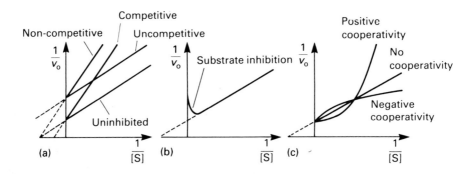

Fig. 4.11. Lineweaver–Burk plots showing (a) the effects of three types of reversible inhibitor, (b) substrate inhibition, and (c) homotropic cooperativity.

and the Eadie–Hofstee equation:

$$v_o = V_{max} - K_m \times \frac{v_o}{[S]}$$
(4.10)

Bisubstrate reactions (Fig. 4.12) such as those catalysed by the transferases, kinases and dehydrogenases, in which two substrates S_1 and S_2 are converted to two products P_1 and P_2 (two substrate–two product, bi-bi, reactions), are inherently more complicated than monosubstrate reactions. They may be sequential, in which case both substrates bind to give a ternary complex before the products are formed. Sequential reactions may be compulsory ordered, in which case the two substrates bind in a definite sequence, or random ordered, in which case either substrate can

(a) $E + S_1 \rightleftharpoons ES_1 \overset{S_2}{\rightleftharpoons} ES_1S_2 \rightleftharpoons EP_1P_2 \rightleftharpoons P_2 + EP_1 \rightleftharpoons P_1 + E$

Sequential compulsory-order mechanism

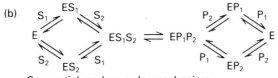
(b)

Sequential random-order mechanism

(c) $E + S_1 \rightleftharpoons ES_1 \rightleftharpoons \epsilon\, P_1 \rightleftharpoons \epsilon + P_1$

$\epsilon + S_2 \rightleftharpoons \epsilon\, S_2 \rightleftharpoons EP_2 \rightleftharpoons E + P_2$

Non-sequential ping-pong mechanism

Fig. 4.12. Possible reaction mechanisms for bisubstrate reactions.

bind first. In both cases there are separate binding sites for each substrate. Alternatively the reaction may be non-sequential, in which case one product is released before the second substrate is bound. One example of this type of mechanism is a ping-pong reaction, which proceeds via a modified form of the enzyme (ϵ) which may take the form of an acylated enzyme. A ping-pong bi-bi mechanism is indicated, but not confirmed, by a series of parallel lines in double reciprocal plots when the variation of initial rate with increasing concentrations of one substrate is investigated in the presence of a series of fixed second substrate concentrations. Double reciprocal plots give a progressively smaller intercept on the 1/[S] axis as the concentration of second substrate is increased. A compulsory-order and a random-order ternary complex mechanism both give non-parallel double reciprocal plots with a progressively smaller intercept on the $1/v_o$ axis as the concentration of fixed second substrate is increased.

For all bisubstrate reactions, by holding constant the concentration of one of the two substrates and studying the influence on the initial rate of varying the concentration of the second substrate, the V_{max} value and a value of K_m for each substrate may be obtained. In these bisubstrate reactions, V_{max} is defined as the maximum initial rate when both substrates are saturating, and the K_m for a particular substrate as the concentration of that substrate that gives $\frac{1}{2} V_{max}$ when the other substrate is saturating. To determine these K_m values, the initial velocity is studied as a function of the concentration of one substrate at a series of fixed second substrate concentrations. A double reciprocal plot is made for each second substrate concentration, giving a series of straight lines called primary plots. A secondary plot is then made of the $1/v_o$ intercepts of the primary Lineweaver–Burk plots against the reciprocal of the second (fixed) substrate. This gives a straight line slope K_m (for the second substrate)/V_{max} and intercept $1/V_{max}$. The study is then repeated reversing the roles of the two substrates. The principle of secondary plots is illustrated in Fig. 4.13.

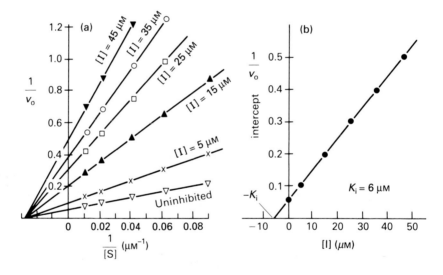

Fig. 4.13. (a) Primary Lineweaver–Burk plots showing the effect of a simple linear non-competitive inhibitor at a series of concentrations and (b) the corresponding secondary plot, which enables the inhibitor constant K_i to be calculated.

The elucidation of the reaction mechanism associated with a particular bisubstrate reaction generally involves a study of the variation of the initial rate with the concentration of one substrate at a series of fixed concentrations of the second substrate, in the absence and presence of the two reaction products, and the application of a series of rules formulated by W. Cleland. Two of these rules are:

(i) the intercept on the $1/v_o$ axis of double reciprocal plots is affected only by an inhibitor that binds reversibly to an enzyme form other than that to which the variable substrate binds;
(ii) the slope of double reciprocal plots is affected by an inhibitor that binds to the same enzyme form as' the variable substrate or to an enzyme form that is connected by a series of reversible steps to that with which the variable substrate binds.

The consequence of rule (i) is that, if characteristic competitive inhibition (Section 4.7.6) behaviour is observed, the inhibitor and the substrate whose concentration is being varied bind at the same site. The consequence of rule (ii) is that, if characteristic uncompetitive inhibition (Section 4.7.6) is observed, there must be no reversible link between the inhibitor and the substrate whose concentration is being varied. Studies applying these rules have, for example, revealed that histamine-N-methyltransferase operates via a compulsory-order mechanism whereas phosphoglycerate mutase operates via a ping-pong mechanism involving a phosphoryl-enzyme intermediate.

4.7.3   Variation of initial rate with enzyme concentration

It can be shown that:

$$v_o = \frac{k_{+2}[\text{E}][\text{S}]}{K_m + [\text{S}]} \text{ and hence that } v_o = \frac{k_{+2}[\text{E}]}{\dfrac{K_m}{[\text{S}]} + 1} \tag{4.11}$$

Thus, when the substrate concentration is very large, the initial rate is directly proportional to the enzyme concentration. This is the basis of the determination of enzyme activity in a particular biological sample (Section 4.8). The importance of the correct measurement of initial rate is illustrated by Fig. 4.8.

4.7.4   Variation of initial rate with temperature

The rate of an enzyme reaction varies with temperature according to the Arrhenius equation:

$$\text{rate} = A e^{-E/RT} \tag{4.12}$$

where A is a constant known as the pre-exponential factor, E is the activation energy (in J mol^{-1}), R is the gas constant (8.2 J mol^{-1} K^{-1}), and T is the absolute temperature.

The equation explains the sensitivity of enzyme reactions to temperature because the relationship between reaction rate and temperature is exponential. The rate of most enzyme reactions approximately doubles for every 10 deg.C rise in temperature (Q_{10} value). At a temperature characteristic of the enzyme and generally in the region 40 to 70 °C, the enzyme is denatured and enzyme activity is lost. The activity displayed in this 40 to 70 °C temperature range depends partly upon the equilibration time before the reaction is commenced. The so-called optimum temperature, at which the enzyme appears to have maximum activity, is therefore time dependent and for this reason is not normally chosen for the study of enzyme activity. Enzyme assays are carried out routinely at 30 or 37 °C (Section 4.8).

Enzymes work by decreasing the activation energy for the reaction relative to the non-enzyme catalysed reaction. Thus catalase decomposes hydrogen peroxide 10^{14} times faster than does the uncatalysed reaction!

4.7.5   Variation of initial rate with pH

The state of ionisation of amino acid residues in the active site of an enzyme is pH dependent. Since catalytic activity relies on a specific state of ionisation of these residues, enzyme activity is also pH dependent. As a consequence the pH–enzyme activity profile is either bell-shaped (two important amino acid residues in the active site), giving a narrow pH optimum, or it has a plateau (one important amino acid residue in the active site). In either case, the enzyme is generally studied at a pH at which its activity is maximal. By studying the variation of log K_m and log V_{max} with pH, it is possible to identify the pK_a values of key amino acid residues involved in the catalytic process. Table 4.2 lists the ionisable groups found in proteins.

4.7.6   Influence of inhibitors on initial rate

Enzyme inhibitors (I) combine specifically with an enzyme to reduce its ability to convert substrate to product. Irreversible inhibitors, such as the organophosphorous and mercuric compounds, cyanide, carbon monoxide and hydrogen sulphide, combine by covalent forces and the extent of their inhibition is dependent upon their reaction rate constant (and hence time) and upon the amount of inhibitor present. The effect of irreversible inhibitors, which cannot be overcome by simple physical techniques such as dialysis, is to reduce the amount of enzyme available for reaction. The inhibition involves reaction with a functional group such as hydroxyl or sulphydryl or with a metal atom in a prosthetic group in the active site or a distinct allosteric site. Reversible inhibitors combine non-covalently with the enzyme and can therefore be readily removed by dialysis. Competitive reversible inhibitors combine at the same site as the substrate and must therefore be structurally related to the substrate. An example is the inhibition of succinate dehydrogenase by malonate:

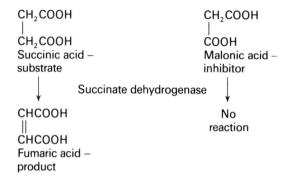

A non-competitive reversible inhibitor combines at a site distinct from that of the substrate but in such a way as to produce a so-called dead-end complex, irrespective of whether or not the substrate is bound. Uncompetitive reversible inhibitors can bind only to the ES complex and not to the free enzyme, so that inhibitor binding may be at a site created by the binding of the substrate to the active site (i.e. a conformational change occurs on substrate binding) or it may be to the bound substrate molecule. The resulting ternary complex, ESI, is also a dead-end complex. All types of reversible inhibitors are characterised by their dissociation constant K_i called the inhibitor constant, which may relate to the dissociation of EI (K_{EI}) or ESI (K_{ESI}).

For competitive inhibition the following two equations can be written:

$$E + S \rightleftharpoons ES \rightarrow E + P$$
$$E + I \rightleftharpoons EI$$

Since the binding of both substrate and inhibitor involves the same site, the effect of a competitive reversible inhibitor can be overcome by increasing the substrate concentration. The result is that V_{max} is unaltered but the concentration of substrate required to achieve it is increased so that when

$$v_o = \frac{1}{2} V_{max}, \quad [S] = K_m \left(1 + \frac{[I]}{K_i} \right) \tag{4.13}$$

where [I] is the concentration of inhibitor. It can be seen from this that K_i is equal to the concentration of inhibitor, which apparently doubles the value of K_m. With this type of inhibition, K_i is equal to K_{EI} and K_{ESI} is infinite because no ESI is formed.

In the presence of a competitive inhibitor the Lineweaver–Burk equation 4.8 becomes:

$$\frac{1}{v_0} = \frac{K_m}{V_{max}} \times \frac{1}{[S]}\left(1 + \frac{[I]}{K_i}\right) + \frac{1}{V_{max}} \tag{4.14}$$

allowing the diagnosis of competitive inhibition (Fig. 4.14). The numerical value of K_i can be calculated from a Lineweaver–Burk plot for the uninhibited and inhibited reactions. In practice, however, a more accurate value is obtained from a secondary plot. The reaction is carried out in the presence of a series of fixed inhibitor concentrations and a Lineweaver–Burk plot for each inhibitor concentration constructed. Secondary plots of the slope of the primary plot against the inhibitor concentration or of the apparent K_m, K'_m, (which is equal to K_m $(1 + [I]/K_i)$ and which can be calculated from the reciprocal of the negative intercept on the 1/[S] axis) against inhibitor concentration will both have intercepts, on the inhibitor concentration axis, of $-K_i$. Sometimes it is possible for two molecules of inhibitor to bind at the active site. In these cases, although all the primary double reciprocal plots are linear, the secondary plot is parabolic. This is referred to as parabolic competitive inhibition to distinguish it from normal linear competitive inhibition.

For non-competitive inhibition the inhibitor may also bind to ES:

$$ES + I \rightleftharpoons ESI$$

Since this inhibition involves a site distinct from the active site, the inhibition cannot be overcome by increasing the substrate concentration. The consequence is that V_{max}, but not K_m, is reduced because the inhibitor and substrate do not affect the binding of each other. With this type of inhibition K_{EI} and K_{ESI} are identical and K_i is numerically equal to both of them.

The Lineweaver–Burk equation 4.8 therefore becomes:

$$\frac{1}{v_o} = \left(\frac{K_m}{V_{max}} \times \frac{1}{[S]} + \frac{1}{V_{max}}\right)\left(1 + \frac{[I]}{K_i}\right) \tag{4.15}$$

Once non-competitive inhibition has been diagnosed (Fig. 4.11) the K_i value is best obtained from a secondary plot of either the slope of the primary plot or $1/V'_{max}$ (which is equal to the intercept on the $1/v_o$ axis) against inhibitor concentration.

Both secondary plots will have an intercept of $-K_i$ on the inhibitor concentration axis (Fig. 4.13).

For uncompetitive inhibition the inhibitor binds *only* to ES:

$$E+S \rightleftharpoons ES \longrightarrow E+P$$
$$-I \upharpoonleft\downharpoonright +I$$
$$ESI$$

As with non-competitive inhibition, the effect cannot be overcome by increasing the substrate concentration, but in this case both K_m and V_{max} are reduced by a factor of $(1 + [I]/K_i)$. An inhibitor concentration equal to K_i will therefore halve the values of both K_m and V_{max}. With this type of inhibitor, K_{EI} is infinite because the inhibitor cannot bind to the free enzyme so K_i is equal to K_{ESI}. The Lineweaver–Burk equation (4.8) therefore becomes:

$$\frac{1}{v_o} = \frac{K_m}{V_{max}} \times \frac{1}{[S]} + \frac{1}{V_{max}}\left(1 + \frac{[I]}{K_i}\right) \tag{4.16}$$

The value of K_i is best obtained from a secondary plot of either $1/V'_{max}$ or $1/K'_m$ (which is equal to the intercept on the $1/[S]$ axis) against inhibitor concentration. Both secondary plots will have an intercept of $-K_i$ on the inhibitor concentration axis.

This classification of the type of inhibition observed is not always straightforward or unambiguous. Thus, it is possible to observe competitive inhibition in cases where the inhibitor binds to a site other than the active site but in such a way that a conformational change follows the binding of the substrate, thereby preventing binding of the inhibitor. Equally, in some cases of bisubstrate reactions, an inhibitor may competitively inhibit the binding of one substrate and non-competitively inhibit the binding of the second (Section 4.7.2). It is also possible for an ESI complex to have some catalytic activity or for K_{EI} and K_{ESI} to be neither equal nor infinite, in which case so-called mixed inhibition kinetics are obtained. Mixed inhibition is characterised by a linear Lineweaver–Burk plot that does not fit any of the patterns shown in Fig. 4.11a. The plots of non-inhibition and inhibition may intersect either above or below the $1/[S]$ axis. The associated K_i can be obtained from a secondary plot of the slope either of the primary plot or of $1/V'_{max}$ for the primary plots against inhibitor concentration. In both cases the intercept on the inhibitor concentration axis is $-K_i$. Non-competitive inhibition may be regarded as a special case of mixed inhibition.

A number of enzymes at high substrate concentration display substrate inhibition characterised by a decrease in initial rate with increased substrate concentration.

The graphical diagnosis of this situation is shown in Fig. 4.11b. It is explicable in terms of the substrate acting as an uncompetitive inhibitor and forming a dead-end complex.

The study of enzyme inhibitors helps our understanding of the mechanisms by which enzymes work. Inhibitors are also used widely in the study of metabolic pathways to help in the identification of intermediates. The whole area of selective toxicity, including the use of antibiotics and insecticides, is based on the exploitation of species differences in susceptibility to enzyme inhibitors.

4.7.7   Cellular control of enzyme activity

The control of the activity of enzymes *in vivo* is vital for the efficient regulation of cellular metabolism. Not all of the factors so far discussed that influence enzyme activity *in vitro* operate within a given organism. Coarse (long-term) control is achieved at the level of protein synthesis by induction or inhibition of enzyme synthesis. Fine (short-term) control is obtained in a number of ways. Fluctuations in the concentration of substrate or reversible inhibitor can influence activity either by a direct effect on the concentration of the ES complex or by an allosteric effect. Studies have also identified the importance of reversible covalent modification. This may involve adenylation or phosphorylation of the enzyme, commonly triggered by cyclic AMP produced by adenylate cyclase and involving active and inactive protein kinases. Examples of such control are found in glycogenolysis and glyconeogenesis. It has also been shown that the activity of some enzymes is dependent upon the cellular concentration of calcium. The calcium is bound to specific binding proteins such as calmodulin and troponin C. Examples of enzymes subject to such control include actomyosin, phosphorylase kinase and cyclic nucleotide diesterase. Another mechanism important for the regulation of some enzymes is irreversible limited proteolysis. Enzymes subject to such control are synthesised in an inactive form called proenzymes or zymogens. These are activated, when required, by the proteolytic cleavage of peptide fragments by enzymes such as trypsin. Examples include chymotrypsinogen (the precursor of chymotrypsin), thrombinogen (the precursor of thrombin), and proelastase (the precursor of elastase).

4.8   Enzyme assay techniques

4.8.1   General considerations

The determination of the activity of an enzyme is based upon the rate of utilisation of substrate or formation of product under controlled conditions. Most assays are carried out at 30 °C but some are performed at 37 °C because of the physiological significance of the temperature. Adequate buffering capacity must be used and care should be taken to ensure that all apparatus is scrupulously clean. Analytical methods may be classified as either continuous (kinetic) or discontinuous (fixed-time). Continuous methods monitor some property change (e.g. absorbance or gas volume) in the reaction mixture, whereas discontinuous methods require samples to be withdrawn from the reaction mixture and analysed by some convenient technique. The inherent greater accuracy of continuous methods commends them whenever they are available.

Irrespective of the principle of the analytical method, enzyme assays require the use of excess substrate (zero-order kinetics; at least equal to $10\,K_m$) and an appropriate control. The control is in all respects the same as the test assay but lacking either enzyme or substrate. Changes in the experimental parameter in the control lacking the test enzyme will give an assessment of the extent of the non-enzymic reaction whereas changes in the control lacking added substrate evaluate any background reaction in the enzyme preparation. All reaction mixtures are incubated at the experimental temperature for at least 2 min before the reaction is started by the addition of either pre-equilibrated enzyme or substrate. It is worth while assaying the enzyme using different volumes of the test solution to confirm linearity between initial rate and enzyme concentration, thereby confirming the absence of activators or inhibitors in the preparation.

4.8.2   Visible and ultraviolet spectrophotometric methods

Many substrates and products absorb light in the visible or ultraviolet region and, provided that the substrate and product do not absorb at the same wavelength and that the Beer–Lambert law (Section 7.4) holds, the change in absorbance can be used as the basis of the assays. Studies are best carried out in a double-beam recording instrument with a temperature-controlled cell housing.

A large number of assays are based on the interconversion of NAD^+ and NADH. Both the oxidised and reduced forms of these two nucleotides absorb at 260 nm but only the reduced form at 340 nm. Enzymes that do not directly involve this interconversion may be assayed by using the concept of coupled reactions. In such reactions, the enzyme to be assayed is linked to one utilising the NAD^+/NADH system by means of common intermediates. The principle is illustrated by the assay of phosphofructokinase (PFK). It can be linked via aldolase to the glyceraldehyde-3-phosphate dehydrogenase (G3PDH) reaction:

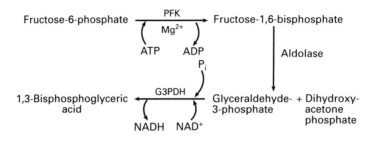

The assay mixture would therefore contain fructose 6-phosphate, ATP, Mg^{2+}, aldolase, G3PDH, NAD^+ and P_i all in excess, so that the rate of NADH production, and hence increase in absorption at 340 nm wavelength, would be determined solely by the concentration of PFK in the known volume of preparation added to the assay mixture. In principle there is no limit to the number of reactions that can be coupled in this way, provided the enzyme under investigation is present in limiting amounts. The number of units of enzyme in the test preparation is calculated as follows:

$$\text{enzyme units} = \frac{\Delta E_{340}}{\varepsilon} \times \frac{a}{1000} \times \frac{1000}{x} \qquad (4.17)$$
$$\text{(kat cm}^{-3})$$

where ΔE_{340} is the control-corrected change in absorption at 340 nm s^{-1}, a is the total volume of reaction mixture (generally about $3\,cm^3$) in a cuvette of 1 cm light path, x is the volume of test preparation included in the reaction mixture, and ε is the molar extinction coefficient for NADH (6.3×10^3 dm^3 mol^{-1} cm^{-1}). A general form of this equation is applicable to all spectrophotometric enzyme assays. In some cases the stoichiometry of the reaction (the number of molecules of compound undergoing the observed change in absorbance) is not unity, in which case a correction for the stoichiometry has to be introduced. The general equation is therefore:

$$\text{enzyme units} = \frac{\Delta E a}{\varepsilon d n} \qquad (4.18)$$

where ε is the molar extinction coefficient of the chromophor, n is the stoichiometry, d is the light path in the cuvette (cm), and ΔE is the change in absorbance at experimental wavelength. By dividing equations 4.17 and 4.18 by C_p, the total concentration of protein in the enzyme preparation, the specific activity of the preparation can be calculated.

The scope of visible spectrophotometric enzyme assays can be extended by the use of artificial substrates and by the production of coloured derivatives of the substrate or product. Many enzymes, especially the hydrolases, will act on synthetic analogues of their natural substrate to release a coloured product such as *p*-nitrophenol and phenolphthalein. An example is the assay of α-glucosidase (maltase):

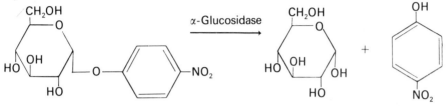

p-Nitrophenyl-α-D-glucopyranoside   D-Glucose   p-Nitrophenol (yellow)

An extension of this approach is the use of synthetic dyes for the study of the oxidoreductases. The oxidised and reduced forms of these dyes are different colours. Examples are the tetrazolium dyes, methylene blue, 2,6-dichlorophenol indophenol and methyl and benzyl viologen. Their use, which is discussed fully in Section 11.3.2, depends upon them having an appropriate oxidation–reduction potential relative to that for the substrate.

Substrates or products containing certain functional groups can be converted to a coloured derivative. Examples are the orange dinitrophenylhydrazone derivatives of aldehydes and of ketones. Thus an assay of isocitrate lyase is based on this reaction:

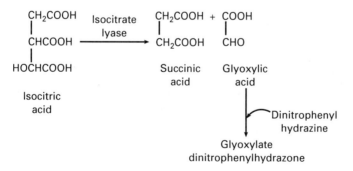

Either samples of the reaction mixture are withdrawn periodically and reacted with dinitrophenylhydrazine or, for a fixed-time assay, the whole reaction mixture is reacted with the reagent at a pre-established time. In some cases it is possible to convert the product of an enzyme reaction to a coloured derivative *in situ* without interfering with the enzyme reaction itself. An example is glucose oxidase, whose product, hydrogen peroxide, can be used to oxidise *o*-dianisidine, incorporated into the assay mixture, to a yellow product.

4.8.3   Spectrofluorimetric methods

Although potentially very sensitive, fluorimetric enzyme assays have the practical limitation that trace impurities in the enzyme preparation can quench the emitted radiation (Section 7.3.2). Additionally, many fluorescent compounds are unstable, especially in the presence of ultraviolet light. Nevertheless, the technique is widely applied to enzyme assays. NAD(P)H is fluorescent so that many enzymes can be assayed by coupling to an appropriate reaction as mentioned above (Section 4.8.2). Equally, synthetic substrates are available that release fluorescent products. Examples include the use of 4-methylumbelliferyl-β-D-glucuronide for the assay of β-glucuronidase (Section 7.3.2).

4.8.4   Luminescence methods

The increasing popularity of bioluminescent reactions (Section 7.11.3), in which the intensity of emitted light is used to study enzyme reactions, is due to the high sensitivity of the technique. One of their problems is occasional lack of reproducibility. Firefly luciferase catalyses the oxidation of luciferin in an ATP-dependent reaction:

$$\text{Luciferin} + \text{ATP} + O_2 \xrightarrow{\text{Luciferase}} \text{Oxyluciferin} + \text{AMP} + PP_i + CO_2 + \textit{Light}$$

The reaction can be used to assay ATP and appropriate enzymes via coupled reactions (Section 4.8.2). The corresponding bacterial luciferase uses reduced FMN to oxidise long-chain aliphatic aldehydes. The resulting FMN can be coupled to NAD(P)H, thus permitting the assay of many enzymes, e.g. malate dehydrogenase:

$$\text{Malic acid} + \text{NAD}^+ \xrightarrow[\text{dehydrogenase}]{\text{Malate}} \text{Oxaloacetic acid} + \text{NADH} + H^+$$

$$\text{NADH} + H^+ + \text{FMN} \xrightarrow{\text{Oxidoreductase}} \text{FMNH}_2 + \text{NAD}^+$$

$$\text{FMNH}_2 + \text{RCHO} + O_2 \xrightarrow{\text{Luciferase}} \text{FMN} + \text{RCOOH} + H_2O + \textit{Light}$$

The luminol/aminophthalic acid system involving microperoxidase can form the basis of the assay of many enzymes, including acetylcholinesterase (ACE) involved in synaptic transmission:

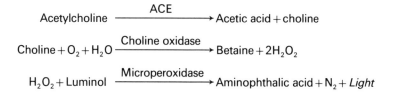

$$\text{Acetylcholine} \xrightarrow{\text{ACE}} \text{Acetic acid} + \text{choline}$$

$$\text{Choline} + O_2 + H_2O \xrightarrow{\text{Choline oxidase}} \text{Betaine} + 2H_2O_2$$

$$H_2O_2 + \text{Luminol} \xrightarrow{\text{Microperoxidase}} \text{Aminophthalic acid} + N_2 + \textit{Light}$$

4.8.5   Radioisotope methods

Although potentially a very sensitive method, the use of radioisotopes (Section 5.5.2) in enzyme assays is restricted to applications where it is possible to separate easily the radiolabelled forms of substrate and product. In those cases where one of the products is a gas, this presents no problem. Thus the assay of glutamate decarboxylase could be based on the evolution of $^{14}CO_2$:

$$\underset{\text{Glutamic acid}}{HOOCCH_2CH_2CH(NH_2)^{14}COOH} \xrightarrow[\text{decarboxylase}]{\text{Glutamate}} {}^{14}CO_2 + \underset{\gamma\text{-Aminobutyric acid}}{HOOCCH_2CH_2CH_2NH_2}$$

The $^{14}CO_2$ evolved could be trapped in alkali and hence the rate of $^{14}CO_2$ evolution measured. In other cases, the substrate and product may be separated by solvent extraction. Thus, in the assay of monoamine oxidase (MAO), samples of the assay mixture could be acidified (thereby converting the monoamine to its salt), the labelled aldehyde removed by ether extraction and the extract added to a scintillation cocktail for counting the radioactivity.

$$\underset{\text{Monoamine}}{R^{14}CH_2NH_2 + O_2 + H_2O} \xrightarrow{\text{MAO}} \underset{\text{Aldehyde}}{R^{14}CHO + H_2O_2 + NH_3}$$

4.8.6   Manometric methods

Enzyme reactions resulting in the net evolution or uptake of a gas such as oxygen and carbon dioxide can be monitored by the Warburg manometer or the Gilson respirometer. Both types of instrument can readily measure small changes in gas volume, provided the temperature is adequately controlled and that corrections are applied for the solubility of the gas in the reaction mixture. The versatility of the method can be extended to examples where one gas is evolved and another taken up, by chemically removing the evolved gas. In the case of CO_2 this would be by absorption in sodium hydroxide solution and in the case of O_2 by absorption by pyrogallol or chromous chloride solution. Enzymes that can be assayed manometrically include glutamate decarboxylase (see above), catalase, malic enzyme and monoamine oxidase (see above):

$$2H_2O_2 \xrightarrow{\text{Catalase}} 2H_2O + O_2$$

$$CH_3COCOOH + CO_2 + NADPH + H^+ \xrightarrow[\text{enzyme}]{\text{Malic}} HOOCCH(OH)CH_2COOH + NADP^+$$
Pyruvic acid   Malic acid

4.8.7   Ion-selective and oxygen electrodes methods

The development of ion-selective electrodes (Section 11.5), such as those for the ammonium ion, and the oxygen electrode, has afforded attractive methods of enzyme assays. The methods are very sensitive, reproducible and can use very small volumes of reaction mixture.

4.8.8   Immunochemical methods

Polyclonal or monoclonal antibodies (Section 2.2) raised to a particular enzyme can be used as the basis for a highly specific ELISA-based assay for the enzyme (Section 2.6). Such systems can distinguish between isoenzymes, which, in the context of clinical measurements of enzyme activities, is of considerable diagnostic value. A monoclonal assay is available for serum prostatic acid phosphatase, which is one of the best means of diagnosing carcinoma of the prostate.

4.8.9   Microcalorimetric methods

Most biological reactions are accompanied by a minute change in heat (enthalpy) that gives rise to a temperature change of the order of 10^{-2} to 10^{-4} deg.C. Measurement of such small changes is possible using thermistors, which are temperature-sensitive metal oxides. The technique, which requires stringent insulation of the reaction vessel, may be improved by coupling the primary reaction to a secondary one which generates a larger heat evolution. Thus reactions releasing protons may be carried out in Tris buffer, which has a large enthalpy change on protonation:

$$\text{Glucose} + \text{ATP} \xrightarrow{\text{Hexokinase}} \text{Glucose 6-phosphate} + \text{ADP} + \text{H}^+ \quad \Delta H^\circ = -28\,\text{kJ mol}^{-1}$$

$$\text{Tris} + \text{H}^+ \longrightarrow \text{TrisH}^+ \qquad\qquad \Delta H^\circ = -47\,\text{kJ mol}^{-1}$$

4.8.10   Automated enzyme analysis

Spectrophotometric methods are the most popular methods for enzyme assays and form the basis of many commercially available reaction rate analysers. These are instruments dedicated to the measurement of enzyme or substrate. Many are automated and based on fixed-change or fixed-time assays. Discrete analysers mix the enzyme and substrate by means of automatic pipettes according to a preprogrammed instruction. Continuous flow analysers pump the substrate continuously through a flow tube and periodically introduce a sample of the enzyme to be analysed into the line, each sample being separated from the next by a bubble. The reactants are mixed in a narrow mixing coil and pumped to the detector. The flow rate ensures that the reactants reach the detector at the correct time. An interesting variant is the fast (centrifugal) analyser. In it the enzyme and substrate are placed in separate wells located near the centre of a horizontal centrifuge plate (rotor) that can accommodate up to 30 separate samples. When the plate is rotated, the reactants are forced centrifugally into a cuvette at the edge of the plate, thus initiating the reaction. The change in absorbance at an appropriate wavelength in each cuvette is continuously recorded, enabling a good estimation to be made of the initial rate.

4.9   Substrate assay techniques

Enzyme-based assays are very convenient methods for the estimation of the amount of substrates present in a biological sample. In theory, the principle of using excess enzyme and relating the substrate concentration to the observed initial rate could be used but in practice this is not popular because of the difficulty of measuring a rapidly decreasing reaction rate at low substrate concentrations. The procedures employed overcome this problem in a variety of ways but basically they are all variants of the so-called end-point technique. In this approach all of the substrate is converted to product and the total change in parameter (e.g. ultraviolet absorption) recorded. This change is then used to compute the amount of substrate originally present. The technique relies on the use of sufficient enzyme to ensure that the reaction goes to completion in a reasonable time. For reversible reactions it is necessary to adjust the position of equilibrium so that the reaction is effectively complete. This can be achieved by such means as adjusting the pH away from the optimum for the enzyme: in the case of bisubstrate reactions, by using a high concentration of the second substrate, or, in the case of reactions using $NAD(P)^+$, by using an analogue such as acetylpyridine adenine dinucleotide (APAD), which has a more favourable oxidation–reduction potential. Coupled reactions are commonly used in substrate assays. In all cases the substrate concentration should be very much smaller than the K_m and the amount of enzyme used should be adjusted so that the reaction is complete in 2–10 min. It can be shown that if the amount of enzyme is such that $V_{max}/K_m \approx 1$, then the reaction will be 99% complete in about 5 min. The detection limits for such assays are determined by the molar extinction coefficient of the compound being monitored. In the case of NAD(P)H, they are of the order of 10^{-2} to 10^{-3} $\mu mol\,cm^{-3}$ in a 1 cm cuvette for ultraviolet spectrophotometric assays.

The sensitivity limits for substrate assays can be improved dramatically by the technique of enzymic cycling. The method, which is particularly valuable in those cases where the substrate is present in very low concentrations, involves the regeneration of substrate by means of a coupled reaction. The product accumulated in a given period of time (30–60 min) is then measured. A precalibration is necessary, using known amounts of the test substrate and with all the other components in the assay present in excess. The resulting 10^4- to 10^5-fold increase in sensitivity lowers the detection limits for visible and ultraviolet spectrophotometry to 10^{-6} to 10^{-8} $\mu mol\,cm^{-3}$. The method is commonly used for the assay of NAD(P)$^+$ and ATP/ADP. In the latter case, pyruvate kinase and phosphoenolpyruvate are used to regenerate ATP. In the case of NAD$^+$ or NADP$^+$, glutamate dehydrogenase may be used. For example in the assay of NADP$^+$, glucose-6-phosphate dehydrogenase (G6PDH) and glutamate dehydrogenase (GDH) may be coupled:

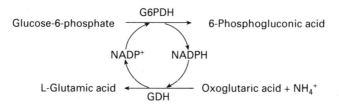

4.10   Pre-steady-state enzyme kinetics

4.10.1   Rapid mixing methods

The experimental techniques so far discussed for steady-state kinetics allow the determination of K_m and V_{max} values, but special techniques must be employed for the determination of the rate constants of the individual steps in the conversion of substrate to product, since the intermediates are transient. Figure 4.14 shows the progress curves in the initial stages of the conversion of substrate to product via ES. The induction period, t, is related to k_{+1}, and k_{+2}.

In the continuous flow method, solutions of the enzyme and substrate are introduced from syringes into a small mixing chamber (typically 100 mm³ capacity) and then pumped at a pre-selected speed through a narrow tube to which is

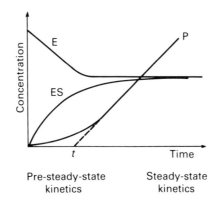

Fig. 4.14. Initial stage progress curve for the reaction $E + S \rightleftharpoons ES \rightarrow E + P$ when $(S) \geqslant [E]$.

attached a light source and a photomultiplier or similar detector. Flow through the tube is fast (typically $10\,\mathrm{m\,s^{-1}}$) so that it is turbulent, thus ensuring homogeneity of the solution. The precise reaction time, after mixing, at the point of observation can be calculated from the known flow of the two solutions. By varying these rates, the reaction time at the observation point can be varied, allowing the extent of reaction to be studied as a function of time. From these data, rate constants can be calculated. The technique, which requires relatively large volumes of reactants, is limited only by the inherent time needed to mix the reactants.

The stopped flow method is a development of the continuous flow method in that, shortly after the reactants emerge from the mixing chamber, the flow is stopped. At this point the recorder is triggered and a continuous recording made of the change in experimental parameter (e.g. absorbance). The attraction of the method is its conservation of reactants. In both methods, the problem of studying the first few hundred microseconds of the reaction can be partially solved by altering the pH or temperature to slow down the reaction or by using alternative substrates with slower turnover times. The versatility of both the continuous flow and stopped flow techniques is increased by the use of synthetic substrates that either generate a chromophoric leaving group or give a chromophoric acyl or phosphoryl intermediate. An example is the use of *p*-nitrophenylacetate in the study of chymotrypsin.

A variant of these rapid flow techniques is the quenching method. In this technique, the reactants from the mixing chamber enter a second chamber where they are mixed with a quenching agent such as trichloroacetic acid, which stops the reaction. The reaction products are then analysed by an appropriate analytical technique. By quenching after a range of reaction times, the progress of the reaction can be followed.

4.10.2   Relaxation methods

The great limitation of the rapid mixing methods is the dead time during which the enzyme and substrate are mixed. In the relaxation method, an equilibrium mixture of the reactants is preformed and the position of equilibrium altered by a change in reaction conditions. The most common procedure for achieving this is the temperature jump technique in which the temperature is raised rapidly by 5 to 10 deg.C by the discharge of a capacitor or infrared laser. The rate at which the system adjusts to its new equilibrium (relaxation time τ) is inversely related to the first-order rate constants involved in the reaction sequence. The rate of return to equilibrium is studied by spectrophotometric techniques. It is often advantageous to use more than one such technique, e.g. ultraviolet absorbance and fluorimetry, since they may yield complementary information. Careful analysis of the total number of relaxation times gives an indication of the number of intermediates involved in the overall process as well as the value of the related rate constants.

Whilst relaxation techniques are best for studying the fastest processes, rapid flow techniques are frequently best for studying the processes involved in the catalytic steps. Rapid kinetic techniques have revealed that enzyme and substrate generally associate very rapidly with first-order rate constants in the range 10^6 to $10^8 \, \text{M}^{-1} \, \text{s}^{-1}$ and dissociate more slowly with rate constants in the range 10 to $10^4 \, \text{s}^{-1}$. The association process is generally slower than that predicted by simple collision theory and indicates the need for specific orientation of the substrate and enzyme and perhaps conformation changes and the involvement of solvation processes.

4.10.3   Detection of intermediates

Although pre-steady-state kinetic measurements permit the calculation of rate constants they do not necessarily directly identify the intermediate(s) predicted by the proposed reaction mechanisms. For this, the intermediate(s) must be isolated and characterised. One way of achieving this is by the use of affinity labels. These are irreversible inhibitors that structurally resemble the substrate but form a covalent bond with an amino acid residue in the active site. The resulting enzyme–inhibitor complex is then isolated and studied by conventional analytical techniques. Examples of affinity labels include organophosphorous compounds, such as diiso-propylfluorophosphate (DFP), which binds irreversibly to acetylcholinesterase and related esterases via a serine residue, and bromohydroxyacetone phosphate, which binds to triosephosphate isomerase. A variant of the technique is the use of photoaffinity labels, which initially bind reversibly to the enzyme but which, on photolysis, are converted to reactive intermediates that bind irreversibly. Examples are diazo compounds, which give carbenes and azides, which give nitrenes:

$$RCOCH_2N_2 \xrightarrow{\textit{Light}} RCO\ddot{C}H_2 + N_2$$
Diazo   Carbene
compound

$$RN_3 \xrightarrow{\textit{Light}} RN{:} + N_2$$
Azide   Nitrene

Affinity labels of course identify specific amino acid residues in the active site in addition to providing information about the chemical nature of the intermediates. It is possible to expand this approach to the identification of amino acid residues in an active site by using reagents that do not resemble the substrate but do react with specific amino acid residues. The success of the method relies upon the fact that certain amino acid residues in the active site are activated, making them more reactive than similar residues elsewhere in the protein. Examples of these reagents include the reaction of N-tosyl-L-phenylalanylchloromethyl ketone (TPCK) with histidine residues, and iodoacetate and iodoacetamide with cysteine residues.

4.11   Protein–ligand binding studies

4.11.1   General principles

The affinity of an enzyme for its substrate is expressed by the dissociation constant K_s of the ES complex. K_s values cannot be directly derived from steady-state kinetics, although under certain circumstances the Michaelis constant, K_m, which is readily measured, may approximate numerically to K_s (Section 4.7.2). In contrast, the dissociation constant for reversible inhibitor–enzyme complexes, K_i, can be obtained directly from steady-state studies (Section 4.7.6) without any assumption being made about the relative values of K_m and K_s. In certain cases K_s can be measured spectrophotometrically by use of substrate binding spectra (Section 7.4.3).

The approach to the direct measurement of K_s is similar to that for the study of other protein–ligand binding studies such as the binding of hormones to plasma membrane receptor proteins and the binding of drugs and hormones to plasma proteins such as albumin, α_1-acid glycoprotein and steroid hormone binding globulins. All such reversible binding can be represented by the general equation:

$$\text{P} + n\text{L} \underset{K_s}{\overset{K_a}{\rightleftharpoons}} \text{PL}_n$$

$$\text{Protein} \quad \text{Ligand} \quad K_s \quad \text{Complex}$$

where n is the number of identical but non-cooperative ligand binding sites, K_a is the protein–ligand binding (affinity) constant, and K_s is the protein–ligand dissociation constant $= 1/K_a$.

Applying the Law of Mass Action to the equilibrium and developing the expression for K_a algebraically, gives the Scatchard equation:

$$\frac{r}{[\text{L}]} = nK_a - rK_a \tag{4.19}$$

where r is the number of moles of ligand bound to each mole of protein,

i.e.:
$$r = \frac{[\text{PL}_n]}{[\text{P}]}$$

where [P] is the molar concentration of protein and [L] is the molar concentration of free (unbound) ligand. To determine the numerical value of n and K_a the equilibrium binding of ligand to protein is studied as a function of ligand concentration at a fixed protein concentration [P]. In cases of enzyme and substrate, the experimental conditions must be such that product formation is not possible. A plot is then made of $r/[\text{L}]$ against r (Fig. 4.15). In some situations the Scatchard plot is biphasic, indicating either that binding is cooperative or that there are two sets of binding sites with different affinities. In this latter case the isolation of the

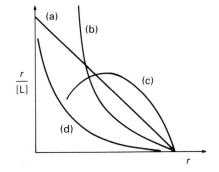

Fig. 4.15. Scatchard plot for (a) a single set of sites with no cooperativity, (b) two sets of sites with no cooperativity, (c) a single set of sites with positive cooperativity, and (d) a single set of sites with negative cooperativity.

two sets of binding constants is difficult. It is worth noting that, although the Scatchard equation is widely used, like the Eadie–Hofstee plot it is inherently unsatisfactory, as one of the variables, r, is incorporated into both axes. A double reciprocal plot of $1/r$ against $1/[L]$ (equation 4.8), analogous to the Lineweaver–Burk plot, lacks this criticism but, like the Lineweaver–Burk equation, gives an uneven distribution of points:

$$\frac{1}{r} = \frac{1}{n} + \frac{1}{n[L]K_a} \tag{4.20}$$

If the numerical value of [P] is not known, for example in a sample of plasma or serum, a plot is constructed of the concentration of bound ligand/concentration of unbound ligand against the concentration of bound ligand i.e. $[PL_n]/[L]$ against $[PL_n]$. It will have a slope of $-K_a$, intercept on the x-axis of $n[P]K_a$ and on the y-axis of $n[P]$. The term $n[P]$ is frequently referred to as the binding capacity.

The binding of some peptide hormones to their membrane receptor proteins displays cooperativity. In such cases the Scatchard plot is concave upwards for negative cooperativity and concave downwards for positive cooperativity but in both cases the intercept on the x-axis is equal to n (Fig. 4.15). If cooperative ligand binding is suspected, it should be confirmed by a Hill plot (equation 4.3), which, in its non-kinetic form, is:

$$\log\left(\frac{Y}{1-Y}\right) = h\log[L] + \log K \tag{4.21}$$

where Y is the fraction of binding sites filled, h is the Hill constant, and K is an overall binding (affinity) constant. For a protein with multiple binding sites that function independently, $h = 1$ and $n = 1,2,3,4\ldots$, whereas for a protein with multiple sites that are interdependent $n = 1,2,3,4\ldots$ and $1 < h < 1$.

Numerous techniques including ultracentrifugation (Section 6.6), exclusion chromatography (Section 10.8), circular dichroism (Section 7.9) and nuclear magnetic resonance spectrometry (Section 7.7) have been used to study ligand binding. In the case of spectrophotometric techniques, information regarding the functional groups in the protein involved in binding the ligand can be obtained in addition to data permitting the calculation of dissociation constants. The principles of these techniques are discussed in the relevant chapters in this book and only details of the two simplest techniques, equilibrium dialysis and ultrafiltration, will be described here.

4.11.2   Equilibrium dialysis

The protein and ligand, each in a buffer of the same pH and concentration, are placed in opposite halves of a dialysis cell. The cell, which generally has a total internal volume of about $3\,cm^3$, is constructed of transparent plastic such as Perspex and unscrews into two halves, which are separated by a cellulose acetate or nitrate semi-permeable membrane mounted on an inert mesh support. Many commercial variants of the cell are available, some consisting of banks of up to six cells. The temperature of the cell is thermostatically controlled and the cell is slowly rotated to help the system to reach equilibrium. The ligand molecules, which are small and diffusable, readily cross the membrane until their unbound concentration is the same on both sides. The protein is confined to one half of the cell. At equilibrium, samples are taken from each half of the cell and analysed for ligand. The sample from the protein half of the cell will give the sum of the bound and unbound ligand concentrations, whereas that in the other half will simply give the unbound ligand concentration, which will be the same as the unbound ligand concentration in the other half of the cell containing the protein. From a knowledge of the protein concentration, the values of r can be calculated and a Scatchard plot constructed. Most commonly the ligand is partly present as a ^3H- or ^{14}C-labelled form so that the assay can be performed by scintillation counting. For reliable results the binding of both the ligand and protein to the membrane must be minimal and the total ion concentration in each half of the cell equalised to minimise any possibility of a charge inequality on either side of the membrane affecting the distribution of ligand (Donnan effect). The limitations of the technique are the relatively long period of time it takes to establish equilibrium and the fact that it cannot be applied to cases where the ligand is a macromolecule, for example in the study of the binding of tRNA to aminoacyl-tRNA synthetase.

4.11.3   Ultrafiltration

The protein and ligand in a buffered solution are contained in a thermostatically controlled cell (generally 1 to $3\,cm^3$ capacity) containing a semi-permeable membrane on an inert mesh support at its base. Since no diffusion across the membrane is required to establish equilibrium, attainment of equilibrium is rapid (20 min). A small sample ($100\,mm^3$) is then forced across the membrane into a collection cup, either by application of a gas pressure to the mixture side or more simply by placing the cell in a low-speed centrifuge and centrifuging at about $3000\,g$ for a few minutes. By analysing the ultrafiltrate (representative of the unbound ligand concentration) and the reaction mixture (bound plus unbound ligand), the influence of ligand concentration on the extent of binding can be studied readily. The speed of the method is its attraction, but binding of the reactants to the membrane must be checked and the volume of the ultrafiltrate kept to a minimum to minimise any possibility that the sampling procedure displaces the position of equilibrium. As with equilibrium dialysis, the method cannot be used to study macromolecular ligands. The binding of such ligands is best studied by ultracentrifugation or exclusion chromatography.

4.12   Immobilised enzymes

Studies of immobilised enzymes have shown that they display kinetic behaviour subtly different from that of free enzymes. The reasons for this are complex but involve the perturbation of the three-dimensional structure of the enzyme by the immobilising matrix and changes in the microenvironment of the active site of the immobilised enzyme. Kinetic studies of immobilised enzymes are therefore valuable for our complete understanding of their action *in vivo*, especially their regulation. Immobilised enzymes, however, are also important because of their considerable analytical and industrial potential. Industrially, solutions of enzymes are commonly used to carry out chemical transformations, especially in the synthesis of antibiotics and steroids, and are used increasingly in preference to the techniques of conventional organic chemistry. Enzymes offer the advantages of both chemical specificity and stereo-specificity and accordingly give purer products. However, they are frequently unstable and difficult to recover from the reaction mixture. In principle, however, the ability to immobilise enzymes, pack them into columns and re-use them many times is commercially most attractive.

Enzymes can be immobilised in many ways, and these can be divided into two types. In physical immobilisation no covalent bonds are formed between the enzyme and the supporting matrix. Adsorption on animal charcoal or alumina was initially used to achieve immobilisation but current approaches include ionic adsorption on to ion-exchangers especially those of the Sephadex (Section 10.8.2) type and adsorption on to controlled-pore glass. Immobilisation by adsorption has the advantages of simplicity, general applicability, high yield and ability to recharge when the catalytic activity of the immobilised enzyme decreases below an unacceptable level. A limitation of the technique is the need strictly to control the working conditions to prevent desorption. Enzyme entrapment in liposomes (artificially produced concentric spheres of phospholipid bilayers) and in water-insoluble polymers such as polyacrylamide and agarose is also a simple and generally applicable method but suffers from poor flow properties, inefficiency and progressive leaching of the enzyme.

Chemical immobilisation results in at least one covalent bond being formed between the enzyme and the matrix. The chemical procedures used to produce immobilisation are similar to those used in affinity chromatography (Section 10.9). Attachment must not involve amino acid residues in the active site of the enzyme. The matrix may be polysaccharide, polymers such as nylon, or inorganic carriers such as glass and titanium dioxide.

Immobilised enzymes are finding an increasing number of analytical applications especially in clinical situations where they offer the potential for fast, sensitive and accurate determinations of analytes such as blood glucose and urea. The most significant development is the combination of immobilised enzymes, with their high specificity, and electroanalytical chemistry with its inherent sensitivity. The so-called enzyme electrode offers the opportunity for accurate analysis without any sample preparation. The principles involved are discussed more fully in Chapter 11.

4.13   Enzymology *in vivo*

Our current understanding of the *in vivo* regulation of metabolic pathways and of the flux of substrates through individual pathways is based on the study of the kinetic properties of isolated enzymes using the principles and techniques discussed in this chapter. The assumption has been made that intracellular compartments such as the cytosol and the mitochondrial matrix can be regarded as homogeneous bags of enzymes in which the individual enzymes, their substrates and products can freely diffuse. Support for this view has come from cell fractionation studies in which the great proportion of total cell proteins is released when cells are disrupted.

Over the past decade, evidence has begun to accumulate that casts doubt on this traditional view of *in vivo* enzymology. For example, studies with high voltage electron microscopy (HVEM) have shown that mammalian cells contain an extensive network of interconnecting strands termed the microtrabecular lattice (MTL), which appears to connect all the intracellular structures. Critics of the existence of the MTL have argued that such structures are artefacts, but there is increasing acceptance that MTL or closely related structures do represent a good approximation of the structure *in vivo*. The consequence of such a model is that previously regarded freely diffusing enzymes are more likely to be loosely bound to the MTL. Such a view gives rise to the concept of the wide existence of multienzyme complexes previously regarded as an exception to the norm. Support for organised arrangements of enzymes has come from the isolation of a fragment of the inner surface of the inner mitochondrial membrane containing all the enzymes of the TCA cycle. The fragment was termed a metabolon. Studies with mutant yeast cells, in which the activity of one of the TCA enzymes was modified, has been interpreted in support of such organised structures.

If these revised views of the arrangement of enzymes *in vivo* in a metabolic pathway are correct, then the concept arises of the channelling of substrates and products from one enzyme to the next. Channelling could occur by a number of mechanisms including the sequential covalent binding of intermediates to active sites; site-to-site transfer of non-covalently bound intermediates (so-called tight channelling), the transfer of intermediates in an unstirred aqueous layer and the prevention of diffusion of intermediates by electrostatic forces. It has been argued that channelling of metabolic intermediates within an organised enzyme complex could lead to increased flux through the pathway and restriction of flux in competing pathways. Whether or not the loose association of enzymes will result in an influence on their individual kinetic properties is still a matter of debate.

The most successful analytical technique for studying enzymology in individual cells and in whole organisms is nuclear magnetic resonance spectroscopy (NMR) (Section 7.7). This non-invasive technique allows the measurement of steady-state metabolite concentrations and of enzyme-catalysed flux using either simple proton NMR, the redistribution of a ^{13}C label among glycolytic intermediates or the use of ^{31}P NMR to measure ATP turnover and flux. Evidence for enzyme–enzyme interaction has been obtained by studying conformational changes in the enzyme protein. This approach requires the protein to be labelled in some appropriate way. One of the most attractive methods is to insert a fluorine atom into the molecule. From an NMR point of view this is an excellent label, since it is a spin $-\frac{1}{2}$ nuclide that is readily studied by NMR. The chemical shift change of the fluorine nucleus is large, making it very sensitive to its local environment in the protein. Moreover, its size is very similar to that of a proton, so that it is unlikely to modify the enzyme's structure. Since fluorine is very rare in biological systems, the NMR signal from the label can be interpreted unambiguously. By studying the relaxation times associated with the nucleus it should be possible to detect restricted motion of the enzyme in a cell due to protein–protein aggregation.

A complementary approach to these NMR studies is that of genetic manipulation. Using molecular biology techniques, it is possible to delete, raise or lower the intracellular concentration of selected enzymes and to study the effect on kinetics and flux. The approach simply requires the availability of cDNA or a genomic clone for the selected enzyme, coupled with gene disruption and anti-sense-RNA methodologies. It is realistic to expect that in the next decade the combination of molecular biology with NMR studies will help to reveal a greater understanding of how enzymes operate in metabolic pathways *in vivo*. Certainly there are many questions to be answered. For example, how is the elaborate structure of the MTL regulated and how are the individual enzymes inserted into it? How are the enzymes ordered and are the enzymes of a given pathway always in the same subcellular location? Do polysomes have spatial specificity and, if so, what determines it for the cytomatrix? Our complete understanding of enzymology still has a long way to go!

4.14   Suggestions for further reading

BOLLAG, D.M. and EDELSTEIN, S.J. (1991) *Protein Methods*. Wiley–Liss, New York. (Contains detailed practical protocols for a large number of protein techniques.)

BRANDEN, C. and TOOZE, J. (1991) *Introduction to Protein Structure*. Garland Publishing Inc., New York. (A beautifully illustrated, colourful and comprehensive account of protein structure.)

CHAPLIN, M.F. and BUCKE, C. (1990) *Enzyme Technology*. Cambridge University Press, Cambridge. (Discusses the use of enzymes in biotechnological applications such as biosensors.)

COLOWICK, S.P. and KAPLAN, N.O. (Eds.) *Methods in Enzymology*. Academic Press, New York. (The most comprehensive reference text on all aspects of practical enzymology.)

CREIGHTON, T.E. (Ed.) (1989) *Protein Structure – A Practical Approach*. IRL Press, Oxford. (Good chapters on estimating molecular weights of polypeptides by SDS polyacrylamide gel electrophoresis, predicting protein structure from amino acid sequences and the use of spectral methods for the study of protein conformation.)

FERSHT, A. (1985) *Enzyme Structure and Mechanism*, (2nd edn). Freeman, New York. (General text on enzymology particularly good on enzyme mechanisms.)

FINDLAY, J.B.C. and GEISOW, M.J. (Eds.) (1989) *Protein Sequencing – A Practical Approach*. IRL Press, Oxford. (Contains specialist practical details on all methods of protein sequencing.)

HARRIS, E.L.V. and ANGAL, S. (Eds.) (1989) *Protein Purification Methods - A Practical Approach*. IRL Press, Oxford. (A good, readable presentation on all aspects of the subject.)

JERVIS, L. and PIERPOINT, W.S. (1989) *Purification Technologies for Plant Proteins. Journal of Biotechnology*, **11**, 161–198. (Excellent review of the subject.)

PRICE, N.C. and STEVENS, L. (1989) *Fundamentals of Enzymology*. Oxford Scientific, Oxford. (A good, readable undergraduate text on enzymology.)

SRERE, P.A., JONES, M.E. and MATTHEWS, C.K. (1990) *Structural and Organisational Aspects of Metabolic Regulation*. Wiley–Liss, New York. (Good coverage of metabolic regulation and of enzymology *in vivo*.)

5

Radioisotope techniques

5.1 The nature of radioactivity

5.1.1. Atomic structure

An atom is composed of a positively charged nucleus that is surrounded by a cloud of negatively charged electrons. The mass of an atom is concentrated in the nucleus, even though it accounts for only a small fraction of the total size of the atom. Atomic nuclei are composed of two major particles, protons and neutrons. Protons are positively charged particles with a mass approximately 1850 times greater than that of an orbital electron. The number of orbital electrons in an atom must be equal to the number of protons present in the nucleus, since the atom as a whole is electrically neutral. This number is known as the atomic number (Z). Neutrons are uncharged particles with a mass approximately equal to that of a proton. The sum of protons and neutrons in a given nucleus is the mass number (A). Thus:

$$A = Z + N$$

where $N =$ the number of neutrons present.

Since the number of neutrons in a nucleus is not related to the atomic number, it does not affect the chemical properties of the atom. Atoms of a given element may not necessarily contain the same number of neutrons. Atoms of a given element with different mass numbers (i.e. different numbers of neutrons) are called

227

isotopes. Symbolically, a specific nuclear species is represented by a subscript number for the atomic number and a superscript number for the mass number, followed by the symbol of the element. For example:

$$^{12}_{6}C \quad ^{14}_{6}C \quad ^{16}_{8}C \quad ^{18}_{8}C$$

However, in practice it is more conventional just to cite the mass number (e.g. ^{14}C). The number of isotopes of a given element varies: there are 3 isotopes of hydrogen, ^{1}H, ^{2}H and ^{3}H, 7 of carbon ^{10}C to ^{16}C inclusive, and 20 or more of some of the elements of high atomic number.

5.1.2   Atomic stability and radiation

In general, the ratio of neutrons to protons in the nucleus will determine whether an isotope of a given element is stable enough to exist in nature. Stable isotopes for elements with low atomic numbers tend to have an equal number of neutrons and protons, whereas stability for elements of higher atomic numbers is associated with a neutron : proton ratio in excess of 1. Unstable isotopes, or radioisotopes as they are more commonly known, are often produced artificially, but many occur in nature. Radioisotopes emit particles and/or electromagnetic radiation as a result of changes in the composition of the atomic nucleus. These processes, which are known as radioactive decay, result, either directly or as a result of a decay series, in the production of a stable isotope.

5.1.3   Types of radioactive decay

There are several types of radioactive decay; only those most relevant to biochemists are considered below.

Decay by negatron emission

In this case a neutron is converted to a proton by the ejection of a negatively charged beta (β) particle called a negatron (β − ve):

Neutron → Proton + Negatron

To all intents and purposes a negatron is an electron, but the term negatron is preferred, although not always used, since it serves to emphasise the nuclear origin of the particle. As a result of negatron emission, the nucleus loses a neutron but gains a proton. The N/Z ratio therefore decreases while Z increases by 1 and A

remains constant. An isotope frequently used in biological work that decays by negatron emission is ^{14}C.

$$^{14}_{6}C \rightarrow \, ^{14}_{7}N + \beta - ve$$

Negatron emission is very important to biochemists because many of the commonly used radionuclides decay by this mechanism. Examples are ^3H and ^{14}C, which can be used to label any organic compound; ^{35}S used to label methionine, for example to study protein synthesis; and ^{32}P, a powerful tool in molecular biology when used as a nucleic acid label.

Decay by positron emission

Some isotopes decay by emitting positively charged β-particles referred to as positrons (β +ve). This occurs when a proton is converted to a neutron:

$$\text{Proton} \rightarrow \text{Neutron} + \text{Positron}$$

Positrons are extremely unstable and have only a transient existence. Once they have dissipated their energy they interact with electrons and are annihilated. The mass and energy of the two particles are converted to two γ-rays emitted at 180° to each other. This phenomenon is frequently described as back-to-back emission.

As a result of positron emission the nucleus loses a proton and gains a neutron, the N/Z ratio increases, Z decreases by 1 and A remains constant. An example of an isotope decaying by positron emission is ^{22}Na:

$$^{22}_{11}Na \rightarrow \, ^{22}_{10}Ne + \beta + ve$$

Positron emitters are detected by the same instruments used to detect γ-radiation. They are used in biological sciences to spectacular effect in brain scanning with the technique positron emission tomography (PET scanning) used to identify active and inactive areas of the brain.

Decay by alpha particle emission

Isotopes of elements with high atomic numbers frequently decay by emitting alpha (α) particles. An α-particle is a helium nucleus; it consists of two protons and two neutrons ($^4He^{2+}$). Emission of α-particles results in a considerable lightening of the nucleus, a decrease in atomic number of 2 and a decrease in the mass number of 4. Isotopes that decay by α-emission are not frequently encountered in biological work. Radium-226 (^{226}Ra) decays by α-emission to Radon-222 (^{222}Rn), which is itself radioactive. Thus begins a complex decay series, which culminates in the formation of ^{206}Pb:

$$^{226}_{88}\text{Ra} \rightarrow {}^{4}_{2}\text{He}^{2+} + {}^{222}_{86}\text{Rn} \rightarrow \rightarrow \rightarrow {}^{206}_{82}\text{Pb}$$

Alpha emitters are extremely toxic if ingested, due to the large mass and the ionising power of the atomic particle.

Electron capture

In this form of decay a proton captures an electron orbiting in the innermost K shell:

$$\text{Proton} + \text{Electron} \rightarrow \text{Neutron} + \text{X-ray}$$

The proton becomes a neutron and electromagnetic radiation (X-rays) is given out.
Example:

$$^{125}_{53}\text{I} \rightarrow {}^{125}_{52}\text{Te}$$

Decay by emission of gamma rays

In contrast to emission of α- and β-particles, γ-emission involves electromagnetic radiation similar to, but with a shorter wavelength than, X-rays. These γ-rays result from a transformation in the nucleus of an atom (in contrast to X-rays, which are emitted as a consequence of excitation involving the orbital electrons of an atom) and frequently accompany α- and β-particle emission. Emission of γ-radiation in itself leads to no change in atomic number or mass.

Gamma radiation has low ionising power but high penetration. For example, the γ-radiation from ^{60}Co will penetrate 15 cm of steel. The toxicity of γ radiation is similar to that of X-rays.
Example:

$$^{131}_{63}\text{I} \rightarrow {}^{131}_{64}\text{Xe} + \beta^{-} + \gamma$$

5.1.4   Radioactive decay energy

The usual unit used in expressing energy levels associated with radioactive decay is the electron volt. (One electron volt (eV) is the energy acquired by one electron in accelerating through a potential difference of 1 V and is equivalent to 1.6×10^{-19}J.) For the majority of isotopes, the term million or mega electron volts (MeV) is more applicable. Isotopes emitting α-particles are normally the most energetic, falling in the range 4.0–8.0 MeV, whereas β- and γ-emitters generally have decay energies of less than 3.0 MeV.

5.1.5   Rate of radioactive decay

Radioactive decay is a spontaneous process and it occurs at a definite rate characteristic of the source. This rate always follows an exponential law. Thus the number of atoms disintegrating at any time is proportional to the number of atoms of the isotope present (N) at that time (t). Expressed mathematically, the exponential curve (Fig. 5.1) gives the equation:

$$\frac{-\mathrm{d}N}{\mathrm{d}t} \propto N$$

or:

$$\frac{-\mathrm{d}N}{\mathrm{d}t} = \lambda N \tag{5.1}$$

where λ is the decay constant, a characteristic of a given isotope defined as the fraction of an isotope decaying in unit time (t^{-1}).

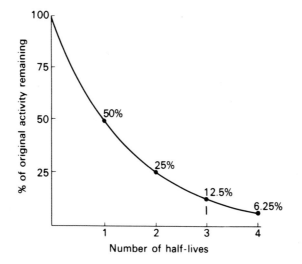

Fig. 5.1. Demonstration of the exponential nature of radioactive decay.

By integrating equation 5.1 it can be converted to a logarithmic form:

$$\ln\frac{N_t}{N_o} = -\lambda t \tag{5.2}$$

Table 5.1. *Half-lives of some isotopes used in biological studies*

Isotope	Half-life
^{3}H	12.26 years
^{14}C	5760 years
^{22}Na	2.58 years
^{32}P	14.20 days
^{35}S	87.20 days
^{42}K	12.40 h
^{45}Ca	165 days
^{59}Fe	45 days
^{125}I	60 days
^{131}I	8.05 days
^{135}I	9.7 h

where N_t is the number of radioactive atoms present at time t, and N_o is the number of radioactive atoms originally present. In practice it is more convenient to express the decay constant in terms of half-life ($t_{\frac{1}{2}}$). This is defined as the time taken for the activity to fall from any value to half that value (Fig. 5.1). If N_t in equation 5.2 is equal to one-half of N_o then t will equal the half-life of the isotope. Thus:

$$\ln \tfrac{1}{2} = -\lambda t_{\frac{1}{2}} \tag{5.3}$$

or:
$$2.303 \log_{10}(1/2) = -\lambda t_{\frac{1}{2}} \tag{5.4}$$

or:
$$t_{\frac{1}{2}} = 0.693/\lambda \tag{5.5}$$

The values of $t_{\frac{1}{2}}$ vary widely from over 10^{19} years for lead-204 (^{204}Pb) to 3×10^{-7} s for polonium-212 (^{212}Po). The half-lives of some isotopes frequently used in biological work are given in Table 5.1. Note that two important elements, oxygen and nitrogen, are missing from the table. This is because the half-lives of radioactive isotopes of these elements are too short for most biological studies (^{15}O has a $t_{\frac{1}{2}}$ of 2.03 min, whereas ^{13}N has a $t_{\frac{1}{2}}$ of 10.00 min).

5.1.6   Units of radioactivity

The Système International d'Unités (SI system) uses the Becquerel (Bq) as the unit of radioactivity. This is defined as one disintegration per second (1 d.p.s.). However, this unit has still not been widely adopted and a commonly used unit is still the Curie (Ci). This is defined as the quantity of radioactive material in which the number of nuclear disintegrations per second is the same as that in 1 g of radium, namely 3.7×10^{10} (or 37 GBq, see Table 5.7). For biological purposes this unit is too large and the microcurie (μCi) and millicurie (mCi) are used. It is important to realise that the Curie refers to the number of disintegrations actually occurring in a sample (i.e. d.p.s.) not to the disintegrations detected by the radiation counter, which will generally be only a proportion of the disintegrations occurring and are referred to as counts (i.e. c.p.s.).

Normally, in experiments with radioisotopes, a carrier of the stable isotope of the element is added. It therefore becomes necessary to express the amount of radioisotope present per unit mass. This is the specific activity. It may be expressed in a number of ways including disintegration rate (d.p.s. or d.p.m.), count rate (c.p.s. or c.p.m.) or Curies (mCi or μCi) per unit of mass of mixture (units of mass are normally either moles or grams). An alternative method of expressing specific activity, which is not very frequently used, is atom percentage excess. This is defined as the number of radioactive atoms per total of 100 atoms of the compound. For quick reference, a list of units and definitions frequently used in radiobiology is provided at the end of the chapter (Table 5.7).

5.1.7 Interaction of radioactivity with matter

Alpha particles

These particles have a very considerable energy (3–8 MeV) and all the particles from a given isotope have the same amount of energy. They react with matter in two ways. Firstly, they may cause excitation. In this process energy is transferred from the α-particle to orbital electrons of neighbouring atoms, these electrons being elevated to higher orbitals. The α-particle continues on its path with its energy reduced by a little more than the amount transferred to the orbital electron. The excited electron eventually falls back to its original orbital, emitting energy as photons of light in the visible or near visible range. Secondly, α-particles may cause ionisation of atoms in their path. When this occurs the target orbital electron is removed completely. Thus, the atom becomes ionised and forms an ion pair, consisting of a positively charged ion and an electron. Because of their size, slow movement and double positive charge, α-particles frequently collide with atoms in their path. Therefore they cause intense ionisation and excitation and their energy is rapidly dissipated. Thus, despite their initial high energy, α-particles are not very penetrating.

Negatrons

Compared with α-particles, negatrons are very small and rapidly moving particles that carry a single negative charge. They interact with matter to cause ionisation and excitation exactly as with α-particles. However, due to their speed and size, they are less likely than α-particles to interact with matter and therefore are less ionising and more penetrating than α-radiation. Another difference between α-particles and negatrons is that, whereas for a given α-emitter all the particles have the same energy, negatrons are emitted over a range of energy, i.e. negatron emitters have a characteristic energy spectrum (Fig. 5.5). The maximum energy level (E_{max}) varies from one isotope to another, ranging from 0.018 MeV for ^{3}H to 4.81 MeV for ^{38}Cl. The difference in E_{max} affects the penetration of the radiation: β-particles from ^{3}H can travel only a few millimetres in air, whereas those from ^{32}P can penetrate over 1 m of air. The reason for negatrons of a given isotope being emitted within an energy range was explained by W. Pauli in 1931, when he postulated that each radioactive event occurs with an energy equivalent to E_{max} but that the energy is shared between a negatron and a neutrino. The proportion of total energy taken by the negatron and the neutrino varies for each disintegration. Neutrinos have no charge and negligible mass and do not interact with matter.

Gamma rays

These rays are electromagnetic and therefore have no charge or mass. They rarely collide with neighbouring atoms and travel great distances before dissipating all their energy (i.e. they are highly penetrating). They interact with matter in many ways. The three most important ways lead to the production of secondary electrons, which in turn cause excitation and ionisation. In photoelectric absorption, low energy γ-rays interact with orbital electrons, transferring all their energy to the electron, which is then ejected as a photoelectron. The photoelectron subsequently behaves as a negatron. In contrast, Compton scattering, which is caused by medium energy γ-rays, results in only part of the energy being transferred to the target electron, which is ejected. The γ-ray is deflected and moves on with reduced energy. Again the ejected electron behaves as a negatron. Pair production results when very high energy γ-rays react with the nucleus of an atom and all the energy of the γ-ray is converted to a positron and a negatron.

5.2   Detection and measurement of radioactivity

There are three commonly used methods of detecting and quantifying radioactivity. These are based on the ionisation of gases, on the excitation of solids or solutions, and the ability of radioactivity to expose photographic emulsions, i.e. auto-radiography.

5.2.1 Methods based upon gas ionisation

The effect of voltage upon ionisation

As a charged particle passes through a gas, its electrostatic field dislodges orbital electrons from atoms sufficiently close to its path and causes ionisation (Fig. 5.2). The ability to induce ionisation decreases in the order:

$$\alpha > \beta > \gamma \qquad (10\,000:100:1)$$

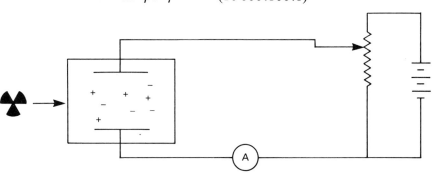

Fig. 5.2. Detection based on ionisation.

Accordingly, α- and β-particles may be detected by gas ionisation methods, but these methods are poor for detecting γ-radiation. If ionisation occurs between a pair of electrodes enclosed in a suitable chamber, a pulse (current) flows, the magnitude of which is related to the applied potential and the number of radiation particles entering the chamber (Fig. 5.3). The various 'regions' shown in Fig. 5.3 will now be considered.

In the ionisation chamber region of the curve, each radioactive particle produces only one ion-pair per collision. Hence the currents are low, and very sensitive measuring devices are necessary. This method is little used in quantitative work, but various types of electroscopes, which operate on this principle, are useful in demonstrating the properties of radioactivity. At a higher voltage level than that of the simple ionisation chambers, electrons resulting from ionisation move towards the anode much more rapidly; consequently they cause secondary ionisation of gas in the chamber, resulting in the production of secondary ionisation electrons, which cause further ionisation and so on. Hence from the original event a whole torrent of electrons reach the anode. This is the principle of gas amplification and is

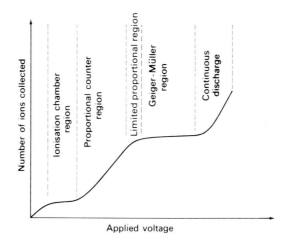

Fig. 5.3. Effect of voltage on pulse flow.

known as the Townsend avalanche effect, after its discoverer. As a consequence of this gas amplification, current flow is much greater. As can be seen in Fig. 5.3, in the proportional counter region the number of ion-pairs collected is directly proportional to the applied voltage until a certain voltage is reached when a plateau occurs. Before the plateau is reached there is a region known as the limited proportional region, which is not often used in detection and quantification of radioactivity and hence will not be discussed.

The main drawback of counters that are manufactured to operate in the proportional region is that they require a very stable voltage supply because small fluctuations in voltage result in significant changes in amplification. Proportional counters are particularly useful for detection and quantification of α-emitting isotopes, but it should be noted that relatively few such isotopes are used in biological work.

In the Geiger–Müller region all radiation particles, including weak β-particles, induce complete ionisation of the gas in the chamber. Thus the size of the current is no longer dependent on the number of primary ions produced. Since maximal gas amplification is realised in this region, the size of the output pulse from the detector will remain the same over a considerable voltage range (the so-called Geiger–Müller plateau). The number of times this pulse is produced is measured rather than its size. Therefore it is not possible to discriminate between different isotopes using this type of counter.

Since it takes a finite time for the ion-pairs to travel to their respective electrodes, other ionising particles entering the tube during this time fail to produce ionisation and hence are not detected, thereby reducing the counting efficiency. This is referred to as the dead time of the tube and is normally 100–200 μs. When the ions reach the electrode they are neutralised. Inevitably some escape and produce their own ionisation avalanche. Thus, if unchecked, a Geiger–Müller tube would tend to give a continuous discharge. To overcome this, the tube is quenched by the addition of a suitable gas, which reduces the energy of the ions. Common quenching agents are ethanol, ethyl formate and the halogens.

Instrumentation

Counters based on gas ionisation used to be the main method employed in the quantification of radioisotopes in biological samples. Currently, scintillation

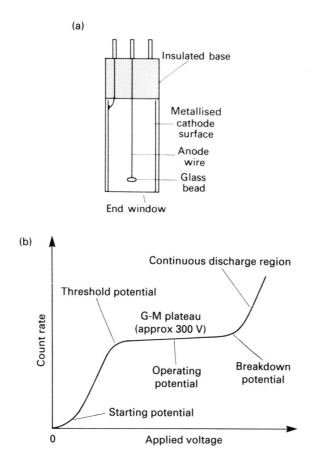

Fig. 5.4. (a) The Geiger–Müller (G-M) tube and (b) the effect of applied voltage on count rate.

counting (Section 5.2.2) has virtually taken over. However, all laboratories use small hand-held radioactivity monitors based on gas ionisation; the end-window design being the most popular type (Fig. 5.4). These counters have a thin end window made from aluminium and can detect β-radiation from high energy (^{32}P) and weak emitters (^{35}C, ^{14}C), but are incapable of detecting ^{3}H because the radiation cannot penetrate the end window. For the same reason they are not very efficient detectors of α-radiation.

End-window ionisation counters are used for routine monitoring of the radioactive laboratory to check for contamination. They are also useful in experimental situations where the presence or absence of radioactivity needs to be known rather than the absolute quantity, for example quick screening of radioactive gels prior to autoradiography or checking of chromatographic fractions for labelled components.

The inability of end-window counters to detect weak β-emitters presents a problem in biosciences because ^{3}H is a very commonly used radioisotope. The problem can be overcome by using a so-called windowless counter where a gas flow is used. These instruments are rather cumbersome and need to be carried around on an object that resembles a golf trolley. They are useful for mass screening of premises for ^{3}H contamination but are rarely used as routine. Most laboratories monitor for ^{3}H by doing a wipe test regularly, i.e. using wet tissues or cotton wool to take swabs for scintillation counting.

5.2.2 Methods based upon excitation

As outlined in Section 5.1.7, radioactive isotopes interact with matter in two ways, causing ionisation, which forms the basis of Geiger–Müller counting, and excitation. The latter effect leads the excited compound (known as the fluor) to emit photons of light. This fluorescence can be detected and quantified. The process is known as scintillation and when the light is detected by a photomultiplier, forms the basis of scintillation counting. The electric pulse that results from the conversion of light energy to electrical energy in the photomultiplier is directly proportional to the energy of the original radioactive event. This is a considerable asset of scintillation counting, since it means that two, or even more, isotopes can be separately detected and measured in the same sample, provided they have sufficiently different emission energy spectra (see below). The mode of action of a photomultiplier is shown in Fig. 5.5.

In summary, scintillation counting provides information of two kinds:

    (i) quantitative, the number of scintillations is proportional to the rate of decay of the sample, i.e. the amount of radioactivity;

    (ii) qualitative, the intensity of light given out and therefore signal from the photomultiplier is proportional to the energy of radiation.

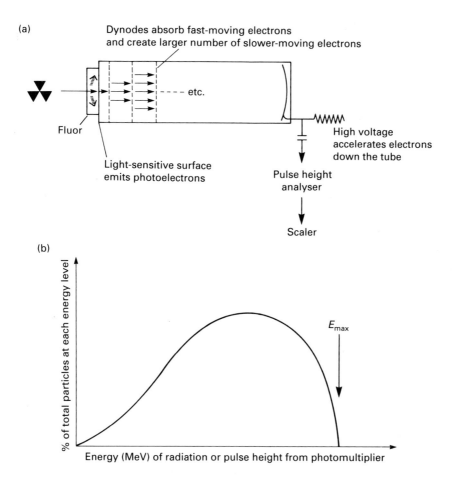

(a)

Dynodes absorb fast-moving electrons
and create larger number of slower-moving electrons

etc.

Fluor

Light-sensitive surface
emits photoelectrons

High voltage
accelerates electrons
down the tube

Pulse height
analyser

Scaler

(b)

% of total particles at each energy level

E_{max}

Energy (MeV) of radiation or pulse height from photomultiplier

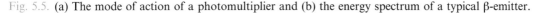

Fig. 5.5. (a) The mode of action of a photomultiplier and (b) the energy spectrum of a typical β-emitter.

Types of scintillation counting

There are two types of scintillation counting, which are illustrated diagrammatically in Fig. 5.6. In solid scintillation counting the sample is placed adjacent to a crystal of fluorescent material. The crystal that is normally used for γ-isotopes is sodium iodide, whereas for α-emitters zinc sulphide crystals are preferred and for β-emitters organic scintillators such as anthracene are used. The crystals themselves are placed near to a photomultiplier, which in turn is connected to a high voltage supply and a scaler (Fig. 5.6a). Solid scintillation counting is particularly useful for γ-emitting isotopes. This is because, as explained in Section 5.1.7, γ-rays are electromagnetic radiation and collide only rarely with neighbouring atoms to cause ionisation or excitation. Clearly, in a crystal the atoms are densely packed, making collisions more likely. Conversely, solid scintillation counting is generally unsuitable

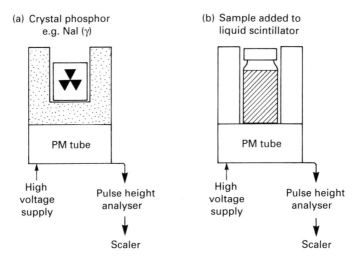

(a) Crystal phosphor
e.g. NaI (γ)

(b) Sample added to
liquid scintillator

PM tube

PM tube

High voltage supply

Pulse height analyser

High voltage supply

Pulse height analyser

Scaler

Scaler

Fig. 5.6. Diagrammatic illustration of solid (a) and liquid (b) scintillation counting methods.

for weak β-emitting isotopes such as ^{3}H and ^{14}C, because even the highest energy negatrons emitted by these isotopes would have hardly sufficient energy to penetrate the walls of the counting vials in which the samples are placed for counting. As many of the isotopes used in radio-immunoassay (Section 2.5) are γ-emitting isotopes, solid scintillation counting is frequently used in biological work.

In liquid scintillation counting (Fig. 5.6b), the sample is mixed with a scintillation cocktail containing a solvent and one or more fluors. This method is particularly useful in quantifying weak β-emitters such as ^{3}H, ^{14}C and ^{35}S, which are frequently used in biological work. For these isotopes, liquid scintillation counting is the usual counting method. Thus the remainder of this section will place particular emphasis on this technique, though it should be pointed out that most of what follows applies equally to solid scintillation counting used in the quantification of γ-emitters.

Energy transfer in liquid scintillation counting

A small number of organic solvents fluoresce when bombarded with radioactivity. The light emitted is of very short wavelength (Fig. 5.7) and is not efficiently detected by most photomultipliers. However, if a compound is dissolved that can accept the energy from the solvent and itself fluoresce at a longer wavelength, then the light can be more efficiently detected. Such a compound is known as a primary fluor and the most frequently used example is 2,5-diphenyloxazole (PPO). Unfortunately the light emitted by PPO is not always detected with very high efficiency (depending on the photomultiplier detector) but this can be overcome by

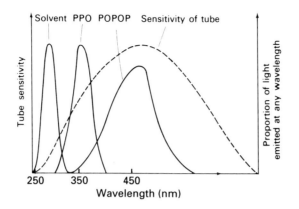

Fig. 5.7. Emission spectra of various fluors in relation to sensitivity of phototubes.

including a secondary fluor or wavelength shifter such as 1,4-bis(5-phenyloxazol-2-yl) benzene (POPOP). Thus, the energy transfer process becomes:

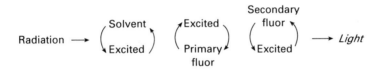

The question obviously arises as to why a primary fluor *and* a secondary fluor are necessary when it is the latter that emits light at the best wavelength for detection. The answer is simply that the solvent cannot transfer its energy directly to the secondary fluor.

PPO and POPOP were among the original fluors used in liquid scintillation counting and remain a favourite choice. However compounds such as 2-(4'-*t*-butylphenyl)-5-(4"-biphenylyl)-1,3,4-oxadiazole (BUTYL-PBD) is a better primary fluor but is quite expensive and is affected by extremes of pH.

Most laboratories now buy their scintillation cocktails already prepared and there are many different makes and recipes on the market. Competition and an increasing awareness of health and safety means that scintillation cocktails are gradually becoming less toxic and have a lower fire hazard. A final point: some cocktails are designed for aqueous samples and others for organic samples; it is important that the appropriate formulation is used.

Advantages of scintillation counting

The very fact that scintillation counting is widely used in biological work indicates that it has several advantages over gas ionisation counting. These advantages are listed below.

(i) The rapidity of fluorescence decay (10^{-9} s), which, when compared to dead time in a Geiger–Müller tube (10^{-4} s), means much higher count rates are possible.

(ii) Much higher counting efficiencies particularly for low energy β-emitters; over 50% efficiency is routine in scintillation counting and efficiency can rise to over 90% for high energy emitters. This is partly due to the fact that the negatrons do not have to travel through air or pass through an end window of a Geiger–Müller tube (thereby dissipating much of the energy before causing ionisation) but interact directly with the fluor; energy loss before the event that is counted is therefore minimal.

(iii) The ability to accommodate samples of any type, including liquids, solids, suspensions and gels.

(iv) The general ease of sample preparation (see below).

(v) The ability to count separately different isotopes in the same sample, which means dual labelling experiments can be carried out (see below).

(vi) Scintillation counters are highly automated, hundreds of samples can be counted automatically and built-in computer facilities carry out many forms of data analysis, such as efficiency correction, graph plotting, radioimmunoassay calculations, etc.

Disadvantages of scintillation counting

It would not be reasonable, having outlined some of the advantages of scintillation counting, to disregard the disadvantages of the method. Fortunately, however, most of the inherent disadvantages have been overcome by improvement in instrument design. These disadvantages include the following.

(i) The cost per sample of scintillation counting is not insignificant; however, other factors including versatility, sensitivity, ease and accuracy outweigh this factor for most applications.

(ii) At the high voltages applied to the photomultiplier, electronic events occur in the system that are independent of radioactivity but contribute to a high background count. This is referred to as photomultiplier noise and can be partially reduced by cooling the photomultipliers. Since temperature affects counting efficiency, cooling also presents a controlled temperature for counting, which may be useful. Low noise photo-multipliers, however, have been designed to provide greater temperature stability in ambient temperature systems. Also the use of a pulse height analyser can be set so as to reject, electronically, most of the noise pulses that are of low energy (the threshold or gate setting). The disadvantage here is that this also rejects the low energy pulses resulting from low energy radioactivity (e.g. ^{3}H). Another method of reducing noise, which is incorporated into most scintillation counters, is to use coincidence counting. In this system two photomultipliers are used. These are set in coincidence such that only when a pulse is generated in both tubes at the same time is it allowed to pass to the scaler. The chances of this happening for a pulse generated by a radioactive event is very high compared to the chances of a noise event occurring in both photomultipliers during the so-called resolution time of the system, which is generally of the order of 20 ns. In general, this system reduces photomultiplier noise to a very low level.

(iii) The greatest disadvantage of scintillation counting is quenching. This occurs when the energy transfer process described earlier suffers interference. Correcting for this quench contributes significantly to the cost of scintillation counting. Quenching can be any one of three kinds.

(a) Optical quenching occurs if inappropriate or dirty scintillation vials are used. These will absorb some of the light being emitted, before it reaches the photomultiplier.

(b) Colour quenching occurs if the sample is coloured and results in light emitted being absorbed within the scintillation cocktail before it leaves the sample vial. When colour quenching is known to be a major problem, it can be reduced, as outlined later.

(c) Chemical quenching, which occurs when anything in the sample interferes with the transfer of energy from the solvent to the primary fluor or from the primary fluor to the secondary fluor, is the most difficult form of quenching to accommodate. In a series of homogeneous samples (e.g. $^{14}CO_2$ released during metabolism of [^{14}C]glucose and trapped in alkali, which is then added to the scintillation cocktail for counting), chemical quenching may not vary greatly from sample to sample. In these cases relative counting using sample counts per minute can be compared directly. However, in the majority of biological experiments using radioisotopes, such homogeneity of samples is unlikely and it is not sufficiently accurate to use relative counting (i.e. counts per minute). Instead, an appropriate method of standardisation must be used. This requires the determination of the counting efficiency of each sample and the conversion of counts per minute to absolute counts (i.e. disintegrations per minute), as described later. It should be noted that quenching is not such a great problem in solid (external) scintillation counting.

(iv) Chemiluminescence can also cause problems during liquid scintillation counting. It results from chemical reactions between components of the samples to be counted and the scintillation cocktail, and produces light emission unrelated to excitation of the solvent and fluor system by radioactivity. These light emissions are generally low energy events and are rejected by the threshold setting of the photomultiplier in the same way as is photomultiplier noise. Chemiluminescence, when it is a problem, can usually be overcome by storing samples for some time before counting, to permit the chemiluminescence to decay. Many contemporaneous instruments are able to detect chemiluminescence and subtract it or flag it on the print-out.

(v) Phospholuminescence results from components of the sample, including the vial itself, absorbing light and re-emitting it. Unlike chemi-luminescence, which is a once-only effect, phospholuminescence will occur on each exposure of a sample to light. Samples that are pigmented are most likely to phosphoresce. If this is a problem, samples should be adapted to dark prior to counting and the sample holder should be kept closed throughout the counting process.

Despite all the complications described above, scintillation counters are universal in biosciences departments. This is because the instruments have automated systems for calculating counting efficiency; in other words the instruments do all the hard work!

Using scintillation counting for dual-labelled samples

A feature of the scintillation process is that the size of electric pulse produced by the conversion of light energy in the photomultiplier is related directly to the energy of the original radioactive event. Because different β-emitting isotopes have different energy spectra, it is possible to quantify two isotopes separately in a single sample, provided their energy spectra are sufficiently different. Examples of pairs of isotopes that have sufficiently different energy spectra are ^{3}H and ^{14}C, ^{3}H and ^{35}S, ^{3}H and ^{32}P, ^{14}C and ^{32}P, ^{14}C and ^{32}P, ^{35}S and ^{32}P. The principle of the method is illustrated in Fig. 5.8, where it can be seen that the spectra of two isotopes (S and T) overlap only slightly. By setting a pulse height analyser to reject all pulses of an energy below X (threshold X) and to reject all pulses of an energy above Y (window Y) and also to reject below a threshold of A and a window of B, it is possible to separate the two isotopes completely. A pulse height analyser set with a threshold and window for a particular isotope is known as a channel (e.g. a ^{3}H channel).

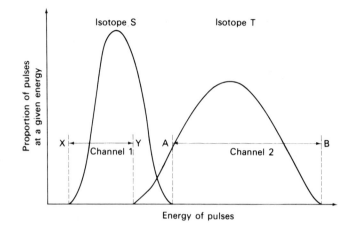

Fig. 5.8. Diagram to illustrate the principle of counting dual-labelled samples.

Most modern counters operate with a so-called multi-channel analyser. These are based on an analogue-to-digital converter; electronic signals from the photo-multiplier are converted to digital signals stored in a computer. Thus, the entire energy spectrum is analysed simultaneously. This greatly facilitates multi-isotope counting and in particular allows the effect of quenching on dual-label counting to be assessed adequately.

Dual-label counting has proved to be useful in many aspects of molecular biology (e.g. nucleic acid hybridisation and transcription), metabolism (e.g. steroid synthesis) and drug development.

Determination of counting efficiency

As outlined above, a major problem encountered in scintillation counting is that of quenching, which makes it necessary to determine the counting efficiency of some, if not all, of the samples in a particular experiment. This can be done by one of several methods of standardisation, all of which apply to both solid and liquid scintillation counting, though again in this section emphasis is placed on the latter method.

Internal standardisation. The sample is counted (and gives a reading of say A c.p.m.), removed from the counter and a small amount of standard material of known disintegrations per minute (B d.p.m.) is added. The sample is then recounted (C c.p.m.) and the counting efficiency of the sample calculated:

$$\text{Counting efficiency} = [100 \ (C - A)/B]\%$$

It is obviously necessary in this method to use an internal standard (the spike) that contains the same isotope as the one being counted and also to ensure that the standard itself does not act as a quenching agent. Suitable ^{14}C-labelled standards include [^{14}C]toluene, [^{14}C]hexadecane, [^{3}H]benzoic acid and $^{3}H_2O$ (benzoic acid and water are themselves quenching agents and must be used in only very small amounts). Internal standardisation is simple and reliable and corrects adequately for all types of quenching. Carefully carried out, it is the most accurate way of correcting for quenching. On the other hand it demands very accurate pipetting when the standard is added, and it is time consuming because each sample must be counted twice. It also means that the sample cannot be recounted in the event of error because it will be contaminated with the standard. Moreover, time elapses between the first and second count and changes in sample quenching characteristics can also occur, which can lead to considerable inaccuracies. However, it is the means by which the following two methods are calibrated.

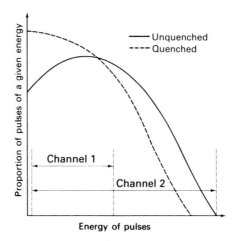

Fig. 5.9. The effect of quenching of a β-energy spectrum.

Channels ratio. When a sample in a scintillation counter is quenched, the scintillation process is less efficient: less light is produced for a given quantum energy of radiation. Thus the energy spectrum for a quenched sample appears to be lower than for an unquenched sample (Fig. 5.9). The higher the degree of quenching, the more pronounced is the resulting decrease in the spectrum. This fact is made use of in the channels ratio method for determining counter efficiency. This method involves the preparation of a calibration curve based on counting in two channels that cover different, but overlapping, parts of the spectrum. As a sample is quenched, and the spectrum shifts to gradually lower apparent energies, the ratio of counts in each channel will vary. To prepare the standard curve, a set of quenched standards is counted: the absolute amount of radioactivity is known and therefore the efficiency of counting in each channel can easily be determined.

The efficiency is then plotted against channels ratio to form the standard curve (Fig. 5.10). Typical data for a set of ^{14}C quenched standards is given in Table 5.2. It is important to realise that a standard curve applies to only one set of

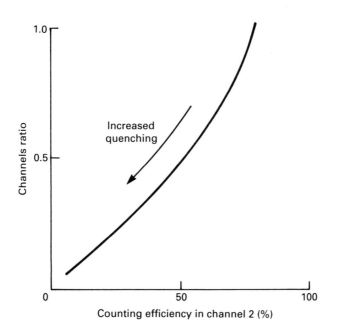

Fig. 5.10. Channels ratio quench correction curve.

circumstances – one radioisotope, counter and scintillation fluid.

Once the standard curve has been prepared, the efficiency of counting experimental samples can be determined. Samples are counted in the same two channels, the ratio is calculated, put into the graph and the efficiency read. In practice all the data can be stored in the counter's computer and corrected values printed automatically.

Multichannel scintillation counters operate on the same principle but the whole shape and position of the spectrum is analysed. This is given a digital parameter that relates to counting efficiency. Manufacturers have developed their own titles for such parameters, e.g. LKB Instruments' Spectral Quench Parameter of Isotope (or SQP(I)) or Packard's Spectral Index Sample.

Table 5.2. *Radioactivity recorded with gradually increasing quench in two channels of a scintillation counter*

Sample	c.p.m.		Ratio	Counting efficiency
	Channel 1	Channel 2	Ch1: Ch2	in channel 2 (%)
^{14}C standard (203 600 d.p.m.) unquenched	171 930	184 250	0.93	90.5
^{14}C standard (203 600 d.p.m.) with increasing quench	146 610	168 840	0.87	82.9
	94 240	135 090	0.70	66.3
	52 260	102 030	0.51	50.1
	16 030	58 320	0.27	28.6
	5920	34 740	0.17	17.1
	2060	20 270	0.10	9.9
	1130	13 260	0.08	6.5

These systems have greater precision than the two-channel approach, as the whole of the spectrum is used for analysis. The channels ratio method is suitable for all types and even high degrees of quenching. Furthermore, counting in more than one channel is simultaneous and, therefore, this method is less time consuming than either internal or external standardisation. It is also, in practice, an acceptably accurate method for determining counting efficiency, provided care is taken in the preparation of the calibration curve. However, it is notoriously inaccurate at low count rates, because the error on the counts per minute is high and there will be a larger error on the channels ratio because it is calculated from two values of the counts per minute. It is also inaccurate for very highly quenched samples. For these reasons the method that follows is most frequently the procedure of choice.

External standardisation. Instruments have a γ-emitting external standard built into the counter. Under the control of the counter each sample to be counted is exposed to this external source, which is automatically shifted from a lead shield to the counting chamber. The γ-radiation penetrates the vial and excites the scintillation fluid. The resulting spectrum is unique to the source and is significantly different from that produced by the sample in the vial. The γ-source used (e.g.

^{137}Cs, ^{133}Ba or ^{226}Ra) varies according to the make of instrument. The spectrum obtained by ^{226}Ra is shown in Fig. 5.11.

Quenching agents present in the scintillation fluid will significantly affect the

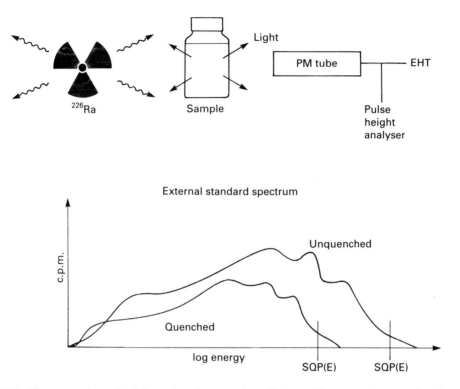

Fig. 5.11. The external standard for estimating counting efficiency. The external source irradiates the sample. The counter analyses the spectrum, which shifts to lower energies if the sample is quenched. The SQP(E), i.e. standard quench parameter (external), expressed without units, is derived from the energy axis and relates to the extent of quench. The greater the quench in the sample, the lower the SQP(E) and the lower the counting efficiency (see Table 5.3).

spectrum obtained. The instrument analyses this spectrum and assigns a quench parameter to it. The precise method used depends on the make of counter; LKB Instruments refer to a standard quench parameter based on a point on the energy axis (Fig. 5.11). Other manufacturers use slightly different approaches but the principle is the same: the spectrum for the external standard varies according to the degree of quench in the vial and, therefore, the efficiency of counting of the internal experimental sample.

As for the channels ratio method, a standard curve is required, i.e. a range of quenched standards is counted and the external standard spectrum analysed in each case. The resulting data (Table 5.3) are used to prepare a standard curve that is held in the instrument's computer. Unknown samples are then counted in the

Table 5.3. *Recorded radioactivity from a ^{14}C standard sample with increasing quench detected by an external standard*

Sample	c.p.m.	External quench parameter[a]	Counting efficiency (%)
^{14}C standard (203 600 d.p.m.)			
unquenched	194 930	810	95.7
^{14}C standard (203 600 d.p.m.)			
with increasing quench	190 411	422	93.5
	181 171	207	89.0
	167 731	126	82.4
	145 879	76	71.6
	126 913	55	62.3
	108 641	42	53.3
	96 103	37	47.2

[a] e.g. SQP(E), see Fig 5.11.

same way, the efficiency read from the standard curve and the sample counts corrected.

The external standard approach is now routine in most laboratories, the main advantage over the channels ratio method being that it is suited to samples with low count rates. However, it is not without disadvantages: a standard curve is required for each set of circumstances (as with the channels ratio) and the user can be lulled into a false sense of security. The system is so highly automated that it is easy to lose sight of the basic principles and the method is not always appropriate. A case in point is the counting of ^{3}H precipitated onto filters counted in scintillation fluid. The external standard method will calculate the degree of quench in the fluid (which will probably be very low) but will not take into account the poor penetration of ^{3}H β-particles from the filter into the scintillation fluid: artificially high efficiencies will be recorded.

In all cases where an automated procedure for calculating counting efficiencies is employed it is prudent to count a few prepared samples in which the true amount of radioactivity is known.

Sample preparation

It is impossible here to give details of all aspects of sample preparation for scintillation counting. However, major considerations are outlined below and the reader is referred to books cited in Section 5.8 for further details.

Sample vials. In solid scintillation counting, sample preparation is easy and only involves transferring the sample to a glass or plastic vial (or tube) compatible with the counter. In liquid scintillation counting, sample preparation is more complex and starts with a decision on the type of sample vial to be used. These may be glass, low potassium glass (with low levels of ^{40}K that reduces background count) or polyethylene. The last of these are cheaper but are not suitable for cleaning and reuse, whereas glass vials can be reused many times provided they are thoroughly cleaned. Polyethylene vials give better light transfer and result in slightly higher counting efficiencies, but are inclined to exhibit more phosphorescence than do glass vials. The recent trend is towards mini-vials, which use far smaller volumes of expensive scintillation cocktails. Modern counters are able to accept many types of vial; the smallest vial possible should be used (within the obvious constraints of sample volume) to save costs and in consideration of environmental issues, as scintillation fluids are toxic. Some counters are designed to accept very small samples in special polythene bags split into an array of many compartments; these are particularly useful, for example, to the pharmaceutical industry with laboratories that do large numbers of receptor binding assays.

Scintillation cocktails. Toluene-based cocktails are the most efficient, but will not accept aqueous samples, because toluene and water are immiscible and massive quenching results. Cocktails based on 1,4-dioxane and naphthalene that can accommodate up to 20% (v/v) water can be used, but they have largely been phased out due to toxicity. Emulsifier-based cocktails are the most frequently used for counting aqueous samples. They contain an emulsifier such as Triton X-100 and can accept up to 50% water (v/v); however, phase transitions occur from single phase to two-phase or gel, as the water content increases. Accurate counting cannot be done if the samples are in the two-phase state. Many ready-made cocktails are on the market and are sold with precise instructions regarding sample condition.

Volume of cocktail. It should be noted that the efficiency of scintillation counting varies with sample volume, though this is less of a problem in modern counters. Nevertheless, care should be taken that sample vials in a given series of counts contain the same volume of sample and that all instrument calibration is done using the same sample volume as for experimental samples.

Overcoming major colour quenching. If colour quenching is a problem it is possible to bleach samples before counting. Care should be taken, however, since bleaching agents such as hydrogen peroxide can give rise to chemiluminescence in some scintillation cocktails.

Tissue solubilisers. Solid samples, such as plant and animal tissues, may be best counted after solubilisation by quaternary amines such as NCS solubiliser or Soluene, etc. Not surprisingly these solutions are highly toxic and great care is required. The sample is added to the counting vial containing a small amount of solubiliser and digestion is allowed to proceed. When digestion is complete, scintillation cocktail is added and the sample counted. Again, chemiluminescence can be a problem with tissue solubilisers.

Combustion methods. A suitable alternative to bleaching of coloured samples or digestion of tissues is the use of combustion techniques. Here samples are combusted in an atmosphere of oxygen, usually in a commercially available combustion apparatus. Thus, samples containing ^{14}C would be combusted to $^{14}CO_2$, which is collected in a trapping agent such as sodium hydroxide and then counted; ^{3}H-containing samples are converted to $^{3}H_2O$ for counting.

As indicated earlier, only important considerations in sample preparation are discussed above and details are not given. However, it is worthy of comment that almost any type of radioactive sample containing β-emitting isotopes can be prepared for counting in a liquid scintillation counter by one method or another, including cuttings from paper chromatographs or membrane filters, again illustrating the versatility and importance of this technique of quantifying radioactivity.

Čerenkov counting

The Čerenkov effect occurs when a particle passes through a substance with a speed higher than that of light passing through the same substance. If a β-emitter has a decay energy in excess of 0.5 MeV, then this causes water to emit a bluish white light usually referred to as Čerenkov light. It is possible to detect this light using a typical liquid scintillation counter.

Since there is no requirement for organic solvents and fluors, this technique is relatively cheap, sample preparation is very easy, and there is no problem of chemical quenching. Table 5.4 lists some isotopes that are suitable for this detection method. Most work has been done on ^{32}P, which has 80% of its energy spectrum above the Čerenkov threshold and which can be detected at around 40% efficiency. It may be noted from Table 5.4 that, as the proportion of the energy spectrum

Table 5.4. *Some isotopes suitable for Čerenkov counting*

Radioisotope	E_{max} (MeV)	% of spectrum above 0.5 MeV	Counting efficiency (%)
^{22}Na	1.39	60	30
^{32}P	1.71	80	40
^{36}Cl	0.71	30	10
^{42}K	3.5	90	80

above 0.5 MeV increases, so too does the detection efficiency.

5.2.3 Methods based upon exposure of photographic emulsions

Ionising radiation acts upon a photographic emulsion to produce a latent image much as does visible light. For a photograph, a radiation source, an object to be imaged and photographic emulsion are required. For an autoradiograph, a radiation source (i.e. radioactivity) emanating from within the material to be imaged (the object) is required, along with a sensitive emulsion. The emulsion consists of a large number of silver halide crystals embedded in a solid phase such as gelatin. As energy from the radioactive material is dissipated in the emulsion, the silver halide becomes negatively charged and is reduced to metallic silver, thus forming a particulate latent image. Photographic developers are designed to show these silver grains as a blackening of the film and fixers to remove any remaining silver halide. Thus, a permanent image of the location of the original radioactive event remains.

This process, which is known as autoradiography, is very sensitive and has been used in a wide variety of biological experiments. These usually involve a requirement to locate the distribution of radioactivity in biological specimens of different types. For instance, the sites of localisation of a radiolabelled drug throughout the body of an experimental animal can be determined by placing whole body sections of the animal in close contact with a sensitive emulsion such as an X-ray plate. After a period of exposure, the plate, upon development, will show an image of the section in tissues and organs in which radioactivity was present (Fig. 5.12). Similarly, radioactive metabolites isolated and separated by chromatographic or electrophoretic techniques during metabolic studies can be located on the chromatograph or electrophoretograph and the radioactive spots can subsequently be recovered for counting and identification (counting may be carried out on the original chromatograph by using a chromatograph scanner, the design of which is

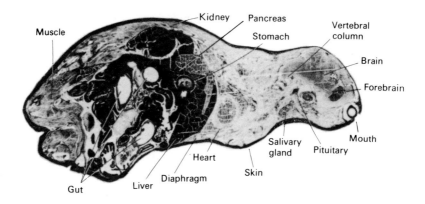

Fig. 5.12. Whole-body autoradiograph of a mouse treated with [L-¹⁴C]dopa. The dark areas indicate the presence of radioactive isotope and show high concentrations in the liver, pancreas, kidney, skin and forebrain.

generally based on Geiger–Müller tubes, or by elution from the paper, plate or gel for counting by liquid scintillation counting).

The techniques of autoradiography have become more important with recent developments in molecular biology (Chapter 3). Consequently more detail is given below on some important aspects of the technique.

Suitable isotopes

In general weak, β-emitting isotopes (e.g. ^{3}H and ^{14}C and ^{35}S) are most suitable for autoradiography, particularly for cell and tissue localisation experiments. This is because, as a result of the low energy of the negatrons, the ionising track of the isotope will be short and a discrete image will result. This is particularly important when radioactivity associated with subcellular organelles is being located. For this, ^{3}H is the best radioisotope since its energy will all be quickly dissipated within the emulsion. Electron microscopy can then be used to locate the image in the developed film. For location within whole organisms or tissues, either ^{14}C or ^{3}H is suitable; whilst more energetic isotopes (e.g. ^{32}P) being less suitable because their higher energy negatrons produce much longer track lengths and result in less discrete images that are not sufficiently discriminatory for microscopic location. Conversely, for location of, for example, DNA bands in an electrophoretic gel, ^{32}P is useful. In this case low energy ^{3}H negatrons would largely dissipate their energy within the gel (and in the wrapping around the gel, which is usually necessary to prevent the gel sticking to the emulsion), thereby reducing sensitivity to a low level. However, the more energetic ^{32}P negatrons will leave the gel and produce a strong image. If very thin gels are prepared, then ^{35}S or ^{14}C can be detected with high resolution, e.g. in DNA sequencing gels where ^{35}S is used as the label.

Choice of emulsion and film

A variety of suitable emulsions is available with different packing densities of the silver halide crystals. Care must be taken to choose an emulsion suitable for the purposes of the experiment, since the sensitivity of the emulsion will affect the resolution obtained. Manufacturers' literature should be consulted and their advice sought if one is in any doubt. X-ray film is generally suitable for macroscopic samples such as whole-body sections of small mammals, chromatographs or electrophoretographs. When light or electron microscopic detection of the location of the image in the emulsion is required (cellular and subcellular localisation of radioactivity), very sensitive films are necessary, as is a very close apposition of sample and film. In these cases a stripping film technique can be used in which the film is supplied attached to a support. It is stripped from this and applied directly to the sample. Alternatively, liquid emulsions are prepared by melting strips of emulsion by heating them to around 60 °C. Then either the emulsion is poured on to the sample or the sample attached to a support is dipped into the emulsion. The emulsion is then allowed to set before being dried. Such a method is often referred to as a dipping-film method and is preferred when very thin films are required.

Background

Accidental exposure to light, chemicals in the sample, natural background radioactivity (particularly ^{40}K in glass) and even pressure applied during handling and storage of film will cause a background fog (i.e. latent image) on the developed film. This can be problematic, particularly in high resolution work (e.g. involving microscopy) and care must be taken at all times to minimise its effect. Background will always increase during exposure time, which should therefore always be kept to a minimum.

Time of exposure and film processing

The time of exposure depends upon the isotope, sample type, level of activity, film type and purpose of the experiment. The same applies to the processing of the film in order to display the image. Generally the process must be adapted to a given purpose, and a great deal of trial and error is often involved in arriving at the most suitable procedures.

Direct autoradiography

In direct autoradiography, the X-ray film or emulsion is placed as close as possible to the sample and exposed at any convenient temperature. Quantitative images are produced until saturation is reached. The approach provides high resolution but limited sensitivity: isotopes of energy equal to, or higher than, ^{14}C ($E_{max} = 0.156$ MeV) are required.

Fluorography

Many of the currently popular methods in molecular biology involve separation of macromolecules or fractions of macromolecules by gel electrophoresis (Section 9.3). The separated macromolecules or fractions form bands in the electrophoretograph that must be located. This is often achieved by radiolabelling the macromolecules with ^{3}H or ^{14}C and subjecting the gel to autoradiography. Because these are weak β-emitters, much of their energy is lost in the gel and long exposure times are necessary even when very high specific activity sources are used. However, if a fluor (e.g. PPO or sodium silicate) is infiltrated into the gel, and the gel dried and then placed in contact with a pre-flashed film (see below), sensitivity can be increased by several orders of magnitude. This is because the negatrons emitted from the isotope will cause the fluor to become excited and emit light, which will react with the film. Thus, use is made of *both* the ionising and the exciting effects of radioactivity in fluorography.

Intensifying screens

When ^{32}P-labelled or γ-isotope-labelled samples (e.g. [^{32}P]DNA or ^{125}I-labelled protein fractions in gels) are to be located, the opposite problem to that presented by low energy isotopes prevails. These much more penetrating particles and rays cause little reaction with the film as they penetrate right through it, producing a poor image. The image can be greatly improved by placing, on the other side of the film from the sample, a thick intensifying screen consisting of a solid phosphor. Negatrons penetrating the film cause the phosphor to fluoresce and emit light, which superimposes its image on the film. There is, therefore, an increase in sensitivity but a parallel reduction in resolution due to the spread of light emanating from the screen.

Low temperature exposure

If the energy of ionising radiation is converted to light (i.e. with fluorography or intensifying screens) the kinetics of the film's response are affected. The light is of low intensity and a back reaction occurs that cancels the forming latent image.

Exposure at low temperature (− 70 °C) slows this back reaction and will therefore provide higher sensitivity. There is no point in doing direct autoradiography at low temperature as the kinetics of the film response are different. There is nothing to be gained by exposing pre-flashed film (see below) at low temperature.

Pre-flashing

As described above, the response of a photographic emulsion to radiation is not linear and usually involves a slow initial phase (lag) followed by a linear phase. Sensitivity of films may be increased by pre-flashing. This involves a millisecond light flash prior to the sample being brought into juxtaposition with the film and is often used where high sensitivity is required or if results are to be quantified.

Quantification

As indicated earlier, autoradiography is usually used to locate rather than to quantify radioactivity. However it is possible to obtain quantitative data directly from autoradiographs by using a densitometer, which records the intensity of the image. This in turn is related to the amount of radioactivity in the original sample. There are many varieties of densitometers available and the choice made will depend on the purpose of the experiment. Quantification is not reliable at low or high levels of exposure because of the lag phase (i.e. the back reaction, as described above) or saturation, respectively; however, pre-flashing combined with fluorography or intensifying screens obviates the problem for small amounts of radioactivity. In this case all photons contribute equally to the image of the pre-exposed film.

5.3 Other practical aspects of counting radioactivity and analysis of data

5.3.1 Counter characteristics

Background count

Radiation counters of all types always register a count, even in the absence of radioactive material in the apparatus. This may be due to such sources as cosmic radiation, natural radioactivity in the vicinity, nearby X-ray generators, and circuit noises. By means of the various methods already outlined and the use of lead shielding, this background radiation may be considerably reduced, but its value must always be recorded and accounted for in all experiments. Some commercial instruments have automatic background subtraction facilities.

Dead time

At very high count rates in Geiger–Müller counting, counts are lost due to the dead time of the Geiger–Müller tube. Correction tables are available and these should be used when necessary to correct for lost counts. Dead time is not a problem in scintillation counting.

Geometry

When samples with an end-window ionisation counter, such as a Geiger–Müller tube, are compared, it is important to standardise the position of the sample in relation to the tube, otherwise the fraction of the emitted radiation entering the tube may vary and hence so will the observed count.

5.3.2 Sample and isotope characteristics

Self-absorption

Self-absorption is primarily a problem with low energy, β-emitters: radiation is absorbed by the sample itself. Self-absorption can be a serious problem in the counting of low energy radioactivity by scintillation counting if the sample is particulate or is, for instance, stuck to a membrane filter. Care should be taken to ensure comparability of samples because the methods of standardisation outlined earlier will not correct for self-absorption effects. Where homogeneity is not possible, particulate samples should be digested or otherwise solubilised prior to counting. Self-absorption is a major problem with Geiger–Müller counting and significantly reduces sensitivity and reliability. It is very difficult to count low energy emitters reliably with these counters and this was a major factor in the switch to scintillation counting.

Half-life

The half-life of an isotope (Section 5.1.5) may be short and, if so, this must be allowed for in the analysis of data.

Statistics

The emission of radioactivity is a random process. This can be demonstrated readily by making repeated measurements of the activity of a long-lived isotope, each for an identical period of time. The resulting counts will not be the same but will vary over a range of values, with clustering near the centre of the range. If a sufficiently large number of such measurements is made and the data are plotted, a normal distribution curve will be obtained. For a single count, therefore, we cannot obtain a true count. Instead, we take the mean of a large number of counts as being very close to the true count. However the accuracy of this mean will depend on the spread or standard deviation (σ) for the data. Statistical theory states that, for a normal distribution such as that shown in Fig. 5.13, 68% of

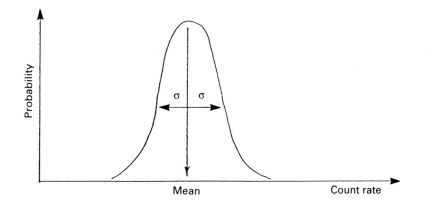

Fig. 5.13. The distribution of count rates around a mean, showing the standard deviation, σ.

values obtained lie ± 1 σ, and 95% lie ± 2 σ from the mean ($\bar{x}$).

Clearly, if we wish to compare samples, and in particular to state that two samples contain different amounts of radioactivity, then we need to take account of the counting statistics. Fortunately Poisson mathematics makes the task relatively easy as:

$$\sigma = \sqrt{\text{total counts taken}}$$

or:

$$\sigma = \sqrt{\left(\frac{\text{count rate}}{\text{time}}\right)}$$

Therefore to quote a figure with 95% certainty, state:

$$\text{total counts } \pm 2\sqrt{\text{total counts}}$$

For example, if 1600 counts are recorded this can be expressed as 1600 ± 80. There is, therefore, a 95% chance that the true figure lies between 1520 and 1680.

When data is expressed as d.p.m. or c.p.m., again using 95% certainty, then:

$$\text{error on count rate} = 2\sqrt{\left(\frac{\text{count rate}}{\text{time}}\right)}$$

Using the same example, if the 1600 counts were obtained in 1 min:

$$\text{error on count rate} = 2\sqrt{\left(\frac{1600}{1}\right)} = 80$$

and therefore 1600 c.p.m. ± 80 c.p.m.

If the 1600 counts were obtained in 10 min:

$$\text{error on count rate} = 2\sqrt{\left(\frac{160}{10}\right)} = 8$$

and therefore 160 c.p.m. ± 8 c.p.m. Note that the error is the same. This is because effectively the same number of measurements, i.e. counts, has been taken in each case.

If we had recorded 160 counts in only 1 min, then:

$$\text{error on count rate} = 2\sqrt{\left(\frac{160}{1}\right)} = 25$$

and therefore 160 c.p.m. ± 25 c.p.m.

Consider these other simple examples for a series of 1 min counts:

counts = 100 $\sigma = \sqrt{100}$,   therefore 10% error at 68% certainty
counts = 1000 $\sigma = \sqrt{1000}$,   therefore 3% error at 68% certainty
counts = 10 000 $\sigma = \sqrt{10\,000}$,   therefore 1% error at 68% certainty

In summary, the counts per minute data become more accurate for *higher count rates* and/or *longer counting times*. It is common practice to count to 10 000 counts or for 10 min, whichever is the quicker, although for very low count rates longer counting times are required.

5.3.3   Supply, storage and purity of radiolabelled compounds

There are several suppliers of radiolabelled compounds, the main ones being Amersham International plc., Du Pont, NEN and ICN. The suppliers usually include details of the best storage conditions and quality control data with their products. This is because several types of decomposition can occur; for example internal decomposition resulting from radioactive decay such as $^{14}C \rightarrow {}^{14}N$, and external decomposition where emitted radiation is absorbed by other radioactive molecules, causing impurities. The extent to which decomposition occurs is dependent on many factors such as temperature, energy of radiation, concentration and the formulation of the compound. It is, therefore, imperative to store radioisotopes by the method recommended by the supplier and to maintain sterility of the stock.

If necessary, chromatographic procedures will be required to check on the purity of the labelled compounds.

5.3.4   Specific activity

The specific activity of a radioisotope defines its radioactivity related to the amount of material (e.g. Bq mol^{-1}, Ci mmol^{-1} or d.p.m. µmol^{-1}). Suppliers offer a range of specific activities for their compound, the highest often being the most expensive. The advantages of using a very high specific activity are as follows:

- Products of a reaction using the labelled precursor can be produced at high specific activity (e.g. for DNA probes, see Section 3.10).
- Small quantities of radiolabelled compound can be added such that the equilibrium of metabolic concentrations is not unduly perturbed.
- Calculating the amount of substance required to make up radioactive solutions of known specific activity is simplified, as the contribution to concentration made by the stock radiolabelled solution is often negligible (see below).

Sometimes, however, it is not necessary to purchase the highest specific activity available. For example, enzyme assays *in vitro* often require a relatively high substrate concentration and so specific activity may need to be lowered. Consider the example below (for definitions of units see Table 5.6):

[^{3}H]Leucine is purchased with a specific activity of 5.55 TBq mmol^{-1} (150 Ci mmol^{-1}) and a concentration of 9.25 MBq 250 mm^{-3} (250 µCi 250 mm^{-3}). A 10 cm^3 solution of 250 mmol dm^{-3} and 3.7 kBq cm^{-3} (0.1 µCi cm^{-3}) is required. It is made up as follows:

- 10 cm^3 at 3.7 kBq cm^{-3} is 37 kBq (1 µCi), therefore pipette 1 mm^3 of stock radioisotope into a vessel (or, to be more accurate, pipette 100 mm^3 of a ×100 dilution of stock in water).
- Add 2.5 cm^3 of a 1 mol dm^{-3} stock solution of cold leucine, and make up to 10 cm^3 with distilled water.

There is no need to take into account the amount of unlabelled leucine in the [^{3}H]leucine preparation; it is a negligible quantity due to the high specific activity. If necessary (for example, to manipulate solutions of relatively low specific activity), however, the following formula can be applied:

$$W = Ma \left[(1/A') - (1/A) \right]$$

where W is the mass of cold carrier required (mg), M is the amount of radioactivity present (MBq), a is the molecular weight of the compound, A is the original specific activity (MBq mmol^{-1}), and A' is the required specific activity (MBq mmol^{-1}).

5.3.5   The choice of radionuclide

This is a complex question depending on the precise requirements of the experiment. A summary of some of the key features of radioisotopes commonly used in biological work is shown in Table 5.5.

Table 5.5. *The relative merits of commonly used β-emitters*

Isotope	Advantages	Disadvantages
^{3}H	Safety High specific activity possible Wide choice of positions in organic compounds Very high resolution in autoradiography	Low efficiency of detection Isotope exchange with environment Isotope effect
^{14}C	Safety Wide choice of labelling position in organic compounds Good resolution in autoradiography	Low specific activity
^{35}S	High specific activity Good resolution in autoradiography	Short half-life Relatively long biological half-life
^{32}P	Ease of detection High specific activity Short half-life simplifies disposal Čerenkov counting	Short half-life affects costs and experimental design External radiation hazard Poor resolution in autoradiography

Taken from *Radioisotopes in Biology, A Practical Approach,* ed. R.J. Slater, Oxford University Press, with permission.

5.4   Inherent advantages and restrictions of radiotracer experiments

Perhaps the greatest advantage of radiotracer methods over most other chemical and physical methods is their sensitivity. For example, a dilution factor of 10^{12} can be tolerated without the detection of ^{3}H-labelled compounds being jeopardised. It is thus possible to detect the occurrence of metabolic substances that are normally present in tissues at such low concentrations as to defy the most sensitive chemical methods of identification. A second major advantage of using radiotracers is that they enable studies *in vivo* to be carried out to a far greater degree than can any other technique.

In spite of these significant advantages, certain restrictions have to be appreciated. Firstly, although they undergo the same reactions, different isotopes may do so at different rates. This effect is known as the isotope effect. The different rates are approximately proportional to the differences in mass between the isotopes. The extreme case is the isotopes ^1H and ^3H, the effect being smaller for ^{12}C and ^{14}C and almost insignificant for ^{31}P and ^{32}P. Secondly, the amount of activity employed must be kept to the minimum necessary to permit reasonable counting rates in the samples to be analysed, otherwise the radiation from the tracer may elicit a response from the experimental organism and hence distort the results. A third consideration is that, in order to administer the tracer, the normal chemical level of the compound in the organism is automatically exceeded. The results are therefore always open to question.

5.5 Safety aspects

The greatest practical disadvantage of using radioisotopes is their toxicity: they produce ionising radiations. When absorbed, radiation causes ionisation and free radicals form that interact with the cell's macromolecules, causing mutation of DNA and hydrolysis of proteins. The toxicity of radiation is dependent not simply on the amount present but on the amount absorbed by the body, the energy of the absorbed radiation and its biological effect. There are, therefore, a series of additional units used to describe these parameters. Originally, radiation hazard was measured in terms of exposure, i.e. a quantity expressing the amount of ionisation in air. The unit of exposure is the roentgen (R), which is the amount of radiation that produces 1.61×10^{15} ion-pairs (kg air)$^{-1}$ (or 2.58×10^{-4} coulombs (kg air)$^{-1}$).

The amount of energy required to produce an ion-pair in air is 5.4×10^{-18} joules (J) and so the amount of energy absorbed by air with an exposure of 1 R is:

$$1.61 \times 10^{15} \times 5.4 \times 10^{-18} = 0.00869 \text{ J (kg air)}^{-1}$$

Although the roentgen has been used as a unit of radiation hazard, it is now considered inadequate for two reasons: first, it is defined with reference to X-rays (or γ-rays) only; and, second, the amount of ionisation or energy absorption in different types of material, including living tissue, is likely to be different from that in air.

The concept of radiation absorbed dose (rad) was introduced to overcome these restrictions. The rad is defined as the dose of radiation that gives an energy absorption of 0.01 J (kg absorber)$^{-1}$; this has now been changed to the gray, an SI unit, representing absorption of 1 J kg^{-1} (i.e. 100 rads).

The gray (Gy) is a useful unit, but it still does not adequately describe the hazard to living organisms. This is because different types of radiation are associated with differing degrees of biological hazard. It is, therefore, necessary to introduce a correction factor, known as the weighting factor (W), which is calculated by comparing the biological effects of any type of radiation with that of X-rays. The unit of absorbed dose, which takes into account the weighting factor is the sievert (Sv) and is known as the equivalent dose. Thus:

$$\text{equivalent dose (Sv)} = \text{Gy} \times W$$

The majority of isotopes used in biological research emit β-radiation. This is considered to have a biological effect that is very similar to X-rays and has a weighting factor of 1. Therefore, for β-radiation, Gy = Sv. Alpha particles, with their stronger ionising power, are much more toxic and have a weighting factor of 20. Therefore, for α-radiation, 1 Gy = 20 Sv. It is likely that, as our knowledge of the biological effectiveness of different forms of radiation progresses, so the quality factor for different types of radiation may change in the future. Absorbed dose from known sources can be calculated from knowledge of the rate of decay of the source, the energy of radiation, the penetrating power of the radiation and the distance between the source and the laboratory worker. As the radiation is emitted from a source in all directions, the level of irradiation is related to the area of a sphere, $4\pi r^2$. Thus the absorbed dose is inversely related to the square of the distance from the source (r); or, put another way, if the distance is doubled the dose is quartered. A useful formula is:

$$\text{dose}_1 \times \text{distance}_1^2 = \text{dose}_2 \times \text{distance}_2^2$$

The relationship between radioactive source and absorbed dose is illustrated in Fig. 5.14.

The rate at which dose is delivered is referred to as the dose rate, expressed in Sv h^{-1}. It can be used to calculate your total dose. For example, a source may be delivering 10 μSv h^{-1}. If you worked with the source for 6 h, your total dose would be 60 μSv.

Currently, the dose limit for workers exposed to radiation is 50 mSv year^{-1} to the whole body. Limits are set for individual organs. The most important of these to know are for hands (500 mSv year^{-1}) and for lens of the eye (150 mSv year^{-1}). Dose limits are constantly under review and the current recommendation is that these be reduced to 40% of their current value. Thus the dose limit to the whole

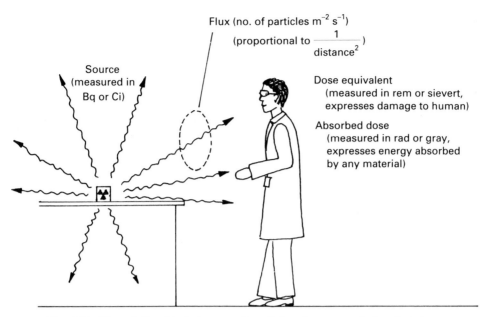

Fig. 5.14. The relationship between radioactivity of source and absorbed dose.

body may soon be reduced to 20 mSv year^{-1}. Although these dose limits are set, it is against the Ionising Radiations Regulations to work to such a limit, i.e. to assume that all is satisfactory if the limit is not exceeded. Instead, the ALARA principle is applied, to work always to a dose limit that is as low as reasonably achievable. Work that may cause a worker to exceed three-tenths or one-tenth of the dose limit must be carried out in a controlled area or a supervised area, respectively. In practice, work in the biosciences rarely involves a worker receiving a measurable dose. Supervised areas are common but not always required (e.g. for ^{3}H or ^{14}C experiment). Controlled areas are required in only certain circumstances, e.g. for isotope stores or radioiodination work. A major problem, however, in biosciences is the internal radiation hazard. This is caused by radiation entering the body, e.g. by inhalation, ingestion, absorption or puncture. This is a likely source of hazard where work involves open sources, i.e. liquids and gases; most work in biology involves manipulations of radioactive liquids. Control of contamination is assisted by:

(i) complying with local rules, written by an employer,
(ii) conscientious personal conduct and strict adherence to good laboratory practice,
(iii) regular monitoring,
(iv) carrying out work in some kind of containment.

Calculating the dose received following the ingestion of a radioisotope is complex. Detailed information is published by the International Commission on Radiological Protection and assessments, e.g. for experiments on human volunteers, can be obtained from the National Radiological Protection Board. However, one relatively simple concept is the annual limit on intake (ALI). The ingestion of one ALI results in a person receiving a dose of either 50 mSv to the whole body or 500 mSv to a particular organ. Some ALIs are shown in Table 5.6.

Table 5.6. *Annual limits on intake (ALI) for some commonly used isotopes*

Nuclide	ALI (MBq)
^{3}H	3000
^{14}C	90
^{32}P	10
^{125}I	1

Management of radiation protection is similar in most countries. In the UK there is the Radioactive Substances Act 1960 and the Ionising Radiations Regulations 1985. Every institution requires certification, monitored by Her Majesty's Inspectorate of Pollution, and employs a Radiation Protection Advisor in compliance with the Act.

When handling radioisotopes the rule is to:

- *Maximise the distance* between yourself and the source,
- *Minimise the time* of exposure and
- *Maintain shielding* at all times.

5.6 Applications of radioisotopes in the biological sciences

5.6.1 Investigating aspects of metabolism

Metabolic pathways

Radioisotopes are frequently used for tracing metabolic pathways. This usually involves adding a radioactive substrate, taking samples of the experimental material at various times, extracting and chromatographically, or otherwise, separating the products. Radioactivity detectors can be attached to gas–liquid chromatography or high-performance liquid chromatography columns to monitor radioactivity coming off the column during separation. Alternatively, radioactivity can be located on paper or thin-layer chromatography with either a Geiger–Müller chromatograph scanner or with autoradiography. If it is suspected that a particular compound is metabolised by a particular pathway, then radioisotopes can also be used to confirm this. For instance, it is possible to predict the fate of individual carbon atoms of [^{14}C]acetate through the tricarboxylic acid (TCA) cycle. Methods have been developed whereby intermediates of the cycle can be isolated and the distribution of carbon within each intermediate can be ascertained. This is the so-called specific labelling pattern. Should the actual pattern coincide with the theoretical pattern, then this is very good evidence for the mode of operation of the TCA cycle.

Another example of the use of radioisotopes to confirm the mode of operation, or otherwise, of a metabolic pathway is in studies carried out on glucose catabolism. There are numerous ways whereby glucose can be oxidised, the two most important ones in aerobic organisms being glycolysis followed by the TCA pathway together with the pentose phosphate pathway. Frequently, organisms or tissues possess the necessary enzymes for both pathways to occur and it is of interest to establish the relative contribution of each to glucose oxidation. Both pathways involve the complete oxidation of glucose to carbon dioxide, but the origin of the carbon dioxide in terms of the six carbon atoms of glucose is different (at least in the initial stages of respiration of exogenously added substrate). Thus, it is possible to trap the carbon dioxide evolved during the respiration of specially labelled glucose (e.g. [6-^{14}C]glucose in which only the atom is radioactive and [1-^{14}C]glucose) and obtain an evaluation of the contribution of each pathway to glucose oxidation.

The use of radioisotopes in studying the operation of the TCA cycle or in evaluating the pathway of glucose catabolism are just two examples of how such isotopes can be used to confirm metabolic pathways. Further details of these and other examples, including use of dual-labelling methods, can be found in the various texts recommended in Section 5.8.

Metabolic turnover times

Radioisotopes provide a convenient method of ascertaining turnover times for particular compounds. As an example, the turnover of proteins in rats will be considered. A group of rats is injected with a radioactive amino acid and left for 24 h, during which time most of the amino acid is assimilated into proteins. The rats are then killed at suitable time intervals and radioactivity in organs or tissues of interest is determined. In this way it is possible to ascertain the rate of metabolic turnover of protein. Using this sort of method, it has been shown that liver protein is turned over in 7 to 14 days, while skin and muscle protein is turned over every 8 to 12 weeks, and collagen is turned over at a rate of less than 10% per annum.

Studies of absorption, accumulation and translocation

Radioisotopes have been very widely used in the study of the mechanisms and rates of absorption, accumulation and translocation of inorganic and organic compounds by both plants and animals. Such experiments are generally simple to perform and can also yield evidence on the route of translocation and sites of accumulation of molecules of biological interest.

Pharmacological studies

Another field where radioisotopes are widely used is in the development of new drugs. This is a particularly complicated process, because, besides showing if a drug has a desirable effect, much more must be ascertained before it can be used in the treatment of clinical conditions. For instance, the site of drug accumulation, the rate of accumulation, the rate of metabolism and the metabolic products must all be determined. In each of these areas of study, radiotracers are extremely useful, if not indispensable. For instance, autoradiography on whole sections of experimental animals (Section 5.2.3 and Fig. 5.12) yields information on the site and rate of accumulation, while typical techniques used in metabolic studies can be used to follow the rate and products of metabolism.

5.6.2   Analytical applications

Enzyme and ligand binding studies

Virtually any enzyme reaction can be assayed using radiotracer methods, as outlined in Section 4.8.5, provided that a radioactive form of the substrate is available. Radiotracer-based enzyme assays are more expensive than other methods, but frequently have the advantage of a higher degree of sensitivity. Radioisotopes have also been used in the study of the mechanism of enzyme action and in ligand binding studies (Section 4.11).

Isotope dilution analysis

There are many compounds present in living organisms that cannot be accurately assayed by conventional means because they are present in such low amounts and in mixtures of similar compounds. Isotope dilution analysis offers a convenient and accurate way of overcoming this problem and avoids the necessity of quantitative isolation. For instance, if the amount of iron in a protein preparation is to be determined, this may be difficult using normal methods, but it can be done if a source of ^{59}Fe is available. This is mixed with the protein and a sample of iron is subsequently isolated, assayed for total iron and the radioactivity determined.

If the original specific activity was 10 000 d.p.m. $(10\ mg)^{-1}$ and the specific activity of the isolated iron was 9000 d.p.m. $(10\ mg)^{-1}$ then the difference is due to the iron in the protein (x), i.e.:

$$\frac{9000}{10} \times \frac{10\,000}{10+x}$$

$$x = 1.1\ mg$$

This technique is widely used in, for instance, studies on trace elements.

Radioimmunoassay

One of the most significant advances in biochemical techniques in recent years has been the development of radioimmunoassay. This technique is discussed in Section 2.5 and is not elaborated upon here.

Radiodating

A quite different analytical use for radioisotopes is in the dating (i.e. determining the age) of rocks, fossil and sediments. In this technique it is assumed that the proportion of an element that is naturally radioactive has been the same throughout time. From the time of fossilisation or deposition the radioactive isotope will decay. By determining the amount of radioisotope remaining (or by examining the amount of a decay product) and from a knowledge of the half-life, it is possible to date the sample. For instance, if the radioisotope normally composes 1% of the element and it is found that the sample actually contains 0.25% then two half-lives can be assumed to have elapsed since deposition. If the half-life is one million years then the sample can be dated as being two million years old.

For long-term dating, isotopes with long half-lives are necessary, such as ^{235}U, ^{238}U and ^{40}K, whereas for shorter-term dating ^{14}C is widely used. It cannot be overemphasised that the assumptions made in radiodating are sweeping and hence palaeontologists and anthropologists who use this technique can only give approximate dates to their samples.

5.6.3   Other applications

Molecular biology techniques

Recent advances in molecular biology that have led to the advances in genetic manipulation have depended heavily upon use of radioisotopes in DNA and RNA sequencing, DNA replication, transcription, synthesis of cDNA, recombinant DNA technology and many similar studies. Many of these techniques are more fully discussed in Chapter 3.

Clinical diagnosis

Radioisotopes are very widely used in medicine, in particular for diagnostic tests. Lung function tests routinely made using xenon-133 (^{133}Xe) are particularly useful in diagnosis of malfunctions of lung ventilation. Kidney function tests using [^{133}I]-iodohippuric acid are used in diagnoses of kidney infections, kidney blockages or imbalance of function between the two kidneys. Thyroid function tests using ^{131}I are employed in the diagnosis of hypo- and hyperthyroidism.

Various aspects of haematology are also studied by using radioisotopes. These include such aspects as blood cell lifetimes, blood volumes and blood circulation times, all of which may vary in particular clinical conditions.

Ecological studies

The bulk of radiotracer work is carried out in biochemical, clinical or pharmacological laboratories, nevertheless, radiotracers are also of use to ecologists. In particular, migratory patterns and behaviour patterns of many animals can be monitored using radiotracers. Another ecological application is in the examination of food chains where the primary producers can be made radioactive and the path of radioactivity followed throughout the resulting food chain.

Sterilisation of food and equipment

Very strong γ-emitters are now widely used in the food industry for sterilisation of prepacked foods such as milk and meats. Normally either ^{60}Co or ^{137}Ce is used, but care has to be taken in some cases to ensure that the food product itself is not affected in any way. Thus doses often have to be reduced to an extent where sterilisation is not complete but nevertheless food spoilage can be greatly reduced. ^{60}Co and ^{137}Ce are also used in sterilisation of plastic disposable equipment such as Petri dishes and syringes, and in sterilisation of drugs that are administered by injection.

Mutagens

Radioisotopes may cause mutations, particularly in micro-organisms (Section 1.6). In various microbiological studies mutants are desirable, especially in industrial microbiology. For instance, development of new strains of a microorganism that produce higher yields of a desired microbial product frequently involve mutagenesis by radioisotopes.

5.7   Radioisotope calculations

Calculations typical of the radioisotope laboratory

1. An experimental sample of ^{3}H on a filter paper in scintillation fluid gave a count rate of 1450 c.p.m. in a liquid scintillation counter. The filter was removed and 5064 d.p.m. added to it. On recounting, the filter gave a reading of 2878 c.p.m. What was the d.p.m. of the experimental sample?

2. Given $\ln N_t/N_o = -\lambda t$ and that the half-life of ^{32}P is 14.2 days, how long would it take a solution containing 42 000 d.p.m. of ^{32}P to decay to 500 d.p.m.?

3. A 1 litre of [^{3}H]uridine with a concentration of 100 $\mu mol\,cm^{-3}$ and 50 000 c.p.m. cm^{-3} is required. If all measurements are made on a scintillation counter with an efficiency of 40%, how would you make up this solution if the purchased supply of [^{3}H]uridine has a specific activity of 5.2 Ci mol^{-1}?

   [NB: M_r uridine = 244; 1 Ci = 22.2×10^{11} d.p.m.]

   Repeat the calculation in bequerels.

4. The efficiency of counting 100 000 d.p.m. of a [^{14}C]leucine solution was estimated in a scintillation counter using two channels, A and B, in scintillation fluid containing increasing amounts of chloroform. The following data were obtained:

Chloroform (cm^3)	c.p.m. A	c.p.m. B
0	48100	54050
1	31612	42150
2	17608	28400
3	7400	15000

An unknown sample of [^{14}C]leucine gave the following data:

Channel A   1890 c.p.m.
Channel B   2700 c.p.m.

How much radioactivity is present in the unknown sample?

5. The efficiency of detecting ^{14}C in a scintillation counter was determined by counting a standard sample containing 105 071 d.p.m. at different degrees of quench analysed by the external standard approach:

c.p.m.	SQP
87451	0.90
62361	0.64
45220	0.46
21014	0.21

SQP, standard quench parameter.

An experimental sample gave 2026 c.p.m. at an SQP of 0.52. What is the true count rate?

6. A sample recording 564 c.p.m. was counted over 10 min. What is the accuracy of the measurement for 95% confidence?

Answers

1. 5142 d.p.m. (from efficiency taken as $(2878 - 1450/5064 \times 100)\%$ used to correct the figure of 1450 c.p.m.).
2. 90.6 days (using equation 5.4 to obtain λ and substitute 42 000 for N_o and 500 for N_t).
3. 56.3 μCi (2.08 MBq) [^{3}H]uridine + 24.4 g uridine; (refer to the example on p. 261).
4. 7130 d.p.m. (plot efficiency in A or B, e.g. $(48\,100 \times 100/100\,000)\%$ v. ratio A/B, use ratio of unknown to compute efficiency from graph, correct appropriate c.p.m., i.e. 1890 or 2700, according to plot of A or B).
5. 4221 d.p.m. (plot efficiency, e.g. $(87\,451 \times 100/105\,071)\%$ v. SQP, obtain efficiency for experimental sample and correct 2026).
6. 564 ± 15 ($\pm 2\sigma = \pm \sqrt{(564/10)}$).

Table 5.7. *Units commonly used to describe radioactivity*

Unit	Abbreviation	Definition
Counts per minute per second	c.p.m. c.p.s.	The *recorded* rate of decay
Disintegrations per minute or second	d.p.m. d.p.s.	The *actual* rate of decay
Curie	Ci	The number of d.p.s. equivalent to 1g of radium (3.7×10^{10} d.p.s.)
Millicurie	mCi	$Ci \times 10^{-3}$ or 2.22×10^9 d.p.m.
Microcurie	μCi	$Ci \times 10^{-6}$ or 2.22×10^6 d.p.m.
Becquerel (SI unit)	Bq	1 d.p.s.
Terabecquerel (SI unit)	TBq	10^{12}Bq or 27.027 Ci
Gigabecquerel (SI unit)	GBq	10^9 Bq or 27.027 mCi
Megabecquerel (SI unit)	MBq	10^6 Bq or 27.027 μCi
Electron volt	eV	The energy attained by an electron accelerated through a potential difference of 1 volt. Equivalent to 1.6×10^{-19} J
Roentgen	R	The amount of radiation that produces 1.61×10^{15} ion-pairs kg^{-1}
Rad	rad	The dose that gives an energy absorption of 0.01 J kg^{-1}
Gray	Gy	The dose that gives an energy absorption of 1 J kg^{-1}. Thus 1 Gy = 100 rad
Rem	rem	The amount of radiation that gives a dose in humans equivalent to 1 rad of X-rays
Sievert	Sv	The amount of radiation that gives a dose in humans equivalent to 1 Gy of X-rays. Thus 1 Sv = 100 rem

5.8   Suggestions for further reading

ARONOFF, S. (1958) *Techniques in Radiobiochemistry*. Iowa State University Press, Ames, IA (Very comprehensive, if dated, account of biochemical uses of radioactivity.)

CHAPMAN, J.M. and AVERY, G. (1981) *The Use of Radioisotopes in the Life Sciences*. George Allen and Unwin, London. (A good undergraduate text on the use and detection of radioactivity.)

SLATER, R.J. (1991) *Radioisotopes in Biology – A Practical Approach*. IRL Press, Oxford. (A more detailed account of the handling and use of radioactivity in biological research.)

6

Centrifugation techniques

▲ ● ■ ▲ ● ■ ▲ ● ■ ▲ ● ■ ▲ ● ■ ▲ ● ■ ▲ ● ■

6.1   Introduction

Centrifugation separation techniques are based upon the behaviour of particles in an applied centrifugal field and assume that the parameters of the molecules under investigation, such as the relative molecular mass, shape and density, may be related to the behaviour of those molecules in a gravitational field.

If a solution of large particles is left to stand, then the particles will tend to sediment under the influence of gravity. For a given particle, the rate or velocity at which it sediments is proportional to the force applied, so that the particles sediment more rapidly when the force applied is greater than the gravitational force of the earth. The basis of centrifugation separation techniques, therefore, is to exert a larger force than does the earth's gravitational field, thus increasing the rate at which the particles sediment. The particles are normally suspended in a specific liquid medium, held in tubes or bottles, which are located in a rotor. The rotor is positioned centrally on the drive shaft of the centrifuge. Particles that differ in density, shape or size can be separated because they sediment at different rates in the centrifugal field, each particle sedimenting at a rate that is directly proportional to the applied centrifugal field.

Centrifugation techniques are of two main types. Preparative centrifugation techniques are concerned with the actual separation, isolation and purification of, for example, whole cells, subcellular organelles, plasma membranes, polysomes, ribosomes, chromatin, nucleic acids, lipoproteins and viruses, for subsequent biochemical investigations. Very large amounts of material may be involved when microbial cells are harvested from culture media, plant and animal cells from tissue culture or plasma from blood. Relatively large amounts of cellular particles may also be isolated in order to study their morphology, composition and biological activity. It is also possible to isolate biological macromolecules, such as nucleic acids and proteins, from preparations that have received some measure of purification by, for example, fractional precipitation (Section 4.5.3). In contrast, analytical centrifugation techniques are devoted mainly to the study of pure, or virtually pure, macromolecules or particles. They are concerned primarily with the study of the sedimentation characteristics of biological macromolecules and molecular structures, rather than with the collection of particular fractions. They require only small amounts of material and utilise specially designed rotors and detector systems to monitor continuously the process of sedimentation of the material in the centrifugal field. Such studies yield information from which the purity, relative molecular mass and shape of the material may be deduced. Since preparative centrifugation techniques are more commonly used in undergraduate courses, this chapter concentrates on these techniques and deals only briefly with analytical centrifugation techniques.

6.2   Basic principles of sedimentation

The rate of sedimentation is dependent upon the applied centrifugal field (G) being directed radially outwards; this is determined by the square of the angular velocity of the rotor (ω, in radians s^{-1}) and the radial distance (r, in centimetres) of the particle from the axis of rotation, according to the equation:

$$G = \omega^2 r \tag{6.1}$$

Since one revolution of the rotor is equal to 2π radians, its angular velocity, in radians s^{-1}, can be readily expressed in terms of revolutions per minute (rev min^{-1}), the common way of expressing rotor speed being:

$$\omega = \frac{2\pi \, \text{rev min}^{-1}}{60} \tag{6.2}$$

The centrifugal field (G) in terms of rev min^{-1} is then:

$$G = \frac{4\pi^2 (\text{rev min}^{-1})^2 \, r}{3600} \tag{6.3}$$

and is generally expressed as a multiple of the earth's gravitational field (g = 981 cm s^{-2}), i.e. the ratio of the weight of the particle in the centrifugal field to the weight of the same particle when acted on by gravity alone, and is then referred to as the relative centrifugal field (RCF) or more commonly as the 'number times g'.

Hence:

$$RCF = \frac{4\pi^2 (\text{rev min}^{-1})^2 \, r}{3600 \times 981} \tag{6.4}$$

which may be shortened to give:

$$RCF = (1.118 \times 10^{-5})(\text{rev min}^{-1})^2 \, r \tag{6.5}$$

When conditions for the centrifugal separation of particles are reported, therefore, rotor speed, radial dimensions and time of operation of the rotor must all be quoted. Since biochemical experiments are usually conducted with particles dissolved or suspended in solution, the rate of sedimentation of a particle is dependent not only upon the applied centrifugal field but also upon the mass of the particle, which may be expressed as the product of its volume and density, the density and viscosity of the medium in which it is sedimenting and the extent to which its shape deviates from spherical. When a particle sediments it must displace some of the solution in which it is suspended, resulting in an apparent upthrust on the particle equal to the weight of liquid displaced. If a particle is assumed to be spherical and of known volume and density, the latter being corrected for the buoyancy due to the density of the medium, then the net outward force (F) it experiences when centrifuged at an angular velocity of ω radians s^{-1} is given by:

$$F = \frac{4}{3} \pi r_p^3 (\rho_p - \rho_m)\omega^2 r \tag{6.6}$$

where $(\frac{4}{3})\pi r_p^3$ is the volume of a sphere of radius r_p, ρ_p is the density of the particle, ρ_m is the density of the suspending medium, and r is the distance of the particle from the centre of rotation. Particles, however, generate friction as they migrate through the solution. If a particle is rigid and spherical and moving at a known velocity, then the frictional force (F_o) opposing motion is given by:

$$F_o = vf \tag{6.7}$$

where v is the velocity or sedimentation rate of the particle, and f is the frictional coefficient of the particle in the solvent.

The frictional coefficient of a particle is a function of its size, shape and hydration, and viscosity of the medium, and by the Stokes equation, for an unhydrated spherical particle, is given by:

$$f = 6\pi\eta r_p \tag{6.8}$$

where η is the viscosity coefficient of the medium.

For asymmetric and/or hydrated particles, the actual radius of the particle in equation 6.8 is replaced by the effective or Stokes radius, r_{eff}. An unhydrated, spherical particle of known volume and density, and present 'in a medium of constant density, therefore accelerates in a centrifugal field, its velocity increasing until the net force of sedimentation equals the frictional force resisting its motion through the medium, i.e.:

$$F = F_o \quad \text{or} \quad \frac{4}{3}\pi r_p^3(\rho_p - \rho_m)\omega^2 r = 6\pi\eta r_p v \tag{6.9}$$

In practice, the balancing of these forces occurs quickly and the particle reaches a constant velocity because the frictional resistance increases with the velocity of the particle. Under these conditions, the net force acting on the particle is zero. Hence, the particle no longer accelerates but achieves a maximum velocity, with the result that it now sediments at a constant rate. Its rate of sedimentation (v) is then given by:

$$v = \frac{dr}{dt} = \frac{2r_p^2(\rho_p - \rho_m)\omega^2 r}{9\eta} \tag{6.10}$$

It can be seen from equation 6.10 that the sedimentation rate of a given particle is proportional to its size, to the difference in density between the particle and the medium and to the applied centrifugal field. It is zero when the density of the particle and the medium are equal; it decreases when the viscosity of the medium increases, and increases as the force field increases. However, since the equation involves the square of the particle radius, it is apparent that the size of the particle has the greatest influence upon its sedimentation rate. Particles of similar density, but only slightly different in size, can therefore have large differences in their sedimentation rate. Integration of equation 6.10 yields equation 6.11, which gives the sedimentation time for a spherical particle in a centrifugal field as a function of the variables involved and in relation to the distance travelled by the particle in the centrifuge tube:

$$t = \frac{9\eta}{2\omega^2 r_p^2(\rho_p - \rho_m)} \ln\frac{r_b}{r_t} \tag{6.11}$$

where t is the sedimentation time in seconds, r_t is the radial distance from the axis of rotation to liquid meniscus, and r_b is the radial distance from the axis of rotation to the bottom of the tube.

It is thus clear that a mixture of heterogeneous, approximately spherical, particles can be separated by centrifugation on the basis of their densities and/or their size, either by the time required for their complete sedimentation or by the extent of their sedimentation after a given time. These alternatives form the basis for the separation of biological macromolecules and of cell organelles from tissue homogenates. The order of separation of the major cell components is generally whole cells and cell debris first, followed by nuclei, chloroplasts, mitochondria, lysosomes (or other microbodies), microsomes (fragments of smooth and rough endoplasmic reticulum) and ribosomes.

Considerable discrepancies exist between the theory and practice of centrifugation. Complex variables not accounted for in equations 6.10 and 6.11, such as the concentration of the suspension, nature of the medium, and the design and handling of the centrifuge, will affect the sedimentation properties of a mixed population of particles. Moreover, aspherical particles exhibit a modified relationship between the sedimentation rate and the particle size, resulting in such particles sedimenting at a slower rate. Equation 6.10 can therefore be modified to give equation 6.12:

$$v = \frac{dr}{dt} = \frac{2r_p^2(\rho_p - \rho_m)\omega^2 r}{9\eta(f/f_o)} \tag{6.12}$$

which now takes into account the frictional effect of varying particle shape on the sedimentation rate of a particle. The frictional ratio, f/f_o, where f is the frictional coefficient of the aspherical and/or hydrated particle, and f_o is the theoretical frictional coefficient of an unhydrated sphere of the same relative molecular mass and density, is a function of the shape and hydration of the particle and, for spherical molecules such as native globular proteins, ranges from about 1 to 1.4. Where there is an appreciable asymmetry, however, as in the case of rod-like molecules such as DNA and proteins such as F-actin and myosin, larger values of f/f_o are found. Hence, particles of a given mass but different shape sediment at different rates. This point is exploited in the study of conformation of molecules by analytical ultracentrifugation (Section 6.10.3).

Although it is convenient to consider the sedimentation of particles in a uniform centrifugal field, in practice this does not occur when preparative rotors are used. Owing to the nature of rotor design (Figs. 6.1a, 6.2a, 6.3a) the effective radial dimension of a given particle will change according to its position in the sample container and will vary between r_{min} and r_{max}. Since the centrifugal field generated is proportional to $\omega^2 r$, a particle will experience a greater field the further away it is from the axis of rotation. The operative centrifugal field, in a fixed-angle rotor for example, can differ by a factor of up to 2 between the top and bottom of the centrifuge tube; thus the sedimentation rate of particles at the bottom of the tube will be twice that of identical particles near the top of the tube. As a result, particles will tend to move faster as they sediment through a non-viscous medium. It is normal, therefore, to record the relative centrifugal field calculated from the average radius of rotation (r_{av}) of the column of liquid in the tube (i.e. the distance

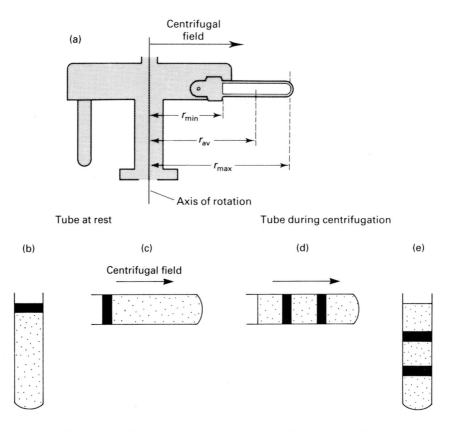

Fig. 6.1. Design and operation of the swinging-bucket rotor. (a) Cross-sectional diagram of a swinging-bucket rotor. (b) The centrifuge tube is initially loaded with gradient, the sample is then layered on top before the tube is placed in the bucket for attachment to the rotor. (c) During acceleration of the rotor the rotor bucket reorients to lie perpendicular to the axis of rotation. (d) Sedimentation and separation of the particles occur during centrifugation. (e) At the end of centrifugation the rotor decelerates, the bucket coming to rest in its original vertical position.

from the centre of rotation to the middle of the liquid column in the centrifuge tube). The average relative centrifugal field (RCF_{av}) is therefore the numerical average of the values exerted at r_{min} and r_{max}. If the sample container is only partially filled then, in the case of fixed-angle and swinging-bucket rotors, the minimum radius (r_{min}) is effectively increased and the particles will therefore start to sediment in a higher gravitational field and have a reduced pathlength to travel. Consequently sedimentation will be quicker. Centrifuge manuals normally provide details of the maximum permitted speed for a rotor, maximum relative centrifugal fields generated and graphs that enable the ready conversion of RCF to rev min^{-1} at r_{min}, r_{av}, and r_{max}.

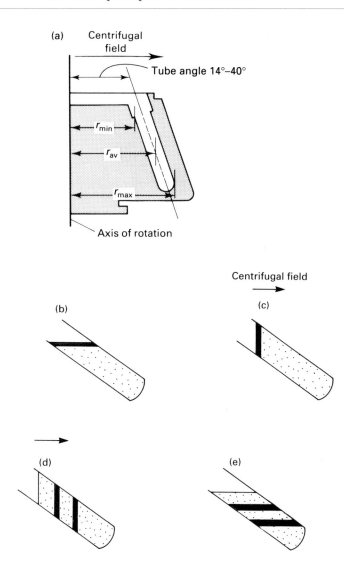

Fig. 6.2. Design and operation of the fixed-angle rotor. (a) Cross-sectional diagram of a fixed-angle rotor. (b) The centrifuge tube after being filled with gradient, is loaded with sample and then placed in the rotor. (c) During rotor acceleration, reorientation of the sample and gradient occur. (d) Sedimentation and separation of the particles occur during centrifugation. (e) Rotor is at rest: the gradient reorients and bands of separated particles appear.

The sedimentation rate or velocity (v) of a particle can also be expressed in terms of its sedimentation rate per unit of centrifugal field, commonly referred to as its sedimentation coefficient, s. From equation 6.12 it can be seen that, if the composition of the suspending medium is defined, then the sedimentation rate is

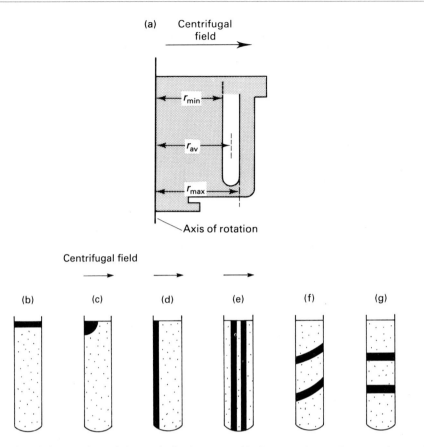

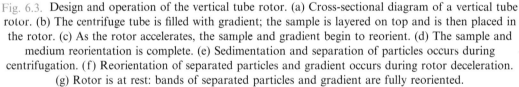

Fig. 6.3. Design and operation of the vertical tube rotor. (a) Cross-sectional diagram of a vertical tube rotor. (b) The centrifuge tube is filled with gradient; the sample is layered on top and is then placed in the rotor. (c) As the rotor accelerates, the sample and gradient begin to reorient. (d) The sample and medium reorientation is complete. (e) Sedimentation and separation of particles occurs during centrifugation. (f) Reorientation of separated particles and gradient occurs during rotor deceleration. (g) Rotor is at rest: bands of separated particles and gradient are fully reoriented.

proportional to $\omega^2 r$, the centrifugal field, and equation 6.12 simplifies to:

$$v = s\omega^2 r \qquad (6.13)$$

or:

$$s = \frac{v}{\omega^2 r} = \frac{dr/dt}{\omega^2 r} \qquad (6.14)$$

Since sedimentation rate studies may be performed using a wide variety of solvent–solute systems, or at different temperatures, the experimentally determined value of the sedimentation coefficient, which is affected by temperature, solution viscosity and density is, by convention, corrected to the sedimentation constant theoretically obtainable in water at 20°C, and by means of equation 6.15, is expressed as the standard sedimentation coefficient or $s_{20,w}$.

$$s_{20,w} = s_{obs} \times \frac{\eta_T}{\eta_{20}} \times \frac{\eta_c}{\eta_o} \times \frac{(1 - \bar{v}\rho_{20,w})}{(1 - \bar{v}\rho_T)} \tag{6.15}$$

where $s_{20,w}$ is the standard sedimentation coefficient, s_{obs} is the experimentally measured sedimentation coefficient, η_T is the viscosity of water at temperature $T(°C)$, η_{20} is the viscosity of water at 20 °C, η_c is the viscosity of the solvent at a given temperature (as near to 20 °C as possible), η_o is the viscosity of water at that temperature, $\rho_{20,w}$ is the density of water at 20 °C, ρ_T is the density of the solvent at temperature $T(°C)$, and $\bar{v}$ is the partial specific volume of the solute (defined as the reciprocal of the non-hydrated density of the particle, i.e. the volume occupied by 1.0 g of the particles).

For many macromolecules including nucleic acids and proteins the sedimentation coefficient usually decreases in value with increase in the concentration of solute, this effect becoming more severe with increase both in the relative molecular mass and the degree of extension of the molecule. Hence $s_{20,w}$ is usually measured at several concentrations and extrapolated to infinite dilution to obtain a value of $s_{20,w}$ at zero concentration, the $s^0_{20,w}$. The sedimentation coefficients of most biological particles are very small, and for convenience its basic unit is taken as 10^{-13} s, which is termed one Svedberg unit (S), in recognition of T. Svedberg's pioneering work in this type of analysis. Therefore, a ribosomal RNA molecule possessing a sedimentation coefficient of 5×10^{-13} s is said to have a value of 5 S.

Examination of equations 6.12 and 6.14 shows that the sedimentation coefficient is influenced by such features as the shape, size and density of the particle, and hence is commonly used to characterise a particular molecule or structure. Generally, the larger the molecule or particle, the larger is its Svedberg unit and hence the faster is its sedimentation rate. Sedimentation coefficients (in Svedberg units) for enzymes, peptide hormones and soluble proteins are 2 to 25 S, nucleic acids 3 to 100 S, ribosomes and polysomes 20 to 200 S, viruses 40 to 1000 S, lysosomes 4000 S, membranes 100 to 100×10^3 S, mitochondria 20×10^3 S to 70×10^3 S, and nuclei between 4000×10^3 S and $40\,000 \times 10^3$ S.

6.3   Centrifuges and their use

Centrifuges may be classified into four major groups: the small bench centrifuges; large capacity refrigerated centrifuges; high speed refrigerated centrifuges; and ultracentrifuges of two types, preparative and analytical.

6.3.1 Small bench centrifuges

These are the simplest and least expensive centrifuges and exist in many types of design. They are often used to collect small amounts of material that rapidly sediment (yeast cells, erythrocytes, coarse precipitates), and generally have a maximum speed of 4000 to 6000 rev min^{-1}, with maximum relative centrifugal fields of 3000 to 7000 g. Most operate at ambient temperature, the flow of air around the rotor controlling rotor temperature. Some of the latest designs, however, incorporate a refrigeration system to keep rotors cool, thus preventing denaturation of proteins. Small microfuges are available, providing virtually instant acceleration to maximum speeds of 8000 to 13 000 rev min^{-1} and developing fields of approximately 10 000 g. These centrifuges have proved extremely useful for sedimenting small volumes (250 mm^3 to 1.5 cm^3) of material very quickly (1 or 2 min). Typical applications include the rapid sedimentation of blood samples, and of synaptosomes used to study the effect of drugs on the uptake of biogenic amines.

6.3.2 Large capacity refrigerated centrifuges

These have a maximum speed of 6000 rev min^{-1} and produce a maximum relative centrifugal field approaching 6500 g. They have refrigerated rotor chambers and vary only in their maximum carrying capacity, all being capable of utilising a variety of interchangeable swinging-bucket and fixed-angle rotors enabling separation to be achieved in 10, 50 and 100 cm^3 tubes. Large total capacity (4 to 6 dm^3) centrifuges are also available that, in addition to accommodating smaller tubes, are also capable of holding bottles, each of 1.25 dm^3 capacity. In all these centrifuges, the rotors are usually mounted on a rigid suspension, hence it is extremely important that the centrifuge tubes and their contents are balanced accurately (to within 0.25 g of each other). Rotors must never be loaded with an odd number of tubes and, where the rotor is only partially loaded, the tubes must be located diametrically opposite each other in order that the load is distributed evenly around the rotor axis. These instruments are most often used to compact or collect substances that sediment rapidly, e.g. erythrocytes, coarse or bulky precipitates, yeast cells, nuclei and chloroplasts.

6.3.3 High speed refrigerated centrifuges

These instruments are available with maximum rotor speeds in the region of 25 000 rev min^{-1}, generating a relative centrifugal field of about 60 000 g. They generally have a total capacity of up to 1.5 dm^3, and a range of interchangeable fixed-angle and swinging-bucket rotors. These instruments are most often used to collect microorganisms, cellular debris, larger cellular organelles and proteins precipitated by ammonium sulphate. They cannot generate sufficient centrifugal force to sediment effectively viruses or smaller organelles such as ribosomes.

6.3.4   Continuous flow centrifuges

The continuous flow centrifuge is a relatively simple high speed centrifuge. The rotor, through which particles suspended in medium flow continuously (usually 1–1.5 dm^3 min^{-1}), is long, tubular, and non-interchangeable. As medium enters the rotating rotor, particles are sedimented against its wall and excess clarified medium overflows through an outlet port. The major application of this type of centrifuge is in the harvesting of bacteria or yeast cells from large volumes of culture medium (10–500 dm^3). Some high speed centrifuges can be adapted to function in the continuous flow mode by being constructed to accept a specially designed rotor (Section 6.4.5).

6.3.5   Preparative ultracentrifuges

Preparative ultracentrifuges are capable of spinning rotors to a maximum speed of 80 000 rev min^{-1} and can produce a relative centrifugal field of up to 600 000 g. The rotor chamber is refrigerated, sealed and evacuated to minimise any excessive rotor temperatures being generated by frictional resistance between the air and the spinning rotor. The temperature monitoring system is more sophisticated than in simpler instruments, employing an infrared temperature sensor that can continuously monitor rotor temperature and control the refrigeration system. An overspeed control system is also incorporated into these instruments to prevent operation of the rotor above its maximum rated speed and there are electronic circuits to detect rotor imbalance. In order to minimise vibration, caused by slight rotor imbalance that may arise due to unequal loading of the centrifuge tubes, ultracentrifuges are fitted with a flexible drive shaft system. Centrifuge tubes and their contents, however, must still be accurately balanced to within 0.1 g of each other. For safety reasons, rotor chambers of high speed and ultracentrifuges are always enclosed in heavy armour plating.

An air-driven, table-top preparative ultracentrifuge, called an airfuge, is available and is capable of accelerating a magnetically suspended 3.7 cm diameter rotor, accommodating 6×175 mm^3 tubes on a virtually friction-free cushion of air in a non-vacuated chamber, to 100 000 rev min^{-1} (160 000 g) in approximately 30 s. The airfuge has found applications in biochemical and clinical research where there are only small volume samples requiring high centrifugal forces. Examples include macromolecule/ligand binding–kinetic studies, steroid hormone receptor assays, separation of the major lipoprotein fractions from plasma, and deproteinisation of physiological fluids for amino acid analysis.

6.3.6 Analytical ultracentrifuges

These instruments are capable of operating at speeds approaching 70 000 rev min^{-1} (500 000 g) and consist of a motor, a rotor contained in a protective armoured chamber that is refrigerated and evacuated, and an optical system to enable the sedimenting material to be observed throughout the duration of centrifugation to determine concentration distributions in the sample at any time during centrifugation (Fig. 6.4a). Three types of optical system are available in the analytical ultracentrifuge: a light absorption system, and the alternative Schlieren system and Rayleigh interferometric system, both of which detect changes in the refractive index of the solution.

The rotor is solid, with holes to hold the cells that contain the samples, and is suspended on a wire coming from the drive shaft of a high speed motor that allows the rotor to find its own axis of rotation. The tip of the rotor contains a thermistor for measuring temperature. Several types of rotor are available, the simplest type of which incorporates two cells – the analytical cell and the counterpoise cell, which counterbalances the analytical cell. Two holes (at distances calibrated from the centre of rotation) are drilled through the counterpoise cell (Fig. 6.4b) to facilitate the calibration of distances in the analytical cell. A wide variety of cells are available and have a capacity of between 0.4 and 1.0 cm^3. Analytical cells used with the ultraviolet light absorption optical system and the Schlieren optical system have a single 2° or 4° sector shape, to prevent convection, and usually have a 12 mm optical pathlength centrepiece, although centrepieces of 1.5 mm to 30 mm are available. Double-sector cells, having two 2.5° sectors and an optical pathlength that can vary from 12 mm to 30 mm, are used with the interference optical system, the absorption optical system when used with the photoelectric scanner, and the Schlieren optical system when a baseline is required. An array of centrepieces is also available for specialised purposes. The centrepiece is so designed that, when correctly aligned in the rotor, the walls will be parallel to the lines of centrifugal force, behaving on the same principle as the swinging-bucket rotor (Section 6.4.2) to give almost ideal conditions for sedimentation. This ensures that there will be no accumulation of material against the wall of the analytical cell during centrifugation. Analytical cells have upper and lower plane windows of quartz or synthetic sapphire, the latter being used with interference optics, as they have less tendency to distort under a high gravitational field. The rotor chamber contains an upper condensing lens and a lower collimating lens; the former, together with a camera lens, focuses light on to a photographic plate, whereas the latter collimates the

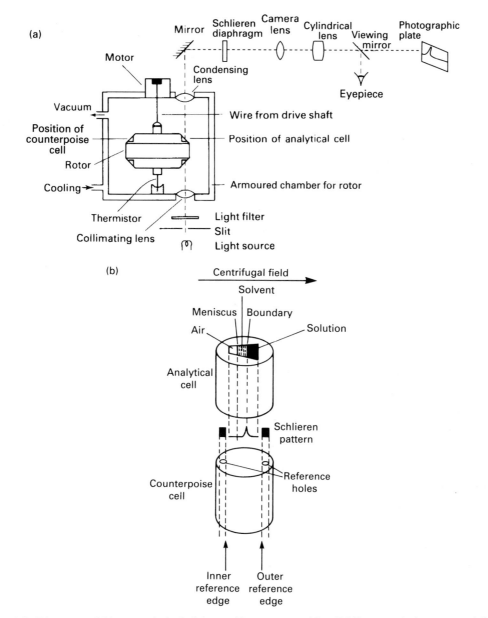

Fig. 6.4. Diagrams of (a) an analytical ultracentrifuge system with a Schlieren optical system and (b) a single sector analytical cell and a counterpoise cell.

light so that the sample cell is illuminated by parallel light. In more advanced instruments the photographic detector system is replaced by an electronic scanning system.

In the ultraviolet light absorption system, light of a suitable wavelength is passed through the moving analytical cell containing the solution under analysis, e.g. protein or nucleic acid, and the intensity of the transmitted light is recorded either on a photographic plate, or by an automatic single or split-beam photoelectric scanning system. The scanner system, unlike the ultraviolet photographic method, has the advantage of allowing direct visualisation of the results during the course of the experiment and can provide a plot of the concentration of the sample at all points in the analytical cell at any particular time. Different wavelengths of light can be selected, enabling the separate movement of single components in a mixture of substances to be monitored, provided that they absorb light at different wavelengths. The Schlieren optical system makes use of the fact that, if light passes through a solution of uniform concentration it does not deviate but, on passing through a solution having different density zones, it is refracted at the boundary between these zones. The optical system records the change in refractive index of the solution, which will vary as the concentration changes. Two Schlieren optical systems are available, differing in the way the deviated light is treated. In the optical Schlieren system the deviated light is passed through an inclined Schlieren diaphragm and a cylindrical lens. In the case of sedimenting materials in an analytical cell, a boundary is formed between the solvent, which has been cleared of particles, and the remainder of the solution containing the sedimenting material. This behaves like a refraction lens, resulting in the production of a peak in the final image on the photographic plate, which is used as the detector system. The peak is an exact record of the refractive index gradient and the area beneath it is proportional to the concentration of solute. As sedimentation proceeds, the boundary, and hence the peak, shifts, and the rate at which the peak moves gives a measure of the rate at which the material is sedimenting (Fig. 6.5). After a period of sedimentation, the peak height diminishes and the width increases, owing to radial dilution of the sample due to the sector shape of the cell. Its area, however, is unchanged. In the bright-field scanning optical system the deviated light is interrupted by a knife edge and the resultant image scanned with a photomultiplier, the resulting refractive index gradient appearing as a dark band against a bright background. The Schlieren optical system plots the refractive index gradient against distance along the analytical cell, which makes it useful for locating boundaries in sedimentation velocity measurements (Section 6.10.1). For some techniques (e.g. the sedimentation equilibrium method for relative molecular mass determinations, (Section 6.10.1)) the Schlieren system is not sufficiently sensitive to detect small concentration differences. Use is therefore made of the more sensitive Rayleigh interference system, which employs a double-sector cell, in which one sector contains the solvent and the other the solution. The optical system measures the difference in refractive index between the reference solvent and the solution by the displacement of interference fringes caused by slits placed behind the two liquid

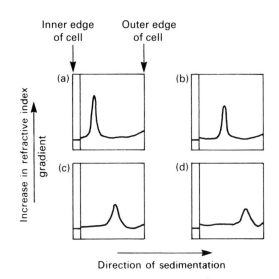

Fig. 6.5. Diagrams of the stages ((a) to (d)) of sedimentation of a macromolecule, using the technique of sedimentation velocity. Sedimentation patterns are obtained using the Schlieren optical system, which measures the refractive index gradient at each point in the cell at varying time intervals.

columns, each fringe tracing a curve of the refractive index gradient against distance in the cell. Since the position of the fringes is determined by solute concentration, it is possible to measure the concentration of solute at any point along the cell.

6.4   Design and care of preparative rotors

6.4.1   Materials used in rotor construction

The centrifugal force created by a spinning rotor generates load or stress on the rotor material. Since rotors used for low speed centrifugation experience a much lower degree of stress in comparison to that of high speed rotors they can be made of brass, steel or Perspex. The higher stress forces generated during high speed centrifugation necessitate the use of aluminium alloy or titanium alloy. Rotors made from titanium alloy have a greater strength-to-weight ratio and are therefore capable of withstanding nearly twice the centrifugal force of rotors made from aluminium alloy. They are also more resistant to chemical corrosion and are less subject to metal fatigue. Aluminium alloy rotors, although less expensive, are far more susceptible to corrosion, being readily attacked by acids and alkalis, and high concentrations of salt solutions (e.g. NaCl and KBr used in lipoprotein fractionations, ammonium sulphate used in protein precipitation, and caesium or rubidium salts used in the preparation of density gradients). All rotors are therefore given a durable protective coating to the metal surface either by anodising in the case of aluminium rotors or applying a black epoxy paint. Concomitant with improvements in rotor safety has been the introduction of materials in rotor construction such as carbon fibre, which is said to produce a lighter rotor with a better strength-to-weight ratio.

6.4.2　Swinging-bucket rotors

The swinging-bucket rotor (Fig.6.1a) has buckets that start off in a vertical position but during acceleration of the rotor swing out to a horizontal position so that during centrifugation the tube, and hence the solution in the tube, is aligned perpendicular to the axis of rotation and parallel to the applied centrifugal field, the tube returning to its original position during deceleration of the rotor. Since the centrifugal field is axial, particles in a centrifugal field fan out radially from the centre of rotation rather than sedimenting in parallel lines. Some particles strike against the wall of the tube and then travel down the wall to the tube base, causing convection currents, referred to as wall effects, that disrupt the sample zones. Undesirable swirling effects that cause mixing of the tube contents are also produced during rotor acceleration and deceleration. Control of convection and swirling effects has been achieved by slowly accelerating and decelerating the rotor, and by the use of density gradient centrifugation techniques (Section 6.6.2), where the particles now sediment to form discrete bands (Fig. 6.1b to e).

6.4.3　Fixed-angle rotors

In fixed-angle rotors (Fig. 6.2a), the tubes are located in holes in the rotor body set at a fixed angle of between 14° and 40° to the vertical. Under the influence of the centrifugal field, which is exerted at an angle to the tube wall, particles move radially outwards and have only a short distance to travel before colliding with, and precipitating on, the outer wall of the centrifuge tube. A region of high concentration is formed that has a density greater than that of the surrounding medium, with the result that the precipitate sinks and collects as a small compact pellet at the outermost point of the tube. The combination of short path to the tube wall and consequent convection caused by the dense layer against the outer wall gives a rapid collection of the sample. Although the strong convective flow produced tends to have an undesirable effect when attempts are made to separate particles of similar sedimentation characteristics, the fixed-angle rotor has proved valuable for the differential separation of particles whose sedimentation rates differ by a significant order of magnitude.

Since the pocket retaining the centrifuge tube is fixed, the solution in the centrifuge tube reorients during rotor acceleration, reorienting back to its original position during deceleration of the rotor (Fig. 6.2b to e). For isopycnic centrifugation (Section 6.6.2) reorientation of the solution in the tubes has the advantage of enhancing the loading capacity of the gradient and increasing the resolution of the sample bands.

6.4.4 Vertical tube rotors

The vertical tube rotor (Fig. 6.3a) can be regarded as a zero-angle fixed-angle rotor in that the tubes are aligned vertically in the body of the rotor at all times. During operation of the rotor, the solution in the tube reorients through 90° during rotor acceleration to lie perpendicular to the axis of rotation and parallel to the applied centrifugal field and returns to its original position during deceleration of the rotor (Fig. 6.3b to g). Since sedimentation of particles occurs across the diameter of the tube, the vertical tube rotor presents the shortest possible pathlength for the particle. Sedimentation of particles can thus be achieved more quickly than by using either the fixed-angle rotor or the swinging-bucket rotor, and, since r_{min} is greater, due to the tubes being at the rotor's edge, a larger minimum centrifugal field is generated. In this type of rotor, however, any pellet formed is deposited along the entire length of the outer wall of the centrifuge tube, which could be a disadvantage because it tends to fall back into the solution at the end of centrifugation.

6.4.5 Zonal rotors

Zonal rotors may be of the batch or continuous flow type, the former being more extensively used than the latter, and are designed to minimise the wall effects that are encountered in swinging-bucket and fixed-angle rotors, and to increase sample size.

Several different types of batch-type rotor are available that differ in the method by which they are loaded and unloaded. Low speed batch rotors, designed to operate near 5000 rev min^{-1} (5000 g) are made of aluminium, having a thick transparent Perspex top and bottom to permit direct examination of particle sedimentation during the course of centrifugation. High speed batch rotors are made of aluminium or titanium alloy and can operate at speeds up to 60 000 rev min^{-1} (256 000 g). The body of a typical batch-type rotor is either a large cylindrical container or a hollow bowl, in which the rotor volume varies with the square of the radial distance from the centre of rotation. The centre of the rotor has a core to which is attached a vane assembly that divides the rotor internally into four sector-shaped compartments and minimises swirling of the rotor contents. The vanes or septa have radial ducts to allow gradient to be pumped to the periphery of the rotor from the centre core. The rotor is enclosed by a threaded lid. The capacity of batch-type rotors range from $300 \, cm^3$ to $2000 \, cm^3$ with the gradient material filling the entire enclosed space. The rotor core may be of two main types. The most commonly used standard core permits the loading and unloading of the rotor while it is spinning (dynamic method), whereas the second core type (the reorienting gradient core) is designed to allow the rotor to be loaded and unloaded with a reorienting gradient while it is at rest (static method).

In the dynamic mode of operation, loading of the standard core type rotor is achieved while the rotor is revolving at approximately 2000 rev min^{-1}. The lighter end of the pre-formed gradient is pumped into the rotor first through a fixed or a removable seal to emerge at the periphery and form a uniform layer held in a vertical orientation against the outer rotor wall by centrifugal force (Fig. 6.6a). The successive addition of denser gradient results in a continuous centripetal displacement of the lighter gradient towards the rotor core (Fig. 6.6b). When the gradient has been pumped into the rotor, fluid cushion, as dense or denser than the heaviest end of the pre-formed gradient, is introduced into the rotor to fill it completely. The sample is then introduced by the fluid line leading to the rotor centre (Fig. 6.6c) from which it is subsequently displaced by the addition of an overlay of low density liquid (Fig. 6.6d), an equal volume of cushion being displaced from the periphery. After removal of the gradient lines to the rotor, the rotor is accelerated to the operating speed, for the required time interval, to give either rate zonal or isopycnic zonal separation (Fig. 6.6e). Recovery of the gradient and separated particles is then accomplished by decelerating the rotor to its original 2000 rev min^{-1} and the rotor contents displaced, lighter end first, by introducing additional cushion to the periphery of the rotor (Fig. 6.6f). A modified rotor core is available that adds versatility to the operation of the zonal rotor by allowing fractions to be recovered at the rotor's edge as well as its centre. Edge loading is accomplished by the introduction of a buffer or distilled water at the rotor centre and collecting the fractions through the edge ports of the core. This modification has the advantage that it is more economic in the use of the displacing gradient.

In the static method of zonal centrifugation, using the reorienting (Reograd) gradient core, the sample solution is layered on top of a density gradient in the rotor while it is at rest (Fig. 6.7a). The rotor is then slowly accelerated to about 1000 rev min^{-1} to prevent mixing of the rotor contents, during which time the sample and gradient layers reorient under centrifugal force (Fig. 6.7b). Near operating speeds the zones approach a vertical orientation, and at very high speeds, where the ratio between the centrifugal force and the acceleration due to gravity in a downward direction is very high, the zones become vertical (Fig. 6.7c). Particle

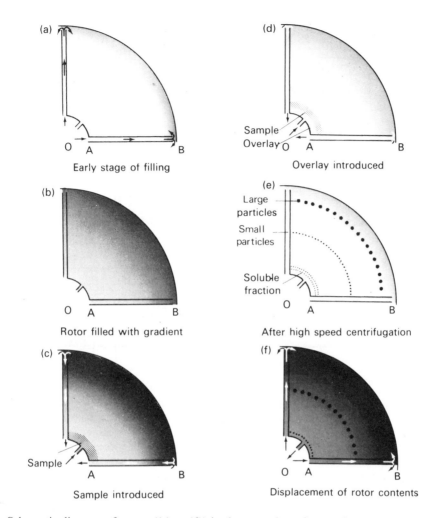

(a) Early stage of filling

(b) Rotor filled with gradient

(c) Sample introduced

(d) Overlay introduced

(e) After high speed centrifugation

(f) Displacement of rotor contents

Fig. 6.6. Schematic diagram of stages ((a) to (f)) in the operation of a zonal rotor. (Reproduced with kind permission of Measuring and Scientific Equipment Ltd.)

separation occurs with the rotor at its operating speed (Fig. 6.7d). At the completion of the separation the rotor is decelerated from its operating speed to 1000 rev min^{-1} and then slowly decelerated to rest. This is to prevent mixing of the rotor contents when sample and gradient layers reorient back to the horizontal position (Fig. 6.7e). With the rotor at rest, the contents can be displaced from it and recovered either by drawing the contents out of the bottom of the rotor or by displacing the gradient out through the top (Fig. 6.7f). Static loading and unloading of the rotor is most suitable for the isolation of long, fragile particles, such as DNA strands, which would otherwise be damaged by the rotating seal assembly used in the dynamic method and has become an accepted method for the separation

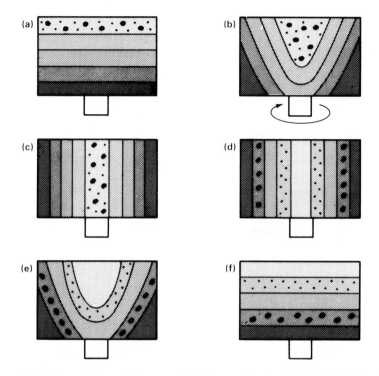

Fig. 6.7. Schematic diagram of a reorienting gradient (Reograd) rotor system. (a) Rotor is filled at rest with density gradient and sample layer. (b) Rotor is accelerated and layers reorient under centrifugal force. (c) Layers become vertical at sufficiently high rotor speed. (d) Particles now separate through the gradient with the rotor at speed. (e) During rotor deceleration, layers containing separated particles reorientate. (f) At rest, rotor contents are displaced and various zones recovered (small dots represent small-sized particles; large dots represent large-sized particles.)

of lipoprotein fractions from large volumes of plasma or serum. The design of all of the batch-type rotor cores enables the zones to be collected without any appreciable loss of resolution achieved during centrifugation. The static and dynamic methods give equally good resolutions. To aid zone isolation, the gradient emerging from the rotor is passed through a suitable monitoring device, for example a photocell to determine protein content by its ultraviolet absorption at 280 nm wavelength (Section 4.4), or a suitable monitor to detect radioactivity (Section 5.2), and then collected into fractions to be analysed for concentration of gradient material (using a refractometer), for specific biological activity, or for some other appropriate property.

Batch-type zonal rotors have been used to remove contaminating proteins from a variety of preparations and for the separation and isolation of hormones, enzymes, macroglobulins, ribosomal subunits, viruses and subcellular organelles from animal or plant tissue homogenates.

Continuous flow zonal rotors are designed for high speed separation of relatively small quantities of solid matter from large volumes of suspension and are particularly useful for the harvesting of cells and the large-scale isolation of viruses. The rotors are similar in shape to batch-type zonal rotors, but differ in the design of the core because of the different fluid flow patterns in the rotor. In operation the suspension is continuously fed into the rotor during centrifugation at a flow rate of up to 9 dm^3 h^{-1}. The sedimenting material moves into the rotor chamber, while the particle-free effluent leaves the rotor through an outlet line. The rotor may be operated without a density gradient, in which case the sample particles are allowed to pellet on the rotor wall and recovered after the rotor has been stopped at the conclusion of the run. Alternatively, a density gradient may be used and the particles separated according to differences in their density. The separated bands may then be recovered in a manner similar to that described for the operation of the batch-type zonal rotor using the standard core.

6.4.6 Elutriator rotors

The elutriator rotor (Fig. 6.8a) is a type of continuous flow rotor that contains recesses to hold a single conical-shaped separation chamber, the apex of which points away from the axis of rotation, and a by-pass chamber on the opposite side of the rotor that serves as a counterbalance and to provide the fluid outlet. The separation chamber may be of two types that differ in capacity and have slightly different shapes. With the aid of a windowed centrifuge door and a stroboscopic lamp, whose flash rate is synchronised with the rotor speed, apertures in the rotor allow the rotor chamber contents to be observed during centrifugation. Particles suspended in a uniform low density medium are pumped into the rotor chamber at its peripheral edge via a rotating seal assembly while the rotor is spinning at a preselected speed (usually between 1000 and 3000 rev min^{-1}). Since the separation chamber is conically shaped (Fig. 6.8b), a gradient of liquid flow velocity, which gradually decreases as the diameter of the chamber increases towards its centripetal end (i.e. towards the axis of rotation), will therefore exist in the chamber that opposes the applied centrifugal field. The tendency of particles of differing sedimentation rate to sediment in the centrifugal field (Fig. 6.8b(i)) is therefore balanced against the controlled flow of liquid being pumped through the separation chamber in the opposite direction towards its centripetal end. Particles then band in the separation chamber at a position where their sedimentation velocity, which in the uniform low density medium used will be proportional to their size, is balanced by liquid flow rate in the opposite direction (Fig. 6.8b(ii)). Larger particles accumulate towards the centrifugal end of the chamber where the liquid flow velocity is high, while the smaller particles accumulate towards the centripetal end

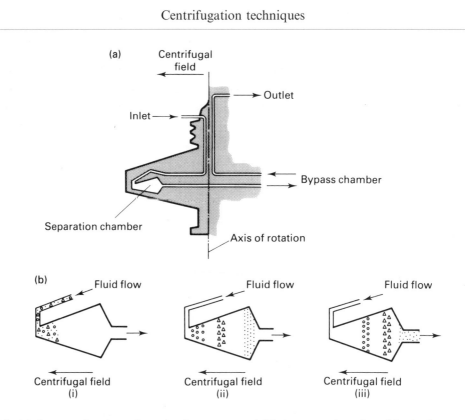

Fig. 6.8. (a) Cross-section through an elutriator rotor and (b) the separation of particles in the separation chamber of an elutriator rotor by centrifugal elutriation.

of the chamber where the liquid flow velocity is low. Either by a stepwise decrease in rotor speed or by a stepwise increase in liquid flow rate through the separation chamber, collection of the separated uniformly sized particles can be made centripetally in order of successively increasing diameter by elutriation from the chamber (Fig. 6.8b(iii)).

With the technique of centrifugal elutriation (Section 6.6.3) the elutriator rotor has been used successfully to separate various cell types from mammalian testis, different types of monocytes and lymphocytes from human blood, to purify Kupffer and endothelial cells from sinusoidal liver cells and fat storing cells from rat liver, and for the bulk separation of rat brain cells and the fractionation of yeast cell populations.

6.4.7   Care of rotors

The protective anodised coating on aluminium rotors is very thin (approximately 0.025 mm) and does not provide a high degree of protection against corrosion, thus rotors should always be handled with care to prevent scratching. The use of acid solutions, strong alkaline detergents (e.g. Decon 90), and salt solutions can also easily damage the protective coating, leading to corrosion and eventually failure of the rotor. After use, therefore, rotors should be thoroughly washed, preferably with deionised water and, because moisture is a potential source of corrosion, left to drain and dry upside down in a warm atmosphere; they should then be stored in a clean, dry environment. Rotor outer surfaces only can be given a protective coat of lanolin or silicone polish. Swinging-bucket rotors, however, should never be completely immersed in water because the bucket hanging mechanism is difficult to dry and can rust. Titanium rotors are essentially resistant to corrosion. However, rotors made of titanium alloy containing aluminium or ferrous compounds are prone to corrosion.

It is important to note that all rotors are designed to carry a maximum load at a maximum speed, which is based on the rotor tubes or bottles being filled with a solution whose density is no greater than 1.2 g cm^{-3}. A reduction in the maximum rated speed (known as derating the rotor) is therefore required if the density of the solution exceeds this value. This reduction can be calculated from the equation:

$$M_n = \sqrt{\left(\frac{1.2 \times M^2}{N}\right)} \qquad (6.16)$$

where M is the usual maximum rotor speed using a solution of density 1.2 g cm^{-3} and M_n is the new maximum rotor speed when a solution with a density of $N \text{ g cm}^{-3}$ is used. Speed reductions are also required when stainless steel caps and/or tubes are used and to prevent recrystallisation of high density salt solutions, which could occur when the salt concentration exceeds the solubility limit of the solution.

To prevent possible damage to the drive shaft of the centrifuge due to vibration caused by rotor imbalance, sample loads should be balanced within the limits specified by the manufacturer – each opposing pair of sample containers being balanced individually, and the total load balanced symmetrically in the rotor. Swinging-bucket rotors should not be run with any buckets or caps removed or individual rotor buckets interchanged, because they form an integral part of the balance of the rotor.

During acceleration and deceleration of the rotor, cyclic stretching and relaxing of the metal can cause metal fatigue, leading to the eventual failure of the rotor. To avoid overstressing the rotor and ensure its continued safe operation an accurate record should be kept of its total usage, i.e. the number of runs (at any speed up to its maximum number of revolutions per minute) and the time of each run, so that the rotor can either be derated after a certain number of runs (e.g. 1000) or hours of centrifugation (e.g. 2500) or replaced after a set period of time, as specified by the manufacturer.

6.5   Sample containers

Centrifuge tubes and bottles are manufactured in a range of different sizes ($100 \, mm^3$ to $1 \, dm^3$), in varying thickness and rigidity and from a wide variety of materials including glass, cellulose esters, polyallomer, polycarbonate, polyethylene, polypropylene, kynar (a high relative molecular mass homopolymer of vinylidene fluoride), nylon and stainless steel.

The correct choice of sample container is important in order to achieve the desired degree of separation of particles from a sample mixture. It is important before commencing centrifugation to consult manufacturers' technical literature to determine the limitations of the material from which the container is made. The type of container used will depend upon such factors as the nature and volume of the sample to be centrifuged, the type of rotor to be used, the available centrifuge, the centrifugal forces to be withstood, its chemical resistance to various solvents, upper and lower temperature limits and physical properties, i.e. whether the tube is transparent or opaque, and can be sliced or punctured for postcentrifugation analysis.

Glass centrifuge tubes are usually suitable only for centrifugation at low speeds because they disintegrate in higher centrifugal fields. Thin-walled tubes may be used in swinging-bucket rotors because the tube is protected from the forces trying to deform it by the surrounding bucket; however, thick-walled tubes are usually required when fixed-angle and vertical tube rotors are used because the large forces exerted on the tube tend to collapse it. Centrifuge tubes and bottles should always be filled to the correct level, and maximum allowable rotor speeds, depending on the particular container, observed. The need to cap bottles and tubes usually depends upon the speed at which the sample is to be centrifuged, the type of container and the nature of the sample. Ideally, tubes used in fixed-angle and vertical tube ultracentrifuge rotors should always be completely filled in order to support the tube against the very high centrifugal forces generated. It is especially important that tubes are capped and sealed with special leak-proof sealing caps when material that may be a biohazard or radioactive is being centrifuged. Capping and sealing of tubes used in vertical tube rotors and most fixed-angle rotors are also important, because the tubes have to withstand the large upward hydrostatic force generated during centrifugation by the liquid in the tube.

6.6   Separation methods in preparative ultracentrifuges

6.6.1   Differential centrifugation

This method is based upon the differences in the sedimentation rate of particles of different size and density. As can be seen from equation 6.12, centrifugation will initially sediment the largest particles. For particles of the same mass but different density, the ones with the highest density (e.g. peroxisomes, $\rho = 1.23\,g\,cm^{-3}$ in sucrose solution) will sediment at a faster rate than the less dense particles (e.g. plasma membranes, $\rho = 1.16\,g\,cm^{-3}$ in sucrose solution) of similar mass. Particles having similar banding densities (e.g. most of the subcellular organelles, where $\rho = 1.1$ to $1.3\,g\,cm^{-3}$ in sucrose solution) can usually be efficiently separated one from another by differential centrifugation or the rate zonal method (Section 6.6.2), provided there is about a ten-fold difference in their mass.

In differential centrifugation, the material to be separated (e.g. a tissue homogenate) is divided centrifugally into a number of fractions by increasing (stepwise) the applied centrifugal field. The centrifugal field at each stage is chosen so that a particular type of material sediments, during the predetermined time of centrifugation, to give a pellet of particles sedimented through the solution, and a supernatant solution containing unsedimented material. Any type of particle originally present in the homogenate may be found in the pellet or the supernatant or both fractions depending upon the time and speed of centrifugation and the size and density of the particle. At the end of each stage, the pellet and supernatant are separated and the pellet washed several times by resuspension in the homogenisation medium followed by recentrifugation under the same conditions. This procedure minimises cross-contamination, improves particle separation and eventually gives a fairly pure preparation of pellet fraction. To appreciate why the pellet is never absolutely pure (homogeneous), however, it is necessary to consider the conditions prevailing in the centrifuge tube at the beginning of each stage.

Initially all particles of the homogenate are homogeneously distributed throughout the centrifuge tube (Fig. 6.9a). During centrifugation, particles move down the centrifuge tube at their respective sedimentation rates (Fig. 6.9b to e) and start to form a pellet on the bottom of the centrifuge tube. Ideally, centrifugation is continued long enough to pellet all the largest class of particles (Fig. 6.9c), the resulting supernatant then being centrifuged at a higher speed to separate medium-sized particles and so on. However, since particles of varying sizes and densities were distributed homogeneously at the commencement of centrifugation, it is evident that the pellet will not be homogeneous, but will contain a mixture of all the sedimented components, being enriched with the fastest (heaviest) sedimenting particles. In the time required for the complete sedimentation of heavier particles,

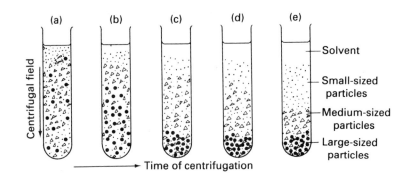

Fig. 6.9. Differential sedimentation of a particulate suspension in a centrifugal field. (a) Particles are uniformly distributed throughout the centrifuge tube. (b) to (e) Sedimentation of particles during centrifugation is dependent upon their size, shape and density.

some of the lighter and medium-sized particles, originally suspended near the bottom of the tube, will also sediment and thus contaminate the fraction. Pure preparations of the pellet of the heaviest particle cannot therefore be obtained in one centrifugation step. It is only the most slowly sedimenting component of the mixture remaining in the supernatant liquid after all the larger particles have been sedimented that can be purified by a single centrifugation step, but its yield is often very low.

The separation achieved by differential centrifugation may be improved by repeated (two or three times) resuspension of the pellet in homogenisation medium and recentrifugation under the same conditions as in the original pelleting, but this will inevitably reduce the yield obtained. Further centrifugation of the supernatant in gradually increasing centrifugal fields results in the sedimentation of the intermediate and finally the smallest and least dense particles. A scheme for the fractionation of rat liver homogenate into the various subcellular fractions is given in Fig. 6.10. In spite of its inherent limitations, differential centrifugation is probably the most commonly used method for the isolation of cell organelles from homogenised tissue.

6.6.2   Density gradient centrifugation

There are two methods of density gradient centrifugation, the rate zonal technique and the isopycnic (isodensity or equal density) technique, and both can be used when the quantitative separation of all the components of a mixture of particles is required. They are also used for the determination of buoyant densities and for the estimation of sedimentation coefficients.

Particle separation by the rate zonal technique is based upon differences in the

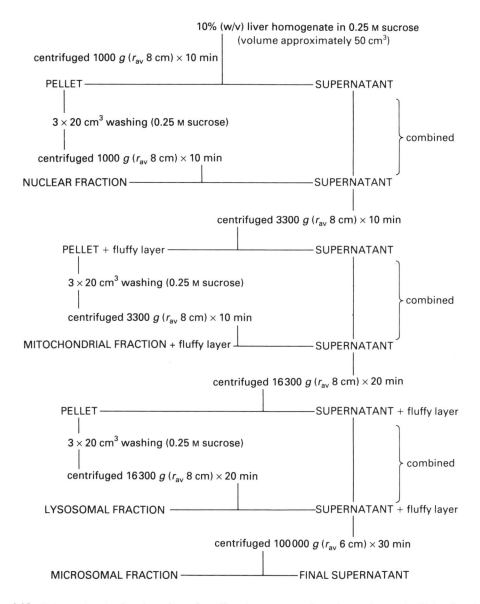

Fig. 6.10. Scheme for the fractionation of rat liver homogenate into the various subcellular fractions.

size, shape and density of the particles, the density and viscosity of the medium and the applied centrifugal field. However, since similar types of biological particle are often similar in shape and their densities fall into a relatively narrow range, and the maximum density of the gradient is chosen not to exceed that of the densest particle to be separated, separation of similar types of particles by the rate zonal technique is based mainly upon differences in their size. Subcellular organelles, therefore, such as mitochondria, lysosomes and peroxisomes, which have different densities but are similar in size, do not separate efficiently using this method, but the separation of proteins of similar density and differing only three-fold in relative molecular mass can be achieved easily. The technique involves carefully layering a sample solution on top of a pre-formed liquid density gradient, the highest density of which does not exceed that of the densest particles to be separated. The function of the gradient is primarily to stabilise the liquid column in the tube against movement resulting from convection currents and secondarily to produce a gradient

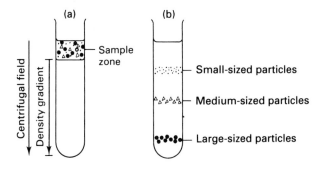

Fig. 6.11. Rate separation and isopycnic separation using a density gradient. (a) Mixture of particles layered on top of a pre-formed liquid density gradient prior to centrifugation; (b) centrifugation of particles. For rate separation, the required fraction does not reach its isopycnic position. For isopycnic separation, centrifugation is continued until the desired particles have reached their isopycnic position in the gradient.

of viscosity that helps to improve the resolution of the gradient. The sample is then centrifuged until the desired degree of separation is effected, i.e. for sufficient time for the particles to travel through the gradient to form discrete zones or bands (Fig. 6.11), which are spaced according to the relative velocities of the particles. Since the technique is time dependent, centrifugation must be terminated before any of the separated zones pellet at the bottom of the tube. The technique has been used for the separation of enzymes, hormones, RNA–DNA hybrids, ribosomal subunits, subcellular organelles, for the analysis of size distribution of samples of polysomes and lipoprotein fractionations and may be adapted for bulk preparative work using a zonal rotor (Section 6.4.5).

Isopycnic centrifugation depends solely upon the buoyant density of the particle and not its shape or size and is independent of time, the size of the particle affecting only the rate at which it reaches its isopycnic position in the gradient. The technique is used to separate particles of similar size but of differing density. Hence soluble proteins, which have a very similar density (e.g. $\rho = 1.3\,\mathrm{g\,cm^{-3}}$ in sucrose solution), can not usually be separated by this method, whereas subcellular organelles (e.g. Golgi apparatus ($\rho = 1.11\,\mathrm{g\,cm^{-3}}$), mitochondria ($\rho = 1.19\,\mathrm{g\,cm^{-3}}$), and peroxisomes

($\rho = 1.23$ g cm^{-3}) in sucrose solution) can be effectively separated.

The methods are a combination of sedimentation and flotation and involve layering the sample on top of a density gradient that spans the whole range of the particle densities that are to be separated. The maximum density of the gradient, therefore, must always exceed the density of the most dense particle. While it is possible to use a discontinuous or stepwise gradient, it is preferable to use a continuous gradient (Section 6.7.1). During centrifugation, sedimentation of the particles occurs until the buoyant density of the particle and the density of the gradient are equal (i.e. where $\rho_p = \rho_m$ in equation 6.12). At this point of isodensity no further sedimentation occurs, irrespective of how long centrifugation continues, because the particles are floating on a cushion of material that has a density greater than their own. Isopycnic centrifugation, in contrast to the rate zonal technique, is an equilibrium method, the particles banding to form zones each at their own characteristic buoyant density (Fig. 6.11). In cases where, perhaps, not all the components in a mixture of particles are required, a gradient range can be selected in which unwanted components of the mixture will sediment to the bottom of the centrifuge tube while the particles of interest sediment to their respective isopycnic positions. Such a technique involves a combination of both the rate zonal and isopycnic approaches.

As an alternative to layering the particle mixture to be separated on to a pre-formed gradient, the sample is initially mixed with the gradient medium to give a solution of uniform density, the gradient self-forming, by sedimentation equilibrium, during centrifugation. In this method (referred to as the equilibrium isodensity method), use is generally made of the salts of heavy metals (e.g. caesium or rubidium), sucrose, colloidal silica or Metrizamide. The sample (e.g. DNA) is mixed homogeneously with, for example, a concentrated solution of caesium chloride (Fig. 6.12a). Centrifugation of the concentrated caesium chloride solution results in the sedimentation of the CsCl molecules to form a concentration gradient and hence a density gradient. The sample molecules (DNA), which were initially uniformly distributed throughout the tube, now either rise or sediment until they reach a region where the solution density is equal to their own buoyant density, i.e. their isopycnic position, where they will band to form zones (Fig. 6.12b). This technique suffers from the disadvantage that often very long centrifugation times (e.g. 36–48 h) are required to establish equilibrium. However, it is commonly used in analytical centrifugation to determine the buoyant density of a particle, the base composition of double-stranded DNA and to separate linear from circular forms of DNA. Many of the separations can be improved by increasing the density differences between the different forms of DNA by the incorporation of heavy isotopes (e.g. ^{15}N) during biosynthesis, a technique used by M. Meselson and F. Stahl to elucidate the mechanism of DNA replication in *Escherichia coli*, or by the

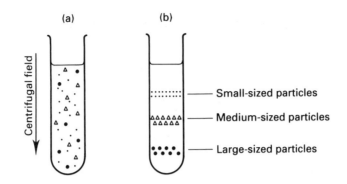

Fig. 6.12. Isopycnic centrifugation using the equilibrium isodensity method. (a) Particles distributed homogeneously throughout the tube prior to centrifugation. (b) During centrifugation the gradient is allowed to establish itself, sample particles redistribute and band in a series of zones at their respective isopycnic positions.

binding of heavy-metal ions or dyes such as ethidium bromide. Isopycnic gradients have also been used to separate and purify viruses and analyse human plasma lipoproteins, to study the variation of buoyant density with pH for proteins and homopolypeptides and has found wide application in the analysis of nucleic acids.

6.6.3   Centrifugal elutriation

In this technique the separation and purification of a large variety of cells from different tissues and species can be achieved by a gentle 'washing action' using an elutriator rotor (Section 6.4.6 and Fig. 6.8). The technique is based upon differences in the equilibrium, set up in the separation chamber of the rotor, between the opposing centripetal liquid flow and applied centrifugal field being used to separate particles mainly on the basis of differences in their size. The technique does not employ a density gradient and has the advantage that any medium totally compatible with the particles can be used, e.g. buffered salt solutions or culture medium, and, because pelleting of the particles does not occur, fractionation of delicate cells or particles, between 5 and 50 μm diameter, can be achieved with minimum damage so that cells retain their viability. Separations can be achieved very quickly, giving high cell concentrations and a very good recovery yield.

6.7 Performing density gradient separations

6.7.1 Formation and choice of density gradients

All density gradient methods involve a supporting column of liquid whose density increases towards the bottom of the centrifuge tube. The function of the density gradient in rate zonal centrifugation is to stabilise the column of liquid in the centrifuge tube, prevent mixing of the separated particles due to convection currents, improve resolution of the separated components by eliminating, or largely alleviating, factors such as mechanical vibration and thermal gradients that disturb the smooth migration of particles through the suspending medium, and permit the quantitative separation of several or all components in a mixture. However, in isopycnic centrifugation the prime function of the density gradient is to band the particles at their buoyant densities.

Density gradients of different shape (i.e. with a concentration profile of gradient medium along the tube) may be produced by techniques that fall into two major groups, the discontinuous or step gradient technique and the continuous density gradient technique.

In the discontinuous technique, two methods are available to produce the initial step gradient. In the overlayering method, known volumes of solutions of decreasing density are allowed to run slowly down the side of the centrifuge tube to form layers over each other in the centrifuge tube. Alternatively, in the underlayering method, a layer of the lightest density solution is introduced into the centrifuge tube; this is then successively underlayered, by means of a pipette placed with its tip at the bottom of the centrifuge tube, with a series of layers (corresponding to the required number of steps) of solution of increasing density, which when added displaces the less dense solution. The discontinuous gradient can be used directly by layering of the sample to be separated as a narrow zone on the top (lowest density) layer. The tube is then centrifuged under the appropriate experimental conditions. Alternatively, if the discontinuous gradient is allowed to stand, the layers slowly merge by diffusion to produce a continuous linear gradient, the time required depending upon such factors as, the concentration range and viscosity of the gradient, thickness of each layer, temperature, and relative molecular mass of the solute.

The continuous density gradient technique is probably the more common of the two, and requires the use of a special piece of apparatus known as a gradient maker, many varieties of which are commercially available and range from the simplest equipment capable of generating either linear, convex- or concave-shaped gradients to sophisticated programmable gradient makers, used mainly in centrifugation studies with zonal rotors, that are capable of producing more complex gradient shapes. Gradient makers capable of forming linear gradients, i.e. where the density of the medium in the centrifuge tube increases linearly with increasing distance from the axis of rotation, consist of two precision-bored cylindrical chambers of identical cross-sectional area that are interconnected at their base by a tube containing a control valve that allows the mixing of the contents of the two chambers to be regulated. One chamber (the mixing chamber) contains a stirrer and possesses an outlet through which the gradient is drawn off, via a length of narrow bore flexible tubing, to the centrifuge tube. If the wall of the centrifuge tube is reasonably wettable by aqueous solutions, as in the case of polypropylene, cellulose nitrate and cellulose acetate butyrate tubes, the outlet pipe from the gradient maker may be allowed to touch the wall of the centrifuge tube near to the top of the tube and the stream of liquid allowed to run slowly down the side of the tube, as it is filled, on to the top of the forming gradient.

Using this top loading method the mixing chamber of the gradient mixer is filled with the most dense solution and the second chamber (the reservoir) filled with an equal amount (by weight) of a less dense solution. The hydrostatic pressures of the two liquid columns need to be equal otherwise liquid will flow through the connecting tube as soon as the control valve is opened. The dense liquid is then allowed to run slowly through the filling pipe, from the mixing chamber and down the wall of the centrifuge tube, and is immediately replaced in the mixing chamber, as a result of hydrostatic pressure, by an equivalent amount of less dense solution via the control valve. This re-establishes hydrostatic equilibrium. As a result the concentration and density of the solution in the mixing chamber, which is maintained homogeneous by constant stirring, constantly and linearly decreases as the device empties. The concentration of the gradient in the centrifuge tube will therefore decrease in a linear manner as the tube is filled.

The top loading method, however, is not suitable for use with centrifuge tubes made from materials that are hydrophobic and not wettable by aqueous solutions, such as polycarbonate or polyallomer, because the gradient material tends to gather in large droplets at the exit of the filling pipe from the gradient maker and then fall into, and disturb, the forming gradient. This problem may be overcome by bottom loading the gradient into the centrifuge tube. In this method, the less dense solution is now placed in the mixing chamber of the gradient maker and the more dense solution placed in the reservoir. By way of a probe that now leads to the bottom of the centrifuge tube less dense solution is allowed to enter the centrifuge tube first and is then displaced by solution of increasing density. Non-linear gradients (i.e. either convex or concave in concentration as a function of volume) may be produced by choosing chambers with varying cross-sectional areas. Alternatively, two mechanically driven syringes containing solutions of different densities may be used. In this case the shape of the gradient may be varied by altering the speed of pumping of one syringe with respect to that of the other.

Gradients of different shape are designed for specific purposes and are important in achieving the desired separation and purification and for the determination of such properties as the buoyant density and sedimentation coefficient of a particle. Discontinuous and continuous gradients are used in isopycnic centrifugation but discontinuous gradients are not often used in rate zonal centrifugation because artefactual bands of material can collect at the interfaces. Discontinuous gradients have been found to be the most suitable for the separation of whole cells or subcellular organelles from plant or animal tissue homogenates and for purification of some viruses. For most purposes linear gradients are used, because the gradual change in density along the gradient has been found to yield a much higher resolution of components such as ribosomal subunits, proteins, enzymes, hormones, and some plant viruses. In general, the greater the slope of the gradient, the better the resolution obtained, because tighter bands are formed due to increasing viscosity. However, for some applications better resolution is achieved using non-linear gradients. Thus, concave gradients can be used to separate light particles by flotation (e.g. serum lipoproteins), whereas heavier particles in the mixture quickly sediment through the less dense upper regions of the gradient to be banded in the region of high viscosity near the bottom of the centrifuge tube. Large particles such as ribosomal subunits, polyribosomes and some viruses usually require steep gradients and long gradient columns to enhance separation. This may be achieved by using rotors with long slender tubes that provide the longer gradient columns, and linear-log gradients in which the logarithm of the gradient column depth is a linear function of the logarithm of the sedimentation coefficient of the particle.

Isokinetic gradients are a specialised type of continuous gradient, generally convex and exponential in shape, and are so designed that the increase in density and viscosity of the gradient balances the increase in the centrifugal field along the tube so that at a constant rotor speed all particles of equal density will move at a constant velocity at all distances along the tube, the distance travelled by each particle being proportional to the relative centrifugal force applied, the time of centrifugation and the sedimentation coefficient of the particle. Hence, if the sedimentation coefficient of a single reference marker particle is known then the sedimentation coefficients of other particles in the same gradient may be calculated, provided each particle has the same density as the particle used as the sedimentation marker.

Most gradient shapes used in zonal rotors are the same as those used in swinging-bucket rotors. However, when preparing gradients for use in zonal rotors, an allowance must be made for the fact that, whereas the parallel walls of the test tube used in swinging-bucket rotors maintain a linear volume-to-radius relationship, the bowl-shaped cavity of a zonal rotor will alter this relationship, since its value varies with the square of the radial distance from the centre of rotation. A gradient will therefore change its shape when loaded into a zonal rotor. Gradients that are prepared linear or convex in concentration with respect to rotor volume will therefore become concave and linear in shape, respectively, when introduced into the zonal rotor.

6.7.2   Sample application to the gradient

Each gradient has a defined sample capacity, which is a function of its density slope, and if exceeded will result in a poor resolution of the sample; therefore, before the sample is applied to the density gradient, its optimum volume and concentration should be determined. The volume of the sample that can be applied to a centrifuge tube is a function of the cross-sectional area of the gradient exposed to the sample. Thus, sample volumes in the range 0.2–$0.5 \, \text{cm}^3$ may be added to tubes 1.0–$1.6 \, \text{cm}$ in diameter, and sample volumes of up to $1 \, \text{cm}^3$ to tubes having a diameter of approximately $2.5 \, \text{cm}$. Effective separation of particles in a multi-component sample would not be achieved with larger sample volumes due to insufficient radial distance in the centrifuge tube. Equally, if the sample concentration on the gradient is too high or too low then either the gradient may become overloaded, resulting in a broadening of the separated zones and loss of resolution, or difficulty may be encountered in the identification of the separated bands. For proteins and nucleic acids, sample concentrations of between $1 \, \mu\text{g}\,\text{cm}^{-3}$ and $1 \, \text{mg}\,\text{cm}^{-3}$ are recommended in sucrose gradients.

For rate zonal separations the correct method of application of the sample to the gradient is of paramount importance, because the resolving power of the gradient is greater the narrower the initial sample zone. The sample is therefore applied, usually using a narrow-bore pipette, as slowly and as gently as possible to reduce mixing with the gradient. The tip of the pipette is touched against the meniscus at the wall of the tube. The pipette is moved slightly upward to leave a thin channel of liquid between the pipette and the meniscus. In order to minimise mixing with the gradient the sample is allowed to run slowly out of the pipette and gently on to the surface of the gradient to form a sharply defined sample layer. The separation achieved in isopycnic separation is independent of the initial distribution of the sample. The volume of the sample and its initial distribution through the gradient are therefore unimportant and the sample can be carefully layered on top of a pre-formed gradient, mixed with the gradient medium in the case of self-forming gradients, or mixed with a dense solution of the gradient medium and layered under the density gradient.

6.7.3   Recovery and monitoring of gradients from centrifuge tubes

After particle separation has been achieved, it is usually necessary to remove the gradient solution in order to isolate the bands of separated material. If, however, the bands can be visually detected, recovery from the gradient can be achieved without having to unload the gradient by placing a Pasteur pipette into the band and slowly withdrawing the required material from the band. Alternatively, the band may be recovered through the side of the centrifuge tube using a hypodermic syringe. Removal of gradients from centrifuge tubes can be achieved by a number of techniques. A method of choice for most applications, however, is that of upward displacement. Either a denser solution of the gradient medium or, for example, a solution that is marketed especially for displacement unloading of gradients (such as Maxidens, an inert, non-viscous, water-immiscible organic liquid with a density of 1.9 g cm^{-3} that does not react with the gradient medium) is either pumped through a long needle passing through the gradient to the bottom of the centrifuge tube or introduced via a hollow needle piercing the bottom of the tube. The gradient is displaced upwards and the fractions removed in sequence using a syringe or pipette, or preferably, by being channelled out through an unloading cap to which is attached a collection pipe leading either to a fraction collector or directly into a flow cell of an ultraviolet spectrophotometer. A variation of this method is to fit a cap to the top of the centrifuge tube and pump either air or a less dense solution of the gradient medium into the top of the tube, the gradient then being displaced through a tube that leads from the bottom of the tube.

Alternatively, provided a pellet of cells is not present, the centrifuge tube may be punctured at its base by using a fine hollow needle. As the drops of gradient pass from the tube through the needle they may be collected by using a fraction collector, and further analysed. Analysis of the displaced gradient, in order to identify and locate the separated components, can be achieved by ultraviolet spectrophotometry (Section 7.4), refractive index measurements, scintillation counting (Section 5.2), and enzymic (Section 4.8) or chemical analysis.

6.7.4   Nature of gradient materials and their use

There is no ideal all-purpose gradient material, the choice of solute depending upon the nature of the particles to be fractionated. The gradient material should:

▲ permit the desired type of separation:
▲ be stable in solution;
▲ be inert towards biological materials and not react with the centrifuge, rotor, tubes or caps;
▲ not absorb light at wavelengths appropriate for spectrophotometric

monitoring (visible or ultraviolet range), or otherwise interfere with assaying procedures;

▲ be sterilisable, non-toxic or flammable;

▲ have negligible osmotic pressure and cause minimum changes in ionic strength, pH and viscosity;

▲ be inexpensive and readily available in pure form and capable of forming a solution covering the density range needed for a particular application without overstressing the rotor;

▲ allow easy separation of the sample material from the gradient medium without loss of the sample or its activity.

Gradient-forming materials that provide the densities required for the separation of subcellular particles include salts of alkali metals (e.g. caesium and rubidium chloride), small neutral hydrophilic organic molecules (e.g. sucrose), hydrophilic macromolecules (e.g. proteins and polysaccharides), and a number of miscellaneous compounds such as colloidal silica (e.g. Percoll and Ludox), and non-ionic iodinated aromatic compounds (e.g. Metrizamide, Nycodenz and Renograffin).

Sucrose solutions, while suffering from the disadvantages of being very viscous at densities greater than 1.1 to 1.2 $g cm^{-3}$ and exerting very high osmotic effects even at very low concentrations (i.e. at approximately 10% (w/v) concentration), have been found to be the most convenient gradient material for rate zonal separations. In some instances, however, glycerol is used, particularly for the separation of enzymes, or alternative media such as Ficoll (a copolymer of sucrose and epichlorohydrin), Metrizamide or Percoll gradients may be utilised. Non-ionic media, such as sucrose, glycerol, Metrizamide, Ficoll and Percoll are generally considered to be more gentle than ionic salts, such as caesium chloride and potassium bromide, and require lower centrifugal fields to achieve an adequate separation of particles. In the case of isopycnic separations, no one medium has proved satisfactory for the isolation of all types of biological particles. Hence a wide range of gradient media has been used for different types of biological sample. Ficoll has been successfully used instead of sucrose for the separation of whole cells and subcellular organelles by rate zonal and isopycnic centrifugation, but, although it is relatively inert osmotically at low concentrations, both osmolarity and viscosity rise sharply at higher concentrations (i.e. above 20% (w/v)). Caesium and rubidium salts are used exclusively for isopycnic separations and have been used most frequently for the separation of high density solutes such as nucleic acids. However, at high concentrations their high ionic strength and osmolarity tend to disrupt intra- and intermolecular bonds. Generally, ionic media have been used for the separation of nucleic acids, proteins and viruses, sucrose gradients for the isolation of organelles and viruses, and non-ionic iodinated aromatic compounds, because of their increased versatility, have been used for a much wider variety of applications. Some of the more commonly used gradient-forming materials and their applications are listed in Table 6.1.

6.8   Selection, efficiency and application of preparative rotors

The selection of a suitable rotor for a particular separation will depend upon such considerations as the capacity and resolving power of the rotor, the quantity and sample type, the time required for the separation, and the centrifugation technique to be used.

In practice, the swinging-bucket rotor is frequently used for rate zonal sedimentation analysis, because the long tube pathlength and minimal wall effects are advantageous in optimising particle separation, particularly with multicomponent samples. It is relatively inefficient, however, for differential centrifugation, due to the relatively long centrifugation time required because of the long pathlength of the tubes. Fixed-angle rotors have proved to be most effective when the rapid, total sedimentation of particles is required by differential centrifugation. Fixed-angle rotors, although giving good sample resolution in isopycnic centrifugation because components are banded over a larger cross-sectional area and also significantly reducing centrifugation time because of their reduced sedimentation pathlength, are seldom used for rate zonal centrifugation because undesirable wall effects can disrupt the sedimentation of the zones, thus limiting particle separation.

Vertical tube rotors are unsuitable for pelleting of particles by differential centrifugation because the pellet is distributed along the entire outer wall of the centrifuge tube and may be disturbed during deceleration of the rotor. They can, however, be used for the rate zonal separation of whole cells and subcellular organelles, but do not achieve the quality of separation obtained when the swinging-bucket rotor is used to separate macromolecules with a relative molecular mass of less than approximately 2×10^5. The main advantage of the vertical tube rotor for rate zonal centrifugation and isopycnic centrifugation is the speed of separation, because particles have a very short sedimentation pathlength across the width of the tube rather than down the length; also the minimum centrifugal field will be higher than in comparable fixed-angle rotors and swinging-bucket rotors because the tubes are located near the edge of the rotor. Although the swinging-bucket rotor, fixed-angle rotor and vertical tube rotor may be used for isopycnic centrifugation, the fixed-angle rotor and vertical tube rotor have been found to be of more use and are often used in preference to the swinging-bucket rotor because larger volumes can be handled and a better degree of resolution achieved in shorter times.

Table 6.1. *Commonly used gradient materials and their applications*

Material	Ionic strength of solution	Maximum density of aqueous solution at 20 °C (g cm⁻³)	Ultraviolet absorbance	Osmotic effect	Common uses
Caesium chloride	+++	1.91	+	+++	Banding of DNA, nucleoproteins, viruses, plasmid isolation
Caesium sulphate	+++	2.01	+	+++	Banding of DNA and RNA, purification of proteoglycans
Sodium bromide	+++	1.53	+	+++	Fractionation of lipoproteins
Sodium iodide	+++	1.90	+++	+++	Banding of DNA and RNA
Glycerol	−	1.26	+	+++	Banding of membrane fragments, protein separation
Sucrose	−	1.32	+	+++	Separation of subcellular particles, proteins, viruses, and membranes
Ficoll (Pharmacia)	−	1.17	+	+(·)	Separation of whole cells, subcellular particles, viruses
Dextran	−	1.13	+	+(·)	Separation of whole cells, banding of microsomes
Bovine serum albumin	−	1.35	+++	+	Separation of whole cells
Percoll (Pharmacia)	−	1.30	+++	+	Separation of whole cells and subcellular particles
Metrizamide (Nyegaard)	−	1.46	+++	++(*)	Separation of whole cells, subcellular particles, nuclei, ribonucleoprotein particles, membranes
Nycodenz (Nyegaard)	−	1.42	+++	++(*)	Separation of whole cells, subcellular particles, nucleoproteins, membranes, viruses

+++, High; ++, medium; +, low; −, non-ionic. (*) Osmotic effect increases almost linearly with concentration.
(·) Very low osmotic effect below 20% (w/v) concentration; increasing almost exponentially above 30% (w/v) concentration.

Table 6.2. *Preparative rotor types and their applications*

Rotor type	Centrifugation technique			
	Differential	Rate zonal	Isopycnic zonal	Elutriation
Swinging-bucket	−	++	+	0
Fixed-angle	+++	−	++	0
Vertical tube	−	+	+++	0
Batch zonal	−	+++	++	0
Continuous flow zonal	+++	−	++	0
Elutriator	0	0	0	+++

Use: +++, excellent; ++, good; +, reasonable; −, poor; 0, not applicable

Batch-type zonal rotors do not differ significantly from conventional rotors in their ability to separate particles, but holding large volumes of gradient make them especially useful for large-scale preparative separations and analytical separations where a number of tests may be required on the separated samples. The continuous flow zonal rotor gives a poor separation with the rate zonal technique, owing to the very short sedimentation pathlength (approximately 1 cm) but can be used for pelleting and isolation by isopycnic banding of particles that might otherwise be damaged by the pelleting technique.

The elutriator rotor provides a means for the rapid, high resolution, large capacity separation of whole cells and larger subcellular organelles, without the need either to use a gradient or to form a pellet. It is, however, of limited use in general teaching laboratories, owing to the requirement for expensive equipment, specialist accessories and an experienced user. The various types of rotor together with their applications are summarised in Table 6.2.

In order to determine the suitability of a preparative rotor for particle sedimentation, use can be made of the k, k' and k^* factors that are related to the dimensions and speed of the rotor and provide a measure of the efficiency of the rotor for the material that is under investigation. The k factor provides an estimate of the time (in hours) that will be required to sediment a particle of known sedimentation coefficient (in Svedberg units, S), at the maximum speed of the rotor and is calculated on the basis that the particles are suspended in a liquid medium with a density and viscosity similar to that of water. The k' and k^* factors are applied to rate zonal separations in a density gradient and provide an estimate of the time required to move a band of particles of known sedimentation coefficient to the bottom of a centrifuge tube through a linear 5% to 20% (w/w) sucrose gradient at 5 °C at the maximum speed of the rotor, the k' factor being calculated on the known density of the particle and the k^* factor calculated on the assumption that the density of the particle in sucrose solution is 1.3 $g cm^{-3}$.

The k factor for the rotor can be either obtained from the manufacturers' data sheets or calculated from the equation:

$$k = 2.53 \times 10^{11} \times \frac{[\ln(r_{max}) - \ln(r_{min})]}{(rev\ min^{-1})^2_{max}} \tag{6.17}$$

Therefore, if the k factor for the rotor and the sedimentation coefficient in Svedberg units (S) of the particle are known, it is possible to estimate the time, in hours, to sediment a particle at the maximum speed of the rotor since:

$$t = \frac{k}{S} \tag{6.18}$$

When the rotor is operated below its maximum speed, the k factor is increased to:

$$k_{revised} = k \times \frac{(rev\ min^{-1})^2_{max}}{(rev\ min^{-1})^2_{selected}} \tag{6.19}$$

The k' factor (or k^* factor if particle density is assumed to be 1.3 $g cm^{-3}$ in sucrose solution) is calculated from the equation:

$$k' = 2.53 \times 10^{11} \times \frac{[I_{z_2} - I_{z_1}]}{(rev\ min^{-1})^2_{max}} \tag{6.20}$$

where Z_1 is the minimum % (w/w) of the sucrose gradient, Z_2 is the maximum % (w/w) of the sucrose gradient, and I is the time integral value for Z_1 and Z_2 for a particle of known density (or an assumed density of 1.3 $g cm^{-3}$ for calculation of the k^* factor) and obtained from tables (supplied by the instrument manufacturer) after determining Z_0 from the equation:

$$Z_0 = \frac{Z_1 r_{max} - Z_2 r_{min}}{r_{max} - r_{min}} \tag{6.21}$$

where Z_0 is the solute concentration corresponding to extrapolation of a linear gradient distribution to zero radius and the time taken, in hours, for the particles to sediment through the gradient calculated from:

$$t = \frac{k'}{S} \tag{6.22}$$

If the time required to achieve a desired degree of separation of particles using a certain rotor (A) is known, then use of equation 6.23 makes it possible to estimate the sedimentation time, in hours, needed to reproduce these results using an alternative rotor (B) if rotor (A) is unavailable:

$$t_1 = t_2(k_1/k_2) \tag{6.23}$$

where t_1 is the sedimentation time using rotor B, t_2 is the sedimentation time using rotor A, $k_1 = k$, k' or k^* factor for rotor B at its maximum speed, and $k_2 = k$, k' or k^* factor for rotor A at its maximum speed.

Since the k, k' and k^* factors are computed from equations (i.e. 6.17, 6.20 and 6.21) that use the r_{min} and r_{max} of a rotor, a new r_{min} has to be determined and hence a revised k, k' or k^* factor calculated should the sedimentation be performed using partially filled tubes.

6.9 Analysis of subcellular fractions

6.9.1 Assessment of homogeneity

It is only when an isolation technique leads to preparations of subcellular particles completely free from contamination by other particles that the properties of the preparations may be attributed to the particles themselves. The evaluation of purity is therefore essential. Light and electron microscopic examination have been used as a means of analysing the success of a particular homogenisation technique and degree of contamination of a fraction after centrifugation. Absence of visible contamination, however, is not conclusive proof of purity because it is often difficult to detect low levels of contaminating material. Quantitative determination of purity has to be obtained by chemical analyses, e.g. protein, DNA, assay of enzyme activity, and possibly immunological properties. Organelles and molecules lacking assayable enzyme activities can be located either by their light absorption or by radioactive labelling techniques.

As a basis for the interpretation of patterns of enzyme distribution in tissue fractionation studies, two general postulates have been put forward. The first presupposes that all members of a given subcellular population have the same enzyme composition. The second assumes that each enzyme is entirely restricted to a single site within the cell. If valid, these postulates would enable enzymes to be used as markers for their respective organelles, e.g. cytochrome oxidase and succinic dehydrogenase as mitochondrial marker enzymes, acid hydrolases as lysosomal marker enzymes, catalase as a marker for peroxisomes and glucose 6-phosphatase as a marker enzyme for microsomal membranes. However, it has been demonstrated that some enzymes are located in more than one fraction, for example malate dehydrogenase, β-glucuronidase, NADPH cytochrome c reductase. Caution must therefore be used in the selection of an enzyme as a marker for a particular subcellular fraction. Further, the absence of marker enzymes cannot be taken as proof of the absence of a particular organelle. It is possible that enzymes released from their respective organelles may have been inhibited or inactivated in some manner during the fractionating process; hence, it is normal practice to assay for at least two marker enzymes in each fraction.

6.9.2   Presentation of results

Enzyme activity and protein content are determined both in the whole homogenate and in each subcellular fraction isolated. The sum of the enzyme activity and protein content of the respective fractions should not differ appreciably from that in the initial homogenate and hence should represent the total recovery. Calculations are then made (Table 6.3) of enzyme activity and protein content in each fraction as a percentage of the total recovery. For this reason there is no need to convert absorbance readings into precise units of enzyme activity or milligrams of protein.

The results obtained from tissue fractionation studies may also be represented graphically, where enzyme distribution patterns, presented in the form of histograms, provide a visual appreciation of the results. In the histogram, each fraction is then presented separately on the ordinate scale, by its own relative specific activity, which is a measure of the degree of purification achieved. On the

Table 6.3. *Distribution pattern of a liver lysosomal enzyme as established by differential centrifugation*

A. *Enzyme assay*

Fraction	Volume (cm³)	Total dilutions	Absorbance (660 nm)	Enzyme activity in fraction (arbitrary units)	% recovered activity in fraction
Whole homogenate	120	1:30	0.50	1800	–
Nuclear	30	1:20	0.22	132	7.4
Mitochondrial	20	1:100	0.33	660	36.9
Lysosomal	16	1:100	0.34	544	30.4
Microsomal	20	1:25	0.44	220	12.3
Supernatant	290	1:20	0.04	232	13.0
				1788	100.0

Table 6.3 contd

B. *Protein assay*

Fraction	Volume (cm^3)	Total dilutions	Absorbance (540 nm)	Protein in fraction (arbitrary units)	% recovered protein in fraction
Whole homogenate	120	1:100	0.16	1920	–
Nuclear	30	1:80	0.13	312	16.4
Mitochondrial	20	1:200	0.13	520	27.4
Lysosomal	16	1:100	0.08	128	6.7
Microsomal	20	1:100	0.18	360	19.0
Supernatant	290	1:25	0.08	580	30.5
				1900	100.0

C. *Relative specific activity calculations*

$$\text{Relative specific activity} = \frac{\%\ \text{enzyme activity in fraction}}{\%\ \text{protein in fraction}}$$

Nuclear $\dfrac{7.4}{16.4} = 0.45$ Mitochondrial $\dfrac{36.9}{27.4} = 1.35$

Lysosomal $\dfrac{30.4}{6.7} = 4.53$ Microsomal $\dfrac{12.3}{19.0} = 0.65$

Supernatant $\dfrac{13.0}{30.5} = 0.43$

abscissa, each fraction is represented cumulatively, from left to right, in the order in which it is isolated, by its percentage of total protein. The area of each rectangle is then equal to the percentage of the enzyme activity in that fraction.

6.10   Some applications of the analytical ultracentrifuge

The analytical ultracentrifuge has found many applications in biology, especially in the fields of protein chemistry and nucleic acid chemistry, yielding information from which the sedimentation coefficient, relative molecular mass, purity and shape of the particle may be deduced.

6.10.1   Determination of relative molecular mass

Two main approaches are available using the analytical ultracentrifuge to determine relative molecular mass of a macromolecule. These are sedimentation velocity and sedimentation equilibrium.

In the sedimentation velocity method the sedimentation coefficient of the molecule is initially determined either by boundary sedimentation or band (zonal) sedimentation. In the boundary sedimentation method the particles are uniformly distributed through the solution in the analytical cell at the start of the experiment. The ultracentrifuge is operated at high speeds, which cause the randomly distributed particles to migrate through the solvent radially outwards from the centre of rotation. A sharp boundary, called the plateau region, is formed between the solvent that has been cleared of particles and the solvent still containing the sedimenting material. The movement of the boundary with time, which is a measure of the rate of sedimentation of the particle, is given by equation 6.14, and is followed using either the ultraviolet absorption optical system or the Schlieren optical system (Section 6.3.6). In the alternative band sedimentation method a small amount of material (about 15 mm^3) is layered on top of a denser solvent, which generates its own density gradient during centrifugation, thus stabilising the solute band against convective disturbances. The movement of the migrating band is then followed using the absorption optical system using a special band-forming centrepiece in the analytical cell.

Rearrangement of equation 6.14 gives:

which, on integration, gives:

$$s\omega^2 t = \ln r \tag{6.25}$$

Therefore, if the logarithm of the distance moved by the boundary (r) is plotted against the time taken in seconds, the sedimentation coefficient (s) for the particle may be calculated from the slope of the line divided by ω^2. The relative molecular mass of the particle may then be determined using the Svedberg equation (equation 6.26):

$$M_r = \frac{RTs}{D(1-\bar{v}\rho)} \tag{6.26}$$

where M_r is the anhydrous relative molecular mass of the molecule, D is the diffusion coefficient of the molecule, $\bar{v}$ is the partial specific volume of the molecule, ρ is the density of solvent at 20 °C, R is the molar gas constant, and T is the absolute temperature in K, and the measured values of s and D are corrected to standard conditions of zero concentration of solute in water at 20 °C. Although the relative molecular mass of a molecule may be calculated using equation 6.26, the calculations are, however, complicated by difficulties encountered in the accurate determination of the diffusion coefficient of the particle and for correction in differences in viscosity and temperature. The determination of the relative molecular mass of a macromolecule using sedimentation velocity analysis is therefore less accurate and invariably more time consuming than determination by sedimentation equilibrium.

Sedimentation equilibrium methods are more versatile, and in the majority of cases the most accurate. They can be used to determine relative molecular mass values ranging from a few hundred to several million. This versatility is due to the large range of centrifugal fields available to the ultracentrifuge. The centrifugal field, which varies with the square of the rotor speed, can cover a 7000-fold range at the rotor speeds of 800 to 68 000 rev min^{-1} utilised by the analytical ultracentrifuge.

In the equilibrium method, the ultracentrifuge is operated until a balance is established between sedimentation, under the influence of the centrifugal field, and diffusion of material in the opposite direction, i.e. until there is no net migration of solute throughout the length of the cell. Relative molecular mass can then be calculated from the concentration gradient of the solute that is set up, using the equation:

$$M_r = \frac{2RT \ln (c_2/c_1)}{\omega^2(1-\bar{v}\rho)(r_2^2-r_1^2)} \tag{6.27}$$

where c_2 and c_1 are the concentrations of solute at distances r_2 and r_1 respectively from the centre of rotation.

The relative molecular mass of the molecule to be studied will dictate the rotor velocity to be used. In general, for low speed sedimentation equilibrium a ratio of approximately 4 : 1 between the ends of the solution column is desirable (i.e. $c_2/c_1 =$ 4, in equation 6.27). Therefore it can be calculated, using equation 6.27, that for a molecule with a relative molecular mass of 50 000, a rotor speed of 10 000 rev min^{-1} should be selected. Molecules of higher relative molecular mass require correspondingly lower rotor speeds. However, it is difficult to perform a good low speed sedimentation equilibrium for molecules with a relative molecular mass greater than 5×10^6 owing to the problem of rotor wobble at low centrifuge speeds (e.g. 1000 rev min^{-1}).

A major disadvantage of low speed sedimentation equilibrium used to be the long periods (several days to several weeks) necessary for equilibrium to be achieved. Modern techniques, however, employ analytical cells using short column depths of liquid, usually 1–3 mm. Since the time taken to reach equilibrium varies with the square of the depth of solution, a great saving of time is possible. The technique of sedimentation equilibrium, unlike that for sedimentation velocity, does not require a knowledge of the diffusion coefficient (compare equations 6.26 and 6.27), making this method more convenient and widely used for the determination of the relative molecular mass of proteins.

High speed equilibrium or meniscus depletion (Yphantis method) methods use short column lengths of liquid (1–3 mm), and in principle a technique similar to that of the low speed method. However, the centrifuge is operated at such a high speed that particles move away from the meniscus, resulting in the concentration of the solute at the meniscus to become essentially zero and producing a concentration gradient due to the solute. The concentrations throughout the cell are then proportional to the difference in refractive index between the meniscus region and any point in the cell. The relative molecular mass is proportional to the slope of a plot of the logarithm of concentration against the square of the distance and can be calculated from:

$$M_r = \frac{2RT}{(1-\bar{v}\rho)\omega^2} \times \frac{\ln \text{concentration}}{(\text{distance})^2} \tag{6.28}$$

Estimates of relative molecular mass, however, are complicated by density heterogeneity in the solution, and possible complex formation in the high concentrations of salts used. The high speed method also suffers from the disadvantage that, for molecules with a relative molecular mass below 10 000, speeds of rotation in excess of 65 000 rev min^{-1} would be required to ensure zero solute concentration at the meniscus. These excessive speeds can produce cell window distortion even when sapphire windows are used (Section 6.3.6). Nevertheless, this technique is extremely useful for the determination of the relative molecular mass of proteins if the material under study is homogeneous.

6.10.2   Estimation of purity of macromolecules

The analytical ultracentrifuge has been used extensively in the investigation of the purity of DNA preparations, viruses and proteins. Sample purity is of course extremely important if an accurate estimation of the relative molecular mass of the molecule is required. The most widely used methods for the determination of the homogeneity of a preparation include the analysis of the sedimenting boundary using the technique of sedimentation velocity. Homogeneity is usually recognised by a single sharp sedimenting boundary. Impurities in the preparation are displayed as additional peaks, shoulders on the main peak, or asymmetry of the main peak.

6.10.3   Detection of conformational changes in macromolecules

Analytical ultracentrifugation has been applied successfully to the detection of conformational changes in macromolecules. DNA, for example, may exist as single or double strands, each of which may be either linear or circular in nature. If exposed to a variety of agents, e.g. organic solvents or elevated temperature, the DNA molecules may undergo a number of conformational changes that may or may not be reversible. Changes in conformation may be ascertained by examining differences in the sedimentation velocity of the sample. The more compact the molecule, the lower would be its frictional resistance in the solvent. The more disorganised the molecule becomes, the greater the frictional resistance, and sedimentation occurs more slowly. Changes in conformation may therefore be detected by differences in sedimentation rates of the sample before and after treatment.

In the case of allosteric proteins (e.g. aspartate transcarbamylase) conformational changes may accompany combination of the protein with substrate and/or small ligands (activators or inhibitors). In addition, treatment of the protein with such reagents as urea and 4-chloromercuribenzoate may result in disaggregation of the protein into its subunits (protomers). All of these changes may readily be studied by analytical ultracentrifugation.

6.11   Safety aspects in the use of centrifuges

Centrifuges can be extremely dangerous instruments if not properly maintained or if incorrectly used. It is therefore essential that all centrifuge users read and understand the operating manual for the particular centrifuge.

The British Standard (BS 4402) governing the manufacture of centrifuges has ensured that operators are safeguarded against accidents by the fitting of effective lid locks that prevent access to the rotor chamber while the rotor is still spinning, imbalance detectors, rotor overspeed devices, and the ability of the centrifuge to contain any failure of the rotor. Since zonal rotors are designed for loading and unloading while the rotor is spinning, the normal safety mechanism that prevents activation of the rotor drive while the lid of the centrifuge chamber is still open is overridden, allowing the rotor to be operated at low speeds only. To prevent possible physical injury when zonal rotors are filled and emptied, and in the operation of continuous flow rotors, care must be taken to ensure that the moving rotor is not touched and that long hair and loose clothing (e.g. ties) do not get caught in any rotating part. This is especially important when using older centrifuges where the lid can be opened before the rotor has stopped rotating.

To minimise the risk of rotor failure, which is one of the more serious hazards likely to arise, manufacturers' instructions regarding rotor use and care should always be followed (Section 6.4.7). It is important when one is centrifuging hazardous materials (e.g. pathogenic microorganisms, infectious viruses, carcinogenic, corrosive or toxic chemicals, radioactive materials), especially in low speed non-refrigerated centrifuges in which rotor temperature is controlled by air flow through the rotor bowl, that the samples are kept in air-tight, leak-proof containers. This is to prevent aerosol formation arising from accidental spillage of the sample, which would contaminate the rotor, centrifuge and possibly the whole laboratory.

6.12   Suggestions for further reading

BIRNIE, G.D. and RICKWOOD, D. (1978) *Centrifugal Separations in Molecular and Cell Biology.* Butterworth, London. (Gives in detail the theory and practice of modern centrifugation techniques in molecular and cell biology.)

FORD, T.C. and GRAHAM, J.M. (1991) *An Introduction to Centrifugation.* Bios Scientific Publishers Ltd, Oxford. (Intended to provide the novice with an introduction to the basic theory and practical applications of centrifugation, the types of centrifuge, rotors, techniques and density gradient media currently available.)

GRIFFITH, O.M. (1983) *Techniques in Preparative, Zonal and Continuous Flow Ultra-*

centrifugation, 4th edn. Beckman Instruments Inc., Palo Alto, CA. (Covering preparative and density gradient ultracentrifugation techniques and their application.)

RICKWOOD, D. (Ed.) (1984) *Centrifugation*, 2nd edn. Published in the Practical Approaches to Biochemistry Series, IRL Press, Oxford/Washington DC. (Covering the theory and practice of centrifugation, important criteria for optimising centrifugal separations and the protocols of illustrative experiments.)

SHARPE, P.T. (1988) *Laboratory Techniques in Biochemical and Molecular Biology*, Vol. 18 *Methods of Cell Separation*. Eds. Burden, R.H., and van Knippenberg, P.H. Elsevier, Amsterdam, New York, Oxford. (Chapters 3 and 5 cover the principles and practice of centrifugation and centrifugal elutriation.)

7

Spectroscopic techniques

▲●■▲●■▲●■▲●■▲●■▲●■▲●■

7.1 Introduction

7.1.1 Properties of electromagnetic radiation

The interaction of electromagnetic radiation with matter is essentially a quantum phenomenon and is dependent on the properties of the radiation and on the appropriate structural parts of the material involved. This is not surprising, as the origin of the radiation is due to energy changes within the matter itself. An understanding of the properties of electromagnetic radiation and its interaction with matter leads to a recognition of the variety of types of spectrum, and consequently spectroscopic techniques, and their application to the solution of biological problems. Also, the transitions that occur within matter (see, for example, Section 7.1.2) are quantum phenomena and the spectra that arise from such transitions are, at least in principle, predictable. Table 7.1 shows the various interactions, with parts of matter, of the electromagnetic spectrum and corresponding wavelength. The various parts of matter give rise to, and are affected by, the radiation in the corresponding region of the spectrum.

Electromagnetic radiation (Fig. 7.1) is composed of an electric vector and a magnetic vector (which give rise to the name), which oscillate in planes at right angles (normal) to each other and both at right angles to the direction of propagation.

Table 7.1. *Interactions of electromagnetic radiation and the various parts or 'structures' of matter*

Phenomenon	Region of spectrum	Wavelength
Nuclear	Gamma	0.1 nm
Inner electrons	X-rays	0.1–1.0 nm
Ionisation	Ultraviolet	10–200 nm
Valency electrons	Near ultraviolet and visible	200–800 nm
Molecular vibrations	Near infrared and infrared	0.8–25 μm
Rotation and electron spin orientation in magnetic fields	Microwaves	400 μm–30 cm
Nuclear spin orientation in magnetic fields	Radiowaves	100 cm and above

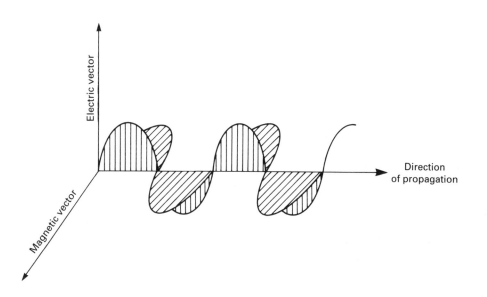

Fig. 7.1. The electric and magnetic vectors or 'oscillations' of electromagnetic radiation and the direction of propagation.

7.1.2   Interaction with matter

Electromagnetic phenomena exhibit energy, frequency, wavelength and intensity. All these are interrelated and can be explained in terms of either waveforms or particles termed photons or quanta. These phenomena are best exemplified by considering electronic spectra. Electrons in either atoms or molecules may be distributed between several energy levels but principally reside in the lowest levels or ground state. In order for an electron to be promoted to a higher level (or excited state), energy must be put into the system and this gives rise to an absorption spectrum if the energy is derived from electromagnetic radiation. Only the exact amount of energy equivalent to the difference in energy level, in accordance with the rules of quantum mechanics, will be absorbed. This is termed one quantum of energy for a single electron transition, and the absolute magnitude of each quantum will differ according to the difference in energy levels involved. When an electron falls from a higher to a lower level then exactly one quantum of energy is emitted from the system, giving rise to an emission spectrum. Energy in other forms may be put into the system, e.g. the heating of metals achieves the promotion of electrons to higher energy levels and, if sufficient energy has been input, when they return to lower levels visible light is emitted. This gives rise to the effect of the glowing of heated metals.

Figure 7.2a is a diagrammatic representation of electron transitions in the sodium atom. These transitions in most atoms give rise to relatively simple line spectra. The situation in molecules is somewhat more complicated, although the same basic principles apply, because more different kinds of energy level exist. Moreover the atoms in molecules may vibrate and rotate about a bond axis, which gives rise to vibrational and rotational sublevels. This situation is shown diagrammatically in Fig. 7.2b but, owing to the subdivision of energy levels in molecules, molecular spectra are usually observed as band spectra.

The energy change for an electron transition is defined in quantum terms by the following simple equation:

$$\Delta E = E_1 - E_2 = h\nu \tag{7.1}$$

where ΔE is the change in energy state of the electron or the energy of electromagnetic radiation absorbed or emitted by an atom or molecule, E_1 is the energy of electron in original state, E_2 is the energy of electron in final state, h is the Planck constant ($= 6.63 \times 10^{-34}$ J s), ν is the frequency of the electromagnetic radiation in hertz (c/λ where c is the speed of electromagnetic radiation (3×10^8 m s^{-1}), and λ is the wavelength of electromagnetic radiation ($1/\bar{\nu}$ or $c\bar{\nu}$, where $\bar{\nu}$ is the wave number of electromagnetic radiation in waves cm^{-1} (kaysers)). Despite the simplicity of the relation, it is of fundamental importance. Figure 7.3 shows some of the inter-relationships. It should be noted that wavelength should be expressed in submultiples of a metre, i.e. nanometre (nm), micrometre (μm), centimetre (cm) etc., not Ångstroms (Å) or mμ or μ; and frequency expressed in hertz (Hz), not cycles per second.

In Fig. 7.2a,b, electron transitions in atoms or molecules give rise to the electronic spectra generally observed as absorption, emission or fluorescence phenomena (Section 7.3) in the ultraviolet and visible regions of the electromagnetic spectrum. The basic quantum relationships hold for other regions also. Of course different

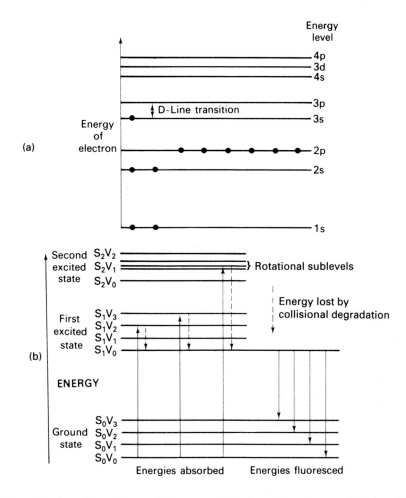

Fig. 7.2. Energy levels and transitions of electrons: (a) in the sodium atom and (b) in a fluorescent organic molecule. Note: for clarity, rotational sublevels have been indicated only for vibrational sublevel S_2V_1.

energy transitions occur in these other regions and these will be indicated as each appropriate part of the system is dealt with.

In the following subsections each region of the electromagnetic spectrum is discussed in terms of the interaction involved, instrumentation used and application to appropriate biological problems. However, as the use in some cases is more limited, the treatment is unequal and some sections are presented in considerably more detail.

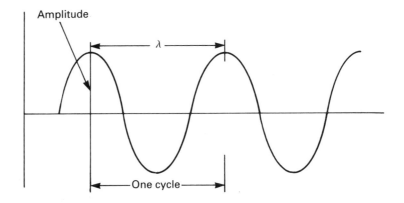

Fig. 7.3. Representation of terms in a single sinusoidal wave form. (The number of cycles occurring in unit time (second) is the frequency measured in hertz.)

7.2 Gamma-ray spectroscopy and γ-ray resonance spectroscopy

7.2.1 Principles

Gamma-rays are of nuclear origin, but they are also part of the electromagnetic spectrum and so in principle it is possible to develop spectroscopic methods involving them. Owing to their considerable penetrating power, the main applications in a biological context are in imaging but also in radiotherapy. The rays arise from energy transitions occurring within the nucleus, the mechanisms of which will not be described here. For further details, see literature cited in Section 7.14.

An important application of the use of γ-ray emission spectroscopy is the use of the element technetium, Tc, which does not occur naturally but is a product of the nuclear industry. This element may be used for medical studies, because, if it is complexed to a compound that is preferentially concentrated in specific biological tissues, particularly bone, liver or brain, its location can be determined by its emission spectrum. The emitted radiation is detected using a device known as a γ-camera, thus enabling the shape and structure of the tissue under study to be investigated. Despite the name given to the instrument, in this type of application the technique is essentially spectroscopic.

Nuclear γ-resonance, the so-called Mössbauer effect, was discovered in 1957. Many isotopes exhibit the effect but the main emphasis appears to have been centred around the ^{57}Fe isotope. Although biological applications have been somewhat limited, compared with some other spectroscopic techniques there is considerable potential for the study of biologically important metal-containing complexes.

7.2.2 Mössbauer spectroscopy

Principles

The γ-ray energy from a radioactive nucleus may be modulated by giving a Doppler velocity to the source. The Doppler effect (observed in all waveforms, sound and electromagnetic) is recognised as the apparent change in frequency that occurs when the source is moving relative to the detector (observer). The change in frequency is proportional to the source velocity and any velocity may be chosen to give the required frequency. Gamma-rays of discrete energy can be resonantly absorbed by appropriate nuclei. The source used is usually ^{57}Co; this emits a range of γ-rays with different energies, an appropriate one of which may be selected. The selected ray is then modulated by the imposed Doppler phenomenon.

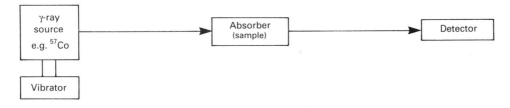

Fig. 7.4. Layout of a simple Mössbauer spectrometer.

Instrumentation

Figure 7.4 is a very simplified diagram of the arrangement required to perform Mössbauer spectroscopy. Usually, because of the energies and wavelengths involved, the Doppler velocity can be imposed by rapidly vibrating the ^{57}Co source.

Applications

The major application of this technique is in the study of the coordination of metal atoms by ligands of an appropriate complexing agent. Model compounds have been investigated that enable a better understanding of how certain metals of biological importance are affected by changes in the binding properties of the ligand, either by chemical modification or by local environmental differences. An example is sickle-cell anaemia, where, compared with normal haemoglobin, the iron atom is distorted out of the plane of the haem moiety.

7.3   X-ray spectroscopy

7.3.1   Principles

Whereas γ-rays are of nuclear origin, X-rays arise from displacement of inner, extranuclear electrons. The electrons with principal quantum numbers 1, 2 and 3 in an atom can be imagined to occupy shells – K, L and M, respectively. Should a bombarding electron from an external source have sufficient energy to displace a K-shell (innermost) electron in a target atom, then this vacancy is filled within a time span of 10^{-4} s by an L-shell electron, and an X-ray of appropriate wavelength is emitted. The energy transition from L to K is of course governed by quantum rules and $E = h\nu$ must be satisfied, hence the frequency and wavelength of the emitted X-ray are determined.

   X-rays can be absorbed by matter and this gives rise to X-ray absorption spectra. The rules applying to the relationship between an incident beam of monochromatic X-radiation (I_0) and the transmitted portion I, are similar to the Beer–Lambert case described in Section 7.4.1. If μ is the linear absorption coefficient of the absorbing material then;

$$I = I_0\, e^{-\mu x} \tag{7.2}$$

where x = the thickness of the absorber.

   If X-rays have a wavelength shorter than the so-called K absorption edge of an atom, then it is possible for the incident radiation to dislodge K electrons. This then results in the emission of X-rays (because of K-electron displacement) of a frequency different from that of the incident ray. The phenomenon is called X-ray fluorescence and gives rise to X-ray fluorescence analysis, XRFA. The general principles of fluorescence will be dealt with in Section 7.8.

7.3.2   Instrumentation and applications

A suitable X-ray source is required that can be focused into the specimen chamber, where the substance under test is excited by the incident beam. A monochromator is required to disperse the fluorescent (emitted) radiation and finally a suitable detector and data-processing facilities are needed. Figure 7.5 is a simple representation of the required layout.

   The technique has wide applications in forensic science and environmental pollution studies, because many elements can be detected and concentrations measured by the technique. Of course the analysis is essentially concerned with elements but can be a useful adjunct to, for example, the detection and measurement of trace elements in fertilisers. Such elements may well find their way into the food

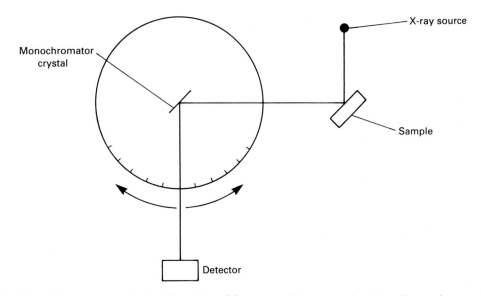

Fig. 7.5. X-ray fluorescence analysis. Dispersion of fluorescent X-rays may be detected at various angles.

chain, with possible toxic consequences if they potentially interfere with normal metabolism. An example of such an application would be the study of the uptake of lead in plants (lethal limit 500 p.p.m.) at various distances from, say, a heavily used thoroughfare.

Absorption and emission spectra are obtained in ways similar to those described below for the ultraviolet/visible region of the electromagnetic spectrum (Section 7.4). X-ray spectrometers obviously require a more rigorous approach to the incorporation of safety features but the essential requirements of source, monochromator and detector are the same.

7.4   Ultraviolet and visible light spectroscopy

These regions of the electromagnetic spectrum and their associated techniques are probably the most widely used for routine analytical work and research in biological problems. The energy transitions that occur here are exactly those described in Section 7.1.2. It is convenient, however, to deal here with the appropriate laws related to the absorption of 'light', that region of the electromagnetic spectrum for which these laws were developed.

7.4.1   Principles

The Beer–Lambert law is a combination of two laws each dealing separately with the absorption of light related to the concentration of the absorber (the substance responsible for absorbing the light) and the path length or thickness of the layer (related to the absolute amount of the absorber). Provided an absorbing substance is partially transparent it will transmit a portion of the radiation incident upon it. The ratio of the intensities of transmitted and incident light gives the transmittance, T, expressed as:

$$T = I/I_o \tag{7.3}$$

where I_o is the intensity of incident radiation, and I is the intensity of transmitted radiation. (Note: Intensity = Number of photons interacting in unit time (seconds)).

A 100% value of T represents a totally transparent substance, with no radiation being absorbed, whereas a zero value of T represents a totally opaque substance that, in effect, represents complete absorption. For intermediate values we can define the absorbance (A) or extinction (E) that is given by the logarithm (to base 10) of the reciprocal of the transmittance:

$$A = E = \log_{10}(1/T) = \log_{10}(I_o/I) \tag{7.4}$$

Absorbance used to be called optical density (O.D.) but continued use of this term should be discouraged. Also, as absorbance is a logarithm it is by definition unitless and has a range of values from 0 ($\equiv 100\%T$) to ∞ ($\equiv 0\%T$).

It is now possible to define the Beer–Lambert law, which, as described above, states that the absorbance is proportional to both the concentration of absorbant and thickness of the layer, as:

$$A = \varepsilon_\lambda cl \tag{7.5}$$

where ε_λ is the molar absorbance coefficient (or molar extinction coefficient) for the absorber at wavelength λ, c is the concentration of absorbing solution, and l is the path length through the solution (or thickness).

In the strictest use of SI units the concentration should be expressed as mol m^{-3} (which is not molar), and the path length in metres. As A is unitless, this would give units for ε_λ as mol^{-1} m^2 (derived from $1/(\text{mol m}^{-3} \text{ m})$, if equation 7.5 is rearranged to give $\varepsilon_\lambda = A/(cl)$). This is to be expected, as the value of the absorbance is also dependent upon the area of illumination by the incident radiation. As this area is identical for both sample and reference, it can be ignored in any calculations. It may also be instructive for the reader to show for him- or herself that mol m^{-3} is equivalent to mmol dm^{-3}. Hence, sometimes a millimolar absorbance coefficient is quoted and this does not contravene strict SI rules. However more practical units for ε_λ are $dm^3\,mol^{-1}\,cm^{-1}$, which conforms to the definition of molarity (despite being incoherent in SI terms) and the common use of cuvettes with 1 cm path lengths. Sometimes molar absorbance coefficients are extremely large and in such cases a more convenient way of expressing values is to quote the absorbance of a 1 cm thick sample of a 1% solution of the absorber. This is distinguished by writing the coefficient as $A_{1\,cm.}^{1\%}$

7.4.2 Instrumentation

The material used in the optical parts of the instruments depends on the wavelength used. In the ultraviolet region it is necessary to use prisms, gratings, reflectors and cuvettes made of silica. Above 350 nm wavelength, borosilicate glass may be used but also there are now some plastic materials (e.g. disposable cuvettes) available that are transparent over virtually the whole of the visible region and into the near ultraviolet.

Wavelength selection is obviously of crucial importance. In the visible region where the analyte may not absorb, but can be readily modified chemically to produce a coloured product, coloured filters may be used that absorb all but a certain limited range of wavelengths. This limited range is known as the bandwidth of the filter. The methods that use filter selectors and depend on the production of a coloured compound are the basis of colorimetry and such methods give moderate accuracy as even the best filters (interference types) do not have particularly narrow bandwidths. The usual procedure is to use two optically matched cuvettes, one containing a blank in which all the materials are mixed except the sample under test, an equivalent volume of solvent being added to this mixture, and the other containing the coloured material to be measured. It is necessary to standardise or zero the instrument using the blank, change cuvettes, and read the absorbance. The best analytical procedure requires the zero to be reset between each measurement as colorimeters, and some filters, are influenced by temperature changes. It is also good practice to work from the most dilute (least colour) to most concentrated, because even if the cuvette is rinsed between each measurement the possibility of carryover should be minimised. Table 7.2 lists a number of commonly used colorimetric assays.

If the wavelength is selected using prisms or gratings, the technique is called spectrophotometry. In both colorimetry and spectrophotometry the most common use is to prepare a set of standards and produce a concentration versus absorbance calibration curve, which is linear because it is a Beer–Lambert plot. Absorbances of unknown compounds are then measured and the concentration interpolated from the linear region of the plot. Interpolation is critical because:

Table 7.2. *Common colorimetric assays*

Substance	Reagent	Wavelength (nm)
Inorganic phosphate	Ammonium molybdate; H_2SO_4; 1,2,4-aminonaphthol; $NaHSO_3$, Na_2SO_3	600
Amino acids	(a) Ninhydrin	570 (proline 420)
	(b) Cupric salts	620
Peptide bonds	Biuret (alkaline tartrate buffer, cupric salt)	540
Phenols, tyrosine	Folin (phosphomolybdate, phosphotungstate, cupric salt)	660 or 750 (750 more sensitive)
Protein	(a) Folin	660
	(b) Biuret	540
	(c) BCA reagent (Bicinchoninic Acid)	562
	(d) Coomassie Brilliant Blue	595
Carbohydrate	(a) Phenol, H_2SO_4	Varies, e.g. glucose 490, xylose 480
	(b) Anthrone (anthrone, H_2SO_4)	620 or 625
Reducing sugars	Dinitrosalicylate, alkaline tartrate buffer	540
Pentoses	(a) Bial (orcinol, ethanol, $FeCl_3$, HCl)	665
	(b) Cysteine, H_2SO_4	380–415
Hexoses	(a) Carbazole, ethanol, H_2SO_4	540 or 440
	(b) Cysteine, H_2SO_4	380–415
	(c) Arsenomolybdate	Usually 500–570
Glucose	Glucose oxidase, peroxidase, *o*-dianisidine, phosphate buffer	420
Ketohexose	(a) Resorcinol, thiourea, ethanoic acid, HCl	520 560
	(b) Carbazole, ethanol, cysteine, H_2SO_4	
	(c) Diphenylamine, ethanol, ethanoic acid, HCl	635
Hexosamines	Ehrlich (dimethylaminobenzaldehyde, ethanol, HCl)	530
DNA	Diphenylamine	
RNA	Bial (orcinol, ethanol, $FeCl_3$, HCl)	665
a-Oxo acids	Dinitrophenylhydrazine, Na_2CO_3, ethyl acetate	435
Sterols	Liebermann–Burchardt reagent (acetic anhydride, H_2SO_4, chloroform)	625
Steroid hormones	Liebermann–Burchardt reagent	425
Cholesterol	Cholesterol oxidase, peroxidase, 4-aminoantipyrine, phenol	500

(i) one should never extrapolate beyond the region for which any instrument has been calibrated, and

(ii) particularly in colorimetry, a phenomenon known as the Job effect (see below) occurs.

If we continue to take measurements beyond the colour reagent limit, it is observed that the linearity of the Beer–Lambert calibration does not continue indefinitely but forms a plateau, at a point which indicates that there is insufficient reagent to produce any more colour. This is the phenomenon known as the Job effect. To extrapolate beyond the linear portion of the curve therefore would potentially introduce enormous errors. Furthermore, if a particular sample gives a very high absorbance reading it is incorrect procedure merely to dilute that sample. This achieves nothing as all the materials in the sample are diluted to the same extent. The correct procedure is to return to the original material and dilute that appropriately and then perform all the steps required to produce colour.

If high precision is not required and the absorbances of the test and standard are close in value, the Beer–Lambert linear relation may be assumed for this experiment and an approximate concentration obtained from the simple relationship:

$$\text{Concentration} = \frac{\text{Test absorbance}}{\text{Standard absorbance}} \tag{7.6}$$

Of course, such an assumption for individual experiments is valid only if the Beer–Lambert relationship has been established for the reaction concerned on a previous occasion.

It is important to note that when plotting calibration curves, despite the facts that the Beer–Lambert relation implies that there is zero absorbance at zero concentration and that the instrument is physically zeroed, it is wrong to force the drawn line through zero. This would be to give greater credence to this point than to any other and assume an unjustified level of precision. The best straight line should be drawn through the points, either by eye or by regression methods.

A further point to note is that the accuracy of the instrument is not uniform throughout the transmission range. The final measurement is an electrical one, involving a galvanometer. The maximum accuracy can be shown to occur at 36.8% transmission, and between 20% and 80% the relative error is about $\pm 2\%$. Owing to the nature of galvanometric measurements the errors at low and high absorbance can be large. This indicates that the analysis should be designed to give absorbance readings in the middle of the range of the transmission scale.

In a colorimeter, the bandwidth of the wavelengths is determined by the filter. A filter that appears red to the human eye is transmitting red light and absorbing almost everything else. This kind of filter would be used to examine blue solutions as they would absorb red light. In general the filter should be of a colour complementary to that of the solution under test.

The arrangement in such an instrument can be very simple, consisting merely of a light source (lamp), filter, cuvette and photosensitive detector to collect the transmitted light. Another detector is required to measure the incident light, or a single detector may be used to measure incident and transmitted light alternately. The latter design is both cheaper and analytically better, because it eliminates variation between detectors.

The spectrophotometer is a much more sophisticated instrument. A photometer is a device for measuring 'light' and 'spectro' implies the whole range of continuous wavelengths that the light source is capable of producing. The detector in the photometer is generally a photocell in which a sensitive surface receives photons and a current is generated that is proportional to the intensity of the light beam reaching the surface. In instruments for measuring ultraviolet/visible light two lamps are usually required: one, a tungsten filament lamp, produces wavelengths in the visible region; the second, a hydrogen or deuterium lamp, is suitable for the ultraviolet. There is a switchover point, usually at 350 nm, although often both lamps are lit all the time an instrument is in use if both ultraviolet and visible light are to be used. The 'switch' in the latter case is then just a mechanical means of directing the appropriate beam along the optical axis, using mirrors or lenses. Mirrors are more frequently used owing to cheapness and the fact that less light is lost, due to chromatic aberration, in a reflectance than in a refraction system. The arrangement may be very simple as in a colorimeter but this really defeats the object of the instrument. Figure 7.6a shows the optical arrangement in a single-beam instrument. Here, first the blank and then the sample must be moved into the beam, adjustments made and readings taken. Figure 7.6b illustrates the double-beam device. In this arrangement the beam is split into two parts, one passing through the blank, or reference, at the same time as the other passes through the sample. This approach obviates any problems of variation in light intensity as both reference and sample would be affected equally. The resultant measured absorbance is the difference between the two transmitted beams of light recorded by the matched detectors. Multi-beam instruments are available that allow the simultaneous recording of absorbance changes at two or more predetermined wavelengths.

The light or radiation emitted from the source lamps covers the whole range of wavelengths that the lamp is capable of producing. In colorimeters, as described above, the filter is used to obtain an appropriate range of wavelengths within the bandwidth it is capable of selecting. In spectrophotometers the bandwidth is selected by the monochromator, which is the optical system used in these devices.

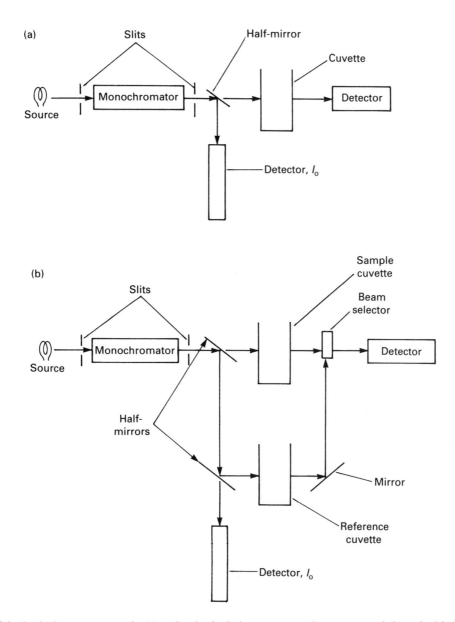

Fig. 7.6. Optical arrangements in (a) a simple single-beam spectrophotometer and (b) a double-beam spectrophotometer.

Theoretically, these systems select a single wavelength of monochromatic radiation, the emergent light being a parallel beam. The bandwidth here is defined as twice the half-intensity bandwidth, which is the range of wavelengths for which the transmitted intensity is greater than half the intensity of the chosen wavelength and it is a function of the slit width.

The optical systems used are usually either prisms, which split the multi-wavelength source radiation into its component parts by the phenomenon known as refraction (an analogy is the natural water-droplet prisms that produce a rainbow), or gratings, which achieve the same thing by diffraction. Refraction occurs because radiation of different wavelengths travels along different paths in the denser medium of the prism material. In order that velocity conservation is maintained overall, a potentially slower moving wavepacket must travel a shorter distance in a dense medium than does a faster one. Diffraction occurs by reflectance at a surface upon which is engraved a series of fine lines. The distance apart of the lines has to be of the same order of magnitude as the wavelength of the radiation being diffracted. The resolution of wavelengths is greater from gratings than from prisms and, originally, gratings were only available in the most expensive research instruments because they were hand engraved. With the advent of photoreproduction in the semiconductor industry, gratings of high quality can be reproduced in large numbers and hence are now relatively cheap.

The optical slit width affects the bandwidth, and the narrower the slit width the more reproducible are measured absorbance values. In contrast, sensitivity is less the narrower the slit, as less radiation gets through to the detector. In the most sophisticated instruments, a high level of control is available to the operator usually via a computer.

The cuvettes used in either spectrophotometry or colorimetry are an integral part of the system. They should be optically matched for the most precise and accurate work, the optical faces parallel and the path lengths identical. In flow cells, used in continuous flow systems, the parallelism of the optical faces and the path length are less critical because the reference (baseline) solution and the sample both occupy the same cell successively in time. Micro-cells are available for limited specimens and the extreme of this is the microscale spectrophotometer, where a very narrow parallel beam of monochromatic radiation passes through the micro-cell and then enters a microscope optical system.

The major advantage of the spectrophotometer, however, is the facility to scan the wavelength range over both ultraviolet and visible light and obtain absorption spectra. These are plots of absorbance versus wavelength and a typical example is shown in Fig. 7.7. This shows the extent of absorbance (absorption peaks) at various wavelengths for reduced cytochrome c. Absorption spectra in the ultraviolet (200–400 nm) and visible (400–700 nm) ranges arise owing to the kinds of electron transition described above (Section 7.1.2), usually the delocalised π-bonding electrons of carbon–carbon double bonds and the lone pairs of nitrogen and oxygen. The wavelengths of light absorbed are determined by the actual electronic transitions occurring and hence specific absorption peaks may be related to known molecular substructures. The term chromophore relates to a specific part

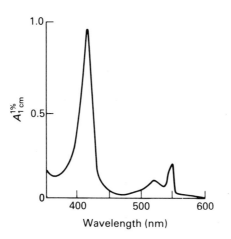

Fig. 7.7. Absolute absorption spectrum of reduced cytochrome c.

of the molecule, which independently gives rise to distinct parts of an absorption spectrum. Conjugation of double bonds lowers the energy (lower frequency) required for electronic transitions and hence causes an increase in the wavelength at which a chromophore absorbs. This phenomenon is termed a bathochromic shift. Conversely a decrease in conjugation (for example, protonation of an aromatic ring nitrogen) causes a hypsochromic shift to lower wavelength. Changes in peak maxima (increase or decrease in absorbance) can also occur. A hyperchromic shift describes an increase and a hypochromic shift a decrease in absorption maximum.

There are a number of specialised types of spectrophotometer available other than those already mentioned above. Recording spectrophotometers are usually capable of scanning a predetermined spectrum (the prism or grating angle is changed by a motor-driven system, thereby emitting a continuously changing bandwidth along the optical axis), or monitoring changes at a predetermined wavelength. Although data are commonly recorded on a chart as hard copy, the more sophisticated devices capture and store data in computer systems and in some cases computer control is an option. Variable chart and scanning speeds and absorbance scale expansion are available. It is also possible to incorporate automatic cell changers and measurement at predetermined intervals for time-dependent changes (e.g. kinetic studies, see Section 4.8.1). Measurement at the temperature of liquid nitrogen ($-196\,°C$) increases the resolution, owing to the reduced thermal motion of the molecules. The absorbance generally increases also as the apparent path length is increased, because of internal reflections occurring in the frozen sample. Reflectance instruments measure the radiation absorbed when a light beam is reflected by the sample, e.g. pastes and suspensions of microorganisms that are too opaque to transmit the radiation. In such cases internal reflection and refraction is occurring and hence the true path length is unknown and the strict Beer–Lambert law is inapplicable, making quantification difficult. A reference reflecting surface is required and magnesium oxide is frequently used.

7.4.3   Applications

Qualitative analysis may be performed in the ultraviolet/visible regions to identify certain classes of compounds both in the pure state and in biological mixtures, e.g. proteins, nucleic acids, cytochromes and chlorophylls. The technique may also be used to indicate chemical structures and intermediates occurring in a system. The most precise analysis, however, is obtained by using infrared methods.

Quantitative analysis may be performed by making use of the fact that certain chromophores, e.g. the aromatic amino acids in proteins and the heterocyclic bases in nucleic acids, absorb at specific wavelengths. Proteins may be measured at 280 nm and nucleic acids at 260 nm, although corrections are usually necessary to account for interfering substances. Such corrections usually require the measurement of the absorbance, by the interfering substance, at a wavelength remote from that for the compound under test, plus a knowledge of the absorbance at the test wavelength. If the ratio of the absorbances of the interfering substance is known for the remote and test wavelengths then the correction is simple, for example (Section 4.4), the $A_{280/260}$ ratio for proteins in the presence of nucleic acid. More sophisticated algebraic techniques are available for the more complicated cases, e.g. R.A. Morton's and D.W. Stubbs' correction for the amount of vitamin A in saponified oils.

The amounts of substances with overlapping spectra, such as chlorophylls a and b in diethylether, may be estimated if their extinction coefficients are known at two different wavelengths. For n components, absorbance data are required at n wavelengths.

A phenomenon known as Rayleigh light scattering (Section 7.10) occurs with moderate concentrations of some biological macromolecules (e.g. large DNA fragments) measured at 260 nm. This introduces an interference leading to error but may be accounted for by measuring the scattering in a region of the spectrum where DNA does not absorb, e.g. 330–430 nm.

A difference spectrum is the difference between two absorption spectra. There are essentially two ways in which difference spectra may be obtained: first, indirectly, by subtraction of one absolute spectrum from another (Fig. 7.8a); second, directly, by placing one compound in the reference cell and the other in the test cuvette (Fig. 7.8b). Fig. 7.8a shows the two absolute spectra of ubiquinone and ubiquinol and differences in absorbance may be calculated at wavelength points with suitable

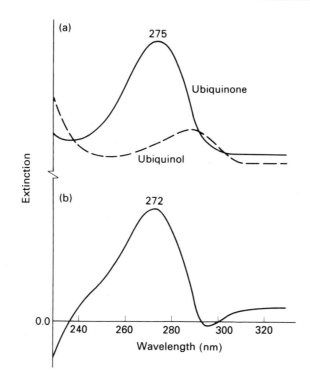

Fig. 7.8. (a) Absolute and (b) difference spectra of ubiquinone and ubiquinol.

regular intervals between them. The resultant absorbance values may then be plotted at the same wavelength points. Figure 7.8b shows this difference spectrum, which is obviously the same, although obtained in a different manner. Difference spectrophotometry has the advantage of enabling the detection of small absorbance changes in a system with a high background absorbance. An example of this kind of investigation is the measurement of changes in the oxidation state of components of the respiratory chain in intact mitochondria and chloroplasts.

The following important observations should be made:

(i)   Difference spectra may contain negative absorbance values.

(ii) Both absorption maxima and minima may be displaced and extinction coefficients are different from those of absolute absorption peaks.

(iii) There are points of zero absorbance in the difference spectrum, equivalent to those wavelengths where both the reduced and oxidised forms of the compound exhibit identical absorbances (isobestic points) and which may be used for checking for the presence of interfering substances.

A more complex example is that of the cytochrome a_3–CO complex minus cytochrome a_3, the difference spectrum being obtained by using anaerobic bacteria in the reference cuvette and the same system complexed with CO in the sample or test cuvette. Cytochrome a_3 is the terminal electron carrier and is the only component in the system that reacts with carbon monoxide.

Frequently the term difference spectrum specifically refers to the absolute reduced spectrum minus the absolute oxidised spectrum, e.g. the difference spectrum of cytochrome c corresponds to the difference between the spectrum for cytochrome c_{red} minus that for cytochrome c_{ox}. A similar difference spectrum may be obtained for a suspension of mitochondria using the so-called reversal technique. This involves measuring the change in absorbance at each wavelength when the preparation passes from the aerobic to the anaerobic state. The resultant spectrum obtained is a combined difference spectrum, for cytochromes a, a_3, b, c, c_1 and NAD^+ and flavoprotein. Shoulders on peaks observed in difference spectra obtained at room temperature may be resolved into distinct peaks at $-196\,°C$ by measuring low temperature difference spectra.

An alternative to low temperature studies of unresolved absorption spectra (difference or absolute) is purely mathematical and is termed differential spectroscopy. If the algebraic relationship that governs the shape of a symmetrical peak is known, then it may be differentiated and the differential plotted against the original variable. An ideal example is shown in Fig. 7.9a and b.

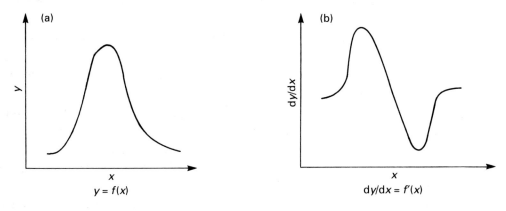

Fig. 7.9. First-differential spectra.

Almost always the algebraic relationship is unknown. However, the results may be readily obtained from digital computer techniques by sampling the curve at small intervals along the x-axis (wavelength). This process gives $\Delta y/\Delta x$ or $\Delta abs/\Delta\lambda$. If the $\Delta\lambda$ intervals were infinitesimally small, the limiting value would be dy/dx. Furthermore, higher-order differential spectra may be obtained by feeding the data back to the processor chip as many times as are required. The value of higher-order calculations is in many cases dubious but second-order differential spectra

(d^2y/dx^2) solve a number of otherwise intractable problems and instruments are commercially available with the facility for the calculations. The binding of a monoclonal antibody to its antigen may be monitored using second-order differential spectroscopy.

Binding spectra or substrate binding spectra may be used to study the extent of interaction between an enzyme and its substrate. The binding of a substrate to a haem group containing a ferric ion in the high spin state perturbs the spectrum by displacing the ligand water from the sixth position of the ferric ion causing it to change to the low spin state. The process may be followed spectrophotometrically. An example of this is the binding of a drug (substrate) to liver microsomal monooxygenase (mixed function oxidase), which causes a blue shift of the cytochrome P450 component of the enzyme from 420 nm to 390 nm (a hypsochromic shift).

Valuable structural studies may be performed on some particular biological macromolecules such as proteins and nucleic acids. In proteins the spectrum of a chromophore depends largely on the polarity of the microenvironment. A change in the polarity of a solvent in which the protein is dissolved changes the spectrum of a particular amino acid chromophore without changing the conformation of the protein. This phenomenon is known as solvent perturbation and obviously, to be accessible to the solvent, the amino acid residue must be on the surface of the protein. Solvents or solutions miscible with water must be used and examples are dimethylsulphoxide, dioxane, glycerol, mannitol, sucrose and polyethylene glycol.

The aromatic amino acids are powerful chromophores in the ultraviolet region. Processes such as denaturation (unfolding) of a polypeptide chain by pH, temperature and ionic strength can be monitored as more of these residues become exposed to the incident radiation.

Many other processes may be followed, particularly if the amino acid residue tyrosine is involved, e.g. protein–protein binding, protein–metal or protein–small molecule interactions. The range may be extended by the use of reporter group techniques, in which an artificial chromophore is attached to the appropriate region of the protein.

In nucleic acid studies solvent perturbation may be used to estimate the number of unpaired bases in RNA. If normal water is replaced by 50% ^2H_2O as solvent, the ^2H_2O changes only the spectral components due to unpaired nucleotides. Also the denaturation of the helical structure of DNA in solution may be investigated when the double-stranded DNA is heated through its transition temperature (Section 3.5.2). The extinction at 260 nm increases (hyperchromic shift) on denaturation and decreases again (hypsochromic shift) on renaturation, which occurs on cooling. Effects on the secondary structure of DNA by pH and ionic strength may be studied in a similar way.

In certain situations an action spectrum may be studied as a plot of physiological (non-extinction) parameter against wavelength. In many complex biological systems such a spectrum often corresponds to the absorption spectrum of a single key compound. An example is the plotting of the rate of oxygen evolution by green plant tissue against the wavelength of light used to irradiate the system. This results in a graph similar to the spectrum of the chlorophylls.

7.5   Infrared and Raman spectroscopy

7.5.1   Principles

These two spectroscopic methods are complementary, giving similar information, but the criteria for the phenomena to occur are different for each type. It is also true that, for asymmetric molecules, absorptions will give rise to both types and virtually the same information could be gained from either. However, for symmetrical molecules having a centre of symmetry, the fundamental frequencies that appear in the Raman do not appear in the infrared and vice versa. The two methods are then truly complementary.

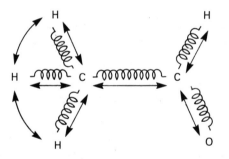

Fig. 7.10. Possible vibrations in acetaldehyde.

The reasons are the two different mechanisms on which the two types depend. Both are indicated in Fig. 7.10 for a simple molecule, acetaldehyde. For the purposes of this discussion the bonds between atoms can be considered as flexible springs so that the atoms are in constant vibrational motion, i.e. the molecule is not fixed and rigid.

Bonds can either stretch or deform (bend) and theory predicts that if a molecule contains n atoms there will be $3n - 6$ fundamental vibrations in total. Of these, $2n - 5$ cause bond deformations and $n - 1$ cause bond stretching. In Table 7.3

Table 7.3. *Some fundamental frequencies associated with vibrations in acetaldehyde*

Functional group	Vibration type	Frequency (cm^{-1})
$-CH_3$	Bending	1460
		1365
$-C-C-$	Stretching	1165
$-C=O$	Stretching	1730
$-C-H$ (in CH_3)	Stretching	2960
		2870
$-C-H$ (in CHO)	Stretching	2720

are listed the most important fundamental frequencies observed in acetaldehyde molecules.

The region of the electromagnetic spectrum ranges from the red end of the visible to the microwave lengths. Energy is input by irradiating in the appropriate region with electromagnetic radiation. The criterion for an infrared spectrum is that there is a change in dipole moment, i.e. a change in charge displacement. Conversely, if there is a change in the polarisability of the molecule, a portion of the scattered radiation will have a frequency different from that of the incident radiation. These different frequencies constitute Raman spectra. It should be noted that more information can be gained about oscillations in molecules by proceeding even further into the microwave region and using microwave spectroscopy.

The fundamental frequencies observed are characteristic of the functional groups concerned and are absolutely specific. This gives rise to the term fingerprint for the infrared pattern obtained. As the number of functional groups increases in more complex molecules, the absorption bands in the infrared patterns become more difficult to assign. However, group frequencies arise that help to simplify interpretation. These groups of certain bands regularly appear near the same wavelength and may be assigned to specific molecular groupings just as particular chromophores absorb in the ultraviolet and visible regions. Such group frequencies are extremely valuable in structural diagnosis. It should be noted that in infrared spectra, which are vibrational spectra, it is usual to work in frequency units, Hz, rather than wavelength.

The frequency associated with a particular group varies slightly, owing to the influence of the molecular environment. This is extremely useful in structural biochemistry studies as it is possible to distinguish between C–H vibrations in methylene (-CH$_2$) and methyl groups (-CH$_3$). Decrease in wavelength also occurs when double bonds are formed as the stretching frequency increases.

7.5.2   Instrumentation

The most common source is a Nichrome alloy coil heated to incandescence. This region of the electromagnetic spectrum contains the heat waves. Samples of solids are either prepared in mulls such as nujol and held as layers between salt plates such as NaCl or pressed into KBr discs. Non-covalent materials must be used for sample containment and also in the optics, as these materials are transparent to infrared.

Detectors are of the heat recognition type. The Golay cell contains a gas or liquid whose expansion is registered when the energy is absorbed. Thermal detectors such as thermocouples can also be employed. Analysis using a Michelson interferometer allows Fourier transform infrared spectroscopy (FT-IR) to be performed. This instrument involves fixed and rotating mirrors that split the incident beam into two. The beams are recombined after passage through the sample but as the two path lengths are different, interference patterns arise that may be analysed by Fourier transform methods. The Beer–Lambert law applies in all cases except for complex mixtures, where more complicated mathematical procedures are required.

7.5.3   Applications

The use of infrared and Raman spectroscopy is mainly in biochemical research for intermediate-sized molecules such as drugs, metabolic intermediates and substrates. Examples are the identification of substances such as penicillin and its derivatives, small peptides and environmental pollutants. It is an ideal and rapid method for measuring certain contaminants in foodstuffs and can be coupled to a gas–liquid chromatograph (GC-IR) when it is also frequently used for the analysis of drug metabolites. Figure 7.11 shows the major bands of an FT-IR spectrum of the drug phenacetin. Gas analysis is rapid, particularly for measuring different concentrations of gases such as CO_2, CO and $CH{\equiv}CH$ (acetylene) in biological samples. Use in the study of photosynthesis and respiration in plants is valuable, particularly for CO_2 metabolism.

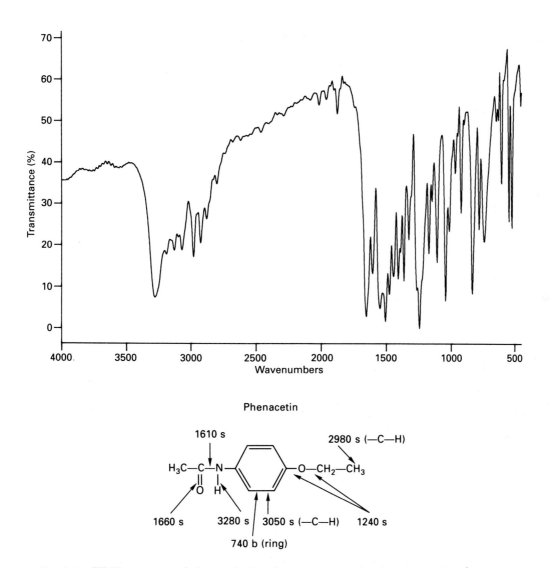

Fig. 7.11. FT-IR spectrum of phenacetin. Bands at the appropriate frequencies (cm^{-1}) are shown indicating the bonds with which they are associated and the type (s, stretching, or b, bending or deformation).

7.6 Electron spin resonance (ESR) spectroscopy

7.6.1 Magnetic phenomena

Prior to any detailed discussion of electron spin resonance (ESR) and nuclear magnetic resonance (NMR) (Section 7.7) methods it is worth while considering the more general phenomena that are applicable to both.

An important consideration is magnetism and how it arises. All substances are magnetic, and magnetism arises from the motion of charged particles. This motion is controlled by internal forces in a system and, for the purposes of this discussion, the major contribution to magnetism in molecules is due to the spin of the charged particle.

Consider the situation in the chemical bonds of a molecule where electrons (negatively charged) have the property of spin controlled by strict quantum rules. The simplest view of the chemical bond is that of paired electrons with opposite spins. It is true that in many chemical systems the electrons may become delocalised, i.e. lose their association with a particular atom, but the essential argument still applies in that for pairs of electrons in molecular orbitals the spins must be opposite (no two electrons may have all the quantum numbers identical: the Pauli exclusion principle). Each of these spinning electronic charges generates a magnetic effect but in electron pairs the effect is almost self-cancelling. The mathematical considerations do not in general apply exactly in molecules, but in atoms a value for magnetic susceptibility may be calculated and is of the order of -10^{-6} g^{-1}. This is diamagnetism, possessed by all substances because all substances contain minuscule magnets, i.e. electrons. Diamagnetism is temperature independent.

If an electron is unpaired there is no counterbalancing opposing spin and the magnetic susceptibility is of the order of $+10^{-3}$ to $+10^{-4}$ g^{-1}. In cases where this possibility arises, the underlying diamagnetism is so small by comparison that it is irrelevant and the free electron case gives rise to paramagnetism. Free electrons can arise in a number of examples, the most notable of which is in the structure of certain metals such as iron, cobalt and nickel. These metals exhibit an extreme case of paramagnetism that is termed ferromagnetism and are the materials from which permanent magnets, with which everyone is familiar, can be made. Some crystal structures allow free electrons to exist, but free radicals (free electron entities) are probably the most important systems in biological investigations.

The way in which a substance behaves in an externally applied magnetic field allows us to distinguish between diamagnetism and paramagnetism. A paramagnetic material is attracted by an external magnetic field and a diamagnetic substance is rejected. This principle is employed in the GUOY balance, which allows the quantification of the magnetic effects. A balance pan is suspended between the poles of a suitable electromagnet (to supply the external field). The substance under test is weighed in air with the current switched off. The same sample is then reweighed with the current on, the result being that a paramagnetic substance apparently weighs more and a diamagnetic substance apparently less.

Exactly similar arguments can be made regarding atomic nuclei. Of course, it is not now the extranuclear electrons but the subnuclear particles that are the spinning charged particles. Strictly speaking (because of interchangeability) it is the number of nucleons (protons plus neutrons) that determine whether a species will exhibit nuclear paramagnetism. It is beyond the scope of this discussion to explore why neutrons (which are neutral and uncharged; Section 5.1) are involved. It is sufficient to note that the hydrogen atoms in a molecule exhibit residual nuclear magnetism and, if some or all are replaced by deuterium, then there is no magnetism from the deuterium. Hydrogen contains a single proton, whereas deuterium contains one proton and one neutron (two nucleons, an even number). Carbon-12 (^{12}C) contains six protons plus six neutrons – even number, no residual magnetism. ^{13}C contains six protons (because it is carbon) but seven neutrons, an odd number of nucleons; hence it exhibits residual nuclear magnetism.

7.6.2   The resonance condition

In both ESR and NMR techniques (Section 7.7) two possible energy states exist for either electronic or nuclear magnetism in the presence of an external magnetic field:

(i) Low energy state E_1: the field generated by the spinning charged particle lies with, or is parallel to, the external field.
(ii) High energy state E_2: the field generated by the spinning charged particle lies against, or is antiparallel to, the external field.

The resonance condition is satisfied when the transition from the low to high energy states occurs and equation 7.1 is satisfied. Energy must be absorbed for these transitions to occur: one quantum or hv (where h is the Planck constant). In the appropriate external magnetic fields it is shown that the frequency of applied radiation, v, occurs in the microwave region for ESR (sometimes called electron paramagnetic resonance (EPR) and the radiofrequency region for NMR (sometimes called nuclear paramagnetic resonance). In both techniques, two possibilities exist for determining the absorption of electromagnetic energy (at the resonance point):

(i) either a constant frequency is employed and the external magnetic field swept,
(ii) or a constant external magnetic field is used and the appropriate region of the spectrum swept.

For technical reasons the more commonly employed option is (i) but the same results would be obtained if either option were chosen.

7.6.3   Principles

The quantum of energy required to cause the resonance condition to be satisfied, and transition between energy states in an ESR experiment, may be quantified as:

$$h\nu = g\beta H \tag{7.7}$$

where g is the spectroscopic splitting factor (a constant), β is the magnetic moment of the electron (termed the Bohr magneton), and H is the strength of the applied external field.

The frequency of the absorbed microwave radiation is a function of the paramagnetic species, β, and the applied magnetic field strength, H (equation 7.7). This indicates that either may be varied to the same effect. The absorption of the energy is recorded as a peak in the ESR spectrum and is indicative of the presence of a paramagnetic species. The area under the peak is proportional to the concentration of that species. Calibration of the instrument with known standards allows the concentration to be calculated. For a delocalised electron (some free radicals), $g = 2.0023$, but for localised electrons, e.g. in transition metal atoms, g varies and its precise value gives information about the nature of the bonding in the environment of the unpaired electron within the molecule. When resonance occurs, the absorption peak is broadened owing to spin–lattice interactions, i.e. the interaction of the unpaired electron with the rest of the molecule. This gives further information about the structure of the molecule.

High resolution ESR may be performed by examining the hyperfine splitting of the absorption peak, which is caused by interaction of the unpaired electron with adjacent nuclei. This yields information about the spatial location of atoms in the molecule. Proton (^{1}H) hyperfine splitting for free radicals occurs in the region of 0 to 3×10^{-3} tesla (1 tesla $= 10^4$ gauss and is a measure of the magnetic field strength) and yields data analogous to those obtained with high resolution NMR.

7.6.4   Instrumentation

Figure 7.12 is a diagram of the main components of an ESR instrument. The field strengths generated by the electromagnets are of the order of 50 to 500 millitesla, and variations of less than 1 in 10^6 are required for highest accuracy. The monochromatic microwave radiation is produced in the Klystron oscillator, the wavelength being of the order of 3×10^{-2} m (9000 MHz).

The samples are required to be in the solid state; hence biological samples are usually frozen in liquid nitrogen. The first-order differential (dA/dH) is usually plotted against H, not A versus H (cf. Section 7.4.3). Hence a plot similar to that in Fig. 7.9 is obtained and this shape is called a 'line' in ESR spectroscopy.

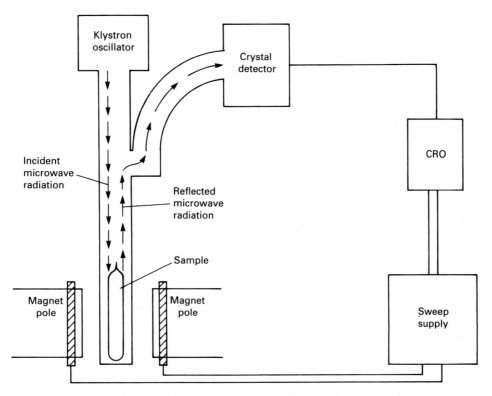

Fig. 7.12. Diagram of an ESR spectrometer. CRO, cathode ray oscilloscope.

Generally there are relatively few unpaired electrons in a molecule, resulting in fewer than 10 lines, which are not closely spaced.

7.6.5   Applications

ESR spectroscopy is one of the main methods used to study metalloproteins, particularly those containing molybdenum (xanthine oxidase), copper (cytochrome oxidase and copper blue enzymes) and iron (cytochrome, ferredoxin). Both copper and non-haem iron, which do not absorb in the ultraviolet/visible regions, possess ESR absorption peaks in one of their oxidation states. The appearance and disappearance of their ESR signals are used to monitor their activity in the multienzyme systems of intact mitochondria and chloroplasts, as well as in isolated enzymes. In metalloproteins there exists a specific stereochemical structure whereby a characteristic number of ligands (frequently amino acid residues of the protein) are coordinated to the metal. ESR studies show that the structural geometry is frequency distorted from that of model systems and the distortion may be related to biological function.

Free radicals are frequently produced when biological material is irradiated, e.g. with γ-rays. For instance -S-S- cross-linkages in proteins may be identified by irradiating the protein and the free radical produced has the free electron localised in the -S-S- region. Another major application in this area is in the examination of irradiated foodstuffs for residual free radicals. The technique can be used to establish whether the packed foodstuff has been irradiated or not.

Spin labelling involves the attachment of a stable and unreactive free radical such as 2,2,6,6-tetramethylpiperidine-1-oxyl (TEMPOL) to biological macro-molecules that lack unpaired electrons. Labelling of glycerophosphatides with TEMPOL allows the lateral diffusion of the labelled molecules in a membrane and also their flip-rate between the inner and outer surfaces of a lipid bilayer to be studied.

7.7   Nuclear magnetic resonance (NMR) spectroscopy

The essential background theory of the phenomena that allow nuclear magnetic resonance to occur has been dealt with in Section 7.6.1. The minuscule magnets involved here are nucleons (in effect protons) rather than electrons. The specific principles, instrumentation and applications are treated below.

7.7.1   Principles

Again there is considerable similarity with ESR. Most studies involve the use of ^{1}H (hence the term proton magnetic resonance (PMR)) but ^{13}C, ^{15}N and ^{31}P isotopes are used in biochemical studies.

The resonance condition in NMR is satisfied in an external magnetic field of several hundred millitesla, with absorptions occurring in the region of radiowave 40 MHz frequency for resonance of the ^{1}H nucleus. The actual field scanned is small compared with the total field applied and the radiofrequencies absorbed are specifically stated on such spectra.

The molecular environment of a proton governs the value of the applied external field at which the nucleus resonates. This is recorded as the chemical shift (τ) and is measured relative to an internal standard, frequently tetramethylsilane (TMS), whose structure $(CH_3)_4Si$ contains 12 identical protons. The chemical shift arises from the applied field inducing secondary fields (15×10^{-4} to 20×10^{-4} tesla) at the proton by interacting with the adjacent bonding electrons. If the induced field opposes the applied field, the latter will have to be at a slightly higher value for resonance to occur. Alternatively if the induced and applied fields are aligned, the latter is required to be at a lower value for resonance. In the opposing field case the nucleus is said to be shielded, the magnitude of the shielding being proportional to the electron-withdrawing power of proximal substituents. In the aligned field case, the nucleus is said to be deshielded. The field axis may be calibrated in units on a scale from 0 to 10, with TMS at the maximum value. The type of proton may thus be identified by the absorption peak position, i.e. its chemical shift and the area under each peak being proportional to the number of such protons in a particular group. Figure 7.13 is a simplified diagram of an ethyl alcohol spectrum in which there are three methyl, two methylene and one alcohol group protons. The peaks appear in the area proportions $3:2:1$.

High resolution NMR yields further structural information derived from the observation of hyperfine splitting. This arises owing to spin–spin splitting or coupling, which arises from the interaction of bonding electrons with like or different spins and may extend to nuclei four or five bonds apart. It is shown as fine structure splitting of peaks already separated by chemical shifts.

NMR spectra are of great value in elucidating chemical structures. Both qualitative and quantitative information may be obtained, and hyperfine splitting yields information about the near-neighbour environment of a nucleus.

7.7.2   Instrumentation

The essential details of an NMR instrument are shown diagrammatically in Fig. 7.14. It will be seen yet again that the layout is almost identical with that of ESR, except that instead of a Klystron oscillator being present to generate microwave radiation, two sets of coils, a transmitter and a receiver, are used for generation and reception respectively, of the appropriate radiofrequency. Samples in solution are contained in sealed tubes that are rotated rapidly in the cavity to eliminate irregularities and imperfections; in this way an average and uniform signal is reflected to the receiver to be processed and recorded. Solid state and high field NMR are more recent and rapidly advancing techniques enabling hitherto difficult or impossible investigations to be carried out. The latest developments allow two-dimensional NMR to be performed, and even more sophisticated structural analyses to be carried out.

7.7.3   Applications

The study of molecular structure, conformational changes and certain types of kinetic investigation are the main uses of NMR in the biological field. Most work is done in solution and in order to eliminate solvent effects the equivalent deuterated solvent (for proton NMR) would be used. The use of the technique in drug metabolism studies is of increasing importance, particularly when coupled with

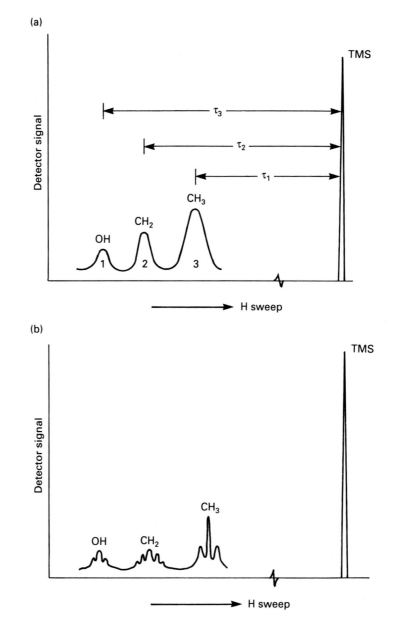

Fig. 7.13. NMR spectrum of ethyl alcohol (a) at low resolution and (b) at high resolution. The latter is resolved only in very pure samples. TMS, trimethylsilane.

infrared and X-ray diffraction data, which can then be used in molecular modelling methods using sophisticated computer techniques to try to elucidate drug action. Figure 7.15 shows a high resolution proton resonance spectrum of phenacetin,

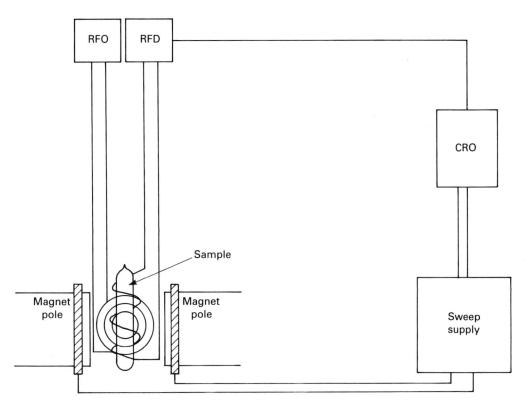

Fig. 7.14. Diagram of an NMR spectrometer. CRO, cathode ray oscilloscope; RFO, radio frequency oscillator; RFD, radio frequency detector.

which together with the FT-IR spectrum shown in Fig. 7.11, yields substantial structural information.

The development of high field magnets and solid state NMR has improved the usefulness of the technique for the biological scientist. The biological action of antibiotics such as gramicidin and valinomycin has been understood largely from NMR studies. The effects of certain lipids on the mobility of specific entities in biological and model artificial membranes have been investigated.

At present there appears to be an upper molecular mass limitation of some 20 000 daltons in the study of biological macromolecules. A major reason for this can be understood if one considers the situation, for instance, in a protein or polypeptide containing large numbers of protons in similar structural environments. All of these ^{1}H nuclei will give absorption peaks in a similar region and even high resolution NMR is difficult to interpret. Increasingly sophisticated computer techniques designed to deconvolute the data are helping to improve the situation. Nevertheless the time-dependent interaction of large and small molecules may be

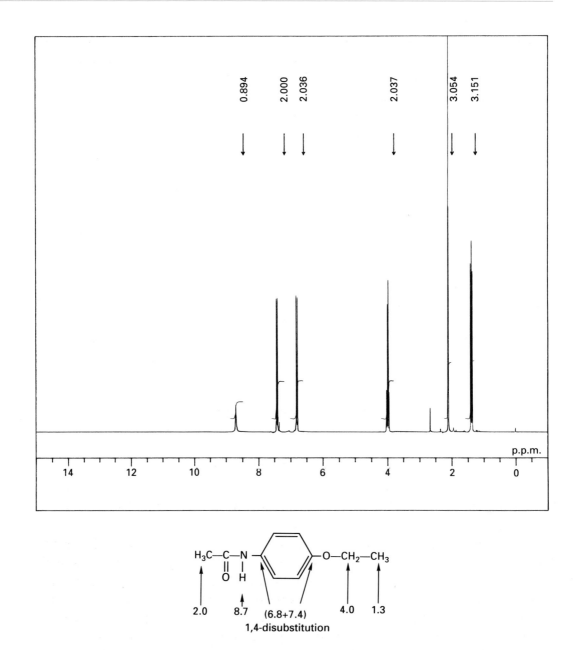

Fig. 7.15. **NMR spectrum of phenacetin.** The values associated with the downward-pointing arrows, shown slightly to the right of each peak in the upper diagram, indicate the approximate number of protons involved. In the lower diagram the shifts in p.p.m. are shown, indicating which proton is involved. The peak at 1.3 p.p.m. is a triplet because it is next to a $-CH_2$ group and that at 4.0 is a quadruplet because it is next to a $-CH_3$ group. The peaks at 6.8 and 7.4 p.p.m. are a pattern characteristic of 1,4-disubstitution in an aromatic ring.

investigated if a sufficiently distinct, unique and variable signal can be identified. This kind of application is invaluable in the study of the small conformational changes accompanying membrane dynamics and the binding of enzymes to substrates, drugs to receptors and antigens to antibodies.

The isotope ^{31}P exhibits nuclear resonance and NMR has been used extensively in studies of phosphate metabolism. The relative and changing concentration of AMP, ADP and ATP can be measured and hence their metabolism studied in living cells and tissues. Intracellular and extracellular inorganic phosphate concentrations may be measured in living cells and tissues also because the chemical shift of inorganic phosphate varies with pH.

The use of NMR in tissue and whole-animal studies is developing rapidly. Some of the physiological-type studies indicated above can be extended to the human subject. ATP metabolism in healthy and unhealthy individuals and changes during exercise are measurable.

One of the body scanners used for the location of tumours is based on NMR using ^1H resonance in water, which appears to be in greater concentration in rapidly dividing cells in a tumour mass. By using appropriate computer methods (tomography) it is possible both to locate the tumour and to identify its shape.

7.8 Spectrofluorimetry

7.8.1 Principles

Fluorescence is an emission phenomenon, the energy transition from a higher to lower state within the molecule concerned being measured by the detection of this emitted radiation rather than the absorption. In order for the transition from higher to lower states to occur, an earlier excitation event must have taken place. This earlier event is due to input of energy by absorption of electromagnetic radiation. The wavelength(s) of absorbed radiation must be at lower values (higher energy) than the emitted (fluoresced) wavelength. The difference, or increase, between these two wavelengths is known as the Stokes shift and in general the best results are obtained from compounds involving large shifts. It is possible for a compound to absorb (be excited) in the ultraviolet region and emit or fluoresce in the visible.

In Fig. 7.2b an example of the various permissible energy levels is shown. Most electrons will occupy the ground state S_0V_0 at room temperature. Elevation to a higher energy level, S_1, S_2, etc., may be achieved by absorption of electromagnetic energy (photons) in less than 10^{-15} s. Energy may be lost very rapidly (as heat) by collision degradation, resulting in minimal vibrational energy in the lowest excited state, S_1V_0. Electrons in this state return to the ground state in less than 10^{-8} s, the emitted energy being manifested as fluorescence. Many organic molecules absorb in the ultraviolet/visible regions but do not fluoresce. Fortunately, of those that do, many are of biological interest. Also, although a knowledge of the structure of an organic molecule may allow predictions about its absorption spectrum, this is not true with fluorescence. Aliphatic molecules, which are usually flexible, tend to photodissociate rather than fluoresce, whereas aromatic compounds with delocalised π-electrons sometimes fluoresce.

The emitted radiation appears as band spectra because there are many closely related values (for the wavelengths) dependent upon the final vibrational and rotational energy levels attained. These band spectra are usually independent of the wavelength of the exciting radiation and have a mirror image relationship with the absorption peak with the greatest wavelength.

An associated phenomenon is phosphorescence, but this emission has long decay times and usually persists when the exciting energy is no longer applied. Phosphorescence arises as a result of intersystem crossing to the lowest triplet state. This light emission usually occurs at longer wavelengths than does fluorescence.

Fluorescence spectra give information about events that occur in less than 10^{-8} s. The ratio:

$$Q = \frac{\text{quanta fluoresced}}{\text{quanta absorbed}} \qquad (7.8)$$

gives Q as the quantum efficiency and is usually independent of the exciting wavelength. At low concentrations, the intensity of fluorescence (I_f) is related to the intensity of the incident radiation (I_o) by:

$$I_r = 2.3 I_o \varepsilon_\lambda c d Q, \text{ i.e. } I_f \propto c \qquad (7.9)$$

where c is the concentration of the fluorescing solution (molar), d is the light path in fluorescing solution (cm), and ε is the molar extinction coefficient for the absorbing material at wavelength λ (dm^3 mol^{-1} cm^{-1}).

Because of the direct relationship between fluorescence intensity and concentration, relatively simple electronics are required. The technique of spectrofluorimetry is most accurate at very low concentrations, whereas, it will be recalled, absorption spectrophotometry is least accurate at these concentrations. In absolute terms, 100 pg of catecholamines or NADH may be measured fluorimetrically, whereas in absorption spectrophotometry 100 μg each of the catecholamines serotonin and adrenaline are required. This is due to increased sensitivity, which is easily adjustable over a large range by amplification of the detector signal. The technique allows great spectral selectivity because, owing to the Stokes shift, two monochromators may be used, one for the exciting wavelength and the other for the emitted fluorescence. Although no reference cuvette is required, a calibration curve must be obtained.

The technique, however, is susceptible to pH, temperature, solvent polarity and the inability to predict whether a particular compound will fluoresce. The major disadvantage is the phenomenon of quenching, which occurs because energy that might have been emitted as fluorescence is lost to other molecules by collisional interaction. This partly explains the increased sensitivity and accuracy in low concentrations because there are fewer molecules, and hence collisions, although the effects of solvent must not be neglected. Many materials such as detergents, stopcock grease, filter paper and some tissues may cause interference by the release of fluorescing agents (Section 6.7.4).

7.8.2   Instrumentation

It was stated above that the electronics required for a spectrofluorimeter were simpler than those required for absorption spectrophotometers. The same is true of the optics. Two monochromators may be employed, the first (M_1) for selecting the excitation wavelength. Fluorescence emission occurs in all possible directions and one direction (90°) is chosen and the second monochromator (M_2) is used for determination of the fluorescence spectrum. The radiation source is generally either a mercury lamp or a xenon arc, excitation wavelengths frequently being selected in the ultraviolet region and the emission wavelengths in the visible region. The detector is usually a sensitive photocell, e.g. a red-sensitive photomultiplier for wavelengths greater than 500 nm. Temperature control is required for accurate work as the intensity of fluorescence may vary between 10% and 50% for a 10 deg.C change at approximately 25 °C.

Two approaches are possible for the illumination of the sample: the simplest is the basic 90° illumination (Fig. 7.16), the alternative approach being front face illumination (FFI; Fig. 7.17) which obviates pre- and post-filter effects. These latter effects are due to the absorption of radiation prior to its reaching the fluorescent molecules (pre-filter absorption) and the reduction in the amount of emitted radiation escaping from the cuvette (post-filter effects). Such effects are more evident in concentrated solutions, and the use of microcuvettes (containing less material) can be of value (Fig. 7.17a). FFI is essential for examining suspensions and cuvettes with only one optical face are required. Excitation and emission occur at the same face but generally the technique is somewhat less sensitive than 90° illumination.

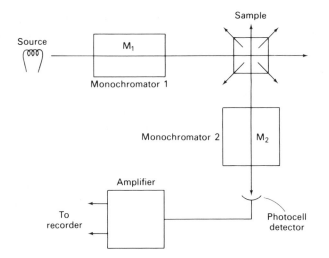

Fig. 7.16. The basic component of a spectrofluorimeter set up for 90° illumination.

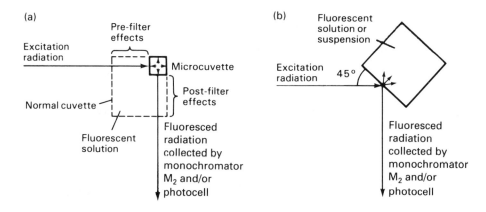

Fig. 7.17. Reduction of filter effects using (a) microcuvettes and (b) front face illumination.

7.8.3 Applications

Applications of the technique are many and varied, despite the fact that relatively few compounds exhibit the phenomenon. A compound may have its fluorescence and absorption spectra compared as an aid to identification; the effects of pH, solvent composition and the polarisation of fluorescence may all contribute to structural elucidation. The measurement of phosphorescence and phosphorescence lifetimes can also be of value in compound identification.

The detection of non-fluorescent compounds may be achieved by coupling a fluorescent probe (or fluor) in a similar way to the use of reporter groups in absorption spectrophotometry (Section 7.4.3). This is termed extrinsic fluorescence as distinct from intrinsic fluorescence, where the native compound exhibits the property. The use of such probes is valuable in both qualitative and quantitative analysis. For instance amino acids and peptides separated by chromatography or electrophoresis may be identified by coupling to their primary amino groups either dansyl chloride or o-phthalaldehyde (Section 4.3.2). The latter conjugates fluoresce intensely blue and the total oligopeptide fingerprint may be determined on only 10^{-5} g of protein. If the separation methods are used in column form then quantification is possible by forming derivatives postcolumn. Acridine orange is an extrinsic fluor that can be used to determine the strandedness of polynucleotides as the Stokes shifts differ between conjugates of single- and double-stranded polynucleotides, which fluoresce red and green, respectively. The fluor should be tightly bound at a specific site, its fluorescence should be sensitive to environmental changes and it should not have adverse effects on the system being studied. Some structures of fluorescent probes are shown in Fig. 7.18.

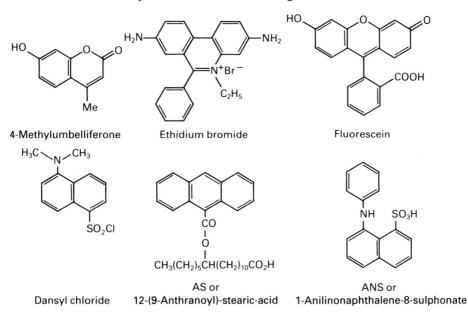

Fig. 7.18. Structure of some fluorescent probes.

The major use of fluorimetry in biochemistry is in quantitative determination of materials present in concentrations too low for absorption spectrophotometry. Assays of vitamin B_1 in foodstuffs, NADH, hormones, drugs, pesticides, carcinogens, chlorophyll, cholesterol, porphyrins and some metal ions indicate the range. Self- and contaminant quenching can be determined by adding a known quantity of a standard to an unknown quantity of a pure compound and measuring the fluorescence before and after the addition.

Calcium ions, Ca^{2+}, may be measured in the cytoplasm by the chelating agent Quin-2, which preferentially binds the metal. The fluorescence increases about five-fold on binding. More sensitive probes for this analysis are Fura-2 and Indo-1. Quin-1 is a chelating agent that may be used as a fluorescent probe to monitor intracellular pH changes in the range 5 to 9. Over this range, there is a 30-fold increase in fluorescence.

There follow some brief accounts of specific methods involving spectro-fluorimetry.

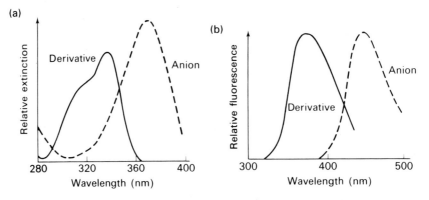

Fig. 7.19. Spectra of the methylumbelliferone anion and derivatives of 4-methylumbelliferone at pH 10. (a) Absorption spectra; (b) fluorescence spectra.

Enzyme assays and kinetic analysis

The general principles of enzyme assays, which are discussed in Section 4.8.3, often rely on the use of spectrophotometric assays. An example is the anion of 4-methylumbelliferone, which fluoresces at 450 nm. Its rate of appearance may be monitored when it is produced as a result of enzymic action on an ether or ester derivative of the fluor. The enzymes used are group-specific hydrolases and their kinetics may be studied by fluorescence measurement; Fig. 7.19a and b shows typical absorption and fluorescence spectra. Irradiation is usually at 350–400 nm wavelength and virtually all the fluorescence measured between 450–500 nm is due to the anion product. It is claimed that one molecule of β-galactosidase may be detected when it acts on fluorescein bis(β-D-galactopyranoside) as substrate, because the sensitivity of the method is so great. Hence actual numbers of molecules in a single bacterial cell may be determined, as may the synthesis of the enzyme in individual cells in a population.

Spectrofluorimetry can be applied widely in metabolic studies where nicotinamide adenine dinucleotide forms are involved as cofactors. This arises because NADH and NADPH fluoresce, whereas the oxidised equivalents do not. Therefore redox processes may be followed kinetically *in vitro* at concentrations similar to those encountered *in vivo*, and also followed in intact cells or organelles (e.g. mitochondria).

Protein structure

The presence of tryptophan and FAD as cofactors allows fluorescence to be measured in proteins. The binding and release of cofactors, inhibitors, substrates, etc., at sites close to the fluor cause changes in the associated fluorescence spectra. Information about conformational changes, denaturation and aggregation may be gleaned. The absence of an intrinsic fluor can be overcome by coupling a suitable extrinsic fluor such as anilino-naphthalene-8-sulphonate (ANS), dansyl chloride and derivatives of fluorescein or rhodamine.

Membrane structure

The fluorescent properties of a molecule are affected by its mobility and environment, particularly the polarity of the latter. These effects in the vicinity of a fluorescent probe may be monitored by measuring changes in fluorescence. Various probes having charged and hydrophobic regions (ANS and *N*-methyl-2-anilino-6-naphthalene sulphonate (MNS)) and hence able to orientate themselves across lipid/aqueous interfaces may be used to study membrane structure and gain information about the properties of such interfaces. Incorporation of phospholipids containing 12-(9-anthroanoyl)-stearic acid and 2-(9-anthroanoyl)-palmitic acid into membranes yields information about the region 0.5 nm and 1.5 nm, respectively, from the phosphate head groups of the lipid bilayer. The basic membrane structure and also the effects of temperature and certain biological phenomena may be studied. Changes in mitochondrial membranes during energy transduction have also been monitored using an ANS probe.

Fluorescence bleaching recovery

If a fluor is exposed to a pulse of high intensity radiation it may be irreversibly bleached, i.e. permanently lose its ability to fluoresce. Molecules such as fluorescently labelled phospholipids incorporated into a biological membrane may be subjected to this treatment and then the motion of such entities (in the membrane) can be studied by monitoring (with low intensity radiation) the re-emergence of fluorescence as the bleached and unbleached molecules interdiffuse. Applications include the lateral motion of extrinsically labelled rhodopsin in the photoreceptor membrane, the study of polymerisation of proteins such as actin and the diffusion of fluorescently labelled proteins microinjected into cells.

Energy transfer studies

In a number of cases energy may be transferred, by resonance energy transfer, from a donor to an acceptor fluor, provided there is overlap between the donor fluorescence spectrum and the acceptor absorption spectrum. The fluors must also be closely situated and transfer efficiency is related to spatial separation. This efficiency may be measured either as quenching of donor fluorescence by acceptor or as the intensities of fluorescence of acceptor when the latter is irradiated both in the presence and in the absence of donor.

Intrinsic fluors such as tryptophan or extrinsic ones attached to amino acids, -SH groups, sugars or fluorescent analogues of substrates, inhibitors, cofactors or phospholipids may be employed in energy transfer experiments to deduce distances within protein molecules. Accuracy is limited to about ± 0.5 nm and determinations include the localisation of metals in metalloproteins, the measurement of the extent of conformational changes in enzymes when substrate binding occurs, the distances between various pairs of proteins in the ribosome and the three dimensional structure of tRNAs.

Fluorescence depolarisation

The excitation wavelengths used may be polarised by introducing a suitable polariser between the first monochromator (M1) and the sample. The emitted radiation may be totally unpolarised or partially polarised and may be detected by using a second polariser between the emission monochromator (M2) and the detector.

Molecular rotations affect fluorescence depolarisation; for instance, the rotation of an absorber chromophore and energy transfer between chromophores increase the depolarisation effect. High concentrations of chromophore and high viscosity of the solvent result in the measurement of mainly energy transfer. At low concentrations and low viscosity the effects of molecular motion predominate.

The mobility of whole molecules, or parts thereof, may be investigated using this technique. The lifetimes of intrinsic fluors of proteins and nucleic acids are usually too short, as these biological macromolecules move relatively slowly. Hence extrinsic fluors are frequently used in these studies. Examples of the use of this method include: the binding of fluorescent substrates; the binding of inhibitors and cofactors to enzymes, which reduces their mobility (increase of overall mass and hence inertia); and the antigen/antibody complexation reaction. The association and dissociation of multi-subunit proteins such as lactate dehydrogenase and chymotrypsin and the viscosity of living cells may also be measured.

An interesting historical aside involves the use of highly viscous glycerol to slow down the rotation and translation of large molecules in depolarisation experiments.

It was knowledge of this totally unconnected fact that gave the clue to the use of glycerol as a matrix in fast atom bombardment mass spectrometry (Section 8.6.2).

Microspectrofluorimetry

In this technique a microscope is combined with a spectrofluorimeter equipped with fibre optics to enable the examination of single bacterial cells binding fluorescent antibodies and also the fluorescent intensity of subcellular structures. The extra amount of nucleic acid that tends to be present in malignant cells will take up more of the fluorescent probe acridine orange than do normal cells. This observation may be used to detect malignant cells in biopsy tissue.

The fluorescence-activated cell sorter

This system, described in Sections 1.5.4 and 2.7.3 makes use of the light emitted by cells carrying a fluorescently labelled antibody to trigger their physical separation from unlabelled cells as they flow through a fine capillary.

Fluorescence immunoassay

These methods are dealt with extensively in Section 2.7 but are worthy of a brief mention here.

Several immunoassays have been developed using fluorescent probes to label either antigen or antibody. The binding of a labelled hapten by an antibody may alter the intensity of fluorescence, thus enabling complex formation to be monitored. Changes in polarisation methods applied to immunoassay have been mentioned above. A major disadvantage of either of these approaches is the high background fluorescence that often accompanies the process and interferes with the measurement. The most promising development in this area is time-resolved fluorescence immunoassay. Two approaches have been combined to reduce the effects of background fluorescence and hence increase the sensitivity. Firstly, europium chelates are usually used as the fluor, as they have large Stokes shifts and long-lived fluorescence. Secondly, a fluorimeter has been designed that delays the measurement of the emitted light by 400 μs, during which time the non-specific background fluorescence has almost completely decayed. Such an approach has led to the development of dissociation-enhanced lanthanide fluoroimmunoassay (DELFIA).

7.9   Circular dichroism (CD) spectroscopy

7.9.1   Principles

It has been known for some time that optical isomers (isomers whose mirror images are non-superimposable) possess the property of allowing the rotation of plane-polarised light. Electromagnetic radiation oscillates in all possible directions and it is possible to select preferentially waves oscillating in a single plane. This is achieved using a polarising material such as Polaroid or a nicol prism. The technique of polarimetry essentially measures the angle through which the plane of polarisation is changed after such light is passed through a solution containing a chiral (optically active) substance. Optical rotary dispersion spectroscopy (ORD) is a technique for measuring this ability to rotate the plane of polarisation, as a function of the wavelength. However, such chiral substances may also absorb the plane-polarised radiation at certain wavelengths. In such cases the chromophore is termed an optically active chromophore or chiral centre as it may only be part of a complex molecule. The technique of ORD has been largely supplanted by circular dichroism spectroscopy (CD), which gives rather better information about the three-dimensional structure of macromolecules containing chiral centres. In CD, circularly polarised light is used and this is obtained by superimposing two plane-polarised light waves of the same wavelengths and amplitudes but differing in phase by one-quarter of a wavelength and in their planes of polarisation by 90°. Just as plane-polarised light may be left (L) or right (R) handed, so can circularly polarised light. Whether R or L circularly polarised light is produced depends on the relative positions of the peaks of the two plane-polarised waves.

The asymmetry inherent in the structure of chiral molecules or centres interacts differently with polarised light. Not only are the R and L waves of plane-polarised light differentially absorbed and refracted, hence resulting in a beam in a different plane (the basis of polarimetry), but in the case of circularly polarised light a similar differential interaction occurs. In the latter case the resultant beam, having passed through the sample, is a recombination of the R and L components to give an emergent beam of elliptically polarised light. In polarimetry the specific rotation $[\alpha]_\lambda$ would be measured whereas in CD spectroscopy it is the ellipticity, θ, which is measured:

$$\theta = 2.303 \Delta A$$
$$= 33 \Delta A \text{ degrees} \tag{7.10}$$

where ΔA is the difference in absorption between R and L components.

A CD spectrum is usually a plot of ellipticity versus wavelength and information regarding the structure of certain entities may be gleaned from it.

7.9.2   Instrumentation

The basic layout of a CD spectrometer is shown in Fig. 7.20. Both L and R circularly polarised light may be produced alternately, from a single monochromator, by

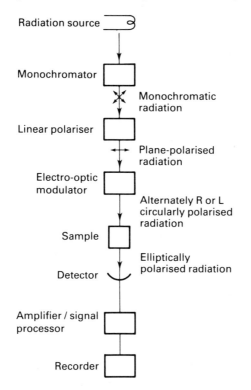

Fig. 7.20. The main components of a CD spectrometer.

the passage of plane-polarised light through an electrooptic modulator. This modulator is a crystal that, when subjected to alternating currents, transmits either the R or the L component, depending on the polarity of the electric field to which it is exposed. The photomultiplier detector produces a voltage proportional to the ellipticity of the resultant beam emerging from the sample container.

7.9.3   Applications

The major application of CD is in the study of conformation of biological macromolecules.

Proteins

Information can be gained about the relative proportions of secondary structure, α, β and random coil, in solution. The application of CD to tertiary structure is limited owing to inadequate theoretical understanding of the influences of different parts of these molecules at this level of structure.

The CD spectra of poly-L-amino acids have been obtained and are used as standards for calculating the percentage of each form of secondary structure in proteins. Curve-fitting procedures using computer processing have been used to apply the method to unknown proteins.

One of the most important benefits to be gained from CD spectroscopy is the study of conformational changes during, or because of, interactions with other entities. Examples are the determination of the binding constants of substrates, cofactors, inhibitors or activators of virtually any enzyme. The binding of the inhibitor 3-cytidylic acid to the active site of pancreatic ribonuclease changes the CD spectrum of a remotely situated tyrosine residue. Hence the binding of this inhibitor must cause a conformational change in a distant part of the enzyme. CD spectroscopy is very sensitive and may be used to monitor the conversion of α- and β-structures to random coil (a major event of the denaturation process).

Nucleic acids

It is possible to calculate the CD spectrum of a single strand of DNA from the known nearest-neighbour frequency. Experimentally determined deviations from this calculated spectrum are indicative of a variation in structure, for example double strandedness. The CD spectrum of double-stranded DNA appears to be independent of the base composition in the range of wavelengths usually used.

A large increase in the CD spectrum of mononucleotides is observed when they link to form even short oligonucleotide chains. This observation provides evidence that hydrophobic interactions between stacked bases are important in stabilising the double-stranded structure of DNA.

All nucleotides exhibit chiral properties, and their CD is greatly increased on the adoption of a helical conformation. Hence the technique may be used to study structural changes in nucleic acids: e.g. loss of helicity of single-stranded DNA as a function of temperature and pH, structural changes on the binding of cations and proteins, tRNA–amino acid binding, transitions between single- and double-stranded DNA, DNA histone interactions in chromatin, the structure of rRNA in ribosomes, and the interaction of double-stranded DNA and intercalating drugs.

7.10   Turbidimetry and nephelometry

The two similar techniques of turbidimetry and nephelometry demand brief consideration owing to their usage in biological sciences. They are both associated with the estimation of the concentrations of dilute suspensions. In turbidimetry, the apparent absorption of radiation by the suspension is measured. The apparent absorption should be measured at a wavelength where *true* absorption is not occurring; hence the Beer–Lambert law does not apply in turbidimetry. These statements require further explanation. When radiation is passed through a transparent medium, e.g. a solution in a cuvette, one or both of two distinct physical phenomena might occur. In the case of extinction, which has been dealt with extensively above, true absorption of energy occurs and allows changes in the energy states of electrons, magnetic conditions, molecular vibrations, etc. When this is the only phenomenon occurring, the medium through which the radiation is passing and in which the absorption is occurring is termed optically empty. However, in the case of a suspension, a quite distinct radiative phenomenon may occur in which the light is scattered by the suspended particles. This scattering is due to reflection and refraction and gives rise to the Tyndall effect; it occurs in all directions and is an example of the more general Rayleigh scattering.

In turbidimetry the incident and transmitted radiation may be measured in an ordinary colorimeter or spectrophotometer but the contribution of true absorption, if any, is small and the Beer–Lambert law is not strictly applicable as it holds only for very thin layers or very dilute suspensions.

The scattered light or Tyndall light may also be measured, usually at right angles (normal) to the incident radiation. This gives the Tyndall ratio, which is the ratio of the Tyndall intensity to that of the incident radiation. If this ratio is measured directly, a Tyndall meter would be used. If, however, the Tyndall intensity is compared with that of a standard suspension of known concentration then the instrument is known as a nephelometer (measures cloudiness). The concentrations of suspensions of microorganisms may be obtained using nephelometry and those of proteins and some other biological macromolecules by turbidimetry (Section 4.4).

These techniques are difficult to use but in experienced hands can be of value. The relationship between energy input (incident radiation) and measured output (transmitted or scattered), however, is complicated and non-linear. It should be noted that these techniques are not strictly spectrophotometric but are included here for completeness.

7.11   Luminometry

7.11.1   Principles

The emission and radiative techniques discussed above all depend on some physical phenomenon within the molecules concerned. The phenomenon also depends on the prior input of energy, frequently obtained from electromagnetic radiation. The radiative phenomenon luminescence arises in a different way. Although it is essentially the emission of electromagnetic radiation in the visible region (i.e. light), it arises as the result of a chemical reaction. Luminometry is the technique used to measure this luminescence and it should be recognised immediately that this is another example of a non-spectrophotometric technique. It is again included for completeness as it represents a method that is increasing in importance as an analytical tool in biological science.

   Chemiluminescence occurs as a result of excited electrons regaining the ground state (Fig. 7.2b). The prior excitation arises as a result of a chemical reaction that yields a fluorescent product, and the chemiluminescent spectrum of a reaction such as luminol with oxygen to produce 3-aminophthalate is the same as the fluorescent spectrum of the product. A similar phenomenon is bioluminescence, so-called because the light emission arises from an enzyme-catalysed reaction (Section 4.8.4) usually involving luciferase. The colour of the light emitted in the latter case depends on the source of the enzyme and varies between 560 nm (greenish yellow) and 620 nm (red) wavelengths. This method has the distinct advantage of high sensitivity as a result of the reaction having a high quantum yield – 100% under favourable conditions.

7.11.2   Instrumentation

As the excitation process is not dependent on the input of energy from electro-magnetic radiation, no monochromator is required. This means that luminometry can be performed with relatively simple photometers. Two minor complications are the need to amplify the output signal prior to recording and the need to maintain fairly strict temperature control. This control is necessary owing to the sensitivity of reactions to temperature, particularly in the case of enzyme-catalysed reactions.

   Figure 7.21 shows the layout of the main components. Some device such as a syringe is required for introducing the reactants into a suitable light-protected reaction vessel in which adequate mixing has to take place. The emitted light is collected by a photomultiplier tube, which is connected to a direct current amplifier with a wide range of sensitivity and linear response.

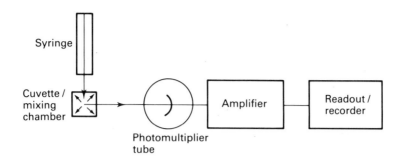

Fig. 7.21. Diagram of the main components of a simple luminometer.

7.11.3 Applications

The firefly luciferase system

Details for the firefly luciferase system are given in Section 4.8.4. ATP concentration may be measured in an assay that is rapid to carry out and whose accuracy is comparable to that of spectrophotometric and fluorimetric assays. The sensitivity is, however, vastly increased, having a limit of detection of 10^{-15} M and a linear range of 10^{-12} to 10^{-6} M ATP. The concentrations of ADP, AMP and cyclic AMP (cAMP) may also be determined using appropriate enzymes, e.g. pyruvate kinase for ADP→ATP, adenylate kinase for AMP→ADP, and phosphodiesterase for cAMP→AMP. In principle, all the enzymes and metabolites involved in ATP interconversion reactions may be assayed by this method. Examples are the enzymes creatine kinase, hexokinase and ATP sulphurase, and the substrates creatine phosphate, glucose, GTP, phosphoenolpyruvate and 1,3-diphosphoglycerate.

The bacterial luciferase system

Details of the bacterial luciferase system also are given in Section 4.8.4. The determination of nicotinamide adenine dinucleotides (and phosphates) and flavin mononucleotides, in their reduced states (i.e. NADH, NADPH and $FMNH_2$), may be made in assays which use this system. A concentration range of 10^{-9} to 10^{-12} M is achievable, which is much more sensitive than the corresponding spectrophotometric and fluorimetric assays, although the NADPH assay is less sensitive than the NADH assay by a factor of about 20. The method can be widely applied to a whole range of coupled enzyme reaction systems of the redox type that involve these nucleotides as coenzymes.

The aequorin system

Despite the development of calcium-specific electrodes the calcium ion concentration may be determined with high sensitivity, intracellularly, using the phosphoprotein aequorin. The protein can be isolated from luminescent Medusae (jellyfish) and is practically non-fluorescent. In the presence of calcium ions, however, it is converted from its natural yellow reflective colour to the blue fluorescent protein BFP. The bioluminescent spectrum of the reaction is identical with the fluorescent spectrum of $BFP:2Ca^{2+}$ but different from that of $BFP:Ca^{2+}$.

The assay is easy to use and has a high sensitivity to, and is relatively specific for, calcium. Aequorin also has the advantage of being non-toxic when injected into living cells. The disadvantages are the scarcity of the protein, its large molecular size, consumption during the reaction and the non-linearity of the light emission relative to calcium concentration. Also the reaction is sensitive to its chemical environment and the limited speed with which it can respond to rapid changes in calcium concentration, e.g. influx and efflux in certain cell types.

Chemiluminescence

Luminol and its derivatives can undergo chemiluminescent reactions with high efficiency. For instance, enzymically generated H_2O_2 may be detected by the emission of light at 430 nm wavelength in the presence of luminol and microperoxidase (Section 4.8.4).

Competitive binding assays may be used to determine low concentrations of hormones, drugs and metabolites in biological fluids. Such assays depend on the ability of proteins such as antibodies and cell receptors to bind specific ligands with high affinity (Section 4.11). Competition between labelled and unlabelled ligand for appropriate sites on the binding compound occurs. If the concentration of the binding compound is known, i.e. the number of available sites is known and a limited but known concentration of labelled ligand is introduced, then under saturation conditions (i.e. all sites are occupied) then the concentration of unlabelled ligand can be determined. Alternatively if labelled ligand only is used then the concentration of binding compound (the number of sites) may be determined. A variety of labels, including radioisotopes, is in common use, enabling the fraction in the bound and free states to be distinguished. Labelling with a luminol derivative and after completion of the binding reaction, separating bound and free fractions allows the protein to be assayed by its chemiluminescence. The system must be calibrated using standards and, under the most favourable conditions, 10^{-12} M of a compound may be determined.

Whilst polymorphonuclear leukocytes are phagocytosing, singlet molecular oxygen is produced that exhibits chemiluminescence. The effects of pharmacological and toxicological agents on these and other phagocytotic cells can be studied by monitoring this luminescence.

7.12   Atomic spectroscopy

All of the methods described above, with the exception of nuclear phenomena in the γ- and X-ray regions, have dealt essentially with molecular spectroscopy. The general theory of electron transitions was discussed in Section 7.1.2 and for simplicity the phenomena were described mainly in atomic terms, although the extension to molecules is not too difficult. It was indicated above (Section 7.1.2) that, in general, molecules give rise to band spectra and atoms to clearly defined line spectra. These lines can be observed by eye either as light, associated with a particular wavelength, which are atomic emission spectra or black lines against a bright background, which are atomic absorption spectra. Some elements, particularly metals, have an important role to play in biological systems whether as simple cofactors in enzymes, the central atom in biological macromolecules such as iron in haemoglobin or magnesium in chlorophyll, or toxic substances that affect metabolism. Use of atomic spectroscopy will enable data to be obtained that are important in the understanding of the biological roles of these elements.

In a spectrum of an element the most obvious wavelengths at which absorption or emission is observed are associated with transitions where the minimal energy change occurs. For example, in Fig. 7.2a is shown the 3s–3p transition in the sodium atom that gives rise to the emission of orange light. This transition is referred to as the D-line transition. When electron transitions occur in an atom they are limited by the availability of an empty orbital or level. An orbital or level could not be overfilled without contravening the Pauli principle (Section 7.1.2). In order for energy changes to be minimal, transitions tend to occur between levels close together in energy terms. This limitation of availability and occupancy of levels means that observed emission and absorption lines are absolutely characteristic of the element concerned. At least for simple atoms it is theoretically possible to deduce their electronic structure from their line spectra. The wavelengths emitted from excited atoms may be identified using a spectroscope, spectrograph or a direct-reading spectrophotometer that use as detectors the human eye, a photographic plate or a photoelectric cell, respectively.

In general, and in contrast to molecular spectroscopy, atom concentrations are not measured directly in solution. The atoms have to be volatilised either in a flame or electrothermally in an oven. In this state the elements will readily emit or absorb monochromatic radiation at the appropriate wavelength. Usually nebulisers (atomisers) will be used to spray the standard or test solution into the flame through which the light is passed. Alternatively the light beam is passed, in an oven, through a cavity containing the vaporised material.

7.12.1   Principles of atomic flame spectrometry

This technique takes advantage of the properties described above to determine the amounts of a specific element that may be present. The emission of light is measured by emission flame spectrophotometry and absorption by atomic absorption flame spectrophotometry.

The energy absorbed or emitted is proportional to the number of atoms in the optical path. In the case of emission it is strictly the number of excited atoms, but under reproducible standard conditions this will be the same as that for a calibrating standard. Standard conditions are not easy to achieve, as flame instability and variation in temperature and composition of the flame affect results. Sodium tends to give high backgrounds and hence should be measured first and then a similar amount added to all other standards. Excess hydrochloric acid is usually added as chloride compounds are often the most volatile salts. Enhancement of emission occurs with alkali metals, and suppression of calcium and magnesium emission by phosphate, silicate and aluminate occurs, the last of these by the formation of non-dissociable salts. This suppression effect may be relieved by the addition of lanthanum and strontium salts. So-called cyclic analysis may be performed that involves the estimation of each interfering substance in a mixture and then the standards for each component in the mixture are doped with each interfering substance. The process is repeated (usually only two to three cycles are necessary) with refined estimates of interfering substance, until self-consistent values are obtained for each component; this implies minimal interference effects resulting from the concentrations approaching those in the unknown sample.

Flame instability requires that assays are carried out in triplicate and it is advantageous to bracket a determination of an unknown with measurements of the same standard to achieve the greatest accuracy. A further improvement in technique is to use internal standards, and lithium is frequently used for this purpose. Polythene bottles should be used for storage if possible, as metal ions are both absorbed and released by glass.

In general it is usual to convert biological samples to ash prior to the determination of metals. Provided sublimation losses are prevented this can be done dry but frequently wet ashing (in solution) is used; this employs an oxidative digestion similar to the Kjeldahl method (Section 4.4).

7.12.2 Instrumentation

Atomic emission spectrophotometry

The nebulisers used are usually of the type that involves passing a stream of air over a capillary tube whose other end dips into the solution under test. Droplets are formed in various sizes and the larger ones may be allowed to settle out in a cloud chamber. This step is required because the larger drops tend not to remain in the hottest part of the flame long enough, in direct injection systems, for their constituents to be volatilised. Combustion of air and natural gas gives a temperature of 1500 °C, which is adequate for sodium determination. Calcium is better assayed at 2000–2500 °C and magnesium and iron require 2500 °C, obtained from an air/acetylene gas mixture. Routine analyses of moderate accuracy may be performed using a filter device for bandwidth selection but more accurate measurements require a monochromator. The best accuracy achieves a resolution of 0.1 to 0.2 nm over the range 200–1000 nm. Table 7.4 lists the wavelengths used for a number of metals together with their detection limits. Detectors are often of the photocell type but flame instability limits their value as their potential accuracy is not realised. Multichannel polychromators allow the emission of up to six elements at one time to be measured. The basic layout of an atomic (flame) emission spectrophotometer is shown in Fig. 7.22.

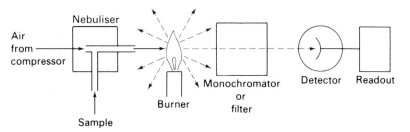

Fig. 7.22. The main components of an atomic emission (flame) spectrophotometer.

Atomic absorption spectrophotometry

In these instruments either a double monochromator with a source of white light or a hollow cathode discharge lamp is used to produce radiation in a very narrow bandwidth. Discharge lamps emit radiation at a wavelength specific for the element being assayed. This specificity can be obtained only from a pure sample of the element that is excited electrically to produce an arc spectrum of that element. Electrodeless discharge lamps that have several advantages are now available. The construction of nebulisers and burners is similar to that of the emission devices

Table 7.4. *The detection limits for various elements in emission and absorption flame spectrophotometry, flameless absorption spectrophotometry, and ion-selective electrodes*

Element	Emission		Absorption				Ion-selective electrode: detection limit (p.p.m.)
	Detection limit (p.p.m.)	Wavelength (nm)	Wavelength (nm)	Detection limit (p.p.m)			
				Flame	Flameless		
Calcium	0.005	442.7	442.7	0.1	0.00007		0.02
Copper	0.1	324.8	324.8	0.1	0.0001		0.0006
Iron	0.5	372.0	248.3	0.2	0.0001		
Lead			283.3	0.5	0.0002		0.21
Lithium	0.001	670.7	670.7	0.03	0.0001		
Magnesium	0.1	285.2	285.2	0.01	0.00001		
Manganese	0.02	403.3	279.5	0.05	0.00004		
Mercury			253.8	10.0	0.018		
Potassium	0.001	766.5	766.5	0.03	0.00003		0.04
Sodium	0.0001	589.0	589.0	0.03	0.00001		0.02
Strontium	0.01	460.7	460.9	0.06	0.0001		

but 10 cm flames are often used to obtain an increased optical length. Both single- and double-beam instruments are available, the latter often incorporating a chopper to give intermittent pulses and prevent stray light from the flame reaching the detector. The most useful wavelength range is 190–850 nm.

Flameless instruments

A flameless atomic absorption spectrophotometer incorporates a graphite tube as an oven, which may be heated electrothermally to 3000 °C. Monochromatic light specific to the element being assayed is produced either by a hollow cathode discharge lamp as described above or in an electrodeless discharge lamp. The graphite tube forms an optical cavity, in which the sample resides and through which the monochromatic radiation is passed. Absorption is measured continuously as the temperature is raised and computer methods allow the superimposition of absorption and temperature profiles, with time, to be produced. This approach allows optimum conditions to be determined for future analyses.

The flameless technique is 100 times more sensitive than flame methods and has the distinct advantage of being able to be automated as the inherent dangers of using combustible gases have been eliminated.

7.12.3 Applications

Sodium and potassium may be assayed at concentrations of a few parts per million (<5), using simple filter photometers. The more sophisticated emission flame spectrophotometers may be used to assay some 20 elements in biological samples, the most common being calcium, magnesium and manganese. Absorption flame spectrophotometers are usually more sensitive than emission types and can usually detect <1 p.p.m. of each of more than 20 elements. Exceptions to this are the alkali metals. Relative precision is about 1% in a working range of 20–200 times the detection limit (Table 7.4).

The techniques are widely used in clinical laboratories, for the determination of metals in body fluids. These determinations aid diagnosis and are valuable in the monitoring of many therapeutic regimes. In physiological and pharmacological research, sodium, potassium, calcium, magnesium, cadmium and zinc may be measured directly, but copper, lead, iron and mercury require prior extraction from the biological source. The methods are also widely used in element determination in soil and plant materials and, after suitable ashing procedures, may be used for metals in macromolecules, organelles, cells and tissues.

7.12.4 Atomic fluorescence spectrophotometry

This method is analogous to molecular fluorescence and depends on the prior excitation of atoms by electromagnetic radiation rather than by thermal energy input. Again the technique requires the atoms to be in the vapour state: the phenomenon is not observed in solution as it is with molecules. The source beam must be intense but less spectrally pure than that required for atomic absorption spectrophotometry, as only the resonant wavelengths will be absorbed and lead to fluorescence. Direct emission from the flame being recorded by the detector must be avoided and this may be achieved by modulation of the detector amplifier to the same frequency as that of the primary source.

The potential for application of this method is considerable, owing to the extreme sensitivity achievable (cf. molecular fluorescence, Section 7.8.1). Although this may be limited to only a few metals, in appropriate cases it achieves a sensitivity better than that of comparable methods. For example, zinc and cadmium may be detected at levels as low as 1 and 2 parts per 10^{-10}, respectively.

7.13 Lasers

Laser is an acronym for light amplification by stimulated emission of radiation. Space here does not permit a detailed explanation of how laser light is generated. A very simple view is that electromagnetic radiation used as the excitation agent can be considered as the input of photons to an absorbing material. This results in elevation of an electron to a higher energy level as described above (Section 7.1.2). If, while the electron is in an excited state, another photon of precisely the correct energy arrives then, instead of the electron being promoted to an even higher energy level, it returns to its original ground state. This return is accompanied by the emission of two coherent photons. These photons have associated wavelengths that are exactly in phase, hence the term coherent. In order to obtain the appropriate conditions, a laser-producing material has to be pumped. This is often achieved by surrounding the material with a rapidly flashing high intensity flash tube that gives an ample supply of suitable photons.

The emitted, coherent light has considerable advantages, but in particular it can be produced with zero bandwidth. This means that unique invariant wavelengths can be selected to excite molecules or atoms in a very precise way. It is also possible to generate, from appropriate sources, groups of selected wavelengths should this be required. Various applications are under development in spectroscopic and spectrophotometric methods that take advantage of the spectral purity of laser light.

An important application is the laser reflectance method for determining complementary DNA (cDNA) in nucleic acid studies. The use of reverse transcriptase and DNA polymerase (Section 3.10.1) allows the nucleotide sequence corresponding to the primary sequence of a peptide fragment or protein to be synthesised. Chain growth occurs at the 3′ end from a primer section, and chain termination occurs when a dideoxynucleotide is incorporated into the growing complementary strand (Chapter 3). Four 'channels' are required, each containing primer, all four deoxynucleoside triphosphates and one of each of the four dideoxy compounds. In each of the four channels, chain termination occurs at different points. Also, at the 5′ end of the primers a different fluorescent label is attached that has no influence on the subsequent reactions but can be used to identify uniquely components of the resulting mixtures in each channel. Mixtures are separated by gel electrophoresis (Section 9.4) in which distance travelled in the gel is effectively inversely proportional to the mass of the fragment. The gel is illuminated with a narrow beam of laser light and fluorescent emission from each label is measured (a different wavelength is emitted from each label). The band on the gel can be identified by including, to interrupt the emitted beam, a rotating filter disc that contains four sectors, each of which allows only one fluorescent wavelength to pass. By design, which fluor relates to which dideoxy terminator is known and mobility, position and amount are determined. The system can be automated and avoids the use of radioisotopes. It is reliable and precise, and data interpretation can be done by computer.

7.14   Suggestions for further reading

BANCROFT, G.M. (1975) *Mössbauer Spectroscopy*. McGraw-Hill, New York. (Early chapters contain details of the instrumentation and theory.)

BROWN, S.B. (1980) *An Introduction to Spectroscopy for Biochemists*. Academic Press, London. (Contains comprehensive sections on the principles, instrumentation and chemical applications of many of the techniques considered in this chapter.)

DELUCA, M.A. (Ed.) (1978) *Methods in Enzymology*, vol. 57 *Bioluminescence and Chemiluminescence*. Academic Press, London. (Contains a comprehensive coverage of luminescence methods.)

GILBERT, B. (1984) *Investigation of Molecular Structure: Spectroscopy and Diffraction Methods*, 2nd edn. Bell and Hyman, London. (A basic introduction to classical spectroscopy techniques.)

HESTER, R.E. and GIRLING, R.B. (Eds.) (1991) *Spectroscopy of Biological Molecules*. Royal Society of Chemistry Special Publication, Cambridge. (Comprehensive and up to date account of applications.)

KNOWLES, P.F., MARSH, D. and RATTLE, H.W.E. (1976) *Magnetic Resonance of Biomolecules*. John Wiley and Sons, London. (Contains a comprehensive coverage of ESR and NMR.)

LAKOWICZ, J.R. (1983) *Principles of Fluorescence Spectroscopy*. Plenum Press, New York. (Starts with basic principles but contains topics of current interest.)

MOORE, G.R., RADCLIFFE, R.G. and WILLIAMS, R.J.P. (1983) NMR and the biochemist. In *Essays in Biochemistry*, vol. 19, pp. 142–195 Academic Press, London. (A comprehensive account of NMR in biochemical studies.)

ROST, F.N.D. (1992, 1994) *Fluorescence Microscopy*, vols. 1 and 2, Cambridge University Press. (An up to date definitive text with a large appendix on flurochromes.)

WHISTON, C. (1987) *X-ray Methods*. ACOL Series. John Wiley and Sons, London. (An open learning text for those needing deeper understanding of the principles.)

WILLIAMS, D.A.R. (1986) *Nuclear Magnetic Resonance Spectroscopy*. ACOL Series. John Wiley and Sons, London. (Comment as for previous reference.)

8

Mass spectrometric techniques

8.1 Introduction

In contrast to the spectroscopic techniques described in Chapter 7, those dealt with here are not dependent on quantum principles. The term mass spectroscopy, though frequently encountered, is in fact incorrect and its use should be discouraged. In spectroscopic techniques it is possible, at least in principle by using quantum mechanical methods, to predict the spectrum. The mass spectrum is essentially dependent upon the thermodynamic stability of the ions produced and collected during a mass spectrometric experiment. Such stability depends essentially on the conditions prevailing during the experiment. In this situation it is difficult to entertain ideas of predicting the mass spectrum, although some attempt has been made to do just this using quasi-equilibrium theory (QET). Usually, to determine the mass spectrum of a compound 'the experiment must be done'.

The variety of mass spectra that may be obtained range from a single peak (generally obtained with so-called soft ionisation methods) to quite complicated patterns of peaks representing various fragments of the original species. Single-peak spectra are of particular value in the determination of very accurate molecular masses. It should be noted that although 'mass' is a universal property (everything has mass), it is not unique. Isomeric compounds (isomers) are chemical entities that have identical masses but different chemical structures, and hence different properties – they are different things. Examples are shown in Fig. 8.1.

In order to use mass spectrometry to identify different chemical structures it is necessary to perform the kind of experiment that causes the molecular entity to

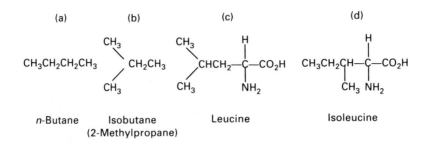

Fig. 8.1. Isomeric pairs: (a) and (b) are isomers as are (c) and (d). Each pair has the same mass but different structures and hence are different compounds.

disintegrate and produce fragment ions, each of which is represented by a peak in the resultant spectrum. It is from these more complicated spectra that skilled interpreters may reconstruct the molecular formulae of the original entity.

8.2   The mass spectrometer

8.2.1   Components

All mass spectrometers (Fig. 8.2) are essentially composed of three parts:

    (i) an ionisation chamber or source,
    (ii) a mass analyser,
    (iii) a detector.

In some cases extra analysers, e.g. energy analysers, may be included in more sophisticated devices.

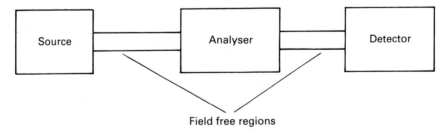

Fig. 8.2. Diagram of a mass spectrometer. The whole system is kept under high vacuum of the order of 10^{-5} torr.

In the simplest mass spectrometers, a single mass analyser is used that involves either magnetic or electric fields. The processes involved in separating entities of different masses are different in each of the cases (magnetic or electric field) and these are explained in detail later (Sections 8.8.1 and 8.8.2). An alternative approach is to measure accurately the times taken for entities of different masses to travel a given distance in space (in a straight line) in the so-called time of flight mass spectrometer (Section 8.6.3).

The essential requirement to obtain a mass spectrum is to produce ions in the gas phase, accelerate them to a specific velocity using electric fields, project them into a suitable mass analyser that separates the entities of different masses, and finally to detect each charged entity of a particular mass sequentially in time.

Any material that can be ionised and whose ions can exist in the gas phase (remember that very low pressures, i.e. high vacua, in the region of 10^{-5} torr are used) can be investigated by mass spectrometry. Almost all compounds of biological interest are 'organic', comprising the elements C,H,N,O and sometimes S and P, although a number of inorganic pollutants are important in environmental studies.

8.2.2   Ionisation techniques

Ions may be produced either by removing an electron from the molecule to produce a positively charged cationic species (cation), or conversely by adding an electron to form an anion. Both positive and negative ion mass spectrometry may be carried out, but clearly not at the same time; only one kind of ion may be accelerated out of the source region at any time, despite the fact that both types of ion may have been produced together in the ionisation process. Cations will be accelerated in either an increasing negative gradient field (attracted towards a negative electrode) or a decreasing positive gradient field (repelled from a positive electrode). Exactly the converse is true for anions.

If electrons (Section 8.3.1) are used to effect the ionisation process, however, cations and anions are not produced in equal amounts, because the removal of an electron from a neutral molecule to form a cation is a much more efficient process than electron capture to form an anion (this efficiency depends on the different ionisation energies and electron affinities involved). Many more cations are produced and it is probably for this reason that positive ion electron impact (EI) mass spectrometry is more common because there is a natural increase in sensitivity. Nevertheless negative ion spectra may be obtained in appropriate cases and produce important and useful complementary data.

Removal or addition of electrons is not, however, the only means of producing ions. The removal or addition of protons (H atoms) will also produce ions. Whereas the mass of an electron removed from, or added to, a molecular entity may be neglected (insignificant relative to the mass being determined), this is not so in the case of protons. In the latter case the mass of the resultant ion will differ by ± 1 from the mass of the original neutral entity. Furthermore it is possible to produce ions by adduct formation with entities such as NH_4^+ or CH_5^+ in a process known as chemical ionisation (Section 8.4) and here the nominal mass difference from the neutral original is 18 or 17, respectively.

The next sections deal with some of the more common ways of producing ions in a mass spectrometer. The range of ionisation sources is quite large and there are many more ways of producing ions in the gas phase than there are for analysing or detecting them.

Whichever ionisation method is chosen, a successful ionisation process requires that energy is put into the material under investigation. This turns out to be extremely important, because the amount of energy input and the way it is applied dramatically affects the thermodynamic conditions and ultimately the type of mass spectrum produced. It also means that there is no universal ionisation source and the type used is determined by the type of information required from the experiment.

8.3   Electron impact ionisation (EI)

8.3.1   Source

EI is probably still the most widely used ionisation source in mass spectrometry. In many biological problems involving metabolic studies, drug studies, pollutants, etc., it would probably be the method of choice. It is not appropriate, however, for the study of biological macromolecules, and other sources are described later that are much more important for these types of investigation. EI is a relatively simple source to understand and therefore a description of it serves the useful purpose of elaborating the principles introduced in Section 8.2.

Many metals, when heated to a sufficiently high temperature (approx. 2000 K) lose electrons by diffusion from their surface. This arises from the structural nature of metals and two that are particularly useful are rhenium and tungsten because they can be readily drawn into thin filaments. Tungsten filaments are sometimes coated with thoria (thorium oxide) to increase the ease with which electrons are emitted. If such heated filaments are subjected to an appropriate potential gradient then electrons are removed from the surface more rapidly and gain an energy directly related to the potential applied. Commonly a 70 V potential is applied and the resultant beam contains 70 eV electrons. The electron volt (eV) is a measure of energy. These electrons stream across an evacuated chamber into which molecules of the substance to be analysed are allowed to diffuse. In the electron impact source the substance to be analysed must be in the vapour state, which obviously limits the applicability to biological materials, although it should be remembered that most substances may have their natural volatility increased by chemical modification.

The stream of 70 eV electrons from the filament interacts with the molecules of the substance to be analysed (which are neutral and in random thermal motion). This interaction results in either loss of an electron from the substance (to produce a cation) or electron capture (to produce an anion). As the positive ionisation potential of most organic molecules is of the order of 20 eV, there is ample excess energy in the beam of bombarding electrons. The electron impact source is shown diagrammatically in Fig. 8.3.

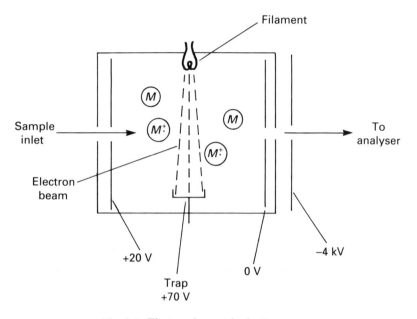

Fig. 8.3. Electron impact ionisation source.

The possible events which may occur are shown below. M represents the neutral molecular species:

$$M + e^- \text{ (bombarding electron)} \longrightarrow M^{\ddot{+}} 2e^-$$

(one bombarding electron
and one electron
removed from M)

or:

$$M + e^- \longrightarrow M^{\ddot{-}}$$

Chemical bonds in organic molecules are formed by the pairing of electrons. Ionisation resulting in a cation requires loss of an electron from one of these bonds (effectively knocked out by the bombarding electrons), but it leaves a bond with a single unpaired electron. This is a radical as well as being a cation and hence the representation as M^{+}, the plus sign $(+)$ indicating the ionic state and the dot $(\cdot)$ a radical. Conversely, electron capture results both in an anion but also the addition of an unpaired electron and therefore a negatively charged radical, hence the symbol M^{-}. Such radical ions are termed molecular ións or parent ions and under the conditions of electron bombardment are relatively unstable. The imparted energy in excess of that required for ionisation has to be dissipated and this latter process results in the parent ion disintegrating into a number of smaller fragments. Also the fragments themselves may be relatively unstable and further fragmentation may occur. This gives rise to a series of daughter ions, which are eventually recorded as the mass spectrum. The whole process is best shown by the purely

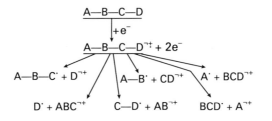

Fig. 8.4. Fragmentation processes in a hypothetical species.

hypothetical case shown in Fig. 8.4. A, B, C and D represent groups in a molecule linked by electron-pair bonds. The example here is for the production of a radical cation. As it is not known where either the positive charge or the unpaired electron actually reside in the molecule it is usual practice to place the dot signs outside the abbreviated bracket sign, $\neg$. This is also true for fragment or daughter ions and will be used throughout the remainder of this chapter.

The first thing to note is that, when the parent ion fragments, one of the possible products carries the charge and the other the unpaired electron, i.e. it splits into a radical and an ion. The daughter ions are therefore true ions and not radical ions. The radicals produced in the fragmentation process are neutral species and therefore do not take any further part in the mass spectrometry but are pumped away by the vacuum system. Only the charged species may be accelerated out of the source and into the mass analyser. It is also important to recognise that almost all possible bond breakages can occur and any given fragment will arise both as an ion and a radical, such as $AB^{\cdot}$ and $AB^{\neg+}$. The distribution of charge and unpaired electron, however, is by no means equal, i.e. it is not usual to get 50% $AB^{\cdot}$ and 50% $AB^{\neg+}$. The distribution depends entirely on the thermodynamic stability of the products of fragmentation. Furthermore, any fragment such as $AB^{\neg+}$ may break down further (until single atoms are obtained) and hence not many ions of a particular type may survive, resulting in a low signal being recorded.

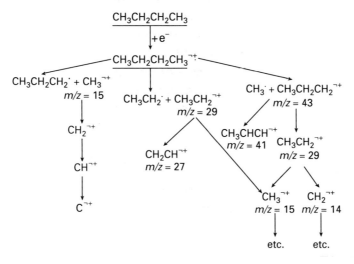

Fig. 8.5. Fragmentation pathways in *n*-butane.

A simple, but non-biological example is given by *n*-butane ($CH_3CH_2CH_2CH_3$) and some of the major fragmentations are shown in Fig. 8.5. The resultant EI spectrum that would be obtained is shown in Fig. 8.6. What is actually recorded is the mass-to-charge ratio spectrum, *m/z*, where *m* is the mass and *z* is the number of charges carried by the ion. In ordinary mass spectrometry, $z = 1$ and hence in effect a mass spectrum is obtained. It will be shown later (Section 8.7.2) that it is possible to produce ions where $z > 1$, enabling much larger species to be analysed.

8.3.2   Isotopic composition

In ordinary mass spectrometry satellite peaks are observed at a particular value of the *m/z* ratio (equation 8.1, p. 391) that arises from the isotopic composition of the species. For instance, 1.108% of all carbon on the planet is the 13-isotope (designated ^{13}C), the remainder being ^{12}C. Hence if an ion, parent or daughter, contained say 12 C atoms, then the approximate percentage probability of at least one of them being the ^{13}C isotope is:

$$12 \times 1.108 = 12.96\%$$

At some *m/z* value (depending on the other atoms present in the ion) there will be a peak in the spectrum corresponding to the ion containing only ^{12}C atoms. The intensity of this peak depends on the thermodynamic stability of the ion. There

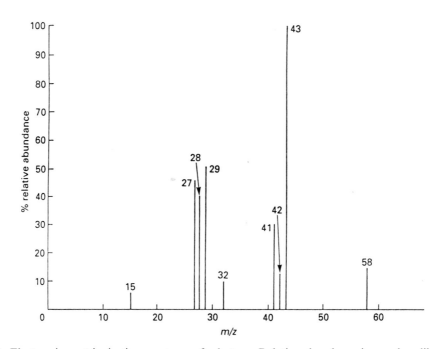

Fig. 8.6. Electron impact ionisation spectrum of *n*-butane. Relative abundance is a scale calibrated from 0 to 100%. The largest peak in the spectrum is set at 100% (base peak) and all others calculated in proportion as a percentage. Spectra produced in this way are said to be normalised, are machine independent and hence directly comparable. Conversely spectra whose ordinates are labelled as intensity are absolute spectra

will also be a satellite isotope peak one atomic mass unit greater owing to the ion containing one ^{13}C atom. The approximate intensity of the satellite peak will be the product of the intensity of the peak representing the ^{12}C only species ($I_{^{12}C}$), the number of carbon atoms in the ion (N) and the natural abundance of the ^{13}C isotope:

$$(I_{^{12}C}) \times (N) \times 1.108\%$$

Of course the intensity of this satellite peak may be greater than this if other species contribute to it, but it is possible that any such contributions may be accounted for using high resolution mass spectrometry (Section 8.8.2).

The possibility of more than one ^{13}C atom being present in an ion is finite but of small value, and to obtain this value one has to perform rather more accurate calculations than the approximations carried out above. To do this one has to solve a binomial expansion:

$$(a+b)^m$$

where *a* is the percentage natural abundance of the light isotope, *b* is the percentage

Table 8.1. *Probability of occurrence of different isotopic compositions*

No of ^{13}C atoms present	% probability of occurrence
0 (all ^{12}C)	95.641
1	4.2863
2	0.0720
3	0.0005
4	0.0000
	99.9998

There have to be four C atoms of one kind or another so that the total percentage adds up to 100.

natural abundance of the heavy isotope, and m is the number of atoms of the element concerned in the molecule.

Consider the earlier simple example of n-butane where $m = 4$. Expansion of $(a + b)^4$ gives:

$$a^4 + 4a^3b + 6a^2b^2 + 4ab^3 + b^4$$

(Note that the binomial coefficients may be found from mathematical tables.) Substituting the appropriate values for the relevant isotopic abundances of carbon isotopes gives:

$$(98.892)^4 + [4 \times (98.892)^3 \times 1.108] + [6 \times (98.892)^2 \times (1.108)^2)]$$
$$+ [4 \times 98.892 \times (1.108)^3] + (1.108)^4$$
$$= (9.5641 \times 10^7) + (4.2863 \times 10^6) + (7.2036 \times 10^4)$$
$$+ (5.3807 \times 10^2) + 1.5071$$

Dividing throughout by 10^6 (the substituted values are actually percentages so the denominators should be 100^4 or 10^8 and this has to be multiplied by 100 to recover a percentage, hence divide by 10^6), the data in Table 8.1 may be constructed. It can be seen that there is now a small but significant probability that natural n-butane will have some molecules containing a ^{13}C atom. The probabilities of there being two, three or four are negligible. However, if the compound contained 100 carbon atoms then there would be a greater chance of any carbon atom being a ^{13}C than a ^{12}C. Biological macromolecules contain several hundred carbon atoms and the isotopic distribution patterns become extremely complicated, the computations unwieldy, and sophisticated computer techniques are necessary to deconvolute the data.

For elements such as chlorine, the isotopic abundances are approximately $3:1$ for $^{35}Cl:^{37}Cl$. If a compound contains a single chlorine atom and can be ionised to give stable molecular ions, then two such species will be observed, with peak intensities in an approximate ratio of $3:1$. If a compound contained two chlorine atoms, then in order to predict the relative abundances of the three possible peaks that would arise (note that if the number of isotopes of a particular element is n then there are $n+1$ possible peaks predicted by the binomial expansion) is obtained from the expansion of:

$$(a+b)^2 = a^2 + 2ab + b^2$$

Substituting $a = 3$ and $b = 1$ for the isotopic abundances of ^{35}Cl and ^{37}Cl, respectively, we obtain for the three possible peaks:

$$^{35}Cl^{35}Cl \quad 9$$
$$^{35}Cl^{37}Cl \quad 6$$
$$^{37}Cl^{37}Cl \quad 1$$

It is left to readers to try as an exercise the calculation of the relative ratios for a compound containing three chlorine atoms. Four peaks are possible with the proportions $27:27:9:1$.

It should be seen that isotope patterns become rather complicated as the possibility of increased numbers of different isotopes being contained within a molecule increases. Nevertheless it is also an aid to interpretation in many cases. Particular interest is being shown by a number of microbiologists in the ability of some microorganisms to metabolise selectively different organohalogen compounds, as these materials are increasingly produced by the heavy-chemical industry. A study of such selective metabolism can be undertaken using even low resolution mass spectrometry, and monitoring the relative consumption of different organohalogen compounds by interpreting the relative isotope abundance patterns obtained.

A rapidly increasing type of mass spectrometry in the biological field is isotope ratio mass spectrometry (IRMS). There is increasing reluctance on the part of clinical research ethical committees to approve metabolic and other studies on human subjects using radioisotopes. The use of the latter in other animals is also diminishing. An alternative approach is to use stable isotopes as labels. Stable isotopes such as ^{13}C, ^{15}N and ^{18}O (for total body water) selectively synthesised into a metabolite may be used as a marker and obviously does not affect the metabolic process in any way that would be different from using the nearest radioisotope. Of course there is no emission of radiation and therefore the concentrations of labelled metabolite cannot be determined by counting methods. Most of the materials used in such investigations, however, can be readily combusted in an atmosphere of oxygen to produce gases such as $^{13}CO_2$, $^{15}N_2O$ and $^{15}NO_2$. These compounds can be readily separated by mass spectrometry and isotope ratios determined from which the metabolic data can be computed.

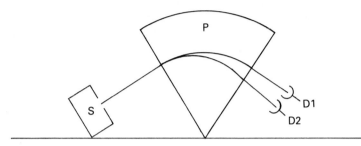

Fig. 8.7. A simple sector mass spectrometer for isotope ratio studies. S, source; P, magnetic pole; D, detectors.

IRMS involves one of the simplest mass spectrometers available, composed of an EI source (all products are already in the vapour state), a magnetic sector mass analyser and a Faraday cup detector (Fig. 8.7). The amplified current from the detector is recorded as a peak in the spectrum, whose height is a measure of the particular ion intensity (ion current). The value of the mass-to-charge ratio is given by:

$$m/z = B^2 R^2 / 2V \qquad (8.1)$$

where B is the magnetic flux density (of the magnetic analyser), R is the radius of the trajectory (of the ion in the magnetic field), and V is the accelerating voltage (used to accelerate ions out of the source).

Consideration of equation 8.1 should indicate that various values of m/z will be obtained if either B or V is varied (R is a constant for a particular magnetic field). In the IRMS technique a permanent magnet is used and hence B is also fixed. By varying V, ions of different masses may be made to traverse the same circular trajectory of radius R slightly separated in time, or for a given appropriate value of V they may be made to traverse two slightly different trajectories of different values of R, i.e. slightly separated in space. In either case, in principle, ions such as $^{13}CO_2^{+}$ and $^{15}NO_2^{+}$ may be separated.

8.3.3   Applications

IRMS analysis of human breath

Helicobacter pylori (*Campylobacter pylori*) infection is linked with 90% of cases of gastric ulcer. It is also estimated that in the developing world some 80% of children are at risk from life-threatening gastroenteritis arising from *H. pylori* infection. As the bacterium contains urease, a non-invasive test for the presence of the organism may be performed. The patient is required to drink a solution (dose) of 99% (v/v) ^{13}C-urea. The ^{13}CO$_2$ produced from the enzyme action is absorbed and eventually exhaled in the patient's breath, which can be collected in a suitable bag. A sample of the breath is then analysed in an IRMS instrument and the amounts of ^{12}CO$_2$ and ^{13}CO$_2$ determined. In a typical set of experiments, all *H. pylori*-infected patients showed an increase of >5% ^{13}C at both 40 and 60 min.

Table 8.2. *Applications of ^{13}C labelling for IRMS measurements*

^{13}C compound	Application
Triolein	Fat malabsorption
Bicarbonate	Energy expenditure
Lactose	Lactase deficiency
Glucose	Carbohydrate metabolism
Palmitate	Fatty acid oxidation
Galactose	Liver enzyme function

Other examples of stable isotope measurements in metabolic studies using ^{13}C labelling are given in Table 8.2. ^{15}N enrichment can also be used to study protein metabolism (turnover) and ^{18}O for total body water and other volume measurements.

8.3.4   Pyrolysis mass spectrometry

Although it has been in use for some time, this technique is now in the process of undergoing something of a renaissance. The principle is simple in that materials are subjected to a precisely controlled high temperature for a fixed and measured time span. Volatile substances are ejected from the material, under vacuum conditions, and can then be ionised by EI, the mass spectrometric analysis being conducted in any of the usual ways.

   The thermal degradation is carried out in a suitable pyrolyser, the most useful
of which is the Curie point device in which a specific temperature can be maintained
very precisely. This is achieved by taking advantage of the fact that in paramagnetic
materials (Section 7.6), the magnetic susceptibility is temperature dependent and
such materials possess a Curie constant. If the temperature varies then the magnetic
susceptibility must vary; conversely, if the susceptibility is maintained by external
means, then the temperature is equally maintained. A disadvantage is that only
certain temperatures are allowed depending upon the Curie point properties of the
heating material used. A continuum of temperatures is not obtainable, only discrete
values.
   Inevitably there are mixtures of pyrolysis products and identification of constituents and relative composition of such mixtures require the use of quite
complicated mathematical and statistical procedures such as factor analysis. Such
methods are readily available on computers so should not be a major impediment
to the use of the technique.
   A major application is in the identification of microorganisms, which, because
such small sample sizes are required, can often be carried out on swab samples.
This reduces the need to culture the cells, other than for confirmation by other
methods, and eliminates the delays involved. Research into the structures and
composition of cell walls can be aided by this technique, particularly in the study
of the effects of different growth conditions and in the presence and absence of a
variety of antibiotics.
   Usually the products of pyrolysis are small-to-medium sized in molecular mass
terms. Hence, large mass range spectrometers of the sector type are not necessary
and small mass range quadrupoles or ion traps (Section 8.8.3) may be used. This
makes the technique portable and it is gaining in importance in field studies.
Provided a suitable power supply is available, which can be obtained from a
generator in a medium-sized off-the-road vehicle, on-site rapid investigations are
facilitated. It can be anticipated that the technique will gain increasing applicability
in environmental science, ecological studies and more general areas of biological
research and monitoring work.

8.4   Chemical ionisation (CI)

CI is essentially based on the EI source but little fragmentation occurs, giving rise
to much cleaner spectra. It is particularly valuable in the determination of molecular
masses, as high intensity molecular or pseudomolecular ions are produced.
   The construction of the source is essentially the same as described above, but
the source is filled, prior to the analytical experiment being carried out, with a

suitable reagent gas such as methane (CH_4) or ammonia (NH_3). The normal generation of ions of these gases by EI will give rise to species such as $CH_4^{\cdot+}$ or $NH_3^{\cdot+}$. However, owing to the relatively high pressure of the reagent gases in the source, the possibility of ion–molecule reactions arises. For example:

or:

$$CH_4^{\cdot+} + CH_4 \longrightarrow CH_5^{\cdot+} + CH_3^{\cdot}$$

$$NH_3^{\cdot+} + NH_3 \longrightarrow NH_4^{\cdot+} + NH_2^{\cdot}$$

The species $CH_5^{\cdot+}$ and $NH_4^{\cdot+}$ have arisen because the original radical ions have abstracted protons from the corresponding neutral species. These resultant ions are powerful proton donors (Lewis acids), in the vapour state, and if a material to be analysed is now introduced into the source (the electron beam is switched off) it will be ionised by protonation, giving rise to a thermodynamically relatively stable parent plus one pseudomolecular ion:

$$RCH_2CH_3 + CH_5^{\cdot+} \longrightarrow (RCH_2CH_3 + H)^{\cdot+} + CH_4$$

This type of ionisation is widely used in the study of drugs and secondary metabolites. Negative chemical ionisation is also made possible by choosing an appropriate reagent gas.

However, an increasingly important method for obtaining an anion is by electron capture chemical ionisation. In order to achieve this with increased efficiency the thermodynamics of the process must be changed, or rather the energetics of the process altered. This is done by slowing down the ionising electrons to thermal energies in the presence of a suitable gas, such as methane, hence making the process of electron capture more probable. The temperature of a gas is a measure of the kinetic energy of the motion of its constituent particles and if they can be slowed down there will be a consequent reduction in temperature. The thermal energies referred to above are those associated with approximately room temperature.

8.5   Field ionisation (FI)

This method of ionisation again requires the sample to be introduced in the vapour state. The molecules are subjected to an intense electric field, of the order of 10^7–10^8 V cm^{-1}. Under such conditions the outer bonding electrons are subject to large forces and the energetics are sufficient to overcome the ionisation potential and an electron is removed to generate a molecular radical cation. These ions can then be accelerated out of the source and into the mass analyser in the usual way.

Such a source has several applications, but they tend to be rather specialised. It is probably the case that for biological applications other forms of ionisation are more appropriate, either EI or the desorption and evaporation methods discussed below (Sections 8.6 and 8.7).

8.6   Ion desorption methods

It is possible to introduce solid samples into the EI source using various devices, the most common of which is the direct insertion probe (DIP). Of course a required property of the substance under test is that the solid would have to be sufficiently volatile (at the low pressures used) to evaporate, or more strictly volatilise, prior to ionisation in the electron beam. Most biological materials do not possess this property, or, if they do, are fragile or thermally labile. Any decomposition that occurs prior to ionisation of course means that what is actually analysed is different from the original.

A number of important desorption methods have been developed that enable solid materials to be introduced into the mass spectrometer, generally with the ionisation process occurring at room temperature or slightly lower temperature, hence introducing some degree of protection for thermally labile compounds. In general, the mechanisms of electron removal or capture in desorption methods are a great deal more complicated than for electron impact, field ionisation or the adducts formed in chemical ionisation. In some cases, the process is not even clearly understood and competing theories exist to explain the observed phenomena. Apart from occasional statements no further attempt will be made in this chapter to delve into the actual processes. Suffice it to say that ions are generated abundantly in these methods, hence making available the power of mass spectrometry to the researcher investigating biological materials.

8.6.1   Field desorption (FD)

FD is an interesting source whose popularity waxes and wanes with time. It is difficult to prepare, even in expert hands and the source is destroyed each time it is used.

The filament of the source has to be prepared in advance. A tungsten wire, 10 μm in diameter, is coated with an appropriate organic material (benzonitrile is commonly used) from which, under suitable conditions, carbon filaments or

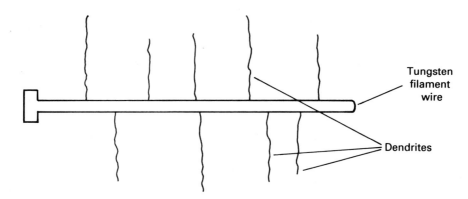

Fig. 8.8. Diagram of field desorption filament.

dendrites may be grown (Fig. 8.8). A solution of the substance to be analysed is then carefully coated on to the dendrites (the 'filament' is extremely fragile) and allowed to dry. This leaves a solid coat and the filament can be fixed into the source housing which is then evacuated to the low pressures required. Intense electric fields, again of the order of 10^7–10^8 V cm^{-1}, can be applied and ions will be desorbed into the evacuated region and accelerated into the mass analyser.

In general the system is run under such conditions that little fragmentation takes place and hence fairly intense molecular ion peaks may be observed.

8.6.2   Fast atom bombardment (FAB) ionisation

It is almost certainly true that the advent of the FAB ionisation method revolutionised mass spectrometry, really opening up the technique to the biologist and may even have been responsible for the renaissance of the analytical method itself. It certainly has generated a massive interest, stimulated developments in other ionisation sources appropriate for biological materials and, since its invention in 1981, has given rise to a whole new field of endeavour that has become known as biological mass spectrometry.

The first important advantage for the investigation of biological materials is that they can be introduced into the ionising beam of neutral atoms, *in solution*. The solution is mixed with a relatively involatile, viscous matrix such as glycerol, thioglycerol or *m*-nitrobenzyl alcohol. It is this admixture, placed on a suitable probe, which is then introduced into the source housing, a vacuum applied and the mixture bombarded with atoms travelling at high velocity. One important theory that attempts to describe the ionisation process states that a very short-lived transient high temperature spike occurs, too brief to cause thermally induced bond breaking but of sufficient length to allow ionisation to occur. Subsequent fragmentation then allows a mass spectrum to be obtained that contains considerable structural information.

Both positive and negative, complementary, mass spectra may be produced, but pseudomolecular species arise as either protonated or deprotonated entities. For example, if M is the molecular entity then $(M + H)^+$ or $(M - H)^-$ is observed. It is assumed that protonation occurs by abstraction of H^+ from the matrix and deprotonation by donation of H^+ to the matrix. Other charged adducts can also arise (such as $(M + Na)^+$ and $(M + K)^+$), but note that in this ionisation mechanism the radical ion does not occur, indicating that the process is quite different from EI. In addition, cluster ions from the matrix arise, e.g. glycerol clusters, $(C_3H_8O_3)^+$, $(C_3H_8O_3)_2^+$, $(C_3H_8O_3)_3^+$ ... etc. Although the peaks that arise from these appear to complicate the spectrum, they serve as accurate markers because their mass is known. Also, for overall calibration of the analyser, mixtures of alkali metal iodides may be used that produce regular peaks to high mass.

The second important advantage in this method arises from the use of liquid matrices. Most solid surfaces are permanently damaged by a beam of high energy atoms, leading to short-lived samples and spectra, but the mobility of the liquid matrices used allows the surface to be continually replenished. This has distinct advantages for the study of many medium-to-large biological molecules.

Despite the important advantages described above there is one major disadvantage in fast atom bombardment mass spectrometry (FAB-MS) and that is suppression effects. The reader may have guessed from the previous paragraph that the surface is important in this method. In fact there is some evidence to suggest that substances that are surface active in the liquid matrix and reside just in the surface are most readily ionised. Those that reside on the surface seem to be less readily ionised and those that are totally dissolved (i.e. they reside in the bulk of the matrix) seem to be least susceptible of all to ionisation.

The problem arises in analyses such as the so-called peptide mapping technique. Ideally, if a protein, for instance, is subjected to an enzyme digestion and the mixture of peptide products analysed by FAB-MS, it should in principle be possible to detect pseudomolecular ions for each product. However, this possibility is not always achieved because, if the different products possess markedly different surface activities, then one or more of the products may be suppressed by being forced to reside in the bulk matrix. The danger that arises should be obvious because, if the experimenter is dealing with an entirely unknown protein, he or she would have no way of knowing whether or not suppression was occurring. It is possible to relate the hydrophobicity (glycerol has some properties in common with water; Section 4.2) of the peptide with surface activity. Hydrophobicity may be expressed in a semiquantitative fashion using, for instance, the Bull and Breeze indices. [Bull and Breeze indices are measures of the water-loving/water-hating (or other solvent) properties of the solute and are expressed as 'hydrophobicities'. The fundamental measure on which the indices are based is thermodynamic and represents the free energy change when a solute crosses a boundary from an aqueous to a non-aqueous medium.] The problem can sometimes be partially overcome by using different matrices, addition of acid, etc., but this approach is somewhat empirical.

The method described above is known as static FAB. Separating the components of the mixed products prior to FAB-MS largely overcomes the suppression problem. High performance liquid chromatography (HPLC; Section 10.4) and electrophoresis (Chapter 9) are powerful methods for achieving such separations. Much greater efficiency can be achieved when these separatory methods are interfaced on-line to the FAB-MS instrument. When this is done, the resulting set up is known as a hyphenated technique. Many ingenious ways have been and are being developed for the crucial interface between separatory devices and mass spectrometers. Much of the technology can be transported from the earlier and highly successful hyphenated technique using gas–liquid chromatography (GLC; Section 10.10) and EI-MS, which is represented as GC-EI-MS or GLC-EI-MS.

In the case of FAB, whatever separatory method is used it gives rise to dynamic or continuous flow FAB (CF-FAB). A small quantity of the appropriate matrix is introduced either pre- or post-column of the separatory method, depending on the application. The eluent diffuses through a frit or sinter at the end of a flow tube in the FAB source. The aim is to match the flow rate of the eluent to the rate of evaporation in order that a steady-state surface is continually presented to the atom beam. Several of these continuous flow systems are being readily adapted to

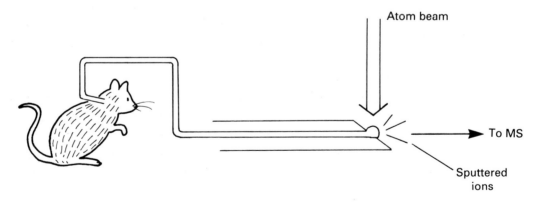

Fig. 8.9. *In vivo* CF-FAB-MS experiment.

other ionisation sources described below, particularly the ion evaporation methods.

Perhaps one of the most remarkable recent developments is *in vivo*-CF-FAB-MS. It should be fairly straightforward to recognise the applications to the analysis of body fluids where these have been removed from the animal, perhaps pretreated in some way and then applied to the separatory device coupled on-line to the mass spectrometer. In the *in vivo* approach, the animal is catheterised, the body fluid sample passed directly into the separatory device and thence to the mass spectrometer (Fig. 8.9). Such small quantities are required for analysis that the animal recovers and suffers no ill effects from the analysis.

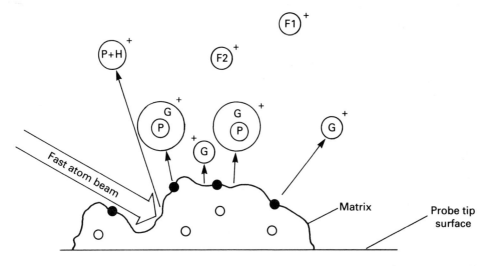

Fig. 8.10. Sputtering phenomenon from liquid matrix during bombardment with fast atoms. G, matrix; $P + G^+$, parent solvate with matrix; $P + H^+$, pseudomolecular ion; F_1, F_2, fragments ions.

During the FAB process, material is said to be sputtered from the surface (Fig. 8.10). The ions of interest are almost certainly solvated with matrix molecules and this helps to stabilise them by dissipating excess energy and limiting fragmentation. This results in reasonable sensitivity for the pseudomolecular ion but still allows sufficient fragmentation to occur as the matrix evaporates.

The generation of the atom beam and the angle of impact is fairly critical in FAB. The atoms, usually argon (Ar) or xenon (Xe), are generated in an atom gun. It is not possible to focus neutral atoms in electric or magnetic fields though the beam may be collimated by passing it through restriction apertures. The first step is to admit the noble gas into what is essentially an EI-type source that produces cations. No radical ions are produced here as the noble gases are monatomic. These ions are accelerated in the usual way and the ion beam focused into a chamber containing neutral atoms of the same gas. Collisions occur between the fast-moving ions and relatively slow-moving atoms (thermal motion only) and, because the same chemical entities are involved, resonant charge exchange takes place. The atoms are knocked out of the collision chamber with virtually the same velocity as the incoming ions, which themselves are slowed down to thermal motion. Just in case any ions manage to get through the chamber without colliding, the emergent atom beam passes between two oppositely charged collector plates. Any positive ions will be deflected from the positively charged plate and attracted to, and collected on, the negatively charged plate. The neutral atoms continue

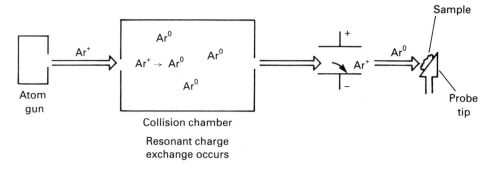

Fig. 8.11. Generation of beams of fast atoms. Ar^0, argon atoms; Ar^+, argon ions.

unimpeded to impact with the matrix mixed with sample on the probe (Fig. 8.11). Atom beams are used rather than the original ions to eliminate undesirable surface effects. In fact it is possible to use beams of fast ions, such as caesium or gold clusters. In these methods (fast ion bombardment (FIB)), we can sacrifice information lost as a result of surface damage of the sample because we gain more from using the more massive species that, owing to their greater mass, have larger kinetic energies. FAB-MS has been applied to the structural elucidation of several different kinds of biological macromolecules including proteins/peptides, nucleic acids, polysaccharides and lipids.

The approach is exemplified by peptide sequencing. It should be recognised that FAB-MS is not meant to be a replacement for conventional Edman-type sequencing methods (Section 4.3.3). There are, however, certain constraints that apply to the Edman method. It does not work at all with peptides containing blocked N-terminal ends, such as pyroglutamic acid. Post-transcriptional and post-translational modifications that remain intact on isolation cannot be determined by Edman-based sequenators, particularly if there are sugar residues present, as the cleavage products are generally too soluble to be extracted and detected. Many of the problems are overcome by using FAB-MS as a complementary technique.

In the case of peptides, although a wide variety of fragmentations may occur, there is a predominance of peptide bond cleavages. This means that the peptide fragments, by losing one amino acid residue at a time, give rise to peaks in the spectrum that differ sequentially by the amino acid residue mass. These sequential mass differences represent exactly the amino acid sequence or primary structure of the peptide. The processes are so reproducible that a set of empirical rules has been devised that enables different series of ions, A, B, C and X, Y, Z, to be

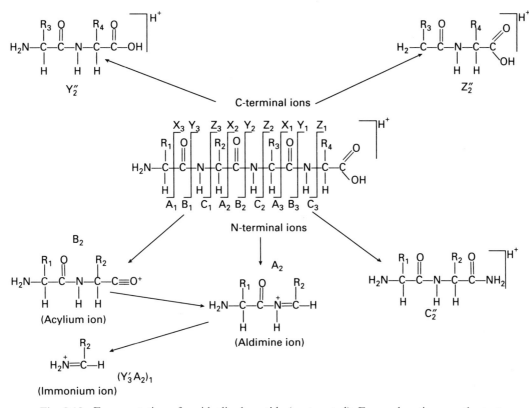

Fig. 8.12. Fragmentation of an idealised peptide (protonated). For explanation, see the text.

recognised, depending upon which mass carries the charge to be recorded in the mass spectrum. Figure 8.12 shows an idealised peptide subjected to FAB. X, Y and Z ions are those that arise by cleavage on the C-terminal side of the peptide bond. For example, the Z_1 ion is the first C-terminal amino acid: Y_1 contains the -NH group as well (15 atomic mass units greater). Z_2 is the first two C-terminal amino acid residues, and so on. The A, B and C ions arise from the N-terminal end, A_1 being the first, A_2 the first two, etc. The superscript double primes represent the number of hydrogen atoms lost or gained by the ion that is collected and recorded. Each series will generally predominate in either positive or negative ion FAB so that, in effect, in two experiments the peptide may be sequenced from both ends by obtaining complementary data (Fig. 8.13). Also it is often possible to leave the sample in the source and merely switch operational modes. Furthermore, D-and W-type ions arise, which enable distinction to be made between isomeric amino acids such as leucine and isoleucine. D-type ions arise from A-type and W-type from Z-type (Fig. 8.14). Substitutions of non-peptidic material may

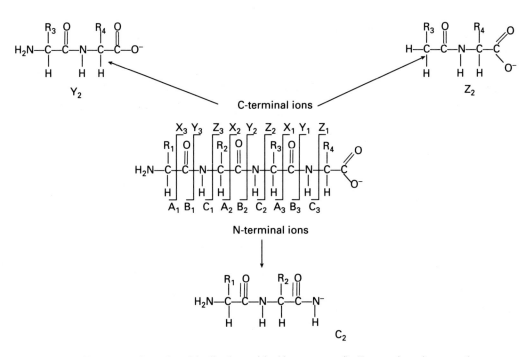

Fig. 8.13. Fragmentation of an idealised peptide (deprotonated). For explanation, see the text.

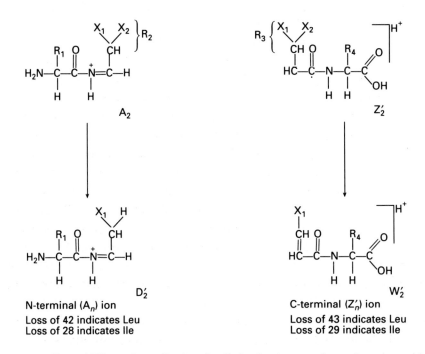

N-terminal (Aₙ) ion
Loss of 42 indicates Leu
Loss of 28 indicates Ile

C-terminal (Z′ₙ) ion
Loss of 43 indicates Leu
Loss of 29 indicates Ile

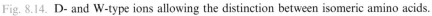

Fig. 8.14. D- and W-type ions allowing the distinction between isomeric amino acids.

also be detected and identified from the same data but this has been omitted from this description for clarity.

Figure 8.15a and b shows the FAB mass spectra of the two isomeric tripeptides Gly-Leu-Ala and Ala-Leu-Gly. The mass losses corresponding to the fragments represented by the most prominent peaks are indicated and show how the two sequences can be distinguished.

The power of the FAB method is enormous and many further developments continue to be published. There are, however, limitations to masses that may be transmitted in sector and quadrupole analysers and other sources have been developed to tackle larger species.

8.6.3   Plasma desorption ionisation (PD)

Plasma, in this context, is used to describe atomic nuclei stripped of electrons. The source of the plasma is radioactive californium, ^{252}Cf, and two typical emission nuclei are the 100 MeV Ba^{20+} and Tc^{18+}, which are ejected in opposite directions, almost colinearly and with equal velocity. This is a pulsed technique, i.e. particles are emitted at discrete time intervals, that requires a different type of mass analysis generally obtained in a time of flight (TOF) mass spectrometer.

Samples of large biological molecules, e.g. haemoglobin, can be coated on to a suitable planchette (nickel, mylar or nafion), sometimes in the presence of other additives such as nitrocellulose (which allows the preferential selection of sample against impurities); it is then placed in front of the source. The emitted plasma particle passes through the support foil and imparts sufficient energy to the sample to cause ionisation and project or desorb the sample ion into the gas phase. The ion drifts down the evacuated tube to a suitable detector and is recorded.

In order to be able to measure accurately the time of flight of the ion, the zero point (time from desorption) must be known. It is here that the nature of the disintegration of ^{252}Cf is important because the plasma particle emitted in the opposite direction to that passing through the sample can be used to trigger a time counter. The desorbed ions can be accelerated electrically as in other sources but for the same terminal velocity, v, ions of different mass will receive different momenta, mv. However, ions of the same momenta but different mass will therefore have higher velocity, i.e. the lower the value of m, the higher the value of v, for constant mv. The lightest (smallest mass) ion therefore travels fastest down the drift tube and arrives at the collector first. Ions of increasing mass (decreasing velocity) arrive successively in time (Fig. 8.16).

In addition to the ability to investigate large ions it is also possible, because of the nature of the pulsed technique and the time intervals involved, to perform kinetic experiments on the foil or planchette, involving enzymic digestion of large

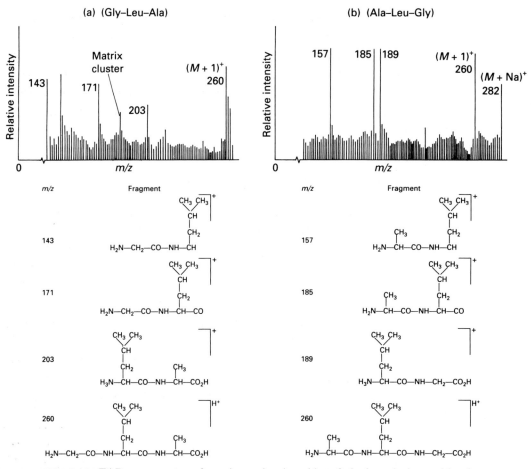

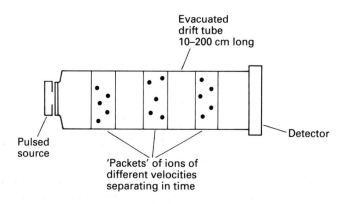

Fig. 8.15. FAB mass spectra of two isomeric tripeptides of alanine, glycine and leucine.

Fig. 8.16. Diagram of the time of flight analyser.

proteins and nucleotides. The parent ions of the digestion products can then be detected. Aberrancies (point substitutions) in haemoglobins have been identified in this way in terms of mass differences (for example, in sickle-cell anaemia), and if sufficient fragmentation can be induced in the enzyme digestion products the position of substitution may be inferred. To confirm the point of substitution, however, other mass spectrometric analyses should be carried out as fragmentation is limited in PD-MS.

8.6.4   Laser desorption ionisation (LD)

Laser beams (being electromagnetic radiation; Section 7.13) can be readily collimated and focused and can be generated with sufficient energy to cause both ionisation and desorption from a sample coated on a suitable probe surface. Either a continuous stream of ions may be produced but the sample is relatively short lived owing to surface damage by the laser, or the laser may be pulsed and the desorbed ions analysed by TOF-MS. Addition of a suitable matrix vastly improves LD and masses of the order of several hundred thousand daltons may be detected.

One of the most recent advances in this method has been the development of matrix-assisted laser desorption ionisation (MALDI), in which the matrix functions as an energy sink (i.e. a substance that absorbs the radiation energy at the wavelength of the coherent laser light used). In the ultraviolet region, substances exhibiting conjugation in their structures make good matrices.

In fact it is possible to choose different matrices which are appropriate for selected wavelengths over the ultraviolet–visible–infrared range of the electromagnetic spectrum. The particular advantage of MALDI is the ability to produce large mass ions, with high sensitivity, the molecular ions being produced with little fragmentation, hence making it a valuable technique for examining mixtures.

8.7   Ion evaporation methods

These are methods that lend themselves readily to interfacing with other separatory systems to produce 'hyphenated' techniques (cf. CF-FAB).

The essential principle in these methods is that a spray of charged liquid droplets is produced by some form of atomisation or nebulisation. The species to be investigated is solvated by the charged drop. As the solvent evaporates in the high vacuum region, the drop size decreases and the charge eventually resides on the entity under study.

8.7.1   Thermospray ionisation source (TSI)

As the name implies, thermal effects are used and this may be disadvantageous for thermally labile compounds. In some designs an electrode and buffers are included to aid the charging of the drop. The mist of drops drifts in an evacuated space where the solvent is removed, by evaporation, and the resulting charged species is then accelerated into the mass spectrometer (Fig. 8.17).

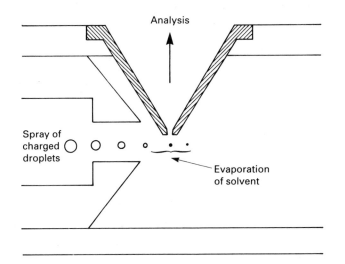

Fig. 8.17. Diagram of the thermospray source. The shading denotes the cross-section through the funnel-shaped orifice.

8.7.2   Electrospray (ESI) and ionspray ionisation (IS)

These two sources are very similar; the latter uses a gas (usually nitrogen) to cause nebulisation and is sometimes referred to as pneumatically assisted electrospray. Two important features of ESI are:

(i) ionisation can occur at atmospheric pressure (note that this method is also sometimes referred to as atmospheric pressure ionisation (API)),
(ii) the ability to impose multiple charges on the molecular species.

Figure 8.18 shows a diagrammatic representation of the ESI source. In one type, a slowly flowing curtain gas (usually nitrogen) is present to aid the evaporation of

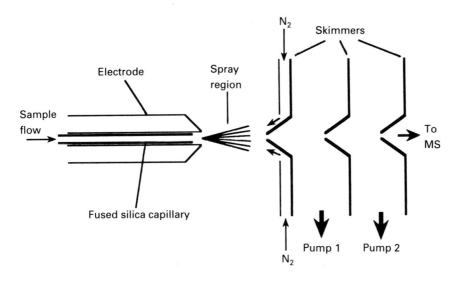

Fig. 8.18. Diagram of the electrospray ionisation source. MS, mass spectrometer.

the solvent at or below room temperature. This has important advantages for thermally labile materials. The resultant charged species are accelerated through differentially pumped regions where the remaining solvent is removed before entry into the mass spectrometer.

It was noted above (Section 8.3.1) that what is generally referred to as the mass spectrum is strictly an m/z spectrum, where m is the mass of the ion and z the number of charges it carries. When $z = 1$ the spectrum is effectively the mass spectrum. The velocity to which the ions can be accelerated depends only on the total charge and the accelerating force. The momentum of an ion is the product of its mass and this velocity. Hence, as the number of charges are increased for the same accelerating force the achievable velocity is greater. Furthermore, as z increases, m/z decreases and the effective result of this is that much more massive species can be mass analysed in this situation. All mass analysers have an upper limit (usually termed the mass range) that is dependent on the design characteristics of the instrument. However if 10 charges can be placed on an appropriate species (e.g. for peptides usually the basic amino acids, arginine, lysine, histidine, would be the main carriers of positive charges) then something with a relative mass of 100 000 would behave, in the mass spectrometer, as if it were a 10 000 mass species ($m = 100\,000$, $z = 10$; $m/z = 10\,000$). This brings biological macromolecules into the range of many existing mass analysers. The main use of ESI at present is to determine very accurate molecular masses (orders of magnitude more accurate than any other method). Little fragmentation occurs but exciting developments are underway that may improve this and produce useful structural information.

The relationship between real relative mass and m/z is shown below. Let m_1, m_2 represent m/z values (peaks in the spectrum) for different ions of the same chemical

$$m_2 = (M + n_2)/n_2$$

$$m_1 = (M + n_1)/n_1$$

$$m_2 > m_1$$

$$n_1 > n_2$$

Fig. 8.19. Diagram of two hypothetical multiply charged peaks in an ESI spectrum. It is assumed that the ions are adducts of neutral molecule and protons. If $n_1 = n_2 + 1$; then $n_2 = (m_1 - 1)/(m_2 - m_1)$ and $M = n_2 (m_2 - 1)$, which is equal to the mass of the neutral molecule. m_1 and m_2 are the recorded masses (equivalent to the m/z values). n_1 and n_2 are the number of charges (z values) or protons added, respectively. By taking peaks in pairs, from the recorded masses, n_2 can be calculated and hence M. A range of values may be obtained for M and an average value calculated. (See example associated with Fig. 8.20.)

entity but carrying different multiple charges (Fig. 8.19). The distribution of peaks in the m/z spectrum resulting from multiple charging of the species M forms fingerprint patterns. The data may be deconvoluted using appropriate computer methods to give an average relative molar mass, M_r, whose accuracy is greater than anything achieved by any other available method. Figures 8.20 and 8.21 show examples of the peak distributions for cytochrome c and insulin B-chain (oxidised). The mass analysers used here are quadrupole devices that will be explained in more detail in Section 8.8.3.

The term relative mass requires some brief explanation. The mass spectrometer does not measure absolute mass. The instrument needs to be calibrated with standard compounds whose M_r values are known accurately. Relative molar mass, however, relates to the relative scale that is used for calculation and is obviously related to relative atomic mass. Three scales are in current use, the hydrogen, carbon and oxygen scales. In mass spectrometry the carbon scale is used exclusively with $^{12}C = 12.000000$. This level of accuracy is achievable in high resolution double-focusing mass spectrometers (Section 8.8.2). At the very high masses that are encountered in biological macromolecules the variations in values between scales is essentially irrelevant and the more familiar dalton is often used. However, mass differences that arise from fragmentations may be small and high resolution enables us to distinguish between groups that have the same nominal mass.

It is worth a brief note about the value of carbon as a primary standard. Pure carbon can be obtained easily (as charcoal). It is remarkably stable, a solid, is insoluble in water, cannot be readily reduced or oxidised and in any case the latter products are gases that may be easily removed.

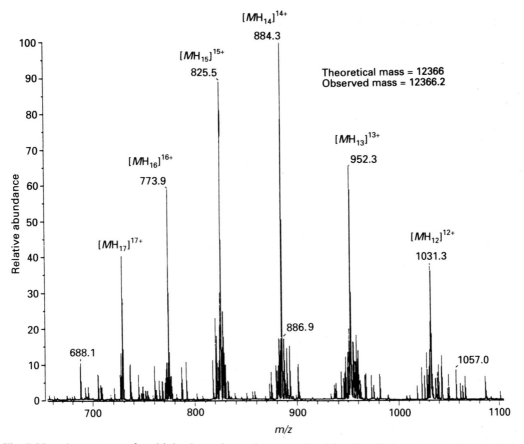

$[MH_{14}]^{14+}$
884.3

$[MH_{15}]^{15+}$
825.5

Theoretical mass = 12366
Observed mass = 12366.2

$[MH_{16}]^{16+}$
773.9

$[MH_{13}]^{13+}$
952.3

$[MH_{17}]^{17+}$

$[MH_{12}]^{12+}$
1031.3

688.1

886.9

1057.0

Relative abundance

m/z

Fig. 8.20. *m/z* spectrum of multiply charged cytochrome *c*. Applying the algebra stated in the caption to Fig. 8.19 and remembering that *z* values must be integers:

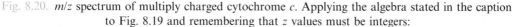

$$m_1 = 952.3 \text{ and } m_2 = 1031.3.$$

Then: $$n_2 = (m_1 - 1)/(m_2 - m_1) = 951.3/(1031.3 - 952.3)$$
$$= 951.3/79 = 12.04 \text{ or } z = 12.$$

(12 positive charges associated with relative mass 1031.3).
Consider the two peaks with relative masses:

$$m_1 = 884.3 \text{ and } m_2 = 952.3$$

then: $$n_2 = 883.3/(952.3 - 884.3) = 883.3/68$$
$$= 12.989 \text{ or } z = 13$$

(13 positive charges associated with relative mass 952.3).

For practice the reader should calculate the *z* values for other pairs of peaks, find the series of associated values for *M* (the mass of the neutral molecule) for each peak (from $M = n_2 (m_2 - 1)$) and find the average value for *M*. (Reproduced by kind permission of JEOL (UK) Ltd, JEOL House, Welwyn Garden City.)

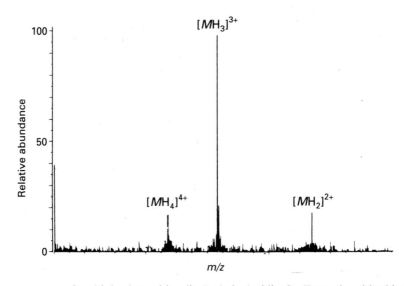

Fig. 8.21. *m/z* spectrum of multiply charged insulin B chain (oxidised). (Reproduced by kind permission of JEOL (UK) Ltd, JEOL House, Welwyn Garden City.)

8.8 Analysers

Considerable emphasis has been placed on ionisation sources and this is justified because most of the important developments of the last decade that have influenced the growth of biological mass spectrometry have been in sources. One analyser, the TOF device (Section 8.6.3), is relatively simple in principle. Analysers are of obvious importance and one of the most widely used of these is the magnetic sector.

8.8.1 Magnetic sector analyser

This is shown diagrammatically in Fig. 8.7. Historically the observation of the phenomenon of charged particles following a circular trajectory in a magnetic field dates back to W. Wien (1898), who was the first to demonstrate that a beam of positively charged particles could be deflected using magnetic and electric fields. J.J. Thomson, in 1912, demonstrated the existence of two stable isotopes of neon using a simple magnetic sector device. More elaborate instruments, developed by A.J. Dempster (1918) and F.W. Aston (1919), could be used for isotopic relative abundance measurements.

The term magnetic sector arises because the beam trajectory traverses only a sector of the circular poles of the magnet. The arrangement shown in Fig. 8.7 indicates only the sector of one pole. In a real system another pole would lie above and parallel to the plane of the diagram. As the whole system must be kept under high vacuum two possibilities exist. With small permanent magnets it is possible to construct the whole in a sealed box that can be evacuated. As indicated in equation 8.1 (p. 391), for given values of B, V and z, ions of different masses will follow different trajectories (different radii). This is satisfactory in IRMS studies because the actual masses of the ions under investigation are quite close together and R will not differ greatly. In fact they are sufficiently close for separate detectors to be mounted within the system. The situation is quite different with mixtures of ions covering a wide range of masses, as it would prove impractical to have enough separate detectors positioned to accommodate all possible radii.

The problem is solved by using the second of the two possibilities, i.e. to make all ions follow the same trajectory of the same radius. There are two ways of achieving this. First, the accelerating voltage V can be varied so as to accelerate ions of different mass to different terminal velocities, v. This type of voltage scanning is used in certain kinds of mass spectrometric experiment. Second, by the use of electromagnets (varying B), ions of different mass (but the same velocity) can be forced to follow the particular trajectory. This is magnetic scanning and is the most commonly used form of analysis. Figure 8.22 shows several hypothetical trajectories in a given magnetic field. Only one, R allows ions to be focused on the detector. If the field is changed, ions travelling along R will be defocused because they do not, in the new field, possess the correct momentum to allow equation 8.1 to be satisfied. A new set of ions will be focused along R at the new value of the field and be collected at the detector. By starting either at the high or low extremes of the magnet range it is possible either to scan down (from high to low mass) or to scan up (from low to high). The whole of this single trajectory resides in a sealed tube, the drift tube, placed between the magnet poles and thus enabling the vacuum to be maintained in the envelope.

In this way the mixture of ions, parent and daughters, are separated according to m/z and the mass spectrum produced. This design results in low resolution mass spectrometry using single-sector or single focusing.

8.8.2 Electric sector analyser

The single-focusing instrument is somewhat limited in application but is still widely used where low resolution is sufficient for the experiment being conducted.

In order to understand how the resolution may be improved the simplest approach is to reconsider the EI source. Assuming that the ions have been formed, it is then necessary to accelerate them out of the source. This is achieved by establishing an electric potential across the source, the ions being attracted towards the plate of opposite charge. The ions will emerge through slits or apertures with

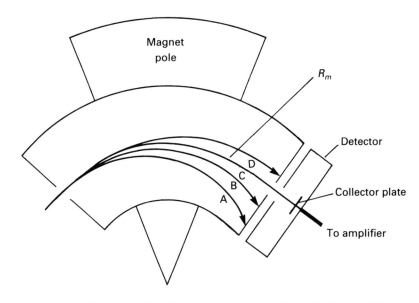

Fig. 8.22. Representation of ion generation between acceleration plates. Only ions following a trajectory of radius R_m will be focused. A, B and D are currently defocused. By altering the field, other ions can be forced to travel along R_m and be focused.

a given terminal velocity directly related to the accelerating force applied. However, the terminal velocities differ because not all the ions are subjected to the same force; it depends entirely on where they arise in space (Fig. 8.23).

Ion A arising at the zero plate will experience the full accelerating force; ion B, half way between the plates, will only experience a force related to -4 kV; and ion C, about 10% of the distance from the -8 kV potential perforated plate, will only experience about 10% of the force. Although this description is considerably oversimplified it should serve to indicate that the ions emerge from the source with varying terminal velocities and hence varying momenta and kinetic energies ($\frac{1}{2}mv^2$). In order to overcome this variation it is necessary to energy analyse the emergent ion beam. This is achieved in the electric sector analyser, which consists of two stainless steel plates bent into segments of concentric circles. The ions follow a circular trajectory, between these plates, whose radius, R_e, is given by:

$$R_e = 2V/E \tag{8.2}$$

where V is the accelerating voltage (in the source), and E is the electrostatic field in the analyser.

The electric sector is usually referred to as the electrostatic analyser (ESA) and packets of ions emerge from this with the whole range of masses but the same velocity. A given packet with the appropriate velocity then enters the magnetic

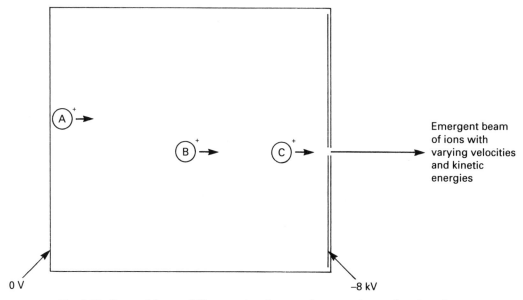

Fig. 8.23. Ions arising at different points in space between the acceleration plates.

sector analyser to undergo mass analysis. The ESA is solely an energy and not a mass analyser. This type of instrument is the two-sector or double-focusing device,

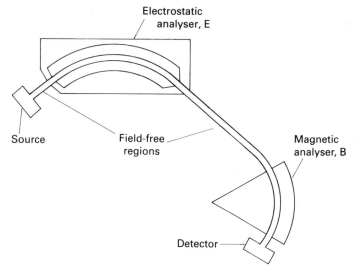

Fig. 8.24. Double focusing mass spectrometer.

which enables high resolution mass spectrometry to be performed (Fig. 8.24). The resolving power of these instruments is such that measurements to accuracies of

Table 8.3. *Relative nominal and accurate masses of some elements*

Element	Nominal mass	Accurate mass
C	12	12.000000
H	1	1.007825
O	16	15.994915
N	14	14.003074
P	31	30.973765
S	32	31.972074
F	19	18.998405

parts per million may be obtained. In Table 8.3 are shown the relative and nominal masses of elements that frequently constitute biological molecules (including drugs etc., which have biological implications). It can be seen from Table 8.3 that adding up the values for the constituent elements of a compound gives different values, depending on whether nominal or accurate atomic masses are used. This difference increases as the compound increases in size (i.e. contains more atoms). High resolution mass spectrometry can be extremely valuable in distinguishing between compounds that have the same nominal M_r but different accurate M_r.

Another important use of high resolution mass spectrometers is the determination, by accurate mass measurement, of the nature of the chemical group that is lost during fragmentation. The reader should recall that much structural information can be gained by considering the mass difference between entities that can show what has to be lost from the greater mass species to give the lesser. Five common groups that may be lost from biological molecules are listed in Table 8.4 together with relative nominal and accurate masses.

Figure 8.24 shows the forward geometry arrangement of what is known as the

Table 8.4. *Nominal and accurate masses of neutral loss groups in fragmentation*

Neutral group lost	Nominal mass	Accurate mass
CO (carbonyl)	28	27.994914
C_2H_4 (double bond fragment)	28	28.031299
N_2 (diazo, in some drugs)	28	28.006158
(CH_2N) (nitrile or isonitrile) (CNH_2) (amino-carbon)	28	28.018723[a]

[a] The last two examples cannot be distinguished by accurate mass measurement.

Nier–Johnson type (after the designers), where the ESA is before the magnetic sector (known as EB; E for electric, B for magnetic). Exactly the same results may be obtained if the reverse geometry (BE) Nier–Johnson type is used. In the latter, the ESA and hence the energy analysis occurs after the mass analysis in the magnetic sector. Different and very sophisticated experiments may be carried out with the different geometries. The regions between the source and analyser, between the analysers and between the last analyser and the collector/detector are the so-called field-free regions. In these regions occur decomposition events; if they can be observed, these give important information regarding structure. Depending on which geometry is employed, certain of the events cannot be observed and there are 'pros' and 'cons' in each case. Also other designs such as the Matteuch–Herzog allow the construction of the mass spectrograph, where the ions all focus in a focal plane (to be recorded by exposure of a photographic plate). Such devices, however, have not found wide application in the biological field.

8.8.3 Quadrupole mass filters (Q)

This type of mass analyser is known as a mass filter. It has the advantage of being a smaller and cheaper device than sector systems and is much lighter in weight, hence its widespread use in benchtop type instruments, particularly many used for hyphenated methods. The disadvantages are the lower mass range and sensitivity.

The theoretical background to these devices is considerably more complicated than that required for sector machines and will not be pursued in any detail here. The underlying idea, however, is a simple one. The device is generally constructed using four solid cylindrical rods, of circular cross-section, to which are applied both direct current (DC) and radiofrequency (RF) voltages (Fig. 8.25). The only component of motion of ions along the linear z-axis of the filter is that derived from the injection velocity. Both the fixed (DC) and oscillating (RF) fields cause the ions to undergo complicated motion in the x–y plane (cross-section). This, together with the component of motion in the z-direction, results in the ions following complicated trajectories through the quadrupole filter. For a given set of field conditions only certain trajectories are stable, allowing ions of specific mass to be transmitted through to the collector/detector. Ions whose mass determines that they travel along unstable trajectories do not get transmitted, hence the term filter. By careful control of the field conditions, ions of different mass can be successively filtered and transmitted. The arrangement of the device is shown in Fig. 8.25. In an ideal situation rods with hyperbolic cross-sections would be used. These give pure quadrupole hyperbolic fields. There are, however, cost implications and manufacturing difficulties with symmetry. The circular cross-section rods are a close approximation and satisfactory for most applications.

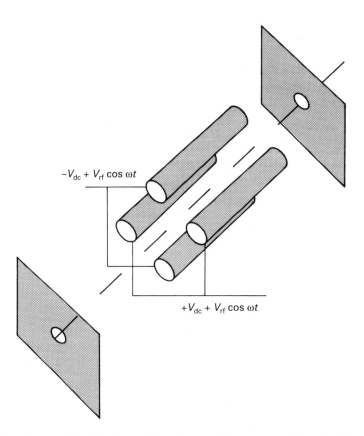

$-V_{dc} + V_{rf} \cos \omega t$

$+V_{dc} + V_{rf} \cos \omega t$

Fig. 8.25. Quadrupole mass filter. In the quadrupole mass filter one opposite pair of rods has a negative DC voltage, $-V_{dc}$, applied and the other pair a positive DC voltage, $+V_{dc}$. There is also a superimposed radiofrequency, RF voltage, $V_{rf} \cos \omega t$, which is 180° out of phase between rod pairs. Mass filtering occurs as these voltages are scanned but the ratio DC to RF is kept constant. In spatial tandem mass spectrometry, Section 8.10.1 and Fig. 8.31, quadrupole collision cells are used which are RF-only devices. No DC voltages are applied and no mass filtering occurs in these cases.

8.8.4   The ion trap

This is an interesting and versatile device that is rapidly gaining in importance in mass spectrometry. It is a relatively small cylindrical device of some 5 cm diameter and 10 cm length. A longitudinal cross-section is shown in Fig. 8.26, the device being constructed from three separate pieces of metal, two of which (mirror images of each other) form a mathematical figure known as a hyperboloid of two sheets and form the end cap electrodes. The third, or ring, electrode is a hyperboloid of one sheet and is like a torus or doughnut except that the cross-section of the ring material is hyperbolic, not circular. The end caps (electrically linked) surround the

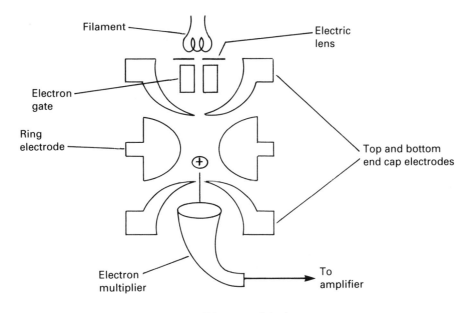

Fig. 8.26. Diagram of the ion trap.

ring electrode (electrically insulated) and together they form the ion trap. Ions are generated, e.g. by EI, and injected into the trap, where they can be constrained and held in constant motion in the space delineated by the electrodes. The trajectories describe Lissajous' figures, i.e. a two-dimensional display of two frequencies perpendicular to each other and normally with a simple ratio between them. Mass analysis is again performed by careful control of the field conditions, the use of which allows ions of increasing mass to be successively ejected from the trap to an appropriate collector/detector. The ion trap belongs to a group of devices known as quadrupole ion storage systems (QUISTORS). The device is readily adapted as an end detector for GLC instruments and is becoming widely used in toxicological studies and pollution monitoring. Like quadrupole rod systems they lend themselves to the design of 'portable' instruments, at least in off the road vehicles, and hence their increasing use in field explorations.

There are several other types of mass spectrometer that have special features: ion cyclotron resonance instruments in which very high mass range and resolution can be obtained, and Fourier transform devices. Space does not permit any further detailed description and, despite their value and importance, their specialised nature is such that they have not gained as widespread use in the biological field as the designs described above.

8.9   Detectors

No information could be gained from sources or analysers without a suitable detector being available. Most detectors are of the impact or ion collection type. All types of detector require a surface on which the ions impinge and the charge is neutralised either by collection or donation of electrons. Hence, electron transfer occurs and an electric current flows that may be amplified and ultimately converted into a signal recorded on a chart or processed by a computer. The total ion current (TIC) is the sum of all the currents carried by all the ions.

8.9.1   The Faraday cup

This is probably the simplest device and works essentially as described above and shown in Fig. 8.27. For example, a positive ion striking the surface of the cup abstracts an electron to neutralise its charge and if enough ions strike then a measurable current flows.

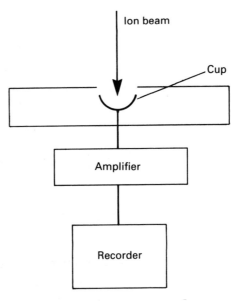

Fig. 8.27. Simplified diagram of the Faraday cup.

8.9.2   Electron multiplier

Greater sensitivity can be achieved with this detector and the degree of amplification is large. The original ions cause a shower of new electrons to be produced. These

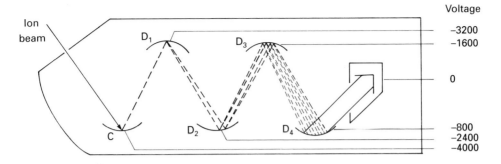

Fig. 8.28. The electron multiplier. The voltages listed on the right-hand side of the diagram are typical. Different manufacturers may produce different values but each dynode has to have a different negative voltage applied to generate the shower of electrons. C, collector; D_1–D_4, dynodes.

electrons impinge on a second dynode and produce yet more electrons. This process continues until a sufficiently large current for normal amplification is obtained. A diagrammatic representation of the electron multiplier is shown in Fig. 8.28.

Modified photomultipliers may also be used, the advantage being that they are cheaper and more robust than electron multipliers. Here the ions generate a shower of electrons that impinge on a surface that produces photons. The remainder of the device is essentially a photomultiplier.

8.9.3   Array detectors

These types of detector are really many detectors arranged in an array or matrix, commonly having 1024 sites for ion collection. Most designs involve the use of microchannel plates (narrow channels along which the ions continually collide with the walls until electrons are produced) as the ion-to-electron converter. Also most array detectors are electrooptical devices in which the electrons strike a suitable surface and produce photons that can be focused and transmitted. Finally, the photons strike a photosensitive plate and an electric current is produced.

Array detectors are ideally suited for use as a focal plane detector and are most easily adapted to the Matteuch–Herzog type of instrument where all the ions would be collected simultaneously. It is possible, however, to arrange the array in a Nier–Johnson instrument where the focal plane would be if it existed. In Nier–Johnson instruments, however, the ions cannot be collected simultaneously. Portions of the mass range (between 4% and 40%, depending on the design) are collected over, say, 900 of the 1024 sites for a given set of field conditions. For example, for 4% of a mass range of 2000 the field conditions would be such that ions with m/z values between $\pm 2\%$ of 2000 ($\pm m/z = 40$) would be detected. The magnetic field then jumps to say 1980 and $\pm 2\%$ of ions are collected at this setting. The process is continued stepwise and it should be noted that an overlap of m/z values is built in at each step. The whole process is under computer control and the data are smoothed so that a continuous spectrum is output and the steps are transparent to the user.

Other specialised variations of detectors are available, but there is nothing like the variety amongst detectors as there is amongst analysers or sources.

8.10   Tandem mass spectrometry

By far the best separatory technique, for charged particles, is the mass spectrometer. However, the observation of a peak in an m/z spectrum does not of itself define the entity it represents in structural terms. Accurate mass measurement may enable us to write down an empirical formula for the species but does not, of itself, confirm any further information. Such difficulties in terms of identifying a peak in a spectrum arise frequently when dealing with mixtures (a situation often encountered with samples from biological sources). In mixed spectra, the facility for identifying the origins of a particular peak are extremely valuable. The experimenter can take several approaches in mass spectrometry, such as linking the scanning modes of the ESA and magnet in sector machines. Space does not permit detailed discussion of these topics but the methods can be found in more advanced texts on mass spectrometry.

A method that must be considered, however, is tandem mass spectrometry. In these methods a particular peak is selected for further investigation. The ions comprising this peak are made to undergo further fragmentation, usually by a method known as collisionally induced decomposition (CID). As the name implies, the ions are allowed to interact collisionally with atoms or molecules (helium, neon, argon and nitrogen have been used as collision gases). Translational energy (the energy of motion) is transferred between the interacting species, up to 50% in elastic collisions. Energy transferred to the ions under investigation can be distributed in a variety of ways. Some of the transferred energy will remain as translational (involved in direction changes, scattering etc.), whilst some will be distributed into vibrational modes of the chemical bonds of the ion. It is this latter energy that, if sufficient, can cause further degradation, the products of which can be analysed in another mass spectrometric experiment. In principle, the process may be continued and multiple mass spectrometry carried out. It will be seen below that this is only feasible in certain types of instrument but such experiments are performed. In view of this, strictly tandem MS is usually abbreviated to MS^2 and in general the abbreviation would be MS^n.

8.10.1   MS² in space

This has nothing to do with extraterrestrial exploration (it should be remembered that a mass spectrometer was used on the Mars probe) but describes the fact that the processes occur sequentially in space. This requires that two mass spectrometers are physically coupled together. Two double-focusing sector machines result in the four-sector device coupled through a collision cell where the CID process occurs. Collision cells are frequently RF-only quadrupoles (no mass filtering occurs here, the RF merely constraining the ions to allow a greater number of collisions to occur), filled with the appropriate collision gas, and such cells are usually designated the symbol q. All geometries are theoretically possible, although not all are in

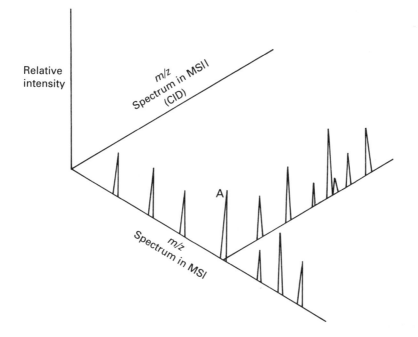

Fig. 8.29. Two-dimensional mass spectrometry spectra obtained in a MSII experiment. The first mass spectrum is generated in mass spectrometer I (MSI) and is shown along the appropriately labelled axis. A particular peak, for example that marked A, is selected and the packet of ions comprising this peak is focused into the collision cell where CID occurs. The *m/z* spectrum corresponding to these decomposition fragments are analysed in MSII. This CID mass spectrum is shown along the isometric axis parallel to the axis labelled spectrum in MSII.

general use: EBqEB, BEqBE, etc. Figure 8.29 shows what has been described as two-dimensional mass spectrometry, the first axis indicating the *m/z* spectrum obtained in the first mass spectrometer, MSI and the isometric axis showing the second *m/z* spectrum obtained in MSII, after CID of the selected peak (packet of ions).

Obviously similar experiments may be performed on other peaks and the amount of information gained is enormous. In Fig. 8.30 is shown the Kratos Concept II H H four-sector instrument. It should be evident to the reader that such instrumentation is extremely expensive and tends to be located only in major research centres.

An alternative approach is to use the triple quadrupole design (Fig. 8.31), which although much cheaper suffers from sensitivity and mass range limitations. The designation here is QqQ or QhQ where, in the latter case, an RF-only hexapole (no mass filtering) is used as the collision cell. The first quadrupole, QI, is used as a mass spectrometer, a selected peak being injected into the collision cell q or h, and the products of decomposition analysed in QII.

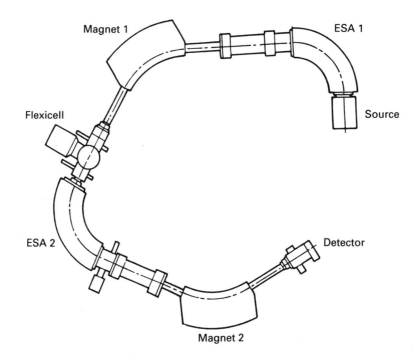

Fig. 8.30. Kratos Concept II H H four-sector instrument. This design includes two forward geometry sector mass spectrometers linked through a collision cell, termed a flexicell by this manufacturer. MSI is represented by ESA 1 and magnet 1 and is where the appropriate first mass spectrum is obtained. Collisionally induced decomposition occurs in the flexicell and the products analysed in MSII, which incorporates ESA 2 and magnet 2. (Reproduced by kind permission of Kratos Analytical, Manchester.)

Hybrid instruments exist and are in quite widespread use. In these designs MSI is a two-sector device and MSII (or QII) is a quadrupole. Geometries such as EBqQ and BEqQ are possible. Note that there is still a sensitivity and mass range limitation but as the masses of the products of CID are less than the original

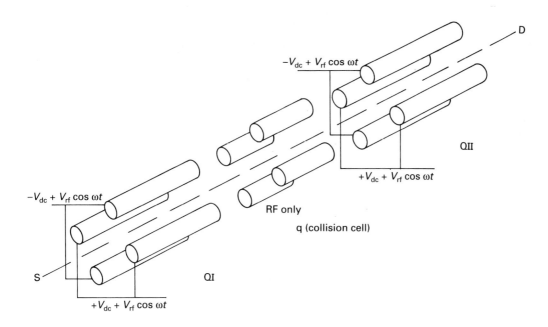

$-V_{dc} + V_{rf} \cos \omega t$

$+V_{dc} + V_{rf} \cos \omega t$

QII

$-V_{dc} + V_{rf} \cos \omega t$

RF only

q (collision cell)

S

QI

$+V_{dc} + V_{rf} \cos \omega t$

Fig. 8.31. Arrangement in triple quadrupole systems. S, source; D, detector; RF, radiofrequency.

sample a convenient compromise is achieved. Note also that using Q as MSI would impose severe mass range constraints and is not therefore a feasible option.

CID is not confined to MS^2 type experiments. Collision cells may be placed in any of the field-free regions leading to a wide variety of experimental methods. Furthermore, in-source CID may be performed in certain situations, e.g. by using the curtain gas nitrogen in an ESI source.

It will be recognised that a collision cell and a detector cannot occupy the same position in space. In four-sector instruments, for instance, the cell position is critical, hence the detectors are set off-axis at appropriate points. To detect ions after MSI and prior to CID in the cell, the ions must be accelerated round a curved path and into the collector/detector. When the peak has been selected the accelerators are switched off and the ions proceed into the cell to undergo CID.

8.10.2   MS^n in time

This can be achieved in the ion trap, where all the processes occur in the *same* region of space and the multiple mass spectrometry is distributed in *time*. Owing to the facility of storing ions in the trap, the procedure is to eject all ions except those corresponding to the selected peak. A suitable collision gas is introduced and the CID occurs and MS^2 can be performed. The process can be repeated successively in time, no extra mass spectrometers or collision cells being required. The limitation is sensitivity as this decreases markedly with each MS experiment, although the claimed world record in an ion trap is currently MS^{14}.

8.11   Suggestions for further reading

DAVIES, R. and FREARSON, M. (1988) *Mass Spectrometry*. ACOL Series. John Wiley and Sons, London. (An open learning text for those needing a deeper understanding of the principles.)

GASKELL, S.J. (1986) *Mass Spectrometry in Biomedical Research*. John Wiley and Sons, London. (An advanced text on specific methods and applications.)

ROSE, M.E. and JOHNSTONE, R.A.W. (1982) *Mass Spectrometry for Chemists and Biochemists*. Cambridge University Press, Cambridge. (Covers instrumentation and applications at an introductory level.)

Many specific applications are available in manufacturers' applications notes, which are generally available free on request. Two of particular interest appear in the VG Monographs in *Mass Spectrometry Series*, published by VG Instruments, Tudor Road, Altrincham, WA14 5RZ, UK:

MELLON, F.A. (1991) *Liquid Chromatography/Mass Spectrometry*, vol. 2, no. 1.

ROSE, M.E. (1990) *Modern Practice of Gas Chromatography/Mass Spectrometry*, vol. 1, no. 1.

9

Electrophoretic techniques

▲ ● ■ ▲ ● ■ ▲ ● ■ ▲ ● ■ ▲ ● ■ ▲ ● ■ ▲ ● ■

9.1　General principles

The term electrophoresis describes the migration of a charged particle under the influence of an electric field. Many important biological molecules such as amino acids, peptides, proteins, nucleotides and nucleic acids, possess ionisable groups and, therefore, at any given pH, exist in solution as electrically charged species either as cations $(+)$ or anions $(-)$. Under the influence of an electric field these charged particles will migrate either to the cathode or to the anode, depending on the nature of their net charge.

The equipment required for electrophoresis consists basically of two items, a power pack and an electrophoresis unit. Electrophoresis units are available for running either vertical or horizontal gel systems. Vertical slab gel units of the type shown in Fig. 9.1 are commercially available and routinely used to separate proteins in acrylamide gels (Section 9.2). The gel is formed between two glass plates that are clamped together but held apart by plastic spacers. Gel dimensions are typically 12 cm×14 cm, with a thickness of 1–2 mm. A plastic comb is placed in the gel solution and is removed after polymerisation to provide loading wells for samples. When the apparatus is assembled, the lower electrophoresis tank buffer surrounds the gel plates and affords some cooling of the gel plates. A typical horizontal gel system is shown in Fig. 9.2. The gel is cast on a glass or plastic sheet and placed on a cooling plate (an insulated surface through which cooling water is passed to conduct away generated heat). Connection between the gel and electrode buffer is made using a thick wad of wetted filter paper (Fig. 9.2): note, however, that

425

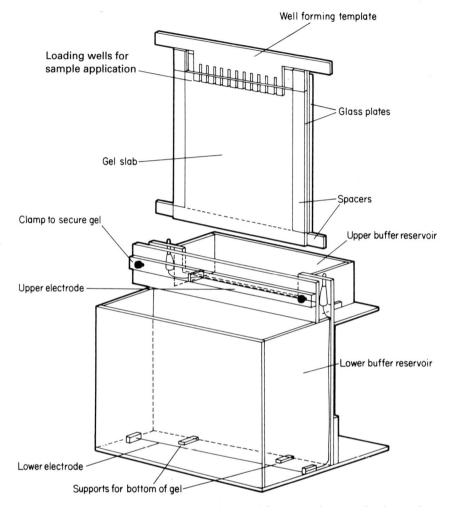

Well forming template

Loading wells for
sample application

Glass plates

Gel slab

Spacers

Clamp to secure gel

Upper buffer reservoir

Upper electrode

Lower buffer reservoir

Lower electrode

Supports for bottom of gel

Fig. 9.1. A typical vertical gel apparatus, such as that used for separating proteins in a polyacrylamide gel. Note that, once the gel has set, the lower spacer is removed prior to the gel being run.

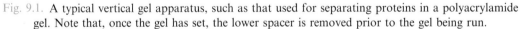
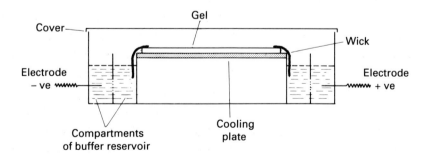

Cover

Gel

Wick

Electrode
− ve

Electrode
+ ve

Compartments
of buffer reservoir

Cooling
plate

Fig. 9.2. A typical horizontal apparatus, such as that used for immunoelectrophoresis, isoelectric focusing and the electrophoresis of DNA and RNA in agarose gels.

agarose gels for DNA electrophoresis are run submerged in the buffer (Section 9.4.1). The power pack supplies a direct current between the electrodes in the electrophoresis unit. All electrophoresis is carried out in an appropriate buffer, which is essential to maintain a constant state of ionisation of the molecules being separated. Any variation in pH would alter the overall charge and hence the mobilities (rate of migration in the applied field) of the molecules being separated.

In order to understand fully how charged species separate it is necessary to look at some simple equations relating to electrophoresis. When a potential difference (voltage) is applied across the electrodes, it generates a potential gradient, E, which is the applied voltage, V, divided by the distance, d, between the electrodes. When this potential gradient E is applied, the force on a molecule bearing a charge of q coulombs is Eq newtons. It is this force that drives a charged molecule towards an electrode. However, there is also a frictional resistance that retards the movement of this charged molecule. This frictional force is a measure of the hydrodynamic size of the molecule, the shape of the molecule, the pore size of the medium in which electrophoresis is taking place and the viscosity of the buffer. The velocity, v, of a charged molecule in an electric field is therefore given by the equation:

$$v = \frac{Eq}{f}$$ (9.1)

where f is the frictional coefficient.

More commonly the term electrophoretic mobility (μ), of an ion is used, which is the ratio of the velocity of the ion to field strength (v/E). When a potential difference is applied, therefore, molecules with different overall charges will begin to separate due to their different electrophoretic mobilities. Even molecules with similar charges will begin to separate if they have different molecular sizes, since they will experience different frictional forces. As will be seen below, some forms of electrophoresis rely almost totally on the different charges on molecules to effect separation, while other methods exploit differences in molecular size and therefore encourage frictional effects to bring about separation.

Provided the electric field is removed before the molecules in the sample reach the electrodes, the components will have been separated according to their electrophoretic mobility. Electrophoresis is thus an incomplete form of electrolysis. The separated samples are then located by staining with an appropriate dye or by autoradiography (Section 5.2.3) if the sample is radiolabelled.

The current in the solution between the electrodes is conducted mainly by the buffer ions with a small proportion being conducted by the sample ions. Ohm's law expresses the relationship between current (I), voltage (V) and resistance (R):

$$\frac{V}{I} = R$$ (9.2)

It therefore appears that it is possible to accelerate an electrophoretic separation by increasing the applied voltage, which would result in a corresponding increase in the current flowing. The distance migrated by the ions will be proportional to both current and time. However, this would ignore one of the major problems for most forms of electrophoresis, namely the generation of heat.

During electrophoresis the power (W, watts) generated in the supporting medium is given by

$$W = I^2 R \tag{9.3}$$

Most of this power generated is dissipated as heat. Heating of the electrophoretic medium can have the following effects:

(i) An increased rate of diffusion of sample and buffer ions leading to broadening of the separated samples.
(ii) The formation of convection currents, which leads to mixing of separated samples.
(iii) Thermal instability of samples that are rather sensitive to heat. This may include denaturation of proteins or loss of activity of enzymes.
(iv) A decrease of buffer viscosity, and hence a reduction in the resistance of the medium.

If a constant voltage is applied, the current increases during electrophoresis due to this decrease in resistance (see Ohm's law, equation 9.2) and this rise in current increases the heat output still further. For this reason, workers often use a stabilised power supply, which provides constant power and thus eliminates fluctuations in heating.

Constant heat generation is, however, a problem. The answer might appear to be to run the electrophoresis at very low power (low current) to overcome any heating problem, but this can lead to poor separations as a result of the increased amount of diffusion resulting from long separation times. Compromise conditions, therefore, have to be found with reasonable power settings to give acceptable separation times and an appropriate cooling system to remove liberated heat. While such systems work fairly well, the effects of heating are not always totally eliminated. For example, for electrophoresis carried out in cylindrical tubes or in slab gels, although heat is generated uniformly through the medium, heat is removed only from the edges, resulting in a temperature gradient within the gel, the temperature at the centre of the gel being higher than that at the edges. Since the warmer fluid at the centre is less viscous, electrophoretic mobilities are therefore greater in the central region (electrophoretic mobilities increase by about 2% for each 1 deg.C rise in temperature), and electrophoretic zones develop a bowed shape, with the zone centre migrating faster than the edges.

A final factor that can effect electrophoretic separation is the phenomenon of electroendosmosis (also known as electroosmotic flow), which is due to the presence of charged groups on the surface of the support medium. For example, paper has some carboxyl groups present, agarose (depending on the purity grade) contains sulphate groups and the surface of glass walls used in capillary electrophoresis

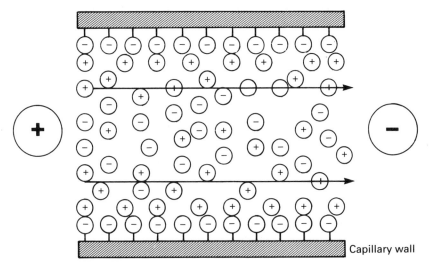

• Acidic silanol groups impart negative charge on wall

• Counter ions migrate toward cathode, dragging solvent along

Fig. 9.3. Electroosmotic flow through a glass capillary. Electrolyte cations are attracted to the capillary wall, forming an electrical double layer. When a voltage is applied, the net movement of electrolyte solution toward the cathode is known as electroendosmotic flow.

(Section 9.5) contains silanol (Si-OH) groups. Figure 9.3 demonstrates how electroendosmosis occurs in a capillary tube, although the principle is the same for any support medium that has charged groups on it. In a fused-silica capillary tube, above a pH value of about 3, silanol groups on the silica capillary wall will ionise, generating negatively charged sites. It is these charges that generate electroendosmosis. The ionised silanol groups create an electrical double layer, or region of charge separation, at the capillary wall/electrolyte interface. When a voltage is applied, cations in the electrolyte near the capillary wall migrate towards the cathode, pulling electrolyte solution with them. This creates a net electroosmotic flow towards the cathode.

9.2 Support media

The pioneering work on electrophoresis by A. Tiselius and co-workers was performed in free solution. However, it was soon realised that many of the problems associated with this approach, particularly the adverse effects of diffusion and convection currents, could be minimised by stabilising the medium. This was achieved by carrying out electrophoresis on a porous mechanical support, which was wetted in electrophoresis buffer and in which electrophoresis of buffer ions and samples could occur. The support medium cuts down convection currents and diffusion so that the separated components remain as sharp zones. The earliest supports used were filter paper or cellulose acetate strips, wetted in electrophoresis buffer. Nowadays these media are less frequently used, although they do still have their uses (see, for example, diagonal electrophoresis, Sections 9.3.6 and 4.3.3). In particular, for many years small molecules such as amino acids, peptides and carbohydrates were routinely separated and analysed by electrophoresis on supports such as paper or thin-layer plates of cellulose, silica or alumina. Although occasionally still used nowadays, such molecules are now more likely to be analysed by more modern and sensitive techniques such as high performance liquid chromatography (Section 10.4). While paper or thin-layer supports are fine for resolving small molecules, the separation of macromolecules such as proteins and nucleic acids on such supports is poor.

However, the introduction of the use of gels as a support medium led to a rapid improvement in methods for analysing macromolecules. The earliest gel system to be used was the starch gel and, although this still has some uses, the vast majority of electrophoretic techniques used nowadays involve either agarose gels or polyacrylamide gels.

9.2.1 Agarose gels

Agarose is a linear polysaccharide (average relative molecular mass about 12 000) made up of the basic repeat unit agarobiose, which comprises alternating units of galactose and 3,6-anhydrogalactose (Fig. 9.4). Agarose is one of the components of agar that is a mixture of polysaccharides isolated from certain seaweeds. Agarose is usually used at concentrations of between 1% and 3%. Agarose gels are formed by suspending dry agarose in aqueous buffer, then boiling the mixture until a clear solution forms. This is poured and allowed to cool to room temperature to form a rigid gel. The gelling properties are attributed to both inter- and intramolecular hydrogen bonding within and between the long agarose chains. This cross-linked structure gives the gel good anticonvectional properties. The pore size in the gel is controlled by the initial concentration of agarose; large pore sizes are formed

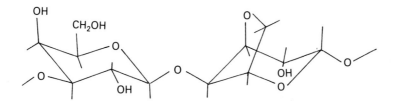

Fig. 9.4. Agarobiose, the repeating unit of agarose.

from low concentrations and smaller pore sizes are formed from the higher concentrations. Although essentially free from charge, substitution of the alternating sugar residues with carboxyl, methoxyl, pyruvate, and especially sulphate groups occurs to varying degrees. This substitution can result in electroendosmosis during electrophoresis and ionic interactions between the gel and sample in all uses, both unwanted effects. Agarose is therefore sold in different purity grades, based on the sulphate concentration – the lower the sulphate content, the higher is the purity.

Agarose gels are used for the electrophoresis of both proteins and nucleic acids. For proteins, the pore sizes of a 1% agarose gel are large relative to the sizes of proteins. Agarose gels are therefore used in techniques such as immuno-electrophoresis (Section 2.4.4) or flat-bed isoelectric focusing (Section 9.3.4), where the proteins are required to move unhindered in the gel matrix according to their native charge. Such large pore gels are also used to separate much larger molecules such as DNA or RNA, because the pore sizes in the gel are still large enough for DNA or RNA molecules to pass through the gel. Now, however, the pore size and molecule size are more comparable and frictional effects begin to play a role in the separation of these molecules (Section 9.4). A further advantage of using agarose is the availability of low melting temperature agarose (62–65°C). As the name suggests, these gels can be reliquified by heating to 65°C and, thus, for example, DNA samples separated in a gel can be returned to solution and recovered.

Owing to the poor elasticity of agarose gels and the consequent problems of removing them from small tubes, the gel rod system sometimes used for acrylamide gels is not used. Horizontal slab gels are invariably used for isoelectric focusing or immunoelectrophoresis in agarose. Horizontal gels are also used routinely for DNA and RNA gels (Section 9.4), although vertical systems have been used by some workers.

9.2.2   Polyacrylamide gels

Electrophoresis in acrylamide gels is frequently referred to as PAGE, being an abbreviation for polyacrylamide gel electrophoresis.

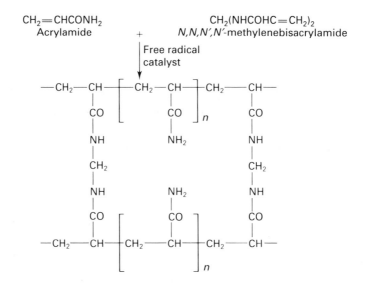

Fig. 9.5. The formation of a polyacrylamide gel from acrylamide and bis-acrylamide.

Cross-linked polyacrylamide gels are formed from the polymerisation of acrylamide monomer in the presence of smaller amounts of N,N'-methylenebisacrylamide (normally referred to as 'bis'-acrylamide) (Fig. 9.5). Note that bis-acrylamide is essentially two acrylamide molecules linked by a methylene group, and is used as a cross-linking agent. Acrylamide monomer is polymerised in a head-to-tail fashion into long chains and occasionally a bis-acrylamide molecule is built into the growing chain, thus introducing a second site for chain extension. Proceeding in this way a cross-linked matrix of fairly well-defined structure is formed (Fig. 9.5). The polymerisation of acrylamide is an example of free-radical catalysis, and is initiated by the addition of ammonium persulphate and the base N,N,N',N'-tetramethylenediamine (TEMED). TEMED catalyses the decomposition of the persulphate ion to give a free radical (i.e. a molecule with an unpaired electron):

$$S_2O_8^{2-} + e^- \longrightarrow SO_4^{2-} + SO_4^- \cdot$$

If this free radical is represented as $R^\cdot$ (where the dot represents an unpaired electron) and M as an acrylamide monomer molecule, then the polymerisation can be represented as follows:

$$R \overset{\frown}{+} M \rightarrow RM^{\bullet}$$

$$RM^{\bullet} \overset{\frown}{+} M \rightarrow RMM^{\bullet}$$

$$RMM^{\bullet} \overset{\frown}{+} M \rightarrow RMMM^{\bullet} \quad \text{etc.}$$

In this way long chains of acrylamide are built up, being cross-linked by the introduction of the occasional bis-acrylamide molecule into the growing chain. Oxygen removes free radicals and therefore all gel solutions are normally degassed (the solutions are briefly placed under vacuum to remove loosely dissolved oxygen) prior to use.

Photopolymerisation is an alternative method that can be used to polymerise acrylamide gels. The ammonium persulphate and TEMED are replaced by riboflavin and when the gel is poured it is placed in front of a bright light for 2–3 h. Photodecomposition of riboflavin generates a free radical that initiates polymerisation.

Acrylamide gels are defined in terms of the total percentage of acrylamide present, and the pore size in the gel can be varied by changing the concentrations of both the acrylamide and bis-acrylamide. Acrylamide gels can be made with a content of between 3% and 30% acrylamide. Thus low percentage gels (e.g. 3%) have large pore sizes and are used, for example, in the electrophoresis of proteins, where free movement of the proteins by electrophoresis is required without any noticeable frictional effect, e.g. in flat-bed isoelectric focusing (Section 9.3.4) or the stacking gel system of an SDS polyacrylamide gel (Section 9.3.1). Low percentage acrylamide gels are also used to separate DNA (Section 9.4). Gels of between 10% and 20% acrylamide are used in techniques such as SDS gel electrophoresis, where the smaller pore size now introduces a sieving effect that contributes to the separation of proteins according to their size (Section 9.3.1).

Proteins were originally separated on polyacrylamide gels that were polymerised in glass tubes, approximately 7 mm in diameter and about 10 cm in length. The tubes were easy to load and run, with minimum apparatus requirements. However, only one sample could be run per tube and, because conditions of separation could vary from tube to tube, comparison between different samples was not always accurate. The later introduction of vertical gel slabs allowed running of up to 20 samples under identical conditions in a single run. Vertical slabs are now used routinely both for the analysis of proteins (Section 9.3) and for the separation of DNA fragments during DNA sequence analysis (Section 9.4). Note, however, that tube gels are still used for the first dimension of two-dimensional gel electrophoresis (Section 9.3.5).

9.3   Electrophoresis of proteins

9.3.1   Sodium dodecyl sulphate polyacrylamide gel electrophoresis (SDS-PAGE)

This form of polyacrylamide gel electrophoresis is the most widely used method for analysing protein mixtures qualitatively. It is particularly useful for monitoring protein purification and, because the method is based on the separation of proteins according to size the method can also be used to determine the relative molecular mass of proteins. SDS (CH_3-$(CH_2)_{10}$-$CH_2OSO_3^-Na^+$) is an anionic detergent. Samples to be run on SDS-PAGE are firstly boiled for 5 min in sample buffer containing β-mercaptoethanol and SDS. The mercaptoethanol reduces any disulphide bridges present that are holding together the protein tertiary structure, and the SDS binds strongly to, and denatures, the protein. Each protein in the mixture is therefore fully denatured by this treatment and opens up into a rod-shaped structure with a series of negatively charged SDS molecules along the polypeptide chain. On average one SDS molecule binds for every two amino acid residues. The original native charge on the molecule is therefore completely swamped by the SDS molecules. The sample buffer also contains an ionisable tracking dye, usually bromophenol blue, that allows the electrophoretic run to be monitored, and sucrose or glycerol, which gives the sample solution density thus allowing the sample to settle easily through the electrophoresis buffer to the bottom when injected into the loading well (see Fig. 9.1). Once the samples are all loaded a current is passed through the gel. The samples to be separated are not in fact loaded directly into the main separating gel. When the main separating gel (normally about 10 cm long) has been poured between the glass plates and allowed to set, a shorter (approximately 1 cm) stacking gel is poured on top of the separating gel and it is into this gel that the wells are formed and the proteins loaded. The purpose of this stacking gel is to concentrate the protein sample into a sharp band before it enters the main separating gel. This is achieved by utilising differences in ionic strength and pH between the electrophoresis buffer and the stacking gel, and involves a phenomenon known as isotachophoresis. The stacking gel has a very large pore size (4% acrylamide), which allows the proteins to move freely and concentrate, or stack, under the effect of the electric field. The band-sharpening effect relies on the fact that negatively charged glycinate ions (in the electrophoresis buffer) have a lower electrophoretic mobility than do the protein–SDS complexes, which, in turn, have lower mobility than the chloride ions (Cl^-) of the loading buffer and the stacking gel. When the current is switched on, all the ionic species have to migrate at the same speed otherwise there would be a break in the electrical circuit. The glycinate ions can only move at the same speed as Cl^- if they are in a region of higher field strength. Field strength is inversely proportional to conductivity, which is proportional to concentration. The result is that the three species of interest adjust their concentrations so that $[Cl^-] > [protein–SDS] > [glycinate]$. There are only a small quantity of protein–SDS complexes, so they concentrate in a very tight band between glycinate and Cl^- boundaries. Once the glycinate reaches the separating gel it becomes more fully ionised in the higher pH environment and its mobility increases. (The pH of the stacking gel is 6.8, that of the separating gel is 8.8.) Thus, the interface between glycinate and Cl^- leaves behind the protein–SDS complexes, which are left to electrophorese at their own rates. The negatively charged protein–SDS complexes now continue to move towards the anode, and, because they have the same charge per unit length, they travel into the separating gel under the applied electric field with the same mobility. However, as they pass through the separating gel the proteins separate, owing to the molecular sieving properties of the gel. Quite simply, the smaller the protein the more easily it can pass through the pores of the gel, whereas large proteins are successively retarded by frictional resistance due to the sieving effect of the gels. Being a small molecule, the bromophenol blue dye is totally unretarded and therefore indicates the electrophoresis front. When the dye reaches the bottom of the gel, the current is turned off, and the gel is removed from between the glass plates and shaken in an appropriate stain solution (usually Coomassie Brilliant Blue, see Section 9.3.8) for a few hours and then washed in destain solution overnight. The destain solution removes unbound background dye from the gel leaving stained proteins visible as blue bands on a clear background. A typical gel would take $1–1\frac{1}{2}$ h to prepare and set, 3 h to run at 30 mA, and have a staining time of 2–3 h with an overnight destain. Vertical slab gels are invariably run, since this allows up to 20 different samples to be loaded on to a single gel. A typical SDS polyacrylamide gel is shown in Fig. 9.6.

Typically, the separating gel used is a 15% polyacrylamide gel. This gives a gel of a certain pore size in which proteins of relative molecular mass (M_r) 10 000 move through the gel relatively unhindered, whereas proteins of 100 000 can only just enter the pores of this gel. Gels of 15% polyacrylamide are therefore useful for separating proteins in the range of 100 000–10 000. However, a protein of 150 000, for example, would be unable to enter a 15% gel. In this case a larger-pored gel (e.g. a 10% or even 7.5% gel) would be used so that the protein could now enter the gel and be stained and identified. It is obvious, therefore, that the choice of gel to be used depends on the size of the protein being studied.

The M_r of a protein can be determined by comparing its mobility with those of

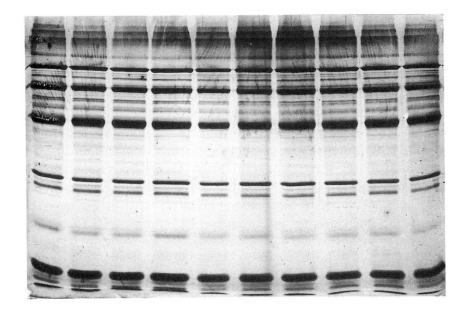

Fig. 9.6. A typical SDS polyacrylamide gel. All 10 wells in the gel have been loaded with the same complex mixture of proteins. (Courtesy of Bio-Rad Laboratories.)

a number of standard proteins of known M_r that are run on the same gel. By plotting a graph of distance moved against log M_r for each of the standard proteins, a calibration curve can be constructed. The distance moved by the protein of unknown M_r is then measured, and then its log M_r and hence M_r can be determined from the calibration curve.

SDS gel electrophoresis is often used after each step of a purification protocol to assess the purity or otherwise of the sample. A pure protein should give a single band on an SDS polyacrylamide gel, unless the molecule is made up of two unequal subunits. In the latter case two bands, corresponding to the two subunits, will be seen. Since only submicrogram amounts of protein are needed for the gel, very little material is used in this form of purity assessment and at the same time a value for the relative molecular mass of the protein can be determined on the same gel run (as described above), with no more material being used.

9.3.2 Native (buffer) gels

While SDS-PAGE is the most frequently used gel system for studying proteins, the method is of no use if one is aiming to detect a particular protein (often an enzyme) on the basis of its biological activity, because the protein (enzyme) is denatured by the SDS-PAGE procedure. In this case it is necessary to use nondenaturing conditions. In native or buffer gels, polyacrylamide gels are again used (normally a 7.5% gel) but the SDS is absent and the proteins are *not* denatured prior to loading. Since all the proteins in the sample being analysed carry their native charge at the pH of the gel (normally pH 8.7), proteins separate according to their different electrophoretic mobilities *and* the sieving effects of the gel. It is not possible to predict the behaviour of a given protein in a buffer gel but, because of the range of different charges and sizes of proteins in a given protein mixture, good resolution is achieved. The enzyme of interest can be identified by incubating the gel in an appropriate substrate solution such that a coloured product is produced at the site of the enzyme (Section 4.8.2). An alternative method for enzyme detection is to include the substrate in an agarose gel that is poured over the acrylamide gel and allowed to set. Diffusion and interaction of enzyme and substrate between the two gels results in colour formation at the site of the enzyme. Often, duplicate samples will be run on a gel, the gel cut in half and one half stained for activity, the other for total protein. In this way the total protein content of the sample can be analysed and the particular band corresponding to the enzyme identified by reference to the activity stain gel.

9.3.3   Gradient gels

This is again a polyacrylamide gel system, but instead of running a slab gel of uniform pore size throughout (e.g. a 15% gel) a gradient gel is formed, where the acrylamide concentration varies uniformly from, typically, 5% at the top of the gel to 25% acrylamide at the bottom of the gel. The gradient is formed via a gradient mixer (Fig. 10.6, p. 479) and run down between the glass plates of a slab gel. The higher percentage acrylamide (e.g. 25%) is poured between the glass plates first and a continuous gradient of decreasing acrylamide concentration follows. Therefore at the top of the gel there is a large pore size (5% acrylamide) but as the sample moves down through the gel the acrylamide concentration slowly increases and the pore size correspondingly decreases. Gradient gels are normally run as SDS gels with a stacking gel. There are two advantages to running gradient gels. First, a much greater range of protein M_r values can be separated than on a fixed-percentage gel. In a complex mixture, very low molecular weight proteins travel freely through the gel to begin with, and start to resolve when they reach the smaller pore sizes towards the lower part of the gel. Much larger proteins, on the other hand, can still enter the gel but start to separate immediately due to the sieving effect of the gel. The second advantage of gradient gels is that proteins with very similar M_r values may be resolved, although they cannot otherwise be resolved in fixed percentage gels. As each protein moves through the gel the pore sizes become smaller until the protein reaches its pore size limit. The pore size in the gel is now too small to allow passage of the protein, and the protein sample stacks up at this point as a sharp band. A similar-sized protein, but with slightly lower M_r will be able to travel a little further through the gel before reaching its pore size limit, at which point it will form a sharp band. These two proteins, of slightly different M_r values therefore separate as two, close, sharp bands.

Nowadays the emphasis in protein electrophoresis centres on the speed with which samples can be analysed. For example, the Phast System sold by Pharmacia LKB comes complete with ready-poured native, SDS-added or gradient minigels and the corresponding buffers in agarose strips. The system is programmed to run 0.5–1.0 µl samples in about 30 min and to stain the gels with either Coomassie Brilliant Blue or silver stain (Section 9.3.8) in approx. 30 min and 2 h, respectively.

9.3.4   Isoelectric focusing (IEF) gels

This method is ideal for the separation of amphoteric substances such as proteins because it is based on the separation of molecules according to their different isoelectric points (Section 4.1). The method has high resolution, being able to separate proteins that differ in their isoelectric points by as little as 0.01 of a pH unit. The most widely used system for IEF utilises horizontal gels on glass plates or plastic sheets. Separation is achieved by applying a potential difference across a gel that contains a pH gradient. The pH gradient is formed by the introduction into the gel of compounds known as ampholytes, which are complex mixtures of synthetic polyamino-polycarboxylic acids (Fig. 9.7). Ampholytes can be purchased in different pH ranges covering either a wide band (e.g. pH 3–10) or various narrow bands (e.g. pH 7–8), and a pH range is chosen such that the samples being separated will have their isoelectric points (pI values) within this range.

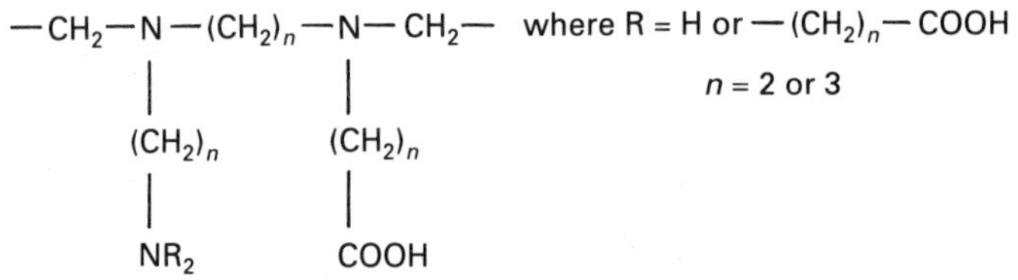

$$-CH_2-N-(CH_2)_n-N-CH_2- \quad \text{where R = H or } -(CH_2)_n-COOH$$

$$n = 2 \text{ or } 3$$

$$(CH_2)_n \qquad (CH_2)_n$$

$$NR_2 \qquad COOH$$

Fig. 9.7. The general formula for ampholytes.

Commercially available ampholytes include Bio-Lyte and Pharmalyte.

Traditionally 1–2 mm thick isoelectric focusing gels have been used by research workers, but the relatively high cost of ampholytes makes this a fairly expensive procedure if a number of gels are to be run. However, the introduction of thin-layer IEF gels, which are only 0.15 mm thick and which are prepared using a layer of electrical insulation tape as the spacer between the gel plates, has considerably reduced the cost of preparing IEF gels, and such gels are now commonly used. Since this method requires the proteins to move freely according to their charge under the electric field, IEF is carried out in low percentage gels to avoid any sieving effect within the gel. Polyacrylamide gels (4%) are commonly used, but agarose is also used, especially for the study of high M_r proteins that may undergo some sieving even in a low percentage acrylamide gel.

To prepare a thin-layer IEF gel, carrier ampholytes, covering a suitable pH range, and riboflavin are mixed with the acrylamide solution, and the mixture is then poured over a glass plate (typically $25 \, cm \times 10 \, cm$), which contains the spacer. The second glass plate is then placed on top of the first to form the gel cassette, and the gel polymerised by photopolymerisation by placing the gel in front of a bright light. The photodecomposition of the riboflavin generates a free radical, which initiates polymerisation (Section 9.2.2). This takes 2–3 h. Once the gel has set, the glass plates are prised apart to reveal the gel stuck to one of the glass sheets. Electrode wicks, which are thick (3 mm) strips of wetted filter paper (the anode is phosphoric acid, the cathode sodium hydroxide) are laid along the long length of each side of the gel and a potential difference applied. Under the effect of this potential difference, the ampholytes form a pH gradient between the anode and cathode. The power is then turned off and samples applied by laying on the gel small squares of filter paper soaked in the sample. A voltage is again applied for about 30 min to allow the sample to electrophorese off the paper and into the gel, at which time the paper squares can be removed from the gel. Depending on which point on the pH gradient the sample has been loaded, proteins that are initially at a pH region below their isoelectric point will be positively charged and will initially migrate towards the cathode. As they proceed, however, the surrounding pH will be steadily increasing, and therefore the positive charge on the protein will decrease correspondingly until eventually the protein arrives at a point where the pH is equal to its isoelectric point. The protein will now be in the zwitterion form with no net charge, so further movement will cease. Likewise, substances that are initially at pH regions above their isoelectric points will be negatively charged and will migrate towards the anode until they reach their isoelectric points and become stationary. It can be seen that as the samples will always move towards their isoelectric points it is not critical where on the gel they are applied. To achieve rapid separations (2–3 h) relatively high voltages (up to 2500 V) are used. As considerable heat is produced, gels are run on cooling plates (10 °C) and power packs used to stabilise the power output and thus to minimise thermal fluctuations. Following electrophoresis, the gel must be stained to detect the proteins. However, this cannot be done directly, because the ampholytes will stain too, giving a totally blue gel. The gel is therefore first washed with fixing solution (e.g. 10% (v/v) trichloroacetic acid). This precipitates the proteins in the gel and allows the much smaller ampholytes to be washed out. The gel is stained with Coomassie Brilliant Blue and then destained (Section 9.3.8). A typical IEF gel is shown in Fig. 9.8. The technique has strong similarities with the technique of chromatofocusing (Section 10.7.3).

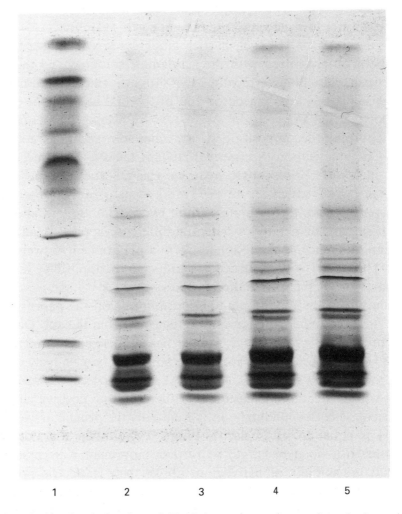

Fig. 9.8. A typical isoelectric focusing gel. Track 1 contains a mixture of standard proteins of known isoelectric points. Tracks 2 to 5 show increasing loadings of venom from Japanese water moccasin snake. (Courtesy of Bio-Rad Laboratories Ltd.)

The pI of a particular protein may be determined conveniently by running a mixture of proteins of known isoelectric point on the same gel. A number of mixtures of proteins with differing pI values are commercially available, covering the pH range 3.5–10. After staining, the distance of each band from one electrode is measured and a graph of distance for each protein against its pI (effectively the pH at that point) plotted. By means of this calibration line, the pI of an unknown protein can be determined from its position on the gel.

IEF is a highly sensitive analytical technique and is particularly useful for studying microheterogeneity in a protein. For example, a protein may show a single band on an SDS-added gel, but may show three bands on an IEF gel. This may occur, for example, when a protein exists in mono-, di- and tri-phosphorylated forms. The difference of a couple of phosphate groups has no significant effect on the overall relative molecular mass of the protein, hence a single band on SDS-added gels, but the small charge difference introduced on each molecule can be detected by IEF.

The method is particularly useful for separating isoenzymes (Section 4.2), which are different forms of the same enzyme often differing by only one or two amino acid residues. Since the proteins are in their native form, enzymes can be detected in the gel either by washing the unfixed and unstained gel in an appropriate substrate or by overlayering with agarose containing the substrate. The approach has found particular use in forensic science where traces of blood or other biological fluids can be analysed and compared according to the composition of certain isoenzymes.

Although IEF is used mainly for analytical separations, it can also be used for preparative purposes. In vertical column IEF, a water-cooled vertical glass column is used, filled with a mixture of ampholytes dissolved in a sucrose solution containing a density gradient to prevent diffusion. When the separation is complete, the current is switched off and the sample components run out through a valve in the base of the column. Alternatively, preparative IEF can be carried out in beds of granulated gel, such as Sephadex G-75 (Section 10.8.2).

9.3.5 Two-dimensional polyacrylamide gel electrophoresis (2-D PAGE)

This technique combines the technique of IEF, which separates proteins in a mixture according to charge (pI), with the size separation technique of SDS polyacrylamide gel electrophoresis. When combined to give 2-D gel electrophoresis, the most sophisticated analytical method for separating proteins available is obtained. The first dimension (isoelectric focusing) is carried out in polyacrylamide gels in narrow tubes (approximately 1–3 mm internal diameter) in the presence of ampholytes, 8 M urea and a non-ionic detergent. The denatured proteins therefore separate in this gel according to their isoelectric points. The gel is then extruded from the tube by applying slight pressure to one end, incubated for 30 min in a buffer containing SDS (thus binding SDS to the denatured proteins), then placed along the stacking gel of an SDS-added gel (either linear or gradient), and fixed in place by pouring molten agarose, in electrophoresis buffer, over the gel. Once the agarose has set, electrophoresis is commenced and the SDS-bound proteins run into the gel, stack, and separate according to size, as described in Section 9.3.1. This method has been shown to be capable of resolving between 1000 and 2000 proteins in a cell extract and some workers claim to be able to resolve as many as 5000 to 10 000 spots, which is comparable to the estimated number of proteins present in a cell. This method can be used, for example, to detect the expression of one extra protein in a cell. mRNA for a given protein can be injected into frog oocytes and the translation of the mRNA detected by comparing 2-D protein profiles of injected and non-injected cell extracts, where the injected cells show the presence of one extra protein spot. It should be noted, however, that running 2-D gels is a technically demanding and labour-intensive procedure. A number of computerised systems now exist for recording and comparing complex 2-D gel patterns. A typical 2-D polyacrylamide gel is shown in Fig. 9.9.

9.3.6   Cellulose acetate electrophoresis

Although one of the older methods, cellulose acetate electrophoresis still has a number of applications. In particular it has retained a use in the clinical analysis of serum samples. Cellulose acetate has the advantage over paper in that it is a much more homogeneous medium, with uniform pore size, and does not adsorb proteins in the way that paper does. There is therefore much less trailing of protein bands and resolution is better, although nothing like as good as that achieved with polyacrylamide gels. The method is, however, far simpler to set up and run. Single samples are normally run on cellulose acetate strips (2.5 cm×12 cm), although multiple samples are frequently run on wider sheets. The cellulose acetate is first wetted in electrophoresis buffer (pH 8.6 for serum samples) and the sample (1–2 μl) loaded as a 1 cm wide strip about one-third of the way along the strip. The ends of the strip make contact with the electrophoresis buffer tanks via a filter paper wick that overlaps the end of the cellulose acetate strip, and electrophoresis conducted at 6–8 V cm^{-1} for about 3 h. Following electrophoresis, the strip is stained for protein (see Section 9.3.8), destained, and the bands visualised. A typical serum protein separation shows about six major bands. However, in many diseased states, this serum protein profile changes and a clinician can obtain information concerning the disease state of a patient from the altered pattern. Although still frequently used for serum analysis, electrophoresis on cellulose acetate is being replaced by the use of agarose gels, which give similar but somewhat better resolution. A typical example of the analysis of serum on an agarose gel is shown in Fig. 9.10. Similar patterns are obtained when cellulose acetate is used.

Enzymes can easily be detected, in samples electrophoresed on cellulose acetate, by using the zymogram technique. The cellulose strip is laid on a strip of filter

Isoelectric focusing

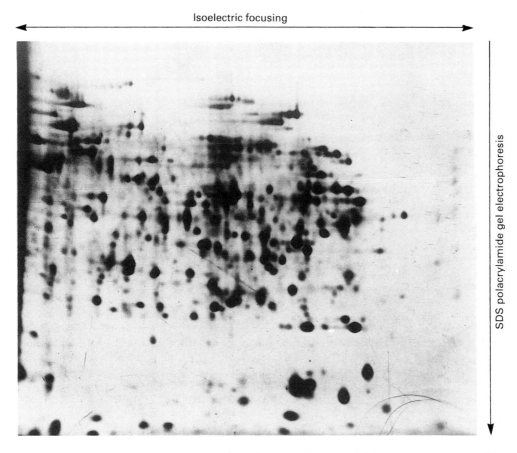

SDS polacrylamide gel electrophoresis

Fig. 9.9. A silver stained 2-D polyacrylamide gel of proteins extracted from *Uromyces viciae faba*. (Courtesy of Bio-Rad Laboratories Ltd; photograph originally provided by Dr Jungblut of the Free University in Berlin.)

paper soaked in buffer and substrate. After an appropriate incubation period, the strips are peeled apart and the paper zymogram treated accordingly to detect enzyme product; hence, it is possible to identify the position of the enzyme activity on the original strip. An alternative approach to detecting and semiquantifying *any* particular protein on a strip is to treat the strip as the equivalent of a protein blot and to probe for the given protein using primary antibody and then enzyme-linked second antibody (Section 9.3.9). Substrate colour development indicates the presence of the particular protein and the amount of colour developed in a given time is a semi-quantitative measure of the amount of protein. Thus, for example, large numbers of serum samples can be run on a wide sheet, the sheet probed using antibodies, and elevated levels of a particular protein identified in certain samples by increased levels of colour development in these samples.

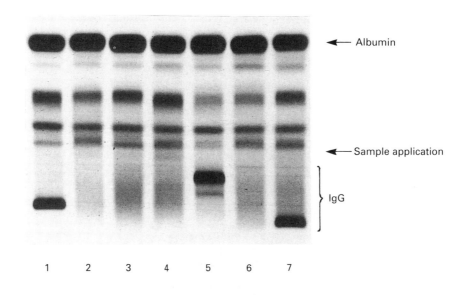

Fig. 9.10. Electrophoresis of human serum samples on an agarose gel. Tracks 2, 3, 4 and 6 show normal serum protein profiles. Tracks 1, 5 and 7 show myeloma patients, who are identified by the excessive production of a particular monoclonal antibody seen in the IgG fraction. (Courtesy of Charles Andrews and Nicholas Cundy, Edgware General Hospital, London.)

9.3.7   Continuous flow electrophoresis

The continuous flow form of electrophoresis is used for separations in free solution in large-scale productions. Electrophoresis takes place continuously as the separating material is carried upwards by a flow of carrier buffer through annular space between two vertical concentric cylinders (Fig. 9.11). The outer cylinder is rotated to maintain a stable laminar flow of the buffer solution. An electrical field is applied between the two cylinders, causing the sample material to separate radially as it is carried upwards by the buffer flow. At the top of the inner cylinder a series of radial slits enables the buffer stream to be separated into as many as 30 individual fractions.

Equipment suitable for large-scale separations is now available commercially. However, it would be fair to say that resolution is certainly no better, and often worse, than can be achieved just as easily and more cheaply by ion-exchange chromatography (Section 10.7), a technique that is also highly amenable to large-scale operation. However, this method does have useful applications in the separation of particles and cells such as human erythrocytes, which otherwise are difficult to separate.

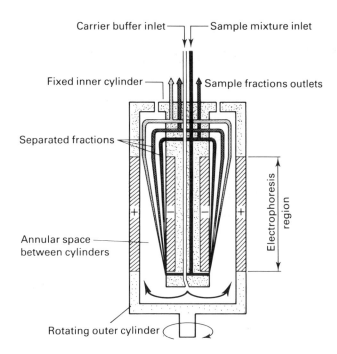

Carrier buffer inlet — — Sample mixture inlet

Fixed inner cylinder — — Sample fractions outlets

Separated fractions —

Electrophoresis region

Annular space between cylinders

Rotating outer cylinder

Fig. 9.11. Continuous flow electrophoresis unit. The sample enters the annular space via the sample inlet and is subject to upward movement due to electrophoresis. The result is radial separation into a series of bands (fractions) that may be collected via a series of sample outlets. (Reproduced by permission of AERE Harwell.)

9.3.8 Detection, estimation and recovery of proteins in gels

The most commonly used general protein stain for detecting protein on gels is the sulphated trimethylamine dye Coomassie Brilliant Blue R-250 (CBB). Staining is usually carried out using 0.1% (w/v) CBB in methanol:water:glacial acetic acid (45:45:10, by vol.). This acid–methanol mixture acts as a denaturant to precipitate or fix the protein in the gel, which prevents the protein from being washed out while it is being stained. Staining of most gels is accomplished in about 2 h and destaining, usually overnight, is achieved by gentle agitation in the same acid–methanol solution but in the absence of the dye. The Coomassie stain is highly sensitive; a very weakly staining band on a polyacrylamide gel would correspond to about 0.1 μg (100 ng) of protein. The CBB stain is not used for staining cellulose acetate (or indeed protein blots) because it binds quite strongly to the paper. In this case, proteins are first denatured by brief immersion of the strip in 10% (v/v) trichloroacetic acid, and then immersed in a solution of a dye that does not stain the support material, e.g. Procion blue, Amido black or Procion S.

Although the Coomassie stain is highly sensitive, many workers require greater sensitivity and use the so-called silver stain. Silver stains are based either on techniques developed for histology or on methods based on the photographic process. In either case, silver ions (Ag^+) are reduced to metallic silver on the protein, where the silver is deposited to give a black band. Silver stains can be used immediately after electrophoresis, or, alternatively, after staining with CBB. With the latter approach, the major bands on the gel can be identified with CBB and then minor bands, not detected with CBB, resolved using the silver stain. The silver stain is about 100 times more sensitive than Coomassie Brilliant Blue, detecting proteins down to 1 ng amounts.

Glycoproteins have traditionally been detected on protein gels by use of the periodic acid–Schiff (PAS) stain. This allows components of a mixture of glycoproteins to be distinguished. However, the PAS stain is not very sensitive and often gives very weak, red-pink bands, difficult to observe on a gel. A far more sensitive method used nowadays is to blot the gel (Section 9.3.9) and use lectins to detect the glycoproteins. Lectins are protein molecules that bind carbohydrates, and different lectins have been found that have different specificities for different types of carbohydrate. For example, certain lectins recognise mannose, fucose, or terminal glucosamine of the carbohydrate side chains of glycoproteins. The sample to be analysed is run on a number of tracks of an SDS polyacrylamide gel. Coloured bands appear at the point where the lectins bind if each blotted track is incubated with a different lectin, washed, incubated with a horseradish peroxidase-linked antibody to the lectin, and then peroxidase substrate added. In this way, by testing a protein sample against a series of lectins, it is possible to determine not only that a protein is a *glyco*protein, but to obtain information about the type of glycosylation.

Quantitative analysis (i.e. measurements of the relative amounts of different proteins in a sample) can be achieved by scanning densitometry. A number of commercial scanning densitometers are available, and work by passing the stained gel track over a beam of light (laser) and measuring the transmitted light. A graphic presentation of protein zones (peaks of absorbance) against migration distance is produced, and peak areas can be calculated to obtain quantitative data. However, such data must be interpreted with caution because there is only a limited range of protein concentrations over which there is a linear relationship between absorbance and concentration. Also, equal amounts of different proteins do not always stain equally with a given stain, so any data comparing the relative amounts of protein can only be semi-quantitative. An alternative and much cheaper way of obtaining such data is to cut out the stained bands of interest, elute the dye by shaking overnight in a known volume of 50% pyridine, and then to measure spectrophotometrically the amount of colour released.

Although gel electrophoresis is used generally as an analytical tool, it can be utilised to separate proteins in a gel to achieve protein purification. Protein bands can be cut out of protein blots and sequence data obtained by placing the blot in a gas phase sequencer. Stained protein bands can be cut out of protein gels and the protein recovered by electrophoresis of the protein out of the gel piece (electroelution). A number of different designs of electroelution cells are commercially available, but perhaps the easiest method is to seal the gel piece in buffer in a dialysis sac and place the sac in buffer between two electrodes. Protein will electrophorese out of the gel piece towards the appropriate electrode but will be retained by the dialysis sac. After electroelution, the current is reversed for a few seconds to drive off any protein that has adsorbed to the wall of the dialysis sac, and then the protein solution within the sac is recovered.

9.3.9   Protein (Western) blotting

Although essentially an analytical technique, PAGE does of course achieve fractionation of a protein mixture during the electrophoresis process. It is possible to make use of this fractionation to examine further individual separated proteins. The first step is to transfer or blot the pattern of separated proteins from the gel on to a sheet of nitrocellulose paper. The method is known as protein blotting, or Western blotting by analogy with Southern blotting (Section 3.5.1), the equivalent method used to recover DNA samples from an agarose gel. Transfer of the proteins from the gel to nitrocellulose can be achieved in one of two ways. In capillary blotting, the gel is placed on a wet pad of buffer-soaked filter paper and a sheet of nitrocellulose placed on the gel. Buffer is then drawn through the gel by placing a pad of dry absorbent material (usually filter paper) followed by a heavy weight on top of the nitrocellulose sheet. Passage of buffer by capillary action through the gel carries the separated proteins on to the nitrocellulose sheet, to which they bind irreversibly by hydrophobic interaction. The process is carried out overnight, but because of the small pore size of the acrylamide gel, only a limited amount of buffer travels through the gel in this time, so that only a fraction (10% to 20%) of each protein in the gel is transferred in this way. A quicker (a few hours) and more efficient method of transfer is achieved by electroblotting. In this method a sandwich of gel and nitrocellulose is compressed in a cassette and immersed, in buffer, between two parallel electrodes (Fig. 9.12). A current is passed at right angles to the gel, which causes the separated proteins to electrophorese out of the gel and into the nitrocellulose sheet. The nitrocellulose with its transferred protein is referred to as a blot. Once transferred on to nitrocellulose, the separated proteins can be examined further. This involves probing the blot, usually using an antibody to detect a specific protein. The blot is firstly incubated in a protein solution, e.g. 10% (w/v) bovine serum albumin, or 5% (w/v) non-fat dried milk (the so-called

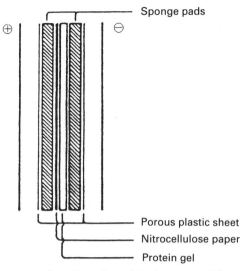

⊕   ⊖

Sponge pads

Porous plastic sheet

Nitrocellulose paper

Protein gel

Fig. 9.12. Diagrammatic representation of an electroblotting set-up. The gel to be blotted is placed on top of a sponge pad saturated in buffer. The nitrocellulose sheet is then placed on top of the gel, followed by a second sponge pad. This sandwich is supported between two rigid porous plastic sheets and held together with two elastic bands. The sandwich is then placed between parallel electrodes in a buffer reservoir and an electrical current passed. The sandwich must be placed such that the immobilising medium is between the gel and the anode for SDS polyacrylamide gels, because all the proteins carry a negative charge.

blotto technique), which will block all remaining hydrophobic binding sites on the nitrocellulose sheet. The blot is then incubated in a dilution of an antiserum (primary antibody) directed against the protein of interest. This IgG molecule will bind to the blot if it detects its antigen, thus identifying the protein of interest. In order to visualise this interaction the blot is incubated further in a solution of a second antibody, which is directed against the IgG of the species that provided the primary antibody. For example, if the primary antibody was raised in a rabbit then the second antibody would be anti-rabbit IgG. This second antibody is appropriately labelled so that the interaction of the second antibody with the primary antibody can be visualised on the blot. Anti-species IgG molecules are readily available commercially, with a choice of a different labels attached. One of the most common detection methods is to use an enzyme-linked second antibody (Fig. 9.13). In this case, following treatment with enzyme-labelled second antibody, the blot is incubated in enzyme-substrate solution, when the enzyme converts the substrate into an insoluble coloured product that is precipitated on to the nitrocellulose. The presence of a coloured band therefore indicates the position of the protein of interest. By careful comparisons of the blot with a stained gel of the same sample, the protein of interest can be identified. The enzyme used in enzyme-

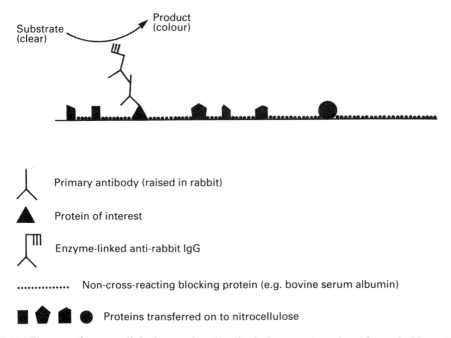

Substrate (clear)

Product (colour)

Primary antibody (raised in rabbit)

Protein of interest

Enzyme-linked anti-rabbit IgG

.... Non-cross-reacting blocking protein (e.g. bovine serum albumin)

Proteins transferred on to nitrocellulose

Fig. 9.13. The use of enzyme-linked second antibodies in immunodetection of protein blots. (1) The primary antibody (e.g. raised in a rabbit) detects the protein of interest on the blot. (2) Enzyme-linked anti-rabbit IgG detects the primary antibody. (3) Addition of enzyme substrate results in coloured product deposited at the site of protein of interest on the blot.

linked antibodies is usually either alkaline phosphatase, which converts colourless 5-bromo-4-chloro-indolylphosphate (BCIP) substrate into a blue product, or horseradish peroxidase, which, with H_2O_2 as a substrate, oxidises either 3-amino-9-ethylcarbazole into an insoluble brown product, or 4-chloro-1-naphthol into an insoluble blue product. An alternative approach to the detection of horseradish peroxidase is to use the method of enhanced chemiluminescence. In the presence of hydrogen peroxide and the chemiluminescent substrate luminol (Fig. 9.14) horseradish peroxidase oxidises the luminol with concomitant production of light, the intensity of which is increased 1000-fold by the presence of a chemical enhancer. The light emission can be detected by exposing the blot to a photographic film. The principle behind the use of enzyme-linked antibodies to detect antigens in blots is highly analogous to that used in enzyme-linked immunosorbent assays (Section 2.6).

Although enzymes are commonly used as markers for second antibodies, other markers can also be used. These include:

▲ ^{125}I-labelled second antibody. Binding to the blot is detected by autoradiography (Section 5.23).

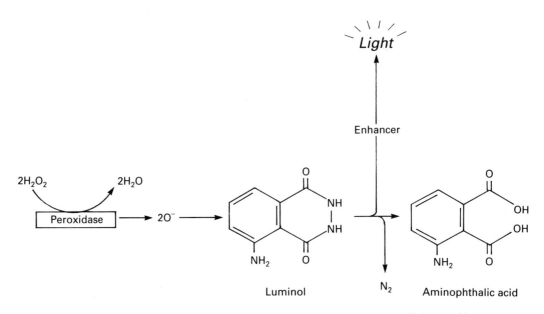

Fig. 9.14. The use of enhanced chemiluminescence to detect horseradish peroxidase.

▲ fluorescein isothiocyanate-labelled second antibody. This fluorescent label is detected by exposing the blot to ultraviolet light.

▲ ^{125}I-labelled Protein A. Protein A is purified from *Staphylococcus aureus* and specifically binds to the Fc region of IgG molecules. ^{125}I-labelled Protein A is therefore used instead of a second antibody, and binding to the blot is detected by autoradiography.

▲ gold-labelled second antibodies. Second antibodies (anti-species IgG) coated with minute gold particles are commercially available. These are directly visible as a red colour when they bind to the primary antibody on the blot.

▲ biotinylated second antibodies. Biotin is a small molecular weight vitamin that binds strongly to the egg protein avidin ($K_d = 10^{-15}$ M). The blot is incubated with biotinylated second antibody, then incubated further with enzyme-conjugated avidin. Since multiple biotin molecules can be linked to a single antibody molecule, many enzyme-linked avidin molecules can bind to a single biotinylated antibody molecule, thus providing an enhancement of the signal. The enzyme used is usually alkaline phosphatase or horseradish peroxidase.

In addition to the use of labelled antibodies or proteins, other probes are sometimes used. For example, radioactively labelled DNA can be used to detect DNA-binding proteins on a blot. The blot is first incubated in a solution of radiolabelled DNA, then washed, and an autoradiograph of the blot made. The presence of radioactive bands, detected on the autoradiograph, identifies the positions of the DNA-binding proteins on the blot.

9.4 Electrophoresis of nucleic acids

9.4.1 Agarose gel electrophoresis of DNA

For the majority of DNA samples, electrophoretic separation is carried out in agarose gels. This is because most DNA molecules and their fragments that are analysed routinely are considerably larger than proteins and therefore, because most DNA fragments would be unable to enter a polyacrylamide gel, the larger pore size of an agarose gel is required. For example, the commonly used plasmid pBR322 has an M_r of 2.4×10^6. However, rather than use such large numbers it is more convenient to refer to DNA size in terms of the number of base-pairs. Although, originally, DNA size was referred to in terms of base-pairs (bp) or kilobase-pairs (kbp), it has now become the accepted nomenclature to abbreviate kbp to simply kb when referring to double-stranded DNA. pBR322 is therefore 4.36 kb. Even a small restriction fragment of 1 kb has an M_r of 620 000. When talking about single-stranded DNA it is common to refer to size in terms of nucleotides (nt). Since the charge per unit length (due to the phosphate groups) in any given fragment of DNA is the same, all DNA samples should move towards the anode with the same mobility under an applied electrical field. However, separation in agarose gels is achieved due to resistance to their movement caused by the gel matrix. The largest molecules will have the most difficulty passing through the gel pores (very large molecules may even be blocked completely), whereas the smallest molecules will be relatively unhindered. Consequently the mobility of DNA molecules during gel electrophoresis will depend on size, the smallest molecules moving fastest. This is analogous to the separation of proteins in SDS polyacrylamide gels (Section 9.3.1), although the analogy is not perfect, as double-stranded DNA molecules form relatively stiff rods and it is not completely understood how they pass through the gel, although it is probable that long DNA molecules pass through the gel pores end-on. While passing through the pores a DNA molecule will experience drag, so the longer the molecule, the more it will be retarded by each pore. Sideways movement may become more important for very small double-stranded DNA and for the more flexible single-stranded DNA. It will be obvious from the above that gel concentrations must be chosen to suit the size range of the molecules to be separated. Gels containing 0.3% agarose will separate double-stranded DNA molecules of between 5 and 60 kb size, whereas 2% gels are used for samples of between 0.1 and 3 kb. Many laboratories routinely use 0.8% gels, which are suitable for separating DNA molecules in the range 0.5–10 kb. Since agarose gels separate DNA according to size, the M_r of a DNA fragment may be determined from its electrophoretic mobility by running a number of standard DNA markers of known M_r on the same gel. This is most conveniently achieved by running a sample of bacteriophage λ DNA (49 kb) that has been cleaved with a restriction enzyme such as *Eco*RI. Since the base sequence of λ DNA is known, and the cleavage sites for *Eco*RI are known, this generates fragments of accurately known size (Fig. 9.15).

4 3 2 1

Fig. 9.15. Photograph showing four tracks from a 0.8% agarose submarine gel. The gel was run at 40 V in Tris/borate/EDTA buffer for 16 h, stained with ethidium bromide and viewed under UV light. Sample loadings were about 0.5 µg of DNA per track. Tracks 1 and 2, λ DNA (49 kb). Track 3, λ DNA cleaved with the enzyme *Eco*RI to generate fragments of the following size (in order from the origin): 21.80 kb, 7.52 kb, 5.93 kb, 5.54 kb, 4.80 kb, 3.41 kb. Track 4, λ DNA cleaved with the enzyme *Hin*dIII to generate fragments of the following size (in order from the origin): 23.70 kb, 9.46 kb, 6.75 kb, 4.26 kb, 2.26 kb, 1.98 kb. (Courtesy of Stephen Boffey, University of Hertfordshire.)

DNA gels are invariably run as horizontal, submarine or submerged gels; so named because such a gel is totally immersed in buffer. Agarose, dissolved in gel buffer by boiling, is poured on to a glass or plastic plate, surrounded by a wall of adhesive tape or a plastic frame to provide a gel about 3 mm in depth. Loading wells are formed by placing a plastic well-forming template or comb in the poured gel solution, and removing this comb once the gel has set. The gel is placed in the electrophoresis tank, covered with buffer, and samples loaded by directly injecting the sample into the wells. Samples are prepared by dissolving them in a buffer solution that contains sucrose, glycerol or Ficoll (Section 6.7.4), which makes the solution dense and allows it to sink to the bottom of the well. A dye such as bromophenol blue is also included in the sample solvent; it makes it easier to see the sample that is being loaded and also acts as a marker of the electrophoresis front. No stacking gel (Section 9.3.1) is needed for the electrophoresis of DNA because the mobilities of DNA molecules are much greater in the well than in the gel, and therefore all the molecules in the well pile up against the gel within a few minutes of the current being turned on, forming a tight band at the start of the run. General purpose gels are approximately 25 cm long and 12 cm wide, and are run at a voltage gradient of about 1.5 V cm^{-1} overnight. A higher voltage would cause excessive heating. For rapid analyses that do not need extensive separation of DNA molecules, it is common to use mini-gels that are less than 10 cm long. In this way information can be obtained in 2–3 h.

Once the system has been run, the DNA in the gel needs to be stained and visualised. The reagent most widely used is the fluorescent dye ethidium bromide. The gel is rinsed gently in a solution of ethidium bromide (0.5 μg cm^{-3}) and then viewed under ultraviolet light (300 nm wavelength). Ethidium bromide is a cyclic planar molecule that binds between the stacked base-pairs of DNA (i.e. it intercalates) (Section 3.5.1). The ethidium bromide concentration therefore builds up at the site of the DNA bands and under ultraviolet light the DNA bands fluoresce orange-red. As little as 10 ng of DNA can be visualised as a 1 cm wide band. It should be noted that extensive viewing of the DNA with ultraviolet light can result in damage of the DNA by nicking and base-pair dimerisation. This is of no consequence if a gel is only to be viewed, but obviously viewing of the gel should be kept to a minimum if the DNA is to be recovered (see below). It is essential to protect one's eyes by wearing goggles when ultraviolet light is used. If viewing of gels under ultraviolet is carried out for long periods, a plastic mask that covers the whole face should be used to avoid 'sunburn'.

9.4.2 DNA sequencing gels

Although agarose gel electrophoresis of DNA is a 'workhorse' technique for the molecular biologist, a different form of electrophoresis has to be used when DNA sequences are to be determined. Whichever DNA sequencing method is used (Section 3.9), the final analysis involves separating single-stranded DNA molecules shorter than about 1000 nt and differing in size by only 1 nt. To achieve this it is necessary to have a small-pored gel and so acrylamide gels are used instead of agarose. For example, 3.5% polyacrylamide gels are used to separate DNA in the range 80–1000 nt and 12% gels to resolve fragments of between 20 and 100 nt. If a wide range of sizes is being analysed it is often convenient to run a gradient gel, e.g. from 3.5% to 7.5%. Sequencing gels are run in the presence of denaturing agents, urea and formamide. Since it is necessary to separate DNA molecules that are very similar in size, DNA sequencing gels tend to be very long (100 cm) to maximise the separation achieved. A typical DNA sequencing gel is shown in Fig. 3.28 (p. 145).

As mentioned above, electrophoresis in agarose can be used as preparative method for DNA. The DNA bands of interest can be cut out of the gel and the DNA recovered by: (a) electroelution, (b) macerating the gel piece in buffer, centrifuging and collecting the supernatant; or, (c) if low melting point agarose is used, melting the gel piece and diluting with buffer. In each case, the DNA is finally recovered by precipitation of the supernatant with ethanol.

9.4.3   Pulsed field gel electrophoresis (PFGE)

The agarose gel methods for DNA described above can fractionate DNA of 60 kb or less. The introduction of pulsed-field gel electrophoresis and the further development of variations on the basic technique now means that DNA fragments up to 2×10^3 kb can be separated. This therefore allows the separation of whole chromosomes by electrophoresis. The method basically involves electrophoresis in agarose where two electrical fields are applied alternately at different angles for defined time periods (e.g. 60 s). Activation of the first electrical field causes the coiled molecules to be stretched in the horizontal plane and start to move through the gel. Interruption of this field and application of the second field force the molecule to move in the new direction. Since there is a length-dependent relaxation behaviour when a long-chain molecule undergoes conformational change in an electrical field, the smaller a molecule, the quicker it realigns itself with the new field and is able to continue moving through the gel. Larger molecules take longer to realign. In this way, with continual reversing of the field, smaller molecules draw ahead of larger molecules and separate according to size. Figure 9.16 shows the separation of yeast chromosomes that vary in size from 260 to 850 kb. Needless to say the physics of designing a PFGE system is complex and in recent years a number of different developments on the same basic theme have resulted in a bewildering array of related techniques. Detailed description of these techniques is beyond the scope of this chapter but the names of a few of these techniques indicates the principles involved, e.g. orthogonal field alternating gel electrophoresis (OFAGE), field inversion gel electrophoresis (FIGE), transverse alternating field gel electrophoresis (TAFE), contour clamped homogeneous electric field electrophoresis (CHEF), and rotating field electrophoresis (RFE).

9.4.4   Electrophoresis of RNA

Like that of DNA, electrophoresis of RNA is usually carried out in agarose gels, and the principle of the separation, based on size, is the same. Often one requires

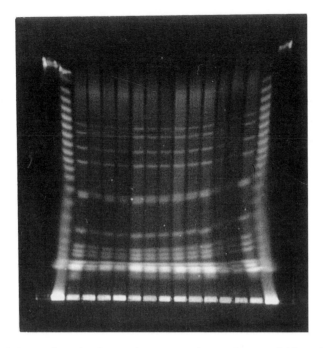

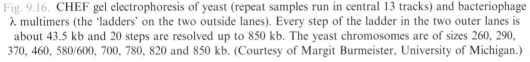

Fig. 9.16. CHEF gel electrophoresis of yeast (repeat samples run in central 13 tracks) and bacteriophage λ multimers (the 'ladders' on the two outside lanes). Every step of the ladder in the two outer lanes is about 43.5 kb and 20 steps are resolved up to 850 kb. The yeast chromosomes are of sizes 260, 290, 370, 460, 580/600, 700, 780, 820 and 850 kb. (Courtesy of Margit Burmeister, University of Michigan.)

a rapid method for checking the integrity of RNA immediately following extraction but before deciding whether to process it further. This can be achieved easily by electrophoresis in a 2% agarose gel in about 1 h. Ribosomal RNAs (18 S and 28 S) are clearly resolved and any degradation (seen as a smear) or DNA contamination is seen easily. However, if greater resolution is required, a smaller-pored acrylamide gel is used to enhance resolution, e.g. to resolve tRNAs (4 S) from 5 S rRNA. This can be achieved on a 2.5% to 5% acrylamide gradient gel with an overnight run. Both these methods involve running native RNA. There will almost certainly be some secondary structure within the RNA molecule owing to intramolecular hydrogen bonding (see e.g. the clover leaf structure of tRNA, Fig. 3.6, p. 117). For this reason native RNA run on gels can be stained and visualised with ethidium bromide. However, if the study objective is to determine RNA size by gel electrophoresis, then full denaturation of the RNA is needed to prevent hydrogen bond formation within or even between polynucleotides that will otherwise affect the electrophoretic mobility. There are three denaturing agents, formaldehyde, glyoxal and methylmercuric hydroxide, that are compatible with both RNA and agarose. Either one of these may be incorporated into the agarose gel and electrophoresis buffer, and the sample is heat denatured in the presence of the denaturant prior to electrophoresis. After heat denaturation, each of these agents forms adducts with the amino groups of guanine and uracil, thereby preventing hydrogen bond reformation at room temperature during electrophoresis. It is also necessary to run denaturing gels if the RNA is to be blotted (Northern blots, Section 3.5.1) and probed, to ensure that the base sequence is available to the probe. Denatured RNA stains only very weakly with ethidium bromide, so acridine orange is commonly used to visualise RNA on denaturing gels. However, it should be noted that many workers will be using radiolabelled RNA and will therefore identify bands by autoradiography. An example of the electrophoresis of RNA is shown in Fig. 9.17.

9.5   Capillary electrophoresis (CE)

The technique has variously been referred to as high performance capillary electrophoresis (HPCE), capillary zone electrophoresis (CZE), free solution capillary electrophoresis (FSCE) and capillary electrophoresis (CE), but the term CE is the one most common nowadays. Capillary electrophoresis can be used to separate a wide spectrum of biological molecules including amino acids, peptides, proteins, DNA fragments (e.g. synthetic oligonucleotides) and nucleic acids, as well as any number of small organic molecules such as drugs or even metal ions. The method has also been applied successfully to the problem of chiral separations (Section 10.6.5).

As the name suggests, capillary electrophoresis involves electrophoresis of samples in very narrow-bore tubes (typically 50 μm internal diameter, 300 μm external diameter). One advantage of using capillaries is that they reduce problems resulting from heating effects. Because of the small diameter of the tubing there is a large surface-to-volume ratio, which gives enhanced heat dissipation. This helps to eliminate both convection currents and zone broadening owing to increased diffusion caused by heating. It is therefore not necessary to include a stabilising medium in the tube and allows free-flow electrophoresis.

Theoretical considerations of CE generate two important equations:

$$t = \frac{L^2}{\mu V} \tag{9.4}$$

where t is the migration time for a solute, L is the tube length, μ is the electrophoretic mobility of the solute, and V is the applied voltage.

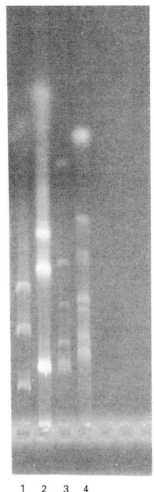

1 2 3 4

Fig. 9.17. Gel electrophoresis of RNA in a 1.4% agarose gel. Track 1 is total RNA from the tobacco plant denatured with glyoxal prior to running. Track 2 is the same sample, *not* denatured. The two faster-running major bands are 18 S and 25 S ribosomal RNA. The slower running major band is nuclear DNA. Tracks 3 and 4 show a mixture of RNA marker fragments, with (track 3) and without (track 4) glyoxal treatment. The sizes of the marker RNA fragments are 0.24, 1.4, 2.4, 4.4, 7.5 and 9.5 kb. Note that, with each sample, denaturation results in lower mobilities for the components of each sample. (Courtesy of Debbie Cook and Robert Slater, Division of Biosciences, University of Hertfordshire.)

The separation efficiency, in terms of the total number of theoretical plates, N, is given by:

$$N = \frac{\mu V}{2D} \tag{9.5}$$

where D is the solute's diffusion coefficient.

From these equations it can be seen firstly that the column length plays no role in separation efficiency, but that it has an important influence on migration time and hence analysis time, and secondly, high separation efficiencies are best achieved through the use of high voltages (μ and D are dictated by the solute and are not easily manipulated).

It therefore appears that the ideal situation is to apply as high a voltage as possible to as short a capillary as possible. However, there are practical limits to this approach. As the capillary length is reduced, the amount of heat that must be dissipated increases owing to the decreasing electrical resistance of the capillary. At the same time the surface area available for heat dissipations is decreasing. Therefore at some point significant thermal effect will occur, placing a practical limit on how short a tube can be used. Also the higher the voltage that is applied, the greater the current, and therefore the heat generated. In practical terms a compromise between voltage used and capillary length is required. Voltages of 10–50 kV with capillaries of 50–100 cm are commonly used.

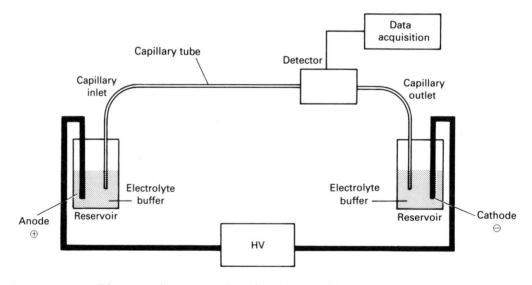

Fig. 9.18. Diagrammatic representation of a typical capillary electrophoresis apparatus.

The basic apparatus for CE is shown diagrammatically in Fig. 9.18. A small plug of sample solution (typically 5–30 μm^3) is introduced into the anode end of a fused silica capillary tube containing an appropriate buffer. Sample application is carried out in one of two ways: by high voltage injection or by pressure injection.

(i) **High voltage injection.** With the high voltage switched off, the buffer reservoir at the positive electrode is replaced by a reservoir containing the sample, and a plug of sample (e.g. 5–30 μm^3 of a 1 mg cm^{-3}

solution) is introduced into the capillary by briefly applying high voltage. The sample reservoir is then removed, the buffer reservoir replaced, voltage again applied and the separation is then commenced.

(ii) *Pressure injection.* The capillary is removed from the anodic buffer reservoir and inserted through an air-tight seal into the sample solution. A second tube provides pressure to the sample solution, which forces the sample into the capillary. The capillary is then removed, replaced in the anodic buffer and a voltage applied to initiate electrophoresis.

A high voltage (up to 20 kV) is then put across the capillary tube and component molecules in the injected sample migrate at different rates along the length of the capillary tube. Electrophoretic migration causes the movement of charged molecules in solution towards an electrode of opposite charge. Owing to this electrophoretic migration, positive and negative sample molecules migrate at different rates. However, although analytes are separated by electrophoretic migration, they are all drawn towards the cathode by electroendosmosis (Section 9.1). Since this flow is quite strong, the rate of electroendosmotic flow usually being much greater than the electrophoretic velocity of the analytes, all ions, regardless of charge sign, and neutral species are carried towards the cathode. Positively charged molecules reach the cathode first because the combination of electrophoretic migration and electroosmotic flow cause them to move fastest. As the separated molecules approach the cathode, they pass through a viewing window where they are detected by a UV monitor that transmits a signal to a recorder, integrator or computer. Typical run times are between 10 and 30 min. A typical capillary electrophoretograph is shown in Fig. 9.19.

This free solution method is the simplest and most widely practised mode of capillary electrophoresis. However, while the generation of ionised groups on the capillary wall is advantageous via the introduction of electroendosmotic flow, it can also sometimes be a disadvantage. For example, protein adsorption to the capillary wall can occur with cationic groups on protein surfaces binding to the ionised silanols. This can lead to smearing of the proteins as it passes through the capillary (recognised as peak broadening) or, worse, complete loss of protein due to total adsorption on the walls. Some workers therefore use coated tubes where a neutral coating group has been used to block the silanol groups. This of course eliminates electroendosmotic flow. Therefore, during electrophoresis in coated capillaries, neutral species are immobile while acid species migrate to the anode and basic species to the cathode. Since detection normally takes place at only one end of the capillary, only one class of species can be detected at a time in an analysis using a coated capillary.

A range of variations on this basic technique also exist. For example, as seen

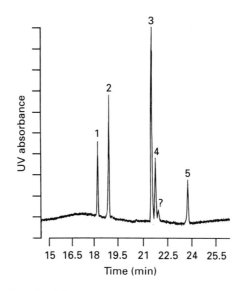

Fig. 9.19. Capillary electrophoresis of five structurally related peptides. Column length was 100 cm and the separation voltage 50 kV. Peptides were detected by their UV absorbance at 200 nm.

Peptide	
1	Lys-Arg-Pro-Pro-Gly-Phe-Ser-Pro-Phe-Arg
2	Met-Lys-Arg-Pro-Pro-Gly-Phe-Ser-Pro-Phe-Arg
3	Arg-Pro-Pro-Gly-Phe-Ser-Pro-Phe-Arg
4	Arg-Pro-Pro-Gly-Phe-Ser-Pro-Phe-Arg
5	Ile-Ser-Arg-Pro-Pro-Gly-Phe-Ser-Pro-Phe-Arg

(Courtesy of Patrick Camilleri and George Okafo, SmithKline Beecham Pharmaceuticals Ltd.)

above, in normal CE neutral molecules do not separate but rather travel as a single band. However, separation of neutral molecules can be achieved by including a surfactant such as SDS with the buffer. Above a certain concentration some surfactant molecules agglomerate and form micelles, which, under the influence of an applied electric field, will migrate towards the appropriate electrode. Solutes will interact and partition with the moving micelles. If a solute interacts strongly it will reach the detector later than one which partitions to a lesser degree. This method is known as micellular electrokinetic capillary electrophoresis (MECC). Since ionic solutes will also migrate under the applied field, separation by MECC is due to a combination of both electrophoresis and chromatography.

9.6   Suggestions for further reading

ANDREWS, A.T. (1986) *Electrophoresis: Theory, Techniques and Biochemical and Clinical Applications*. Oxford University Press, Oxford. (Very comprehensive text on theory and practical details.)

DUNN, M.J. (1993) *Gel Electrophoresis: Proteins*. Bias Scientific, Oxford, UK. (A good introduction to protein electrophoresis.)

HAMES, B.D. and RICKWOOD, D. (Ed.) (1990) *Gel Electrophoresis of Proteins – A Practical Approach*, 2nd edn. IRL Press, Oxford. (A detailed text, with practical details, primarily for research workers.)

SLATER, R.J. (Ed.) (1986) *Experiments in Molecular Biology*. Humana Press, Totowa, N.J. (Detailed laboratory protocols for training undergraduates.)

10

Chromatographic techniques

▲ ● ■ ▲ ● ■ ▲ ● ■ ▲ ● ■ ▲ ● ■ ▲ ● ■ ▲ ● ■

10.1   General principles

10.1.1   Distribution coefficients

The basis of all forms of chromatography is the partition or distribution coefficient (K_d), which describes the way in which a compound distributes itself between two immiscible phases. For two such immiscible phases A and B, the value for this coefficient is a constant at a given temperature and is given by the expression:

$$\frac{\text{Concentration in phase A}}{\text{Concentration in phase B}} = K_d$$

The term effective distribution coefficient is defined as the total amount, as distinct from the concentration, of substance present in one phase divided by the total amount present in the other phase. It is in fact the distribution coefficient multiplied by the ratio of the volumes of the two phases present. If the distribution coefficient of a compound between two phases A and B is 1, and if this compound is distributed between $10 \, cm^3$ of A and $1 \, cm^3$ of B, the concentration in the two phases will be the same, but the total amount of the compound in phase A will be 10 times the amount in phase B.

Basically, all chromatographic systems consist of the stationary phase, which may be a solid, gel, liquid or a solid/liquid mixture that is immobilised, and the mobile phase, which may be liquid or gaseous and which flows over or through the stationary phase. The choice of stationary and mobile phases is made so that the compounds to be separated have different distribution coefficients. This may be achieved by setting up:

(i) an adsorption equilibrium between a stationary solid phase and a mobile liquid phase (adsorption chromatography; hydrophobic interaction chromatography);

(ii) a partition equilibrium between a stationary liquid phase and a mobile liquid or gas phase (partition chromatography; reversed-phase chromatography; ion-pair chromatography; chiral chromatography; gas–liquid chromatography; countercurrent chromatography);

(iii) an ion-exchange equilibrium between a stationary ion exchanger and mobile electrolyte phase (ion-exchange chromatography; chromatofocusing);

(iv) an equilibrium between a liquid phase trapped inside the pores of a stationary porous structure and a mobile liquid phase (exclusion or gel chromatography);

(v) an equilibrium between a stationary immobilised ligand and a mobile liquid phase (affinity chromatography; immunoaffinity chromatography; lectin affinity chromatography; metal chelate chromatography; dye–ligand chromatography; covalent chromatography).

In practice it is quite common for two or more of these equilibria to be involved simultaneously in a particular chromatographic separation.

10.1.2 Modes of chromatography

Chromatographic separations may be achieved using three contrasting modes:

Column chromatography in which the stationary phase attached to a suitable matrix (an inert, insoluble support) is packed into a glass or metal column and the mobile phase passed through the column either by gravity feed or by use of a pumping system or applied gas pressure. This is the most commonly used mode of chromatography.

Thin-layer chromatography in which the stationary phase attached to a suitable matrix is coated thinly on to a glass, plastic or metal foil plate. The mobile liquid phase passes across the thin-layer plate, held either horizontally or vertically, by capillary action. This mode of chromatography has the practical advantage over column chromatography that a large number of samples can be studied simultaneously.

Paper chromatography in which a stationary liquid phase is supported by the cellulose fibres of a paper sheet. As in thin-layer chromatography, with which paper chromatography has several similarities, the mobile phase passes along the paper sheet either by gravity feed or by capillary action. This is one of the older forms of chromatography and, although it is still used to demonstrate the principles of chromatography, it has few serious biochemical applications today.

10.1.3   Performance of column chromatography

The principle of a column chromatographic separation may be depicted by considering a column packed with a solid granular stationary phase to a height of 5 cm, surrounded by the mobile liquid phase of which there is 1 cm³ per cm of column, as shown in Fig. 10.1. If 32 μg of a compound is added to the column in 1 cm³ of mobile phase, then as this 1 cm³ moves on to the column to occupy position A, 1 cm³ of mobile phase will leave the base of the column. If the compound added has an effective distribution coefficient of 1, it will distribute itself equally between the solid and liquid phases (stage 1). If a further 1 cm³ of mobile phase is introduced on to the column, the mobile phase in section A will move down to B, taking 16 μg of the compound with it, leaving 16 μg at A (stage 2). At both A and B a redistribution of the compound will occur so that there is 8 μg in the mobile phase and 8 μg in the solid phase. The addition of a further 1 cm³ of mobile phase to the column displaces the mobile phase in A to B and that in B to C giving the distribution of the compound as shown in stage 3. Addition of a further 1 cm³ of mobile phase leads to the distribution shown at stage 4, and a further 1 cm³ of mobile phase leads to the distribution shown at stage 5.

It is apparent that after a relatively small number of equilibrations the compound

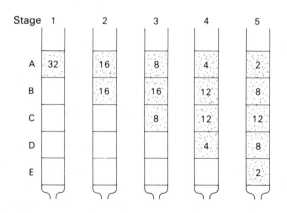

Fig. 10.1. Principle of column chromatographic separation.

distributes itself symmetrically within a band. It should equally be apparent that if a mixture of two compounds, one having a distribution coefficient of 1 the other a distribution coefficient of 100, was added to the column it would separate rapidly into distinct bands. In a real chromatographic column a very large number of equilibrations occur as the mobile phase passes down the column as a result of more mobile phase being added constantly to the top of the column.

A typical column chromatographic system using a liquid mobile phase consists of the column, a mobile phase reservoir and delivery system, a detector for identifying the separated compounds (analytes) as they emerge in the effluent from

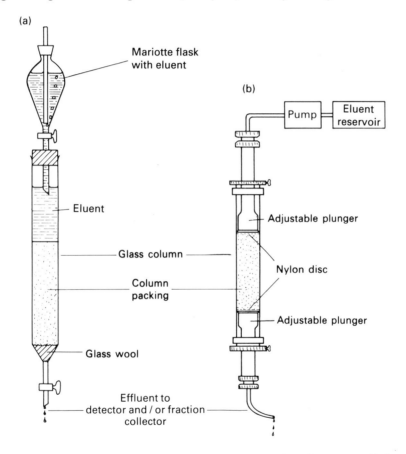

Fig. 10.2. Equipment for column chromatography: (a) simple version; (b) more sophisticated version.

the column, a recorder and a fraction collector (Fig. 10.2). The detector and recorder give a continuous record of the presence of the analytes in the effluent, as measured by a physical parameter such as ultraviolet absorption. Each separated analyte is represented by a peak on the chart recorder and the fraction collector allows each individual analyte to be collected separately and studied further if necessary. Liquid column chromatography can be subdivided according to the pressure generated within the column during the separation process. Low pressure liquid chromatography (LPLC) generates pressures of less than 5 bar (1 bar = 14.5 lbf in.$^{-2}$ = 0.1 MPa), since there is little resistance to solvent flow owing to the physical nature of the stationary phase. Medium pressure liquid chromatography (MPLC) generates pressures of between 6 and 50 bar and high pressure liquid chromatography (HPLC) pressures in excess of 50 bar. In practice the distinctions between MPLC and HPLC are often blurred and their equipment and procedures are virtually identical. Both give excellent resolutions and hence the term high performance liquid chromatography is preferred for both of them, since it better describes the chromatographic characteristics of the techniques and avoids the misconception that it is the high pressure that is fundamentally responsible for the high performance chromatography.

The time taken for each analyte peak to emerge from the column is referred to as its retention time, t_R (Fig. 10.3). Under defined chromatographic conditions, t_R is a characteristic of the analyte. The volume of mobile phase required to elute the analyte under these conditions is referred to as the elution or retention volume, V_R. Elution volume and retention time are related by the flow rate, F_c, of the mobile phase through the column:

$$V_R = t_R F_c \tag{10.1}$$

The flow rate is influenced by the physical dimensions of the column (internal diameter, length), by the physical characteristics of the particles of the column packing (particle size, shape, porosity) and the viscosity of the mobile phase. For a liquid partition chromatographic column, the elution volume is related to the volume of the stationary phase, V_S, the distribution coefficient, K_d, of the analyte between the stationary and mobile phases and to the void volume or dead space, V_M, of the mobile phase around and within the packed stationary phase particles, by the equation:

$$V_R = V_M + K_d V_S \tag{10.2}$$

For adsorption column chromatography V_S is replaced by the surface area, A_S, of the adsorbent.

One of the most important parameters in column chromatography is the partition ratio or capacity ratio, k'. This ratio, which has no units, is a measure of the time spent by the analyte in the stationary phase relative to the time spent in the mobile phase:

$$k' = \frac{C_S V_S}{C_M V_M} + K_d \frac{V_S}{V_M} \tag{10.3}$$

where C_S and C_M are the concentrations of the analyte in the stationary (C_S) and mobile (C_M) phases, respectively.

The ratio of V_S to V_M is referred to as the volumetric phase ratio, β. Hence $k' = K_d\beta$. The capacity ratio is actually a measure of the additional time the analyte takes to elute from the column relative to an unretained or excluded analyte that does not partition into the stationary phase and for which $k' = 0$. Hence:

$$k' = \frac{t_R - t_M}{t_M} = \frac{V_R - V_M}{V_M} \tag{10.4}$$

where t_M is the transit time, through the column, of an unretained compound. t_M is determined by the length of the column, L, and the linear velocity of the mobile phase through the column, $\bar{u}$, according to the equation:

$$\bar{u} = \frac{L}{t_M} \tag{10.5}$$

Capacity ratios, which characterise column performance, are most commonly reported in liquid chromatography as they are easily determined. They are reported infrequently in gas–liquid chromatography (GLC) due to the practical difficulty of measuring t_M or V_M. For most good, acceptable liquid chromatographic systems, k' values range from 1 to 10.

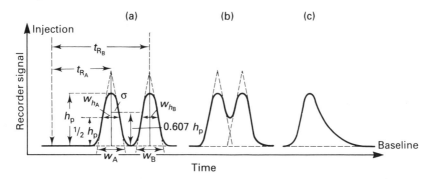

Fig. 10.3. (a) Chromatograph of two compounds showing complete resolution and the calculation of retention times; (b) two compounds giving incomplete resolution and the production of fused peaks; (c) a compound showing excessive tailing.

As already seen, as each analyte passes down the column it becomes distributed in a band that has a Gaussian distribution about the mean (Fig. 10.3a). The width of the base of such a peak is taken to be four standard deviations (4σ) and the extent of peak broadening as the variance (σ^2). For symmetrical Gaussian peaks the base width is equal to 4σ and the peak width at the point of inflexion, w_i, is equal to 2σ (Fig. 10.3a). In some cases, however, the peak is asymmetrical, displaying either fronting (extended front portion of peak) or tailing (Fig. 10.3c). Peak asymmetry has many causes, including the application of too much analyte to the column, poor packing of the column, poor application of the sample to the column or solute–support interactions.

The success of any chromatographic procedure is measured by its ability to separate completely (resolve) one analyte from a mixture of similar compounds. Peak resolution (R_S) is related to the properties of the peaks (Fig. 10.3) such that:

$$R_S = \frac{2(t_{R_B} - t_{R_A})}{w_A + w_B} \qquad (10.6)$$

where t_{R_A} and t_{R_B} are the retention times of compounds A and B, respectively, and w_A and w_B are base widths of the peaks for A and B, respectively. It can be shown that when $R_S = 1.5$ the separation of the two peaks is 99.7% complete. In most practical cases, R_S values of 1.0, corresponding to 98% separation, are adequate for quantitative analysis, provided the detector responses of the two compounds are similar. According to the definition of resolution, $R_S - 1$ peaks fit between any two peaks in question.

Quite often in chromatographic separations it is not possible completely to resolve two peaks. As a result they produce fused peaks and for their analysis the assumption has to be made that the peak characteristics of the two unresolved compounds do not affect each other (Fig. 10.3b).

Chromatography columns are considered to consist of a number of adjacent zones in each of which there is sufficient space for the solute to achieve complete equilibration between the mobile and stationary phases. Each zone is called a theoretical plate and its length in the column is called the plate height (H), which has dimensions of length. The more efficient the column, the greater the number of theoretical plates that are involved. The way in which the number of theoretical plates affects the distribution of a solute with an effective distribution coefficient of 1 is shown in Fig. 10.4.

The number of theoretical plates (plate number) (N) involved in the elution of a particular analyte is given by:

$$N = 16 \left(\frac{t_R}{w} \right)^2 \qquad (10.7)$$

or:

$$N = 5.54 \left(\frac{t_R}{w_h} \right)^2 \qquad (10.8)$$

where w is the peak base width and is equal to 4σ, and w_h is the peak width at half the peak height and is equal to 2.355σ.

The value of N can be calculated from data on the chart recorder paper. The plate number can be increased by simply increasing the column length (L) but there is a limit to this because the retention time and peak width increase

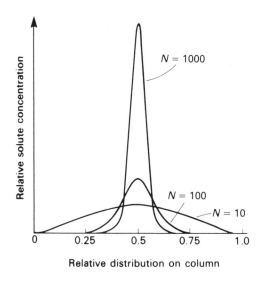

Fig. 10.4. Diagrammatic effect of the number of theoretical plates (N) on the shape of the solute.

proportionally with L whereas the peak height decreases as the square root of N. Although N is a measure of the efficiency of the column, the plate height, which is also called the height equivalent to a theoretical plate (HETP), is useful for comparative purposes such as operating the column under different conditions. It can be shown that:

$$\text{HETP} = \frac{L}{N} = H \tag{10.9}$$

The maximum number of peaks that can be separated by a specific chromatographic system is called the peak capacity (n). It is related to the retention volumes of the first and last peaks (V_α and V_ω, respectively) and to the plate number:

$$n = 1 + \sqrt{\frac{N}{16}\left(\ln\frac{V_\omega}{V_\alpha}\right)} \tag{10.10}$$

Peak capacity is determined by acceptable separation times and by detection sensitivity. In practice it can be increased either by the procedure of gradient elution, as in liquid chromatography (Section 10.3.6), or by temperature programming, as in gas–liquid chromatography (Section 10.10.3).

The success of a particular chromatographic system is judged by its ability to achieve good resolution and is determined by the following three functions.

Selectivity. This is a measure of the inherent ability of the system to discriminate between structurally related compounds. This is reflected in their K_d and k' values such that the ratio of the partition coefficients for two compounds gives their relative retention ratio, α. It is influenced by the chemical nature of the mobile and stationary phases. Some chromatographic systems are inherently highly selective. A good example is affinity chromatography (Section 10.9) and chiral chromatography (Section 10.6.5). These techniques are commonly used towards the end of a purification protocol.

Efficiency. This is a measure of the diffusion effects that occur in the column to cause peak broadening and overlap. It is influenced by the physical parameters of the mobile and stationary phases and by the quality of the packing of the column.

Capacity. This is a measure of the amount of material that can be resolved without causing peaks to overlap, irrespective of such action as gradient elution and temperature programming. Ion-exchange chromatography (Section 10.7) and chromatofocusing (Section 10.7.3) are high capacity procedures, which is why they are commonly used to concentrate impure materials.

Resolution, R_S, is related to capacity ratio, relative retention ratio, and plate number by the equation:

$$R_S = \frac{N^{1/2}}{4}\left(\frac{\alpha-1}{\alpha}\right)\left(\frac{k'}{k'+1}\right) \tag{10.11}$$

10.1.4   Quantification, internal and external standards

When the column effluent has been monitored to produce a chart recording, then generally the area of each peak can be shown to be proportional to the amount of a given analyte eluting from the column. The area of the peak may be determined by measuring the height of the peak (h_P) and its width at half the height (w_h). The product of these dimensions is taken to be equal to the area of the peak. Alternatively the peak may be cut out of the chart paper and weighed and the assumption made that area and weight are linearly related. These procedures are very time consuming when complex and/or a large number of analyses are involved. The calculations are best performed by dedicated integrators or microcomputers. These can be programmed to compute retention time and peak area and to relate them to those of standards (pure reference compounds), enabling relative retention ratios and relative peak area ratios to be calculated. These may be used to identify a particular analyte and to quantify it using previously obtained and stored calibration data from standards. The data system can also be used to correct problems inherent in the chromatographic system. Such problems can arise either from the characteristics of the detector or from the efficiency of the separation process. Problems that are attributable to the detector are baseline drift, where the detector signal gradually changes with time, and baseline noise, which is a series of rapid minor fluctuations in detector signal, commonly the result of the operator using too high a detector sensitivity or possibly an electronic fault.

When the new peak area has been determined, the amount of the analyte present may be determined by use of a calibration curve obtained by chromatographing, under identical conditions, known amounts of the pure form of the analyte. To aid this assay by attempting to compensate for variations in chromatographic conditions and of any preliminary extraction procedure, use is made of an internal standard. The internal standard is a compound that has physical properties as similar as possible to the test analytes and which chromatographs near to, but distinct from, them. A known amount of the standard is introduced into the test sample as early as possible in the extraction, and is therefore taken through any preliminary procedures with it. Any loss of standard during the analysis will be identical with the loss of the test analytes. The peak area associated with the fixed amount of internal standard is used to calculate the relative peak area ratio for each peak in both the calibration data and the sample under analysis. A calibration curve therefore consists of a plot of relative peak area ratio against the known amount of the analyte, thereby enabling the amount of the analyte in the test sample to be calculated. An alternative procedure is to use an external standard. In this method the standard is added to the test sample immediately before the sample is chromatographed and separate analyte and calibration standard solutions are produced independently. It is therefore not taken through any preliminary extraction procedure and cannot compensate for variations in the efficiency of the extraction procedure. This method is valid only in those cases where the recovery of the analyte from the test sample is virtually quantitative and in those cases where there are no short-term fluctuations in detector response.

10.2 Sample preparation

Whilst chromatographic techniques are designed to separate mixtures of compounds this does not mean that no attention has to be paid to the preliminary purification (clean up) of the test sample. On the contrary, it is clear that, for quantitative work using HPLC techniques in particular, such preliminary action is essential particularly if the test compound(s) is in a complex matrix such as plasma, urine, cell homogenate or microbiological culture medium. The extraction and purification of the components from a cell homogenate is often a complex multistage process.

The associated principles for protein purification are discussed in Section 4.5. For some forms of analysis, e.g. the analysis of drugs in biological fluids, sample preparation is relatively much simpler. The simplest and most commonly used clean-up technique is solvent extraction. This is based on the fact that organic compounds can usually be extracted from aqueous mixtures by extraction with a low boiling water-immiscible solvent such as diethylether or dichloromethane. The technique is another example of the application of the principle of partition coefficients. Organic compounds that are weak electrolytes, such as acids and bases, can exist in ionised or unionised forms depending upon their pK_a and the prevailing pH. For extraction into an organic solvent the unionised species is generally required and hence the pH of the test sample must be adjusted to the appropriate value. Organic solvents such as diethylether and dichloromethane also extract significant quantities of water and, in general, this should be removed, e.g. by the addition of an anhydrous salt such as sodium sulphate or magnesium sulphate, before the extract is evaporated to dryness (often under nitrogen or *in vacuo*), dissolved in an appropriate solvent such as methanol or acetonitrile, and subjected to chromatographic separation. This solvent extraction procedure tends to lack selectivity and is often unsatisfactory for the HPLC analysis of compounds in the $ng\,cm^{-3}$ or less range. It can sometimes be improved by the technique of ion-pairing (Section 10.6.4).

The alternative to solvent extraction is solid phase extraction. Its advantage over simple solvent extraction is that it exhibits greater selectivity mainly because it is a form of chromatography. The test solution is passed through a small (few millimetres in length) disposable column packed with relatively large particles of a bonded silica similar to those used for HPLC (Section 10.4.3). These selectively adsorb the analyte under investigation and ideally allow interfering compounds to pass through. Preliminary thought has to be given to the particular bonded silica selected and the test sample should be treated with agents such as trichloroacetic acid, perchloric acid or organic solvents such as acetonitrile to deproteinise it so that the opportunity for protein binding of the analyte is minimised. The pH of the test solution should also be adjusted to maximise the retention of the analyte. Once the test solution has been passed through the column, either by simple gravity feed or by the application of a slight vacuum to the receiver vessel, the column is washed with water and the adsorbed analyte recovered by elution with an organic solvent such as methanol or acetonitrile. The minimum volume of elution solvent is used because the analyte is recovered by evaporating the solution to dryness (under nitrogen or *in vacuo*) and the residue dissolved in the minimum volume of an appropriate solvent prior to chromatographic analysis. Several commercial forms of this solid phase extraction technique are available that facilitate the simultaneous treatment of a large number of test samples.

A more sophisticated procedure for sample preparation, particularly suited to the analysis of analytes in very low concentrations in complex mixtures by HPLC, is the technique of column switching. In this technique, the test solution is applied to a preliminary short column similar to the type used in solid phase extraction. Once the test analyte has been adsorbed and impurities washed through the column, the analyte is eluted with a suitable organic solvent and the column effluent transferred directly to an analytical HPLC column. Technically, this is not easy to achieve and requires several pumps and switching valves and is therefore expensive. One of the main problems with the technique is that unless all interfering compounds are eluted from the preliminary column before the adsorbed analyte is switched to the analytical column they will eventually accumulate in the analytical column and reduce its resolving power. Nevertheless, the technique has achieved many very difficult resolutions.

In recent years, a new approach to sample preparation has been investigated. Termed supercritical fluid extraction (SFE), it exploits the fact that gases such as carbon dioxide exist as a liquid under certain critical conditions. In the case of carbon dioxide, these conditions are $31.1°C$ and 7.38 MPa (10.7 lbf in.$^{-2}$) and the resultant liquid carbon dioxide can be used as the extraction solvent, behaving as a low polarity solvent comparable to hexane. By altering the physical conditions of the extract, the carbon dioxide can be made to revert to a gas, thus simplifying the recovery of the extracted analytes. Whether or not SFE will become a general laboratory procedure depends upon the success of its further evaluation. The technique of analyte pre- or post-column derivatisation may facilitate better chromatographic separation and detection. Derivatisation is designed to improve separation and analyte detection using reagents of the type shown in Table 10.1.

10.3   Low pressure column chromatography

10.3.1   Columns

The glass column used should have a means of supporting the stationary phase as near to the base of the column as possible in order to minimise the dead space below the column support in which post-column mixing of separated analytes could occur. Commercial columns possess either a porous glass plate fused on to the base of the column or a suitable device for supporting a replaceable nylon net, which in turn supports the stationary phase. A cheaper alternative to these commercial column supports is to use a small plug of glass wool together with a minimal amount of quartz sand or glass beads. Capillary tubing normally leads the effluent from the column to the detector and/or fraction collection system

Table 10.1. *Examples of derivatising reagents*

Analyte	Reagent
A. Pre-column	
Ultraviolet detection	
Alcohols, amines, phenols	3,5-Dinitrobenzoyl chloride
Amino acids, peptides	Phenylisothiocyanate, dansyl chloride
Carbohydrates	Benzoyl chloride
Carboxylic acids	1-p-Nitrobenzyl-N,N'-diisopropylisourea
Fatty acids, phospholipids	Phenacyl bromide, naphthacyl bromide
Electrochemical detection	
Aldehydes, ketones	2, 4-Dinitrophenylhydrazine
Amines, amino acids	o-Phthalaldehyde, fluorodinitrobenzene
Carboxylic acids	p-Aminophenol
Fluorescent detection	
Amino acids, amines, peptides	Dansyl chloride, dabsyl chloride, fluorescamine, o-phthalaldehyde
Carboxylic acids	4-Bromomethyl-7-methoxycoumarin
Carbonyl compounds	Dansylhydrazine
B. Post-column	
Ultraviolet detection	
Amino acids	Ninhydrin
Carbohydrates	Orcinol and sulphuric acid
Penicillins	Imidazole and mercuric chloride
Fluorescent detection	
Amino acids	o-Phthalaldehyde

(Fig. 10.2). For some chromatographic separations, it is necessary for the temperature of the column to be maintained constant during the separation. This is most simply achieved by jacketing the column so that liquid from a thermostatically controlled bath (set at the required working temperature, which may be below room temperature) may be pumped around the outside of the column. More sophisticated methods include placing the column in a heating block or in a thermostatically controlled oven.

10.3.2 Matrix materials

The matrix is the material used to support the stationary phase. The selection of a matrix for a particular stationary phase is vital to the successful chromatographic use of the phase. Generally speaking, a matrix needs to have: high mechanical stability to encourage good flow rates and to minimise pressure drop along the column; good chemical stability; functional groups to facilitate the attachment of the stationary phase; and high capacity, i.e. density of functional groups to minimise bed volume. It also needs to be available in a range of particle sizes. In addition some forms of chromatography require a matrix with a porous structure, in which case the pores need to be of the correct size and shape. Finally, the surface of a matrix needs to be inert to minimise the non-selective adsorption of analytes.

In practice, the six most commonly used types of matrices are as follows:

- *Agarose* – a polysaccharide made up of D-galactose and 3,6-anhydro-1-galactose units (Fig. 9.4). The unbranched polysaccharide chains are cross-linked with agents such as 2,3-dibromopropanol to give gels that are stable in the pH range 3–14. They have good flow properties and high hydrophilicity but should never be allowed to dry out otherwise they undergo irreversible change. Commercial examples include Sepharose and Bio-Gel A.
- *Cellulose* – a polysaccharide of β-1–4-linked glucose units. For matrix use it is cross-linked with epichlorohydrin, the extent of cross-linking dictating the pore size. It is available in bead, microgranular and fibrous forms, has good pH stability and flow properties, and is highly hydrophilic.
- *Dextran* – a polysaccharide consisting of α-1–6-linked glucose units. For matrix use it is cross-linked with epichlorohydrin but is less stable to acid hydrolysis than are cellulose matrices. It is stable up to pH 12 and is hydrophilic. Commercial examples include Sephadex.
- *Polyacrylamide* – a polymer of acrylamide cross-linked with N,N'-methylene-bisacrylamide (Fig. 9.5). It is stable in the pH range 2–11. Commercial examples include Bio-Gel P.
- *Polystyrene* – a polymer of styrene cross-linked with divinylbenzene. Polystyrene matrices have good stability over all pH ranges and are most commonly used for exclusion and ion-exchange chromatography. They have relatively low hydrophilicities.
- *Silica* – a polymeric material produced from orthosilicic acid. The numerous silanol (Si-OH) groups make it hydrophilic. When derivatised, excess silanol groups can be removed by treatment with trichloromethylsilane. The stability of silica matrices is confined to the

pH range 3–8. Closely related to the silica matrices is controlled pore glass. It is chemically inert but, like the silicas, tends to dissolve above pH 8.

10.3.3   Stationary phases

The chemical nature of the stationary phase depends upon the particular form of chromatography to be carried out. Full details are given in later sections of this chapter. Most stationary phases are available attached to the matrices in a range of sizes and shapes. Both properties are important because they influence the flow rate and resolution characteristics. The larger the particle, the faster the flow rate but, conversely, the smaller the particle the larger the surface area-to-volume ratio and potentially the greater their resolving power. In practice a balance has to be struck. The best packing characteristics are given by spherical particles and most stationary phases now have a spherical or approximately spherical shape. Particle size is commonly expressed by a mesh size, which is a measure of the openings per inch in a sieve; hence the larger the mesh size, the smaller is the particle. A 100–120 mesh is most common for routine use, whereas a 200–400 mesh is used for higher resolution work.

10.3.4   Column packing

This is one of the most critical factors in achieving a successful separation by any form of column chromatography. Packing a column is normally carried out by gently pouring a slurry of the stationary phase in the mobile phase into a column that has its outlet closed, whilst the upper part of the slurry in the column is stirred and/or the column is gentle tapped to ensure that no air bubbles are trapped and that the packing settles evenly. Poor column packing gives rise to uneven flow (channelling) and reduced resolution. The slurry is added until the required height is obtained. Once the required column height has been obtained, the flow of mobile phase through the packed column is started by opening the outlet, and continued until the packing has completely settled. This whole process generally requires considerable practice to achieve reproducible results. To prevent the surface of the packed material from being disturbed either by the addition of mobile phase to the column or during the application of the sample to the column, it is normal to place a suitable protection device, such as a filter paper disc or nylon or rayon gauze, on the surface of the column. Some commercial columns possess an adaptor and plunger, which serve the dual purpose of protecting the surface of the column and providing an inlet (often capillary tubing) to carry the mobile phase to the column surface. Once a column has been prepared, it is imperative that no part of it should be allowed to run dry; hence a layer of mobile phase should always be maintained above the column surface.

It is difficult to generalise about the ideal column height-to-diameter ratio and the total bed volume. They both influence the amount of material that can be separated on the column and in practice will need to be determined by systematic trial and error. Experience frequently provides a guide; thus, for example, in exclusion chromatography, a height-to-diameter ratio of 10:1 to 20:1 is normally suitable.

10.3.5 Application of sample

Several methods are available for the application of the sample to the top of the prepared column. A simple way is to remove most of the mobile phase from above the column by suction and *just* to drain the remainder into the column bed. The sample is then carefully applied by pipette and it too is allowed just to run into the column. A small volume of mobile phase is then applied in a similar manner to wash final traces of the sample into the bed. More mobile phase is then carefully added to the column to a height of 2–5 cm. The column is then connected to a suitable reservoir that contains more mobile phase so that the height of the phase in the column can be maintained at 2–5 cm. An alternative procedure, which avoids the necessity to drain the column to the surface of the bed, is to increase the density of the sample by addition of sucrose to a concentration of about 1%. When this solution is layered on to the liquid above the column bed, it will automatically sink to the surface of the column and hence be quickly passed into the column. This method of sample application is satisfactory, provided that the presence of sucrose in no way interferes with the separation and subsequent analysis of the sample. A third method involves the use of capillary tubing and/or syringe or peristaltic pump to pass the sample directly to the column surface. This latter method is the most satisfactory of the three.

In all cases, care must be taken to avoid overloading the column with sample, otherwise irregular separation will occur. It is also advantageous to apply the sample in as small a volume of mobile phase as possible because this ensures an initial tight band of material when the separation commences.

10.3.6 Column development and sample elution

The components of the applied sample are separated by the continuous passage of the mobile phase through the column. This is known as elution development.

During the elution process it is essential that the flow of mobile phase is maintained at a stable rate and this is most simply achieved by gravity feed. The flow rate may be regulated by adjusting the operating pressure, which corresponds to the difference between the level of the liquid in a reservoir situated above the column and the level at the outlet site of the column. An ordinary open reservoir is not satisfactory because the operating pressure will drop in the course of the experiment owing to the drop in level in the reservoir as the mobile phase runs into the column. This can be overcome by the use of a Mariotte flask, which will keep the operating pressure constant (Fig. 10.2). An alternative and more effective method of maintaining stable flow rates is to use a peristaltic pump. The most commonly used pumps are of the roller type (Fig. 10.5). These deliver the mobile liquid phase through silicon, polyvinylchloride or fluororubber tubing by compressing the tubing as the revolving disc is rotated at a rate predetermined according to the flow rate required. Pumps and the tubing of known diameter must be precalibrated to determine the flow rate. Care must be exercised when pumps are used, to ensure that the operating pressure does not compact the column excessively and cause the column structure to change.

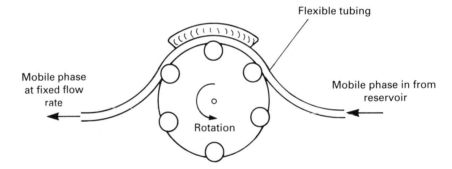

Fig. 10.5. Simple peristaltic pump commonly used in low pressure liquid chromatography.

Column development using a single liquid as the mobile phase is known as an isocratic elution. However, in many cases in order to increase the resolving power of the mobile phase, it is necessary continuously to change its pH, ionic concentration or polarity. This is known as gradient elution. In order to produce a suitable gradient, two eluents have to be mixed in the correct proportions prior to their entering the column. This may be achieved by use of commercially available gradient mixers or more simply as follows. The two eluents are placed in separate chambers, a recipient chamber linked to the column and a donor chamber, linked to the recipient chamber by a siphon (Fig. 10.6a). As eluent enters the column from the recipient chamber, the eluent in the donor chamber replaces it and is mixed by a stirrer. The relative pH, ionic strength or polarity of the eluent in the

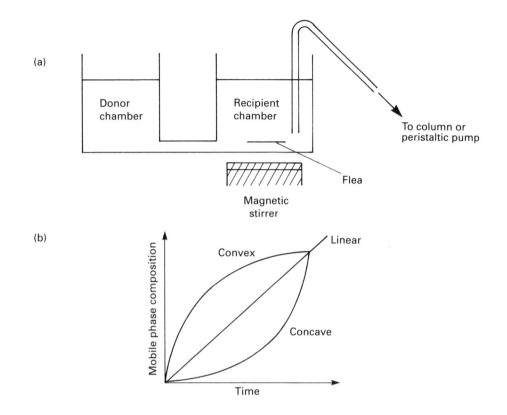

Fig. 10.6. (a) Simple apparatus for producing gradient elution and (b) gradient shapes commonly used.

donor with respect to that in the recipient chamber will determine the direction in which the gradient will be formed. Moreover, the relative diameters and shapes of the two chambers will determine whether the gradient varies with time in a linear, convex, or concave manner (Fig. 10.6b). In more sophisticated systems, two pumps are used to deliver the two eluents at predetermined rates into a mixing area, before application to the column. Convex gradients give better resolution initially, whereas concave gradients give better resolution at the end.

10.3.7   Detectors and fraction collection

As the resolved analytes emerge in the effluent from the column it is necessary to detect their presence. For coloured analytes this can be achieved simply by visual observation but for colourless compounds alternatives are necessary. Detection may be based on ultraviolet absorption, fluorescence spectroscopy, changes in the refractive index of the effluent, the presence of a radioactive emission atom or on the ease of oxidation or reduction of the analytes as measured by an electrochemical detector (Section 11.7).

Ultraviolet detectors are probably the most common form of detector in biochemical analysis. The best instruments allow proteins to be detected and quantified at 190–220 nm and 260–280 nm wavelengths (Section 4.4). Fluorescence detectors generally give greater sensitivity if they can be used but tend to be more sensitive to impurities in the mobile phase. All spectrophotometric detectors use continuous flow cells with a small internal volume (typically $8\,mm^3$) (Section 7.4.2).

For studies in which the analyte in the effluent is to be collected and studied further, the effluent has to be divided into fractions. Two approaches are available to achieve this objective: either the effluent can be continuously monitored and the fraction containing a particular analyte collected, or the effluent can be divided into small ($1–10\,cm^3$) fractions, which are subsequently analysed and those containing a particular analyte bulked together.

A range of automatic fraction collectors is available commercially. They are designed to collect a certain amount of effluent in each tube before a new tube is placed in position automatically. The amount of effluent in each fraction may be determined in one of several ways. There may be a siphoning or similar system to deliver a predetermined volume into each tube, or there may be an electronic means of allowing a predetermined number of drops of effluent to enter each tube. This latter method has the slight disadvantage that, if the composition of the effluent changes (e.g. during gradient elution), so too may its surface tension and hence droplet size, so that the actual volume collected also changes. A further possibility is that the effluent is allowed to enter each tube for a fixed time interval. In this case, if the flow rate through the column varies, so too will the volume of each fraction, but this is unusual and in practice fixed-time collectors are the most common.

10.4   High performance liquid chromatography (HPLC)

10.4.1   Principle

As can be seen from equations 10.1 to 10.11, the resolving power of a chromatographic column increases with column length and the number of theoretical plates per unit length, although there are limits to the length of a column owing to the problem of peak broadening. As the number of theoretical plates is related to the surface area of the stationary phase, it follows that the smaller the particle size of the stationary phase, the better the resolution. Unfortunately, the smaller the particle size, the greater is the resistance to flow of mobile phase. This creates a back-pressure in the column that is sufficient to damage the matrix structure of the stationary phase, thereby actually reducing eluent flow and impairing resolution.

In the recent past there has been a dramatic development in column chromatography technology that has resulted in the availability of new, smaller particle size, stationary phases that can withstand these pressures. This development, which has occurred in adsorption, partition, ion-exchange, exclusion and affinity chromatography, has resulted in faster and better resolution and explains why HPLC has emerged as the most popular, powerful and versatile form of chromatography.

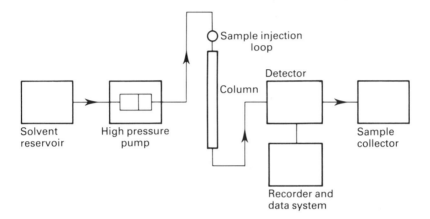

Fig. 10.7. Diagram of the components of an isocratic HPLC system.

10.4.2   Columns

The columns (Fig. 10.7) used for HPLC are generally made of stainless steel and are manufactured so that they can withstand pressures of up to 5.5×10^7 Pa ($\sim$8000 lbf in.$^{-2}$). Straight columns of 15–50 cm length and 1–4 mm diameter are generally used, although smaller microbore columns are available. Microbore columns have an internal diameter of 1–2 mm and are generally 25 cm long. They can sustain flow rates of 0.05–0.20 cm^3 min^{-1} as opposed to the 2 cm^3 min^{-1} of conventional HPLC columns. Preparative columns are also available commercially, with internal diameters of up to 25 mm; they can sustain flow rates of up to 100 cm^3 min^{-1}. The best columns are precision bored, with an internal mirror finish that allows efficient packing of the column. Porous plugs of stainless steel or Teflon are used in the ends of the columns to retain the packing material. These plugs must be homogeneous to ensure uniform flow of liquid through the column. It is advantageous in some separations involving liquid partition chromatography and ion-exchange chromatography to maintain the column temperature slightly above room temperature (up to 60 °C) during the analysis.

10.4.3   Matrices and stationary phases

Three forms of column packing material are available, based on a rigid solid (as opposed to gel) structure. These are as follows:

- ■ *Microporous supports* in which micropores ramify through the particles which are generally 5–10 μm in diameter.
- ■ *Pellicular (superficially porous) supports* in which porous particles are coated on to an inert solid core such as a glass bead of about 40 μm in diameter.
- ■ *Bonded phases* in which the stationary phase is chemically bonded on to an inert support such as silica.

For adsorption chromatography, adsorbents such as silica and alumina are available as microporous or pellicular forms with a range of particle sizes. Pellicular systems generally have a high efficiency but low sample capacity and therefore microporous supports are preferred when available.

In partition chromatographic systems, the stationary phase may be coated on to the inert microporous or pellicular support. One disadvantage of supports coated with liquid phases is that the developing mobile phase may gradually wash off the liquid phase. To overcome this problem, bonded phases have been developed in which the supporting material is silica.

In normal phase liquid chromatography, the stationary phase is a polar compound such as an alkyl nitrile or an alkylamine, and the mobile phase is a non-polar solvent such as hexane. For reversed-phase liquid chromatography, the stationary phase is a non-polar compound such as octasilane (OS) or octadecylsilane (ODS) and the mobile phase is a polar solvent such as water/acetonitrile or water/methanol.

Many different types of ion-exchanger suitable for HPLC are available, of which the cross-linked microporous polystyrene resins are widely used. Pellicular resin forms are also available, as are bonded-phase exchangers covalently bonded to a cross-linked silicone network. These resins are classed as hard gels and readily withstand the pressures generated during analysis.

The stationary phases for exclusion separations are generally porous silica, glass, polystyrene or polyvinylacetate beads and are available in a range of pore sizes. They are generally used where the eluting solvent is an organic system. Semi-rigid gels such as Sephadex or Bio-Gel P, and non-rigid gels such as Sepharose and Bio-Gel A, are of only limited use in HPLC because they can withstand only low pressures. The supports for affinity separations are similar to those for exclusion separations. The spacer arm and ligand are attached to these supports by chemical means similar to those used in conventional low pressure affinity chromatography

Table 10.2. *Some examples of HPLC stationary phases*

Chromatographic separation principle	Commercial name	Nature of stationary phase	Type of support
Adsorption	Partisil C$_8$	Octylsilane	Porous
	Corasil	Silica	Pellicular
	Pellumina	Alumina	Pellicular
	Partisil	Silica	Microporous
	MicroPak A1	Alumina	Microporous
Partition	Bondapak-C$_{18}$/Corasil	Octadecylsilane (ODS)	Pellicular
	μBondapak-C$_{18}$	Octadecylsilane	Porous
	ULTRApak TSK ODS	Octadecylsilane	Porous
	μBondapak-NH$_2$	Alkylamine	Porous
	ULTRApak TSK-NH$_2$	Alkylamine	Porous
Ion-exchange	Partisil-SAX	Strong base	Porous
	MicroPak-NH$_2$	Weak base	Porous
	Partisil-SCX	Strong acid	Porous
	AS Pellionex-SAX	Strong base	Pellicular
	Zipak-WAX	Weak base	Pellicular
	Perisorb-KAT	Strong acid	Pellicular
Exclusion	Bio-Glas	Glass	Rigid solid
	Styragel	Polystyrene-divinylbenzene	Semi-rigid gel
	Superose	Agarose	Soft gel
	Fractogel TSK	Polyvinylchloride	Semi-rigid gel

(Section 10.9). Table 10.2 lists some examples of commonly used HPLC stationary phases.

10.4.4 Column packing

HPLC columns may be purchased already packed with specified packing material, structure and dimensions. Many workers, however, prefer to pack their own columns as this is cheaper. Several methods are available for packing columns and the method used will depend on the nature of the packing material and the dimensions of the particles. The major priority in the packing of a column is to obtain a uniform bed of material with no cracks or channels. Rigid solids and hard gels should be packed as densely as possible, but without fracturing the particles during the packing process. The most widely used technique for column packing is the high pressure slurrying technique. A suspension of the packing is made in a solvent of density equal to that of the packing material. The slurry is then pumped rapidly at high pressure into a column with a porous plug at its outlet. The resulting bed of packed material within the column can then be prepared for use by running the developing mobile phase through the column. When hard gels are to be used, it is necessary for them to be allowed to swell in the solvent to be used in the chromatographic process before they are packed under pressure. Soft gels cannot be packed under pressure and have to be allowed to pack from a slurry in a way similar to that of the packing of columns for LPLC (Section 10.3.4).

10.4.5   Mobile phases and pumps

The choice of mobile phase to be used in any separation will depend on the type of separation to be achieved. Isocratic separations may be made with a single pump, using a single solvent or two or more solvents premixed in fixed proportions. Gradient elution (Section 10.3.6) generally uses separate pumps to deliver two solvents in proportions predetermined by a gradient programmer. All solvents for use in HPLC systems must be specially purified because traces of impurities can affect the column and interfere with the detection system. This is particularly the case if the detection system is measuring absorbance below 200 nm. Purified solvents for use in HPLC systems are available commercially, but even with these solvents a 1–5 μm microfilter is generally introduced into the system prior to the pump. It is also essential that all solvents are degassed before use otherwise gassing (the presence of air bubbles in the solvent) tends to occur in most pumps. Gassing, which tends to be particularly bad for aqueous methanol and ethanol, can alter column resolution and interfere with the continuous monitoring of the effluent. Degassing may be carried out in several ways; by warming the solvent, by stirring it vigorously with a magnetic stirrer, subjecting it to a vacuum, by ultrasonication, or by bubbling helium gas through the solvent reservoir.

Pumping systems for delivery of the mobile phase are one of the most important features of HPLC systems. The main features of a good pumping system are that it is capable of outputs of at least 5×10^7 Pa (~7200 lbf in.$^{-2}$) and ideally there must be no pulses (i.e. cyclical variations in pressure) as this may affect the detector response. There must be a flow capability of at least 10 cm^3 min^{-1} and up to 100 cm^3 min^{-1} for preparative separations. Various pumping systems are available that operate on the principle of constant pressure or constant displacement. Constant pressure pumps produce a pulseless flow through the column, but any decrease in the permeability of the column will result in lower flow rates for which the pump will not compensate. These pumps operate by the introduction of high pressure gas into the pump and the gas in turn forces the solvent from the pump chamber on to the column. The use of an intermediate solvent between the gas and the eluting solvent reduces the chances of dissolved gas directly entering the eluting solvent and causing problems during the analysis. Such pumps are seldom used in contemporaneous liquid chromatography.

Constant displacement pumps maintain a constant flow rate through the column irrespective of changing conditions within the column. One form of constant displacement pump is a motor-driven syringe-type pump that delivers a fixed volume of solvent on to the column by a piston driven by a motor. The reciprocating pump is the most commonly used form of constant displacement pump. The piston is moved by a motorised crank, and entry of solvent to the column is regulated by check valves. On the compression stroke solvent is forced from the pump chamber on to the column. During the return stroke the exit check valve closes and solvent is drawn in via the entry valve to the pump chamber ready to be pumped on to the column on the next compression stroke. Such pumps produce small pulses of flow and pulse dampeners are usually incorporated into the system to minimise this pulsing effect. All constant displacement pumps have in-built safety cut-out mechanisms so that if the pressure within the chromatographic systems changes from preset limits the pump is inactivated automatically.

10.4.6   Application of sample

The correct application of the sample on to an HPLC column is a particularly important factor in achieving successful separations and one of two methods is generally used. The first method makes use of a microsyringe to inject the sample either directly on to the column packing or on to a small plug of inert material immediately above the column packing. This injection can be done while the system is under pressure or it may require the pump to be turned off and, when the pressure has dropped to near atmospheric, the injection is made and the pump switched on again. This latter approach is termed a stop flow injection. The second method of sample introduction that retains the column pressure is by use of a loop injector (Fig. 10.8). This consists of a metal loop, of fixed small volume, that can be filled with the sample. By means of an appropriate valve switching system, the eluent from the pump is channelled through the loop, the outlet of which leads directly on to the column. The sample is thus flushed on to the column by the eluent without interruption of flow to the column.

Repeated application of highly impure samples such as sera, urine, plasma or whole blood, which have preferably been deproteinated, may eventually cause the column to lose its resolving power. To prevent this occurrence, a guard column is

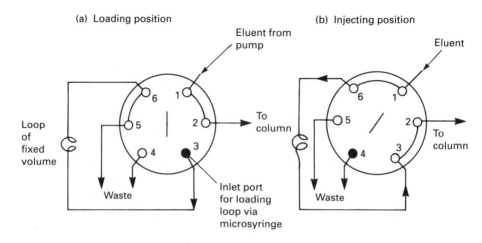

Fig. 10.8. HPLC loop injector: the loop is loaded (a) via port 3 with excess sample going to waste via port 5. In this position the eluent from the pump passes to the column via ports 1 and 2. In the injecting position (b), eluent flow is directed through the loop (ports 1 and 6) and then on to the column.

frequently installed between the injector and the analytical column. This guard column is a short (1–2 cm) column of the same internal diameter and packed with material similar to that present in the analytical column. The packing in the guard column preferentially retains contaminating material and can be replaced at regular intervals.

10.4.7   Detectors

Since the quantity of material applied to the column is frequently very small, it is imperative that the sensitivity of the detector system is sufficiently high and stable to respond to the low concentrations of each analyte in the effluent. Most commonly the detector is a variable wavelength detector based upon ultraviolet–visible spectrophotometry. This type of detector is capable of measuring absorbances down to 190 nm wavelength and has sensitivities as low as 0.001 absorbance units for full-scale deflection (AUFS). Scanning wavelength detectors have the facility to record the complete absorption spectrum of each analyte, thus aiding identification. Such opportunities are possible either by temporarily stopping the effluent flow or by the use of diode array techniques, which allow the simultaneous measurement of absorbance at many or all wavelengths within 0.01 s (Fig. 10.9).

Fluorescence detectors are extremely valuable for HPLC because of their sensitivity but the technique is limited by the fact that relatively few compounds fluoresce. Electrochemical detectors, which are selective for electroactive species (Section 11.7), are potentially highly sensitive. Two types are

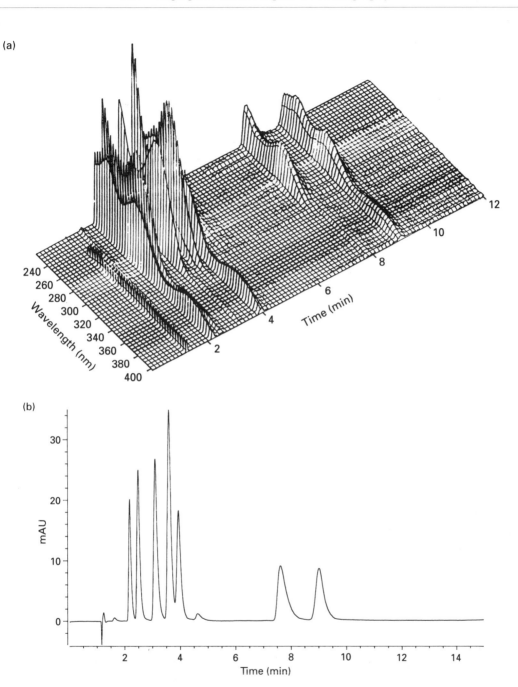

Fig. 10.9. Separation by HPLC of the dihydropyridine calcium channel blocker lacidipine and its metabolites. Column: ODS Hypersil. Eluent: Methanol/acetonitrile/water (66%, 5%, 29%, by vol.) acidified to pH 3.5 with 1% formic acid. Flow rate: $1 \, cm^3 \, min^{-1}$. Column temperature: 40 °C. (a) As recorded by diode array detector; (b) as recorded by an ultraviolet detector. (Reproduced with permission of Glaxo Group Research, Greenford.)

available, amperometric and coulometric, the principles of which are similar. A flow cell is fitted with two electrodes – a stable counter electrode (Ag/AgCl or calomel) and a working electrode, which is highly polarisable. A constant potential is applied to the working electrode at such a value that, as an analyte flows through the flow cell, molecules at the electrode surface undergo either an oxidation or a reduction, resulting in a current flow between the two electrodes. The potential applied to the counter electrode is sufficient to ensure that the current detected gives a full-scale deflection on the recorder. Compounds capable of undergoing oxidation include hydrocarbons, amines, amides, phenols, di- and triazines, phenothiazines, catecholamines and quinolines. Compounds capable of undergoing reduction include olefins, esters, ketones, aldehydes, ethers, azo and nitro compounds. The mobile phase should of course be free from compounds capable of responding to the detector. The electrochemical detector has been particularly successful in the assay of catecholamines, vitamins and antioxidants.

Perhaps the greatest advance in detection using HPLC has been made by the coupling of the HPLC to a mass spectrometer. The technical problems associated with the logistics of removing the bulk of the mobile phase before the sample is introduced into the mass spectrometer have been resolved in a number of ways. The direct liquid insertion interface (DLI) consists of a flow of effluent via a capillary into a direct insertion probe in the mass spectrometer. The analyte enters the ionisation source together with an excess of mobile phase molecules and as a consequence the ionisation is weak (soft) and CI (Section 8.3) rather than EI spectra (Section 8.4) predominate. However the $(M + H)^+$ ion of the analyte is clearly visible and this facilitates structure identification. Thermospray ionisation (TSI) (Section 8.7.1) normally requires the use of an aqueous mobile phase commonly containing ammonium acetate. The end of a heated (400 °C) capillary from the effluent flow enters a chamber exposed to electron bombardment and to a vacuum pumping system. The high temperature creates a supersonic expanding jet of droplets that are ionised and gradually reduced in size due to the pumping system. Eventually ions of the analyte are ejected from the droplet into the mass analyser. CI spectra are again recorded as $(M + H)^+$ or $(M + NH_4)^+$ ions. In electrospray ionisation (ESI) (Section 8.7.2) the capillary containing the effluent is held at a high voltage and the effluent mixed with nitrogen gas. The voltage creates a charge on the liquid surface that causes it to disperse into a very fine spray. The solvent in the fine droplets evaporates and is flushed away by the nitrogen gas. The ions of the analyte are swept as a supersonic stream into a vacuum chamber and eventually into the mass analyser.

The sensitivity of ultraviolet absorption, fluorescence and electrochemical detectors can often be increased significantly by the process of derivatisation, whereby the analyte is converted pre- or post-column to a chemical derivative. Examples are given in Table 10.1.

10.4.8 Applications

The wide applicability, speed and sensitivity of HPLC has resulted in it becoming the most popular form of chromatography and virtually all types of biological molecule have been assayed or purified using the technique. HPLC has had a big impact on the separation of oligopeptides and proteins. Instruments dedicated to the separation of proteins have given rise to the technique of fast protein liquid chromatography (FPLC). There are no unique principles associated with FPLC, it is simply based on reversed-phase, affinity, exclusion, hydrophobic interaction and ion-exchange chromatography and on chromatofocusing. Microbore glass-lined stainless steel columns enable very small amounts of sample to be used, with separation taking as little as 10 min. The technique enables such complex mixtures as tryptic digests of proteins and the culture supernatant of microorganisms to be applied directly to the column, but protein mixtures from cell extracts still need some form of preliminary fractionation (Section 4.5.3) prior to study.

10.5 Adsorption chromatography

10.5.1 Principle

This is the classic form of chromatography first introduced by M. Tswett at the beginning of the century. It is based upon the principle that certain solid materials, collectively known as adsorbents, have the ability to hold molecules at their surface. This adsorption process, which involves weak, non-ionic attractive forces of the van der Waals' and hydrogen-bonding type, occur at specific adsorption sites. These sites have the ability to discriminate between molecules and in the adsorption chromatographic process are occupied by molecules of the eluent or of the analytes present in the mixture in proportions depending upon the relative strengths of their interaction. As eluent is constantly passed down the column, differences in these binding strengths eventually lead to the separation of the analytes. The strength of binding of a particular analyte depends upon the functional groups present in its structure. Hydroxyl groups and aromatic groups tend to increase interaction with the adsorption surface, whereas aliphatic groups of different size usually differ only slightly in their interaction. In general, adsorption chromatography is influenced more by the presence of specific groups than by simple molecular size because only a specific group rather than the whole molecule can interact with the adsorption site.

A typical adsorbent is silica, which has silanol (Si-OH) groups on its surface. These groups, which are slightly acidic, can interact with polar functional groups of the analyte or eluent. The topology (arrangement) of these silanol groups in different commercial preparations of silica explains their different separation properties. Other commonly used adsorbents are alumina and carbon. Materials based on carbon, alumina or silica are available for low pressure chromatography and for HPLC (Table 10.2). Whereas the silicas are acidic and good for the separation of basic materials, the aluminas are more basic and better suited for the resolution of acidic materials.

The selection of the correct eluent (mobile phase) is essential for good resolution because it influences the capacity factor, k', of the analytes (Section 10.1.3). In general, an eluent with a polarity comparable to that of the most polar analyte in the mixture is chosen. Thus, alcohols would be selected if the analytes contained hydroxyl groups, acetone or esters would be selected for analytes containing carbonyl groups, and hydrocarbons such as hexane, heptane and toluene for analytes that are predominantly non-polar. Mixtures of solvents are commonly used in the context of gradient elution (Section 10.3.6). The presence of small amounts of water in the mobile phase is often beneficial when silica is used as the stationary phase, as the water molecules selectively block the more active silanol groups leaving a more selective population of weaker binding sites.

Adsorption chromatography, which can be carried out in the thin-layer mode as well as the column mode, is most commonly used to separate non-ionic, water-insoluble compounds such as triglycerides, PTH amino acids (Section 4.3.3), vitamins and many drugs.

10.5.2 Hydroxylapatite chromatography

Crystalline hydroxylapatite ($Ca_{10}(PO_4)_6(OH)_2$) is an adsorbent used to separate mixtures of proteins or nucleic acids. The mechanism of adsorption is not fully understood but is thought to involve both the calcium ions and phosphate ions on the surface and to involve dipole–dipole interactions and possibly electrostatic attractions. One of the most important applications of hydroxylapatite chromatography is the separation of single-stranded from double-stranded DNA. Both forms of DNA bind at low phosphate buffer concentrations but as the buffer concentration is increased single-stranded DNA is selectively desorbed. As the buffer concentration is increased further, double-stranded DNA is released. This behaviour is exploited in the technique of Cot analysis (Section 3.5.2). The affinity of hydroxylapatite for double-stranded DNA is so high that the latter can be selectively removed from RNA and proteins in cell extracts by use of this form of chromatography.

Hydroxylapatite is available commercially in a range of forms suitable for LPLC and HPLC. These forms include crystalline or spheroidal hydroxylapatite and forms bonded to an agarose matrix. The adsorption capacity of all these forms is maximum around neutral pH and such conditions are usually employed using 20 mM phosphate buffer for the adsorption process. Elution is achieved by increasing the phosphate buffer concentration to 500 mM.

10.5.3   Hydrophobic interaction chromatography (HIC)

This form of chromatography was developed to purify proteins by exploiting their surface hydrophobicity, which is related to the presence of non-polar amino acid residues (Section 4.2). Groups of hydrophobic residues are scattered over the surface of proteins in a way that gives characteristic properties to each protein. In aqueous solution, these hydrophobic regions on the protein are covered with an ordered film of water molecules that effectively masks the hydrophobic groups. These groups can, however, be exposed by the addition of salt ions, which preferentially take up the ordered water molecules. The exposed hydrophobic regions can then interact with each other and this is the basis of salting-out (Section 4.5.3) using ammonium sulphate. In HIC, rather than facilitating protein–protein interaction by the exposure of the hydrophobic groups, the presence of hydrophobic groups attached to a suitable matrix facilitates protein–matrix interaction. The most commonly used stationary phases are alkyl (hexyl, octyl) or phenyl groups attached to an agarose matrix. Commercial materials include Phenyl Sepharose and Phenyl SPW, both for low pressure HIC, and Bio-Gel TSK Phenyl and Spherogel TSK Phenyl for HPLC.

Since HIC requires the presence of salting-out compounds such as ammonium sulphate to facilitate the exposure of the hydrophobic regions on the protein molecule, it is commonly used immediately after fractionation with ammonium sulphate (Section 4.5.3) as ammonium and sulphate ions are already present in the protein sample. To maximise the process, it is advantageous to adjust the pH of the protein sample to that of its isoelectric point. Once the proteins have been adsorbed to the stationary phase, selective elution can be achieved in a number of ways, including the use of an eluent of gradually decreasing ionic strength or of increasing pH (this increases the hydrophilicity of the protein) or by selective displacement by a displacer that has a stronger affinity for the stationary phase than has the protein. Examples include non-ionic detergents such as Tween 20 and Triton X-100, aliphatic alcohols such as butanol and ethylene glycol, and aliphatic amines such as butylamine. One of the potential problems with HIC is that some of these elution conditions may cause protein denaturation. The other practical problem with the technique is its non-predictability in that it works well for some proteins but not for others and a trial study is invariably necessary. Proteins purified by the technique include aldolase, transferrin, cytochrome c and thyroglobulin.

10.6   Partition chromatography

10.6.1   Principle

This form of chromatographic separation is based on differences in capacity ratios, k', and distribution coefficients, K_d, of the analytes using liquid stationary and mobile phases. It can be subdivided into liquid–liquid chromatography, in which the liquid stationary phase is attached to a supporting matrix by purely physical means, and bonded-phase liquid chromatography, in which the stationary phase is covalently attached to the matrix. An example of liquid–liquid chromatography is one in which a water stationary phase is supported by a cellulose, starch or silica matrix, all of which have the ability to bind physically as much as 50% (w/v) water and remain free-flowing powders. The advantages of this form of chromatography are that it is cheap, has a high capacity and has broad selectivity. Its disadvantage is that the elution process may gradually remove the stationary phase, thereby altering the chromatographic conditions. This problem is overcome by the bonded phases and this explains their more widespread use. Most bonded-phases use silica as the matrix, which is derivatised to immobilise the stationary phase by reaction with an organochlorosilane:

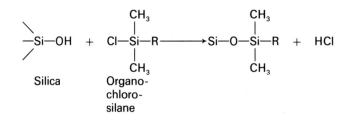

Surplus silanol groups are removed by capping with chlorotrimethylsilane to improve the quality of the chromatography by decreasing tailing (Fig. 10.3c). There are two commonly used modes of partition chromatography that differ in the relative polarities of the stationary and mobile phases and give rise to normal phase liquid chromatography and reversed-phase liquid chromatography.

10.6.2 Normal phase liquid chromatography

In this form of chromatography, the stationary phase is polar and the mobile phase relatively non-polar. The most popular stationary phase is an alkylamine bonded to silica (Table 10.2). The mobile phase is generally an organic solvent such as hexane, heptane, dichloromethane and ethylacetate. The mechanism of separation exploits the ability of the analyte to displace molecules of the mobile phase adsorbed as a monolayer on the surface of the stationary phase, as well as the ability of the analyte to compete with mobile phase molecules in the formation of a bilayer on the stationary phase surface. The order of elution of analytes is such that the *least* polar is eluted *first* and the *most* polar *last*. Indeed, polar analytes generally require gradient elution with a mobile phase of *increasing* polarity, generally achieved by the use of methanol or dioxane. The main advantages of normal phase liquid chromatography are its ability to separate analytes that have low water solubility and those that are not amenable to reversed-phase liquid chromatography.

10.6.3 Reversed-phase liquid chromatography

In this form of liquid chromatography, the stationary phase is non-polar and the mobile phase relatively polar. By far the most commonly used type is the bonded-phase form, in which alkylsilane groups are chemically attached to silica. Butyl (C_4), octyl (C_8) and octadecyl (C_{18}) silane groups are invariably used (Table 10.2). The mobile phase is commonly water or aqueous buffers, methanol, acetonitrile, or tetrahydrofuran or mixtures of them. The organic solvent is referred to as an organic modifier. Reversed-phase liquid chromatography differs from most other forms of chromatography in that the stationary phase is essentially inert and only non-polar (hydrophobic) interactions with analytes are possible. Chromatographic separation of analytes is determined principally by the characteristics of the mobile phase. No simple model has been described to explain reversed-phase chromatography but the solvophobic theory is the one most widely considered. It is based on the consideration of the balance of free energy and entropy changes associated with bonding of the analyte with the stationary phase and with the mobile phase. The attraction of the reversed-phase technique is that small changes in the mobile phase composition such as the addition of salts, change of pH or the amount of organic solvent, profoundly affect the separation characteristics. Moreover, the technique is sensitive to temperature change such that a 10 deg.C increase approximately halves the capacity factor, k' (Section 10.1.3).

In reversed-phase chromatography, *polar* analytes elute *first* and *non-polar* analytes *last*. Non-polar analytes may need gradient elution using increasing proportions of a low polarity solvent such as hexane.

Reversed-phase HPLC is probably the most widely used form of chromatography mainly because of its flexibility and high resolution. It is widely used to analyse drugs and their metabolites, insecticide and pesticide residues, and amino acids. It is also now widely applied to proteins by using FPLC. Octadecylsilane (ODS) phases bind proteins more tightly than do octyl- or methylsilane phases and are therefore more likely to cause protein denaturation because of the more extreme conditions required for the elution of the protein. In non-aqueous form, reversed-phase chromatography can be used to separate lipophilic compounds such as fats.

10.6.4   Ion-pair reversed-phase liquid chromatography

The separation of some highly polar compounds, such as amino acids, peptides, organic acids and the catecholamines, which are difficult to resolve adequately by conventional reversed-phase chromatography, can often be improved by one of two possible approaches. The first is ion suppression in which the ionisation of the compound is suppressed by chromatographing at an appropriately high or low pH. Weak acids, for example, can be chromatographed using an acidified mobile phase. The second is ion-pairing in which a counter ion with charge opposite to that to be separated is added to the mobile phase so that the resulting ion-pair has sufficient lipophilic character to be retained by the non-polar stationary phase of a reversed-phase system. Thus, to aid the separation of acidic compounds, which would be present as their conjugate anions, a quaternary alkylamine ion such as tetrabutylammonium would be used as the counter ion, whereas for the separation of bases, which would be present as cations, an alkyl sulphonate such as sodium heptanesulphonate would be used:

$$RCOO^- \quad + R'_4N^+ \quad \rightleftharpoons [RCOO^- \overset{+}{N}R_4]$$

Carboxylic   Counter cation   Ion-pair
acid anion

$$R\overset{+}{N}H_3 \quad + R'SO_3^- \quad \rightleftharpoons [R\overset{+}{N}H_3^-O_3SR]$$

Conjugate acid   Counter anion   Ion-pair
of weak base

The mechanism by which ion-pairing results in better separation is not clear but two theories have been proposed. The first suggests that the ion-pair behaves as a single neutral species, whilst the second suggests that an active ion-exchange surface is produced in which the counter ion, which has considerable lipophilic properties, and the ions to be separated are adsorbed by the hydrophobic, non-polar stationary phase. In practice, the success of the ion-pairing approach is variable and somewhat empirical. The size of the counter ion, its concentration and the pH of the solution are all factors that may profoundly influence the outcome of the separation.

Octyl- and octadecylsilane-bonded phases are used most commonly in conjunction with a water/methanol or water/acetonitrile mobile phase. One of the advantages of ion-pair reversed-phase chromatography is that, if the sample to be resolved contains a mixture of non-ionic and ionic compounds, the two groups of compounds can be separated simultaneously because the ion-pair reagent does not affect the chromatography of the non-ionic species. This is not true of ion-exchange chromatography.

10.6.5   Chiral chromatography

Chiral compounds contain either at least one asymmetric carbon atom or they are molecularly asymmetric. They exist in two enantiomorphic forms, related as object and mirror images, have the same physical and chemical properties and differ only in their interaction with plane-polarised light such that one is dextrorotatory ($+$) and the other laevorotatory ($-$). There are a number of conventions for indicating the spatial configuration, as opposed to optical properties, of enantiomers. The classical D and L system for monosaccharides and amino acids cannot be applied easily to other structures and the Cahn–Ingold–Prelog systems, which assigns R (*rectus*) or S (*sinister*) configurations to an enantiomer, is of more general use. Until recently it has not been possible to resolve mixtures of enantiomers and this has created problems for the pharmaceutical industry in its development and clinical use of drugs, many of which are chiral, for although enantiomers have identical chemical and physical properties they are distinguishable biologically. Thus they differ in their ability to interact with the receptors involved in a range of physiological responses and they are often metabolised and excreted at different rates.

Chromatographic techniques have now been developed that allow mixtures of enantiomers to be resolved. One of these techniques is based on the fact that diastereoisomers, which are optical isomers that do not have an object–image relationship, do differ in physical properties even though they contain identical functional groups. They can therefore be separated by conventional chromatographic techniques, most commonly reversed-phase chromatography (Section 10.6.3). The diastereoisomer approach requires that the enantiomers contain a function group that can be derivatised by a chemically and optically pure chiral derivatising agent (CDA) to convert them to a mixture of diastereoisomers:

$$(R + S) \quad + \quad R' \quad \rightarrow \quad RR' + SR'$$

| Mixture of enantiomers | Chiral derivatising agent | Mixture of diastereoisomers |

Examples of CDAs include the R or S form of the following:

For amines	N-trifluoroacetyl-L-prolylchloride, α-phenylbutyric anhydride
For alcohols	2-Phenylpropionyl chloride, 1-phenylethylisothiocyanate
For ketones	2,2,2-Trifluoro-1-pentylethylhydrazine
For aliphatic and alicyclic acids	1-Menthol, Desoxyephedrine

Although this approach to chiral resolution is relatively simple, it is essential that the derivatisation process should be rapid and quantitative. Very often this is not the case and this has restricted its use. An alternative approach to the resolution problem is to use a chiral mobile phase. In this technique a transient diastereomeric complex is formed between the enantiomers and the chiral mobile phase agent. Examples of chiral mobile phase agents include albumin, $α_1$-acid glycoprotein, α, β- and γ-cyclodextrins, 10-camphorsulphonic acid and N-benzoxycarbonylglycyl-L-proline, all of which are used with a reversed-phase chromatographic system.

The most successful approach to chiral chromatography, however, has been the use of a chiral stationary phase. These are based upon the principle that the need for a three-point interaction between the stationary phase (working as a chiral discriminator) and the enantiomer would allow the resolution of racemic mixtures due to the different spatial arrangement of the functional groups at the chiral centre in the enantiomers. One such successful approach uses Pirkle phases, based on dinitrobenzoyl derivatives of amino acids such as phenylglycine that are bonded to silica. These phases are thought to function by allowing transient formation of enantiomer–stationary phase complexes by bonding such as hydrogen bonding and van der Waals' forces. Elution is generally by the reversed-phase technique. Alternative chiral stationary phases include triacetylcellulose and various cyclodextrins bonded to silica. These cyclodextrins are cyclic oligosaccharides that have an open truncated conical structure 6–8 Å (0.6–0.8 nm) wide at their base. Their inner surface is predominantly hydrophobic, but secondary hydroxyl groups are located around the wide rim of the cone. β-Cyclodextrin has 7 glucopyranose units and contains 35 chiral centres and α-cyclodextrin has 6 glucopyranose units, 30 chiral centres and is smaller than β-cyclodextrin. Collectively they are referred to as chiral cavity phases because they rely on the ability of the enantiomer to enter the three-dimensional cyclodextrin cage while at the same time presenting functional groups and hence the chiral centre for interaction with hydroxyl groups on the cone rim. Enantiomers possessing a five-, six- or seven-membered aromatic ring have been resolved by this approach in conjunction with reversed-phase elution.

Since proteins are optically active, they can in principle be used as a chiral stationary phase. Bovine serum albumin and $α_1$-acid glycoprotein (AGP) have

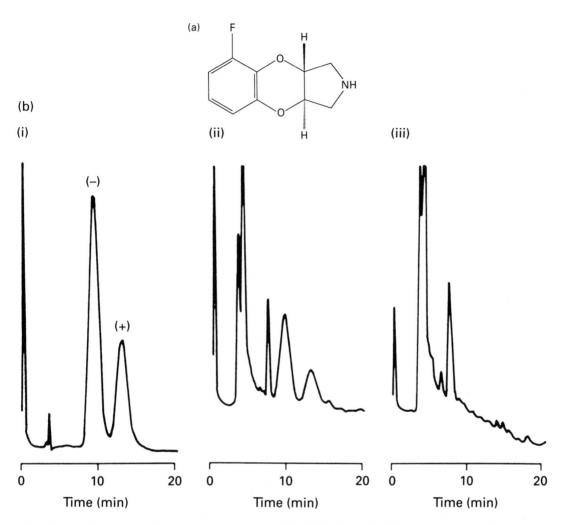

Fig. 10.10. Chiral separation of the enantiomers of GR50360 ((a), M_r 195) in pure solution (b(i)) and in human plasma (b(ii)). (b(iii)) A blank sample of plasma. Resolution was achieved on a bovine serum albumin (Resolvosil 7) column using 0.25 M phosphate buffer, pH 6.2, flow rate 0.8 cm^3 min^{-1} at 35 °C and ultraviolet detection (220 nm wavelength). (Reproduced with permission of Glaxo Group Research, Ware.)

been evaluated and found to be successful for a wide range of separations, but their mechanism of chiral separation is poorly understood. Both albumin and α_1-acid glycoprotein occur in plasma and have long been known to bind drugs (Fig. 10.10). Albumin has at least two distinct binding sites to which acidic and basic drugs may bind. α_1-Acid glycoprotein has a single drug-binding site restricted to the binding of basic drugs such as propranolol. These protein chiral phases are used in conjunction with aqueous buffers and cannot be used at extremes of pH or in the presence of organic solvents.

10.6.6 Countercurrent chromatography (CCC)

This separation process is based upon the distribution of a compound between two immiscible liquid phases. These phases may be mixtures of organic solvents, buffers, salts and various complexing agents. The technique is atypical of normal partition chromatographic techniques in that neither of the phases is supported by an inert support and the process is not conducted in a simple column. Nevertheless, the separation of compounds is based upon the different distribution coefficients between two immiscible phases and therefore the principle of the separation is the same as that of conventional liquid–liquid partition chromatography.

The apparatus most commonly used is the Craig Countercurrent Distribution apparatus. It consists of between 30 and 1000 interconnected H-shaped vessels (the so-called train), each of which retains a fixed volume of the stationary liquid phase. The solute mixture is introduced to the first vessel in the train and equilibrated with the immiscible and less dense mobile phase by the repeated rocking of the vessel through 90°. After equilibration is complete (1 to 2 min), the mobile phase is transferred to the next vessel as a result of the complete tipping of the first vessel. When this returns to its original position, fresh upper phase is introduced automatically. The whole process is repeated so that the mobile phase is transferred progressively along the series of lower phases. The solutes are transferred at a rate determined by their distribution coefficients and the relative volumes of the two solvents in each vessel. Each solute eventually accumulates in a specific group of vessels, the resolution being determined by the total number of transfers and their differences in partition coefficient. A number of miniaturised versions of the technique are now available commercially.

CCC is one of the few forms of chromatography that have been used successfully for cell organelle fractionation. It has also been used for cell fractionation and membrane receptor isolation.

10.7 Ion-exchange chromatography

10.7.1 Principle

This form of chromatography relies on the attraction between oppositely charged particles. Many biological materials, e.g. amino acids and proteins, have ionisable groups and the fact that they may carry a net positive or negative charge can be utilised in separating mixtures of such compounds. The net charge exhibited by these compounds is dependent on their pK_a and on the pH of the solution in accordance with the Henderson–Hasselbalch equation (Section 1.3.1).

Ion-exchange separations are carried out mainly in columns packed with an ion-exchanger. There are two types of ion-exchanger, namely cation and anion exchangers. Cation exchangers possess negatively charged groups and these will attract positively charged cations. These exchangers are also called acidic ion-exchange materials because their negative charges result from the ionisation of acidic groups. Anion exchangers have positively charged groups that will attract negatively charged anions. The term basic ion-exchange materials is also used to describe these exchangers, as positive charges generally result from the association of protons with basic groups.

The ion-exchange mechanism is thought to be composed of five distinct steps:

(i) Diffusion of the ion to the exchanger surface. This occurs very quickly in homogeneous solutions.

(ii) Diffusion of the ion through the matrix structure of the exchanger to the exchange site. This is dependent upon the degree of cross-linkage of the exchanger and the concentration of the solution. This process is thought to be the feature that controls the rate of the whole ion-exchange process.

(iii) Exchange of ions at the exchange site. This is thought to occur instantaneously and to be an equilibrium process.

Cation exchanger

$$RSO_3^- \ldots Na^+ + \overset{+}{N}H_3R' \rightleftharpoons RSO_3^- \ldots \overset{+}{N}H_3R' + Na^+$$

| Exchanger | Counter ion | Charged molecule to be exchanged | | Bound molecular ion | Exchanged ion |

Anion exchanger

$$(R)_4\overset{+}{N} \ldots Cl^- + {}^-OOCR' \rightleftharpoons (R)_4\overset{+}{N} \ldots {}^-OOCR' + Cl^-$$

The more highly charged the molecule to be exchanged, the tighter it binds to the exchanger and the less readily it is displaced by other ions;

(iv) Diffusion of the exchanged ion through the exchanger to the surface.

(v) Selective desorption by the eluent and diffusion of the molecule into the external eluent. The selective desorption of the bound ion is achieved by changes in pH and/or ionic concentration or by affinity elution, in which case an ion that has greater affinity for the exchanger than has the bound ion is introduced into the system.

10.7.2 Materials and applications

Low pressure ion-exchange chromatography can be carried out using a variety of matrices and ionic groups. Matrices used include polystyrene, cellulose and agarose. Functional ionic groups include sulphonate ($-SO_3^-$) and quaternary ammonium ($-\overset{+}{N}R_3$), both of which are strong exchangers because they are totally ionised at all normal working pH values, and carboxylate ($-COO^-$) and diethylammonium ($-HN(CH_2CH_3)_2$), both of which are termed weak exchangers because they are ionised over only a narrow range of pH values. Examples are given in Table 10.3. Bonded phase ion-exchangers suitable for HPLC, containing a wide range of ionic groups, are now available in pellicular and porous forms. The porous variety, which are based on polystyrene, porous silica or hydrophilic polyethers, are particularly valuable for the separation of proteins. They have a particle diameter of 5–25 μm. Most HPLC ion-exchangers are stable up to 60°C and separations are often carried out at this temperature owing to the fact that the raised temperature decreases the viscosity of the mobile phase and thereby increases the efficiency of the separation.

All exchangers are characterised by a total exchange capacity, which is defined as the number of milliequivalents of exchangeable ions available, either per gram of dried exchanger or per unit volume of hydrated resin. Sometimes available capacity is also used to express the available capacity for an arbitrarily chosen molecule such as haemoglobin. These exchange capacities give an indication of the degree of substitution of the exchanger and are therefore a helpful guide in deciding on the scale of a particular application.

The choice of the ion-exchanger depends upon the stability of the sample components, their relative molecular mass and the specific requirements of the separation. Many biological components, especially proteins, are stable within only a fairly narrow pH range so the exchanger selected must operate within this range. Generally, if the sample is most stable below its isoionic point, giving it a net positive charge, a cation exchanger should be used, whereas if it is most stable above its isoionic point, giving it a net negative charge, an anion exchanger should be used. Compounds that are stable over a wide range of pH may be separated by either type of exchanger. The choice between a strong and weak exchanger also depends on sample stability and the effect of pH on sample charge. Weak electrolytes requiring a very low or high pH for ionisation can be separated only on strong exchangers, as only they operate over a wide pH range. In contrast, for strong electrolytes, weak exchangers are advantageous for a number of reasons, including a reduced tendency to cause sample denaturation, their inability to bind weakly charged impurities and their enhanced elution characteristics. Although the degree of cross-linking of an exchanger does not influence the ion-exchange

Table 10.3. *Examples of commonly used ion-exchangers*

Type	Matrices	Functional groups	Functional group name
Weakly acidic (cation exchanger)	Agarose Cellulose Dextran Polyacrylate	$-COO^-$ $-CH_2COO^-$	Carboxy Carboxymethyl
Strongly acidic (cation exchanger)	Cellulose Dextran Polystyrene	$-SO_3^-$ $-CH_2SO_3^-$ $-CH_2CH_2CH_2SO_3^-$	Sulpho Sulphomethyl Sulphopropyl
Weakly basic (anion exchanger)	Agarose Cellulose Dextran Polystyrene	$-CH_2CH_2\overset{+}{N}H_3$ $-CH_2CH_2\overset{+}{N}H(CH_2CH_3)_2$	Aminoethyl Diethylaminoethyl
Strongly basic (anion exchanger)	Cellulose Dextran Polystyrene	$-CH_2\overset{+}{N}(CH_3)_3$ $-CH_2CH_2\overset{+}{N}(CH_2CH_3)_3$	Trimethylaminomethyl Triethylaminoethyl
		$-CH_2\overset{+}{N}(CH_3)_2$ $\quad\vert$ $\quad CH_2CH_2OH$	Dimethyl-2-hydroxyethylaminomethyl
		$-CH_2CH_2\overset{+}{N}(CH_2CH_3)_2$ $\quad\vert$ $\quad CH_2CHCH_3$ $\qquad\vert$ $\qquad OH$	Diethyl-2-hydroxypropylaminoethyl

mechanism, it does influence its capacity. The relative molecular mass and hence size of the sample component therefore determines which specific exchanger should be used.

The pH of the buffer used should be at least one pH unit above or below the isoionic point of the compounds being separated. In general, cationic buffers such as Tris, pyridine and alkylamines are used in conjunction with anion exchangers, and anionic buffers such as acetate, barbiturate and phosphate are used with cation exchangers. The precise initial buffer pH and ionic strength should be such as just to allow the binding of the sample components to the exchanger. Equally, a buffer of the lowest ionic strength that effects elution should initially be used for the subsequent elution of the components. This ensures that, initially, the minimum number of undesired substances bind to the exchanger and that, subsequently, the maximum number of these impurities remain on the column. The amount of sample that can be applied to a column is dependent upon the size of the column and the capacity of the exchanger. Generally, if the starting buffer is to be used throughout the development of the column (isocratic elution), the sample volume should be 1% to 5% of the bed volume. If, however, gradient elution is to be used, the initial conditions chosen are such that the entire sample is bound by the exchanger at the top of the column. In this case the sample volume is not important and large volumes of dilute solution can be applied, thereby effectively introducing a concentration stage.

Gradient elution is far more common than isocratic elution. Continuous or stepwise pH and ionic strength gradients may be employed but continuous gradients tend to give better resolution with less peak tailing (Section 10.1.3). Generally with an anion exchanger the pH gradient decreases and the ionic strength increases, whereas for cation exchangers both the pH and ionic gradients increase.

The separation of amino acids (e.g. in a protein hydrolysate) is usually achieved using a strong acid cation exchanger. The sample is introduced on to the column at a pH of 1–2, thus ensuring complete binding of all of the various types of amino acid (Section 4.1). Gradient elution using increasing pH and ionic concentration results in the sequential elution of amino acids. The acidic amino acids, aspartic and glutamic, are eluted first, followed by the neutral amino acids such as glycine and valine. The basic amino acids such as lysine and arginine retain their net positive charge up to pH values of 9–11 and are eluted last. These principles are embodied in automatic amino acid analysers. The effluent from the column is mixed with ninhydrin colour reagent (Fig. 4.2, p. 172) and nitrogen is also introduced to break the effluent stream into discrete segments and thus avoid axial mixing. The mixture is heated to 105°C to develop the colour, the intensity of which is then determined by two colorimeters, one set at 570 nm wavelength to monitor the majority of the amino acids and a second set at 440 nm wavelength to monitor specifically the colour produced by proline and hydroxyproline. Alternatively, the amino acids may be detected by conversion to derivatives that undergo fluorescence. Whilst this dispenses with the need for two detectors, the method is more tedious and generally less reproducible than the ninhydrin method. In all cases, the system is calibrated for quantitative work by the use of standards of each amino acid. The separation of amino acids in a protein hydrolysate is increasingly being achieved by reversed-phase chromatography (Section 4.3.2) rather than ion-exchange, because of its greater speed and convenience and also because of the fact that it dispenses with the need for a dedicated amino acid analyser.

10.7.3   Chromatofocusing

The technique of chromatofocusing, the principle of which is similar to that of isoelectric focusing (Section 9.3.4), is particularly suitable for protein separations. A linear pH gradient is generated in the column by exploiting the high buffer capacity of an ion-exchanger pre-equilibrated to a particular pH. Using an amphoteric buffer that has even buffering capacity over a range of pH, and with a starting pH lower than that at which the ion-exchange column has been pre-equilibrated, the pH gradient, which is 3–4 pH units higher at the top of the column than at the bottom, is formed by running the buffer through the column for a predetermined time. When the protein is added to this pH gradient, with a

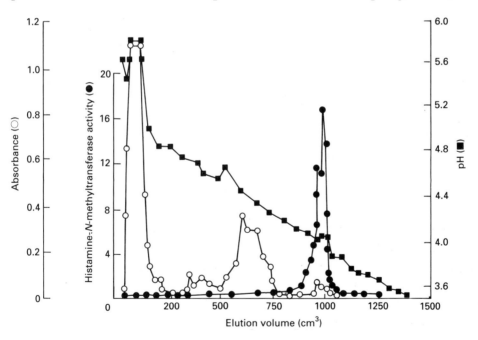

Fig. 10.11. Chromatofocusing elution profile of rat kidney histamine-N-methyl transferase. The partially purified sample (approximately 130 mg of protein) was chromatographed on a PBE 94 column (62 cm × 1.6 cm), previously equilibrated with 25 mM piperazine-HCl at pH 5.5. Five cm³ of the eluent, Polybuffer™74, pH 3.5, preceded the sample. Elution was carried out at a flow rate of 20 cm³ h^{-1}. The fractions (5.0 cm³) were assayed for pH (■), histamine-N-methyl transferase activity (●) expressed as extracted d.p.m. × 10^{-4} per 2.5 cm³ chloroform, and absorbance at 600 nm (○) measured after reaction with Coomassie Brilliant Blue. (Reproduced with permission of Glaxo Group Research, Ware.)

buffer whose pH is similar to that prevailing at the top of the column, it will migrate down the column, encountering a rising pH, until it reaches a pH corresponding to its isoelectric point. Just beyond this point it will be able to bind to the positive groups of the exchanger (in this example, an anion exchanger would be in use). As the elution continues with the starting buffer, the prevailing pH will be lowered, causing the binding of the protein to cease. The protein will continue its movement down the column until once again it encounters a pH slightly below its isoelectric point, when again it will bind. This process is repeated continuously until the protein is eluted at a pH slightly below its isoelectric point (Fig. 10.11). In a mixture of proteins, each molecule would elute in the order of its isoelectric point. More protein added to the top of the column during this elution process would automatically catch up with the initial protein, thereby producing a focusing effect and enabling large volumes to be applied to the column, with no deleterious effect. Thus the technique has a high capacity. Chromatofocusing gives a good resolution of quite complex mixtures of proteins, provided that there are discrete differences in their isoelectric point. Proteins possessing very similar isoelectric points tend to be poorly resolved.

10.8   Exclusion (permeation) chromatography

10.8.1   Principle

The separation of molecules on the basis of their molecular size and shape utilises the molecular sieve properties of a variety of porous materials. Probably the most commonly used of such materials is a group of polymeric organic compounds that possess a three-dimensional network of pores that confers gel properties upon them. The term gel filtration is used to describe the separation of molecules of varying molecular size utilising these gel materials. Porous glass granules have also been used as molecular sieves and the term controlled-pore glass chromatography introduced to describe this separation technique. The terms exclusion or permeation chromatography describe all molecular separation processes using molecular sieves. This section is devoted mainly to gel filtration, as its principles and applications are best documented, but it must be appreciated that controlled-pore glass chromatography has much in common with gel filtration.

The general principle of exclusion chromatography is quite simple. A column of gel particles or porous glass granules is in equilibrium with a suitable mobile phase for the molecules to be separated. Large molecules that are completely excluded from the pores will pass through the interstitial spaces and will appear in the effluent first. Smaller molecules will be distributed between the mobile phase inside and outside the molecular sieve and will then pass through the column at a slower rate, hence appearing last in the effluent. Three stages in such a column are represented diagrammatically in Fig. 10.12.

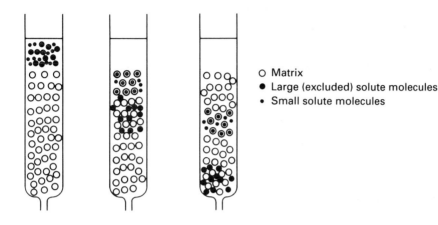

Fig. 10.12. Diagram of separation by exclusion chromatography.

The mobile phase absorbed by a gel is available to an analyte to an extent that is dependent upon the porosity of the gel particle and the size of the analyte molecule. Thus, the distribution of an analyte in a column of a gel is determined solely by the total volume of mobile phase, both inside and outside the gel particles, that is available to it. For a given type of gel, the distribution coefficient, K_d, of a particular analyte between the inner and outer mobile phase is a function of its molecular size. If the analyte is large and completely excluded from the mobile phase within the gel, $K_d = 0$, whereas, if the analyte is sufficiently small to gain complete access to the inner mobile phase, $K_d = 1$. Due to variation in pore size between individual gel particles, there is some inner mobile phase that will be available and some that will not be available to analytes of intermediate size; hence K_d values vary between 0 and 1. It is this complete variation of K_d between these two limits that makes possible the separation of analytes within a narrow molecular size range on a given gel.

For two substances of different relative molecular mass and K_d values, K_d' and K_d'' the difference in their elution volumes, V_S, can be derived from equation 10.2 and shown to be:

$$V_S = (K_d' - K_d'')V_i \tag{10.12}$$

where V_i is the inner volume within the gel available to a compound whose $K_d = 1$.

In practice, deviations from ideal behaviour, e.g. owing to poor packing of the column, make it advisable to reduce the sample volume below the value of V_S because the ratio between sample volume and inside gel volume affects both the sharpness of the separation and the degree of dilution of the sample.

10.8.2 Materials

Gels that are commonly used include cross-linked dextrans (e.g. Sephadex), agarose (Sepharose, Bio-Gel A, Sagavac), polyacrylamide (Bio-Gel P), poly-acryloylmorphine (Enzocryl Gel) and polystyrene (Bio-Beads S). Examples of materials for LPLC are shown in Table 10.4, and examples of materials suitable

Table 10.4. *Some gels commonly used for low pressure liquid exclusion chromatography*

Polymer	Trade name		Fractionation range[a] $(M_r \times 10^{-3})$
Dextran	Sephadex	G10	<0.7
		G25	1.0–5
		G50	1.5–30
		G100	4.0–150
		G200	5.0–600
	Sephacryl	S200	5.0–250
		S300	10.0–1500
		S400	20.0–8000
Agarose	Sepharose	2B	10.0–4000
		4B	60.0–20 000
		6B	70.0–40 000
	Bio-Gel	A5m	10.0–5000
		A15m	40.0–15 000
		A50m	100.0–50 000
		A150m	1000.0–150 000
Polyacrylamide	Bio-Gel	P2	0.1–1.8
		P6	1.0–6.0
		P30	2.5–40.0
		P100	5.0–100.0
		P300	60.0–400.0

[a]Determined for globular proteins. The range is approximately the same for single-stranded nucleic acids and smaller for fibrous proteins and double-strandard DNA.

for HPLC are shown in Table 10.2. The latter type are semi-rigid cross-linked polymers or rigid controlled-pore glasses or silicas.

Exclusion chromatography requires a single mobile phase and isocratic elution. It is most commonly used with ultraviolet absorption spectrophotometric detectors.

Exclusion chromatography columns tend to be longer than those for other forms of chromatography in order to increase the amount of stationary phase and hence pore volume.

10.8.3   Applications

Purification. The main application of exclusion chromatography is in the purification of biological macromolecules by facilitating their separation from larger and smaller molecules. Viruses, proteins, enzymes, hormones, antibodies, nucleic acids and polysaccharides have all been separated and purified by use of appropriate gels or glass granules.

Relative molecular mass determination. The elution volumes of globular proteins are determined largely by their relative molecular mass (M_r). It has been shown that, over a considerable range of relative molecular masses, the elution volume is an approximately linear function of the logarithm of M_r. Hence the construction of a calibration curve, with proteins of a similar shape and known M_r, enables the M_r values of other proteins, even in crude preparations, to be estimated.

Solution concentration. Solutions of high M_r substances can be concentrated by the addition of dry Sephadex G-25 (coarse). Water and low M_r substances are absorbed by the swelling gel, whereas the high M_r substances remain in solution. After 10 min the gel is removed by centrifugation, leaving the high M_r material in a solution whose concentration has increased but whose pH and ionic strength are unaltered.

Desalting. By use of a column of Sephadex G-25, solutions of high M_r compounds may be desalted. The high M_r substances move with the void volume, whereas the low M_r components are distributed between the mobile and stationary phases and hence move slowly. This method of desalting is faster and more efficient than dialysis. Applications include removal of phenol from nucleic acid preparations, ammonium sulphate from protein preparations and salt from samples eluted from ion-exchange chromatography columns.

Protein-binding studies. Exclusion chromatography is one of a number of methods commonly used to study the reversible binding of a ligand to a macromolecule such as a protein, including receptor proteins (Section 4.11). A sample of the protein/ligand mixture is applied to a column of a suitable gel that has previously been equilibrated with a solution of the ligand of the same concentration as that in the mixture. The sample is eluted with buffer in the standard way and the concentration of ligand and protein in the effluent determined. The early fractions will contain unbound ligand, but the subsequent appearance of the protein will result in an increase in the total mount of ligand (bound plus unbound). If the experiment is repeated at a series of ligand concentrations, the appropriate binding constants can be calculated (Section 4.11.1).

10.9   Affinity chromatography

10.9.1   Principle

Purification by affinity chromatography is unlike most other forms of chromatography and such techniques as electrophoresis and centrifugation in that it does not rely on differences in the physical properties of the molecules to be separated. Instead, it exploits the unique property of extremely specific biological interactions to achieve separation and purification. As a consequence, affinity chromatography is theoretically capable of giving absolute purification, even from complex mixtures, in a single process. The technique was originally developed for the purification of enzymes, but it has since been extended to nucleotides, nucleic acids, immunoglobulins, membrane receptors and even to whole cells and cell fragments.

The technique requires that the material to be isolated is capable of binding reversibly to a specific ligand that is attached to an insoluble matrix:

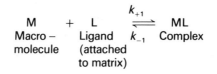

Under the correct experimental conditions, when a complex mixture containing the specific compound to be purified is added to the immobilised ligand, generally contained in a conventional chromatography column, only that compound will bind to the ligand. All other compounds can therefore be washed away and the compound subsequently recovered by displacement from the ligand (Fig. 10.13).

The method requires a detailed preliminary knowledge of the structure and biological specificity of the compound to be purified so that the separation conditions that are most likely to be successful may be carefully planned. In the case of an enzyme, the ligand may be the substrate, a reversible inhibitor or an allosteric activator. The conditions chosen would normally be those that are optimal for enzyme–ligand binding. Since the success of the method relies on the

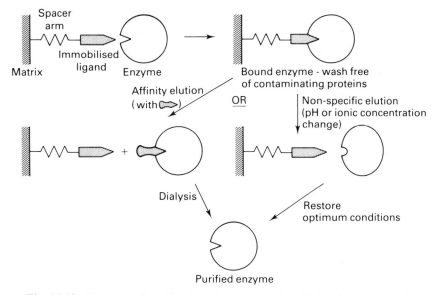

Fig. 10.13. Diagram of purification of an enzyme by affinity chromatography.

reversible formation of the complex and on the numerical values of the first-order rate constants k_{+1} and k_{-1}, as the enzyme is added progressively to the insolubilised ligand in a column, the enzyme molecules will be stimulated to bind and a dynamic situation develops in which the concentration of the complex and the strength of the binding increase. It is because of this progressive increase in effectiveness during the addition of the sample to the column that column procedures are invariably more successful than batch-type methods. Nevertheless, alternative forms have been developed and are particularly suitable for large-scale work. They include: (a) affinity precipitation, in which the ligand is attached to a soluble carrier that can be subsequently precipitated by, for example, a pH change; and (b) affinity partitioning in which the ligand is attached to a water-soluble polymer such as polyethylene glycol and which, with the ligand bound, preferentially partitions into an aqueous polymer phase that is in equilibrium with a pure aqueous phase. In all cases, for effective chromatography, the association constant, K_a, for the complex should be in the region of 10^4–10^8 M^{-1}.

10.9.2   Materials

Matrix

An ideal matrix for affinity chromatography must possess the following characteristics:

(i) It must contain suitable and sufficient chemical groups to which the ligand may be covalently coupled and it must be stable under the conditions of the attachment.

(ii) It must be stable during binding of the macromolecule and its subsequent elution.

(iii) It must at the most interact only weakly with other macromolecules to minimise non-specific adsorption.

(iv) It should exhibit good flow properties.

In practice, particles that are uniform, spherical and rigid are used. The most common ones are the cross-linked dextrans (Sephacryl S), agarose (Sepharose, Bio-Gel A), polyacrylamide gels (Bio-Gel P), polystyrene (Bio-Beads S), cellulose and porous glass and silica.

Selection and attachment of ligand

The chemical nature of the ligand is determined by the prior knowledge of the biological specificity of the compound to be purified. In practice it is frequently possible to select a ligand that displays absolute specificity in that it will bind exclusively to one particular compound. Alternatively, it is possible to select a ligand that displays group selectivity in that it will bind to a closely related group of compounds that possess a similar in-built chemical specificity. An example of the latter type of ligand is 5' AMP, which can bind reversibly to many NAD^+ dependent dehydrogenases because it is structurally similar to part of the NAD^+ molecule. It is essential that the ligand possesses a suitable chemical group that will not be involved in the reversible binding of the ligand to the macromolecule, but which can be used to attach the ligand to the matrix. The most common such groups are $-NH_2$, -COOH, -SH and -OH (phenolic and alcoholic). To prevent the attachment of the ligand to the matrix interfering with its ability to bind the macromolecule, it is generally advantageous to interpose a spacer arm between the ligand and the matrix. The optimum length of this spacer arm is six to ten carbon atoms or their equivalent. In some cases, the chemical nature of this spacer is critical to the success of the separation. Some spacers are purely hydrophobic, most commonly consisting of methylene groups, others are hydrophilic, possessing carbonyl or imido groups. Spacers are most important for small immobilised ligands but generally are not necessary for macromolecular ligands (e.g. in immunoaffinity chromatography, Section 10.9.5) as their binding site for the mobile macromolecule is well displaced from the matrix.

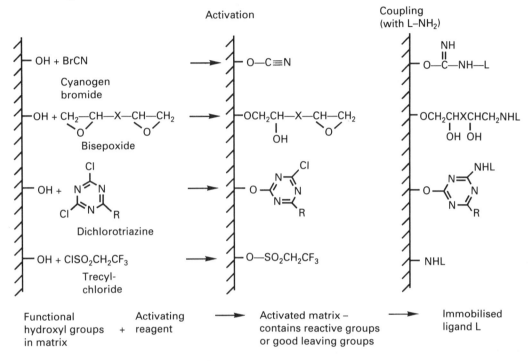

Fig. 10.14. Examples of coupling reactions used to immobilise ligands (L) for affinity chromatography. If a spacer arm is to be introduced between the immobilised ligand and the matrix, the coupling chemistry is similar.

The most common method of attachment of the ligand to the matrix involves the preliminary treatment of the matrix with cyanogen bromide (CNBr) (Fig. 10.14). The reaction conditions and the relative proportion of the reagents will determine the number of ligand molecules that can be attached to each matrix particle. Alternative coupling procedures involve the use of bis-epoxides, N, N'-carbonyldiimidazole (CDI), sulphonyl chloride, sodium periodate and dichloro-triazines. Many pre-activated matrices, prepared using these coupling reagents, are available commercially.

A number of different spacer arms are used. Examples include 1,6-diaminohexane, 6-aminohexanoic acid and 1,4-bis-(2,3-epoxypropoxy)butane. They must possess a second functional group to which the ligand may be attached by conventional organosynthetic procedures, which frequently involve the use of succinic anhydride and a water-soluble carbodiimide. A number of supports of the agarose, dextran

Table 10.5. *Examples of group-specific ligands commonly used in affinity chromatography*

Ligand	Affinity
Nucleotides	
5' AMP	NAD^+–dependent dehydrogenases, some kinases
2',5' ADP	$NADP^+$–dependent dehydrogenases
Calmodulin	Calmodulin-binding enzymes
Avidin	Biotin–containing enzymes
Fatty acids	Fatty-acid-binding proteins
Heparin	Lipoproteins, lipases, coagulation factors, DNA polymerases, steroid receptor proteins
Proteins A and G	Immunoglobulins
Concanavalin A	Glycoproteins containing α–D-mannopyranosyl and α–D-gluco-pyranosyl residues
Soybean lectin	Glycoproteins containing *N*-acetyl-α-(or ß)-D-galactopyranosyl residues
Phenylboronate	Glycoproteins
Poly(A)	RNA containing poly(U) sequences, some RNA-specific proteins
Lysine	rRNA
Cibacron Blue F3G-A	Nucleotide-requiring enzymes, coagulation factors

and polyacrylamide type are commercially available, with a variety of spacer arms and ligands pre-attached ready for immediate use. Examples of ligands are given in Table 10.5.

Practical procedure

The procedure for affinity chromatography is similar to that used in other forms of liquid chromatography. The ligand-treated matrix is packed into a column in the normal way for the particular type of support. The buffer used must contain any cofactors, such as metal ions, necessary for ligand–macromolecule interaction. Once the sample has been applied and the macromolecule bound, the column is eluted with more buffer to remove non-specifically bound contaminants.

The purified compound is recovered from the ligand by either specific or non-specific elution. Non-specific elution may be achieved by a change in either pH or ionic strength. pH shift elution using dilute acetic acid or ammonium hydroxide results from a change in the state of ionisation of groups, in the ligand and/or the macromolecule, that are critical to ligand–macromolecule binding. A change in ionic strength, not necessarily with a concomitant change in pH, also causes elution due to a disruption of the ligand–macromolecule interaction; 1 M NaCl is frequently used for this purpose. If elution is achieved by a pH change, the pH of the collected fractions must be readjusted to the optimum value to minimise the opportunity for protein denaturation. Affinity elution involves the addition of a high concentration of substrate, or reversible inhibitor of the macromolecule if it is an enzyme, or the addition of ligands for which the macromolecule has a higher affinity than it has for the immobilised ligand. The purified material is eventually recovered in a buffered solution that may be contaminated with specific eluting agents or high concentrations of salt and these must be removed by such techniques as exclusion chromatography before the isolation is complete.

10.9.3   Applications

A wide range of enzymes and other proteins, including receptor proteins and immunoglobulins, has been purified by affinity chromatography. The application of the technique is limited only by the availability of immobilised ligands. The principles have been extended to nucleic acids and have made a considerable contribution to developments in molecular biology. Messenger RNA, for example, is routinely isolated by selective hybridisation on poly(U)-Sepharose 4B by exploiting its poly(A) tail. Immobilised single-stranded DNA can be used to isolate complementary RNA and DNA. Whilst this separation can be achieved on columns, it is usually performed using single-stranded DNA immobilised on nitrocellulose filters. Immobilised nucleotides are useful for the isolation of proteins involved in nucleic acid metabolism.

A valuable development of affinity chromatography is its use for the separation of a mixture of cells into homogeneous populations. The technique relies on the antigenic properties of the cell surface or the chemical nature of exposed carbohydrate residues on the cell surface or on a specific membrane receptor–ligand interaction. The immobilised ligands used include Protein A, which binds to the Fc region of IgG, a lectin or the specific ligand for a membrane receptor.

10.9.4   Lectin affinity chromatography

The lectins are a group of proteins produced by animals, plants and slime moulds that have the ability to bind carbohydrate and hence glycoproteins. They have a polymeric structure, most being tetrameric. Their subunits may be either identical, in which case they recognise a single specific saccharide, or of two types in which case they recognise two different saccharides. They all have an M_r in the range 40 000–400 000. Their ability to recognise and bind specific saccharides has made them highly valuable in the purification of glycoproteins, particularly membrane receptor proteins.

The most widely used lectins for lectin chromatography are those from leguminous plants (pea, castor bean, soybean) due to their abundance. They can be immobilised to agarose matrices by conventional techniques and many are now available commercially. If the nature of the saccharide component of a glycoprotein is not known, the lectin of choice is selected by a simple screening procedure. Once the glycoproteins have been bound to the immobilised lectin, elution can be achieved in a number of ways, including affinity elution using the simple monosaccharide for which the lectin has an affinity; use of a borate buffer, which forms a complex with glycoprotein; by the careful change of pH (not below pH 3 or above pH 10) or by the addition of a reagent such as ethylene glycol to reduce ligand hydrophobic interaction. One of the attractions of lectin affinity chromatography is that it can be carried out in the presence of relatively high salt concentrations because it does not rely on ionic interactions. In principle, therefore, it can be applied directly after salt fractionation. It has also been used to separate mixtures of cells by taking advantage of the saccharide components of their outer membranes. Most lectin affinity chromatography has been carried out using conventional LPLC.

10.9.5   Immunoaffinity chromatography

The use of antibodies as the immobilised ligand has been exploited in the isolation and purification of a range of proteins including membrane proteins of viral origin. Monoclonal antibodies may be linked to agarose matrices by the cyanogen bromide technique. Protein binding to the immobilised antibody is achieved in neutral buffer solution containing moderate salt concentrations. Elution of the bound protein quite often requires forceful conditions because of the very tight binding with the antibody ($K_d = 10^{-8}$–10^{-12}M) and this may lead to protein denaturation. Examples of elution procedures include the use of high salt concentrations with or without the use of a detergent and the use of urea, sodium dodecylsulphate or guanidine hydrochloride, all of which cause denaturation. The use of chaotropic agents such as thiocyanate, perchlorate and trifluoracetate or lowering the pH to about 3 may avoid denaturation.

10.9.6   Metal chelate chromatography (immobilised metal affinity chromatography)

This is a special form of affinity chromatography in which an immobilised metal ion such as Cu^{2+}, Zn^{2+}, Hg^{2+} or Cd^{2+} or a transition metal ion such as Co^{2+}, Ni^{2+}, or Mn^{2+} is used to bind proteins selectively by reaction with imidazole groups of histidine residues, thiol groups in cysteine residues and indole groups in tryptophan residues. The immobilisation of the protein involves the formation of a coordinate bond that must be sufficiently stable to allow protein attachment and retention during the elution of non-binding contaminating material. The subsequent release of the protein can be achieved either by simply lowering the pH, therefore destabilising the protein–metal complex, or by the use of complexing agents such as EDTA. Most commonly the metal atom is immobilised by attachment to an iminodiacetate- or tris(carboxymethyl)-ethylenediamine-substituted agarose. Proteins purified by this technique include fibrinogen, superoxide dismutase and the non-histone nuclear proteins.

10.9.7   Dye–ligand chromatography

A number of triazine dyes that contain both conjugated rings and ionic groups, fortuitously have the ability to bind to some proteins. The term pseudo-ligands has therefore been used to describe the dyes. It is not possible to predict whether a particular protein will bind to a given dye as the interaction is not specific but is thought to involve interaction with ligand-binding domains via both ionic and hydrophobic forces. Dye binding to proteins enhances their binding to materials such as Sepharose 4B and this is exploited in the purification process. The attraction of the technique is that the dyes are cheap, readily couple to conventional matrices and are very stable. The most widely used dye is Cibacron Blue F3G-A. Dye selection for a particular protein purification is empirical and is made on a trial-and-error basis. Attachment of the protein to the immobilised dye is generally achieved at pH 7–8.5. Elution is most commonly brought about either by a salt gradient or by affinity elution.

10.9.8   Covalent chromatography

This form of chromatography has been developed specifically to separate thiol (-SH) containing proteins by exploiting their interaction with an immobilised ligand containing a disulphide group. The principle is illustrated in Fig. 10.15. The most commonly used ligand is a disulphide 2′-pyridyl group attached to an agarose matrix such as Sepharose 4B. On reaction with the thiol-containing protein, pyridine-2-thione is released. This process can be monitored spectrophotometrically at 343 nm wavelength, thereby allowing the adsorption of the protein to be followed. Once the protein has been attached covalently to the matrix, non-thiol containing contaminants are eluted and unreacted thiopyridyl groups removed by use of 4 mM dithiothreitol or mercaptoethanol. The protein is then released by displacement with a thiol-containing compound such as 20–50 mM dithiothreitol, reduced glutathione or cysteine. The matrix is finally regenerated by reaction with 2,2′-dipyridyldisulphide. The method has been used successfully for many proteins

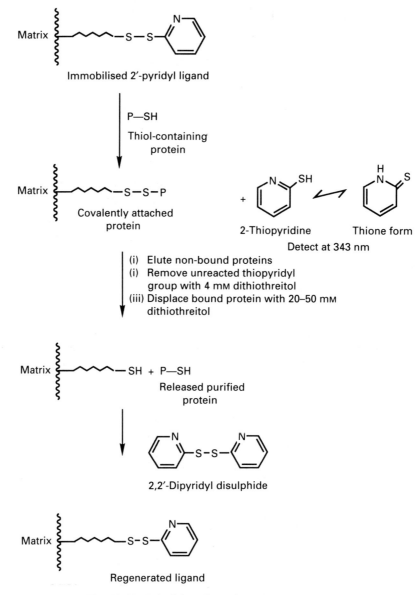

Fig. 10.15. Principles of covalent chromatography.

but its use is limited by its cost and the rather difficult regeneration stage. It can, however, be applied to very impure protein preparations.

10.10   Gas–liquid chromatography (GLC)

10.10.1   Apparatus and materials

This technique, which is based upon the partitioning of compounds between a liquid and a gas phase, is a widely used method for the qualitative and quantitative analysis of a large number of compounds because it has high sensitivity, reproducibility and speed of resolution. It has proved to be most valuable for the separation of compounds of relatively low polarity. A stationary phase of a high boiling point liquid material such as a silicone grease is supported on an inert granular solid. This material is packed into a narrow coiled glass or steel column 1–3 m long and 2–4 mm internal diameter, through which an inert carrier gas (the mobile phase) such as nitrogen, helium or argon is passed. The column is maintained in an oven at an elevated temperature, which ensures that the compounds to be separated are kept in the vapour state and that analysis times are reasonable (Section 10.10.3). The basis for the separation is the difference in the partition coefficients of the volatilised compounds between the liquid and gas phases as the compounds are carried through the column by the carrier gas. As the compounds leave the column they pass through a detector that is linked via an amplifier to a chart recorder, which, in turn, records a peak as each analyte passes through the detector (Fig. 10.16).

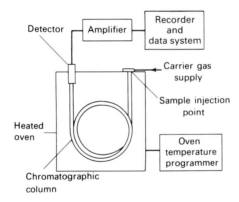

Fig. 10.16. Diagram of a GLC system.

GLC may also be performed using capillary columns, which are made of glass or metal with diameters of between 0.03 and 1.0 mm and which may be up to 100 m in length. There are two types of capillary column system known as wall-coated open tubular (WCOT) columns and support-coated open tubular (SCOT), also known as porous layer open tubular (PLOT) columns, for adsorption work. In WCOT columns the stationary phase is coated directly on to the walls of the capillary tubing. As there is only a small amount of stationary phase present, only very small amounts of sample may be chromatographed. Consequently a splitter system has to be used at the sample injection port so that only a small fraction of the injected sample reaches the column. The remainder of the sample is vented to waste. The design of the splitter is critical in quantitative analyses in order to ensure that the ratio of sample chromatographed to sample vented is always the same. Some instruments are equipped with on-column injectors that require considerable skill in their operation.

In SCOT columns a support material is bonded to the walls of the capillary column, and the stationary phase is coated on to the support. The capacity of SCOT columns is considerably higher than that of WCOT columns and consequently small samples can be injected directly on to such columns without the need for a splitter system. SCOT systems are therefore considerably simpler to use for quantitative analyses than are WCOT systems. Their efficiency is less than that of WCOT systems but considerably better than that of conventional GLC columns. Generally speaking, PLOT columns have little biochemical use.

The efficiency of a GLC column is determined by the principles outlined in Section 10.1.3. There is an optimum carrier gas flow rate for maximum column efficiency (minimum HETP). For capillary columns, the maximum number of theoretical plates that can be obtained is independent of the carrier gas used. In these cases, a decrease in column diameter should give a proportional increase in the number of plates per unit length, i.e. HETP. As the length of these columns is very much greater than that of conventional columns, very high efficiencies are obtained (equation 10.9) and these systems are very useful for the analysis of complex mixtures.

Matrix

Since this is used to provide a supporting surface on which is coated the film of stationary phase, it is important that the support should be inert to the sample. This is generally no problem when the support is holding a high percentage coating of stationary phase, but when the percentage coating is low, exposure of the support to the sample often hinders separation. The most commonly used support is Celite (diatomaceous silica), which because of the problem of support–sample interaction is often treated so that the hydroxyl groups that occur in the Celite are modified. This is normally achieved by silanisation of the support with such compounds as hexamethyldisilazane. In addition to the support, the glass column, the glass wool plug located at the base of the column and any other surface that may come into contact with the sample are also silanised. The support particles have an even size, which, for the majority of practical applications, is 60–80, 80–100 or 100–120 mesh (Section 10.3.3).

Stationary phase

The requirements for any GLC stationary phase are that it must be involatile and thermally stable at the temperature used for analysis. Often the phases used are high boiling point organic compounds, and these are coated on to the support to give from 1% to 25% loading, depending upon the analysis. Such phases are of two types, either selective, where separation occurs by utilisation of different chemical characteristics of components, or non-selective, where separation is achieved on the basis of differences in boiling points of the sample components. The operating temperature for the analysis must be compatible with the phase chosen for use. Too high a temperature results in excessive column bleed owing to the phase being volatilised off, contaminating the detector and giving an unstable recorder baseline. The choice of phase for analysis depends on the compound under investigation and is best chosen after reference to the literature. Commonly used stationary phases include the polyethylene glycols, methylphenyl- and methyl-vinylsilicone gums (so-called OV phases), Apiezon L and esters of adipic, succinic and phthalic acids.

The columns are dry-packed under a slight positive gaseous pressure and after packing must be conditioned for 24–48 h by heating to near the upper working temperature limit, whilst the carrier gas at normal flow rates is passed through the column. During this conditioning, the column should not be connected to the detector, to prevent contamination. With good-quality liquid phases, column conditioning can be simplified to flushing with carrier gas at 100 °C.

10.10.2   Preparation and application of sample

The majority of non- and low-polar compounds are directly amenable to GLC, but other compounds possessing such polar groups as -OH, -NH$_2$, -COOH are generally retained on the column for excessive periods of time if they are applied directly. This excessive retention is inevitably accompanied by poor resolution and peak tailing (Section 10.1.3). This problem can be overcome by derivatisation of these polar groups. This increases the volatility and effective distribution coefficients of the compounds. Methylation, silanisation and perfluoracylation are common derivatisation methods for fatty acids, carbohydrates and amino acids.

The sample for chromatography is dissolved in a suitable solvent such as acetone, heptane or methanol. Chlorinated organic solvents are generally avoided as they contaminate the detector. The sample is injected on to the column using a microsyringe through a septum in the injection port that is attached to the top of the column. Normally 0.1–10 mm^3 of solution is injected. It is common practice to maintain the injection region of the column at a slightly higher temperature than the column itself as this helps to ensure rapid and complete volatilisation of the sample. Sample injection is automated in many commercial instruments.

10.10.3 Separation conditions

Nitrogen, helium and argon are the three most commonly used carrier gases. They are passed through the column at a flow rate of 40–80 cm^3 min^{-1}. The column temperature must be within the working range of the particular stationary phase and is chosen to give a balance between peak retention time and resolution. In GLC, partition coefficients are particularly sensitive to temperature so that analysis times may be regulated by adjustment of the column oven, which can be operated in either of two modes. In isothermal analysis a constant temperature is employed. In the separation of compounds of widely differing polarity or M_r it may be advantageous to increase the temperature gradually. This is referred to as temperature programming. This, however, often results in excessive bleed of the stationary phase as the temperature is raised, giving rise to baseline variation. Consequently some instruments have two identical columns and detectors, one set of which is used as a reference. The currents from the two detectors are opposed; hence, assuming equal bleed from both columns, the resulting current gives a steady baseline as the column temperature is raised.

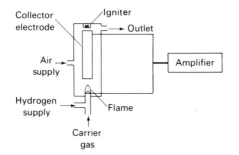

Fig. 10.17. Diagram of a flame ionisation detector.

10.10.4 Detectors

By far the most widely used detector is the flame ionisation detector (FID). It responds to almost all organic compounds, can detect as little as 1 ng and has a wide linear response range (10^6). A mixture of hydrogen and air is introduced into the detector to give a flame, the jet of which forms one electrode, whilst the other electrode is brass or platinum wire mounted near the tip of the flame (Fig. 10.17). When the sample components emerge from the column they are ionised in the flame, resulting in an increased signal being passed to the recorder:

The carrier gas passing through the column and the detector gives a small background signal, which can be offset electronically to give a baseline. A FID has a minimum detection quantity of the order of $5 \times 10^{-12}\,\mathrm{g\,s^{-1}}$ and an upper temperature of 400 °C.

The nitrogen–phosphorus detector (NPD), which is also called a thermionic detector, is similar in design to a FID but has a crystal of a sodium salt fused on to the electrode system, or a burner tip embedded in a ceramic tube containing a sodium salt, or a rubidium chloride tip. The NPD has excellent selectivity towards nitrogen- and phosphorus-containing compounds and shows a poor response to compounds possessing neither of these two elements. Its linearity (10^4), upper temperature limit (300 °C) and detection limits ($10^{-11}\,\mathrm{g\,s^{-1}}$) are not quite as good as a FID. It is widely used in organophosphorus pesticide residue analysis.

The electron capture detector (ECD) responds only to substances that capture electrons, particularly halogen-containing compounds. This detector is widely used in the analysis of polychlorinated compounds, such as the pesticides DDT, dieldrin and aldrin. It has very high sensitivity ($10^{-12}\,\mathrm{g\,s^{-1}}$) and an upper temperature limit of 300°C but its linear range (10^2 to 10^4) is much lower than that of the FID. The detector works by means of a radioactive source (^{63}Ni) ionising the column gas (e.g. $N_2 \rightarrow N_2^+ + e^-$), the electrons so produced giving a current across the electrodes to which a suitable voltage is applied. When an electron-capturing compound (generally one containing a halogen atom) emerges from the column, the ionised electrons are captured, the current drops and this change in current is recorded. The carrier gas most commonly used in conjunction with an ECD is nitrogen or an argon + 5% methane mixture.

The volatile solvent used to introduce the test sample gives rise to a solvent peak at the beginning of the chromatograph. The three main forms of detector respond to this solvent with varying sensitivity, thereby affecting the detection and resolution of rapidly eluting analytes. In cases where authentic samples of the test compounds are not available for calibration purposes or in cases where the identity of the analytes is not known, the detector may be replaced by a mass spectrometer. Special separators are available for removing the bulk of the carrier gas from the sample emerging from the column and prior to its introduction in the mass spectrometer (Section 8.7). More recently, GLC has been linked to other types of detector, including an infrared spectrophotometer, and to a nuclear magnetic resonance spectrometer, the resulting spectra aiding in the identification of unknown compounds.

10.10.5   Applications

Until the development of HPLC, GLC was probably the most commonly used form of chromatography. Its use nowadays is confined to volatile, non-polar compounds that do not need derivatisation. Analytes are characterised by their retention time or preferably by their retention time relative to a standard reference compound. In the analysis of compounds that form a homologous series, e.g. the methyl esters of the saturated fatty acids, there is a linear relationship between the logarithm of the retention time and the number of carbon atoms. There are similar but parallel lines for mono- and di-unsaturated fatty acids. This can be exploited, for example, to identify an unknown fatty acid ester in a fat hydrolysate. A widely used system for quantitative analysis is the retention index (RI), which is based on the retention of a compound relative to n-alkanes. The compound is chromatographed with a number of n-alkanes and a semi-logarithmic plot constructed of retention time against number of carbon atoms. Each n-alkane is assigned an RI of 100 times the number of carbon atoms it contain (pentane therefore has an RI of 500), allowing the RI for the compound to be calculated. Many commercially available GLC systems with data-processing facilities have the capacity to calculate RI values automatically.

10.11   Thin-layer chromatography (TLC)

10.11.1   Principle

A thin layer of the stationary phase is formed on a suitable flat surface such as a glass, foil or plastic plate. Since the layer is so thin, the movement of the mobile phase across the layer, generally by simple capillary action, is rapid, there being little resistance to flow. As the mobile phase moves across the layer from one edge to the opposite, it transfers any analytes placed on the layer at a rate determined by their distribution coefficients, K_d, between the stationary and mobile phases. In practice, the principle of the distribution process may be based on that of adsorption, partition, ion-exchange or exclusion chromatography. Analyte movement ceases either when the mobile phase (solvent front) reaches the end of the layer and capillary action flow ceases or when the plate is removed from the mobile phase reservoir.

The movement of the analyte is expressed by it retardation factor, R_F such that:

$$R_F = \frac{\text{distance moved by analyte from origin}}{\text{distance moved by solvent front from origin}} \qquad (10.13)$$

The efficiency of a thin-layer plate is expressed by its number of theoretical plates, N, and plate height, H (Section 10.1.3) such that:

$$N = 16 \left(\frac{d_A}{w} \right)^2 \tag{10.14}$$

where d_A is the distance moved by the analyte from origin and w is the width of the spot (Fig. 10.17), and:

$$H = \frac{d_A}{N} \tag{10.15}$$

The capacity factor, k', for the analyte, is given by:

$$k' = \frac{d_m}{d_A} = \frac{1 - R_F}{R_F} \tag{10.16}$$

where d_m is the distance moved by the solvent front from the origin.

10.11.2   Thin-layer preparation

A slurry of the stationary phase, generally in water, is applied to a glass, plastic or foil plate, generally 20 cm square, as a uniform thin layer by means of a plate spreader starting at one end of the plate and moving progressively to the other. The thickness of the slurry layer used is dictated by the nature of the desired chromatographic separation. For analytical separations the layer is of the order of 0.25 mm thick and for preparative separations it may be up to 2 mm. Where the stationary phase is to be used for adsorption chromatography, a binding agent such as calcium sulphate is incorporated into the slurry in order to facilitate the adhesion of the adsorbent to the plate. With the exception of thin-layer exclusion chromatography (Section 10.8.1), once the slurry layer has been prepared, the plates are dried to leave the coating of stationary phase. In the case of adsorbents, drying is carried out in an oven at 100–120 °C. This also serves to activate the adsorbent. A range of preprepared plates is available commercially. So-called polyamide layer sheets, which consist of poly-ε-caprolactam coated onto *both* sides of a solvent-resistant polyester sheet, are unusual in that they are semi-transparent, allowing unknowns and standards run on opposite sides of the plate to be compared. They can also be reused if cleaned immediately with ammonia-acetone, and are widely used in protein sequencing studies by the dansyl-Edman method (Section 4.3.3).

10.11.3   Sample application

The sample is applied to the plate 2.0–2.5 cm from the edge by means of a micropipette or microsyringe. It is possible for this process to be automated. The solvent may be removed from the spot by gentle heating or by use of an air blower, care being taken in the case of volatile or thermolabile compounds. It is then possible to apply more sample to the spot if necessary. In the case of adsorption chromatography, diffusion of the sample from the applied spot may be minimised by using a solvent in which components have a low R_F value. For preparative thin-layer chromatography, the sample is applied as a band across the plate rather than as a single spot.

10.11.4   Plate development

Separation most commonly takes place in a glass tank that contains the developing solvent (mobile phase) to a depth of about 1.5 cm. This is allowed to stand for at least 1 h with a lid over the top of the tank to ensure that the atmosphere within the tank becomes saturated with solvent vapour (equilibration). Unless this is done, irregular running of the solvent will occur as it ascends the plate by capillary action, resulting in poor separations being achieved. After equilibration, the lid is removed, and the thin-layer plate is then placed vertically in the tank so that it stands in the solvent. The lid is replaced and separation of the compounds then occurs as the solvent travels up the plate. It is also possible to develop the plate in a horizontal plane by connecting the sample end of it to a reservoir of mobile phase by means of a suitable wick. It is preferable to keep the system at a constant temperature whilst the development is occurring, to avoid anomalous solvent-running effects. One of the biggest advantages of TLC is the speed at which separation is achieved. This is commonly about 30 min and is hardly ever greater than 90 min.

   In order to improve the resolution of partition and adsorption separations, the technique of two-dimensional chromatography may be used. The material to be chromatographed is placed towards one corner of the plate as a single spot and the plate developed in one direction and then removed from the tank and allowed to dry. It is then developed by another solvent system, in which the compounds to be separated have different K_d values, in a direction at right angles to the first development.

10.11.5   Analyte detection

Several detection methods are available. Examination of the plate under ultraviolet light will show the position of ultraviolet-absorbing or fluorescent compounds. Many commercially available thin-layer adsorbents contain a fluorescent dye so that, when the plate is examined under ultraviolet light, the separated compounds show up as blue, green or black areas against a fluorescent background. Subjecting the plate to iodine vapour is useful if unsaturated compounds are being investigated. Spraying of plates with specific colour reagents will stain certain compounds, e.g. ninhydrin will locate amino acids and peptides. If the compounds are radiolabelled, the plates may be subjected to autoradiography (Section 5.2.3), which will detect the spots as dark areas on X-ray film, or the plate may be scanned by a radiochromatograph scanner (Section 5.2.3). A general, non-specific technique is to spray the plate with 50% (v/v) sulphuric acid or 25% (v/v) sulphuric acid in ethanol and heating at 110 °C, which will result in most compounds becoming charred and showing up as brown spots. Great care has to be taken when this latter method is performed.

Although the movement of compounds on TLC may be characterised by specific R_F values, these measurements are not always reproducible. Component identification is therefore most commonly made on the basis of a comparison of the movement of the components with those of reference compounds chromatographed alongside the sample on the TLC plate.

The amount of compound present in a given spot may be determined in a number of ways. On-plate quantification may be achieved by use of radio-chromatograph scanning in the case of radiolabelled compounds or more generally by means of densitometry (Fig. 10.18). Precision densitometers are commercially available that measure the ultraviolet or visible absorption of the compound as well as simultaneously giving a complete absorption spectrum of the compound for identification purposes. Off-plate quantification may be carried out by scraping off the spot and the immediate surrounding stationary phase from the plate and eluting the compound with a suitable solvent. The amount of compound in solution can then be determined by standard methods, most commonly colorimetry or fluorimetry.

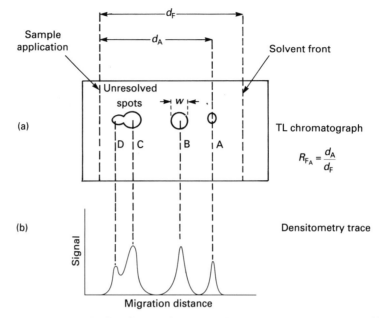

Fig. 10.18. TLC chromatograph of a mixture of compounds A–D (a) and the corresponding densitometer trace (b) from which quantitative data can be calculated.

10.12   Paper chromatography

10.12.1   Principle

The principle of paper chromatography is very similar to that of thin-layer chromatography. The cellulose fibres of chromatography paper act as the supporting matrix for the stationary phase. The stationary phase may be water, a non-polar material such as liquid paraffin or impregnated particles of a solid adsorbent. Chromatography papers are available that have different running characteristics, e.g. slow, medium and fast. Others have been washed with acid to remove traces of impurities that may affect certain analyses. Paper suitable for adsorption and normal phase partition paper chromatography is commercially available. Paper for reversed-phase chromatography must be prepared immediately before use.

10.12.2   Experimental procedure

After the sample has been applied to the sheet by techniques that are similar to those for thin-layer chromatography (Section 10.11.3) there are two methods which may be employed for the development of paper chromatographs – ascending or descending. In both cases the solvent is placed in the base of a sealed tank or glass jar to allow the chamber to become saturated with the solvent vapour. In the ascending method, the procedure is identical with that described above for thin-layer chromatography (Section 10.11.4). The sample spots should be in a position just above the surface of the solvent so that, as the solvent moves vertically up the paper by capillary action, separation of the sample is achieved. In the descending technique, the end of the paper near which the sample spots are located is held in a trough at the top of the tank and the rest of the paper allowed to hang vertically but not in contact with the solvent in the base of the tank. When the analysis is to be started, solvent is added to the trough. Separation of the sample then occurs as the solvent moves downwards under gravity and capillary action. Although ascending chromatography is often preferred because of the simplicity of the set-up, the flow of solvent is faster in the descending technique. Two-dimensional chromatography may be used for paper systems in a manner similar to that described for TLC.

The methods used for component detection are similar to those described for TLC, but spraying with sulphuric acid is not recommended, as this causes the paper to disintegrate. The identification of a given compound may be made on the basis of its R_F value. Paper chromatography has few current serious biochemical applications.

10.13   Selection of a chromatographic system

It is possible to rationalise to some extent the type of system most likely to be applicable to the separation of compounds for which the physical characteristics are known (Fig. 10.19).

The majority of chromatographic procedures exploit differences in physical properties of compounds, the exception being affinity chromatography, which is based upon the specific ligand-binding properties of biological macromolecules. If this form of chromatography can be applied it is the most likely to be successful. Volatile compounds are best separated by GLC, whereas non-volatile compounds that are soluble in organic solvents are generally best separated either by adsorption or normal phase liquid chromatography. If the compounds have different functional groups, adsorption chromatography on silica with non-polar solvent is probably the better method. To separate low polarity compounds in a homologous series, normal phase liquid systems are preferred. If water soluble compounds are non-ionic or weakly ionic, reversed-phase liquid chromatography is preferable where a non-polar stationary phase such as a hydrocarbon is used together with a polar

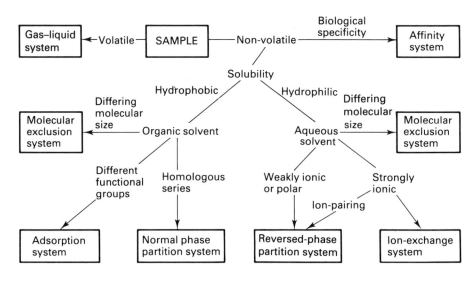

Fig. 10.19. Rationale for the choice of a chromatographic system.

mobile phase such as water/acetonitrile or water/methanol mixtures. Water-soluble compounds that are strongly ionic are best chromatographed by an ion-exchange system, using either an anionic or cationic resin, together with a suitable buffer system for elution. Ionic compounds can, however, be chromatographed by reversed-phase partition systems by the technique of ion-pairing. Compounds differing in molecular size are best separated by exclusion chromatography.

Whatever form of liquid chromatography is chosen for a particular biochemical study, the decision to use LPLC or HPLC depends on many factors including the availability of apparatus, cost, the scale of the separation, and whether the separation is to be qualitative or quantitative. The modern trend is to select HPLC, which is certainly capable of giving fast, accurate and precise data. Reversed-phase HPLC, in particular, is proving to be an extremely versatile technique. The application of HPLC techniques to protein separations, via FPLC, is also proving to be a quick, robust technique, particularly in cases where protein denaturation is not a problem. The simplicity of TLC, especially for qualitative work, with its facility for concurrent investigation of many samples including standards, remains a considerable attraction. Equally, the recent developments in capillary gas chromatography make it a fast and sensitive system for volatile compounds.

10.14   Suggestions for further reading

CLEMENT, R.E. (Ed.) (1990) *Gas Chromatography – Biochemical, Biomedical and Clinical Applications.* Wiley Interscience, Toronto. (An excellent, comprehensive coverage of principles, apparatus and applications.)

HANCOCK, W.S. (Ed.) (1990) *HPLC in Biotechnology.* Wiley Interscience, New York. (Good review of the applications of HPLC to biotechnology problems.)

HEARN, M.T.W. (Ed.) (1991) *HPLC of Proteins, Peptides and Polynucleotides.* VCH Publishers Inc., New York. (Specialist applications of HPLC to these important biomolecules.)

HUNT, B.J. and HOLDING, S.R. (Eds.) (1989) *Size Exclusion Chromatography.* Blackie, New York. (Detailed review of technique that also includes chapter on supercritical fluid chromatography.)

LOUGH, W.J. (Ed.) (1990) *Chiral Liquid Chromatography.* Blackie, New York. (Excellent review of this increasingly important technique.)

TREIBER, L.R. (Ed.) (1987) *Quantitative TLC and its Industrial Applications.* Marcel Dekker, New York. (Comprehensive account of all aspects of the principles and applications of TLC.)

11

Electrochemical techniques

11.1 Introduction

11.1.1 Biological interest in electrochemistry

Frequently biologists are interested in the electrical properties and behaviour of biological substances or with the transformation of chemical energy into electrical energy (and vice versa), i.e. the electrochemistry of living systems (Section 11.1.2). In addition, biologists may often utilise electrochemical techniques to measure the concentrations of a great number and diversity of biologically important substances such as oxygen, catecholamines and glucose, and to determine important parameters of biological environments such as pH (Section 11.1.3).

Electrochemistry is, in fact, one of the oldest specialities of classical physical chemistry, tracing its origins to the mid-nineteenth century. However, electrochemical techniques have developed over the years through advances in a number of scientific disciplines including chemistry, physics, electronics and most recently biology (in the development of biosensors). Consequently the terminology surrounding the subject may, at times, be confusing, or clearly inconsistent. In addition, the sheer number of techniques can be bewildering, and whereas fundamental differences might exist between some groups of techniques (which need to be appreciated before such techniques can be sensibly applied to biological situations) numerous techniques may differ from others only in minor detail. The names of the techniques themselves have been applied historically with little or no semblance of a systematic approach, and only recently has there been any effort

530

to bring some sort of coherence to the nomenclature. This has been done, not by renaming techniques, but by grouping them together in logical sets. This approach has been used in Section 11.2.3. However, this may still leave the student of the subject at a loss to understand why, for example, the Clark oxygen electrode (Section 11.6) may be called an amperometric device by some authors, a voltammetric device by others and a polarographic device by yet others (including Clark himself).

11.1.2 Electrochemistry and energy transduction

The energy contained in many molecules can be released by living organisms through oxidation to produce free energy (i.e. that available to do work). The rates of such oxidation reactions are controlled by enzymes in order to supply the energetic demands of the organism. The relationship between the oxidised and reduced forms of a molecule may be expressed by the general equation:

$$\text{Reduced form} \rightleftharpoons \text{Oxidised form} + ne^-$$

Oxidation in this sense thus means the loss of electrons from a substance, and this loss can occur in three main ways:

(i) *Direct loss.* Electrons may be lost directly and passed on to a second electron acceptor molecule, e.g.:

$$\text{Cytochrome}(Fe^{2+}) \rightarrow \text{Cytochrome}(Fe^{3+}) + e^- \text{ (to acceptor)}$$

This may make up part of an electron transport chain, where the electrons are shuttled down the chain, passing from one carrier to the next. Alternatively single reactions of this type may occur, catalysed by oxidases, where oxygen is not incorporated into the molecule but is used as the electron acceptor to form hydrogen peroxide or occasionally water.

(ii) *Removal of hydrogen.* Electrons are lost during dehydrogenation. Most biological oxidations occur in this fashion rather than by the addition of oxygen. In nearly all cases two electrons plus two protons $(2e^- + 2H^+)$ are removed, e.g.:

$$CH_3CH_2OH + NAD^+ \underset{\text{Alcohol dehydrogenase}}{\rightleftharpoons} CH_3CHO + NADH + H^+$$
$$\text{Ethanol} \qquad\qquad\qquad\qquad \text{Acetaldehyde}$$

though occasionally loss may occur as a hydride ion and a proton $(H^- + H^+)$. Dehydrogenases have a coenzyme requirement (e.g. NAD^+

or $NADP^+$) or have a flavoprotein (FAD or FMN) prosthetic group to serve as the electron and hydrogen acceptor.

(iii) *Addition of oxygen.* In terms of movement of electrons, it is more difficult to understand why addition of oxygen is in fact an oxidation. Closer inspection of the mechanism of such reactions reveals that the first stage of such reactions involves the loss of electrons to molecular oxygen, though this subsequently shares its electrons with the donor molecule when the covalent bond forms. Alternatively, it may be argued that the addition of a highly electronegative oxygen atom to a molecule results in electrons being pulled away from the rest of the atoms in the molecule, i.e. in a partial form of oxidation. Oxidations of this type are catalysed by oxygenases. Dioxygenases catalyse the incorporation of both atoms of molecular oxygen into a structure, whereas mono-oxygenases (often called hydroxylases) introduce only one atom. The oxidation of natural compounds by oxygenases is rather uncommon.

Within living cells, oxidation reactions in the glycolytic pathway and in the citric acid cycle are catalysed by dehydrogenases. Such reactions yield NADH and $FADH_2$ which then pass their electrons on to the respiratory electron transport chain found on the inner membrane of the mitochondria. This chain is represented in Fig. 11.1. Electron transport from NADH to oxygen results in the phosphorylation of three molecules of ADP to ATP per atom of oxygen consumed, whereas electron transport from $FADH_2$ bypasses the first phosphorylation site and thus produces only two molecules of ATP per atom of oxygen. The number of molecules of ATP produced from a particular substrate can be found by comparing the rate of phosphorylation with the rate of electron transport, i.e. with the rate of oxygen uptake. Phosphorylation can be measured by the disappearance of either ADP or inorganic phosphorus, whilst oxygen consumption can be measured using a Clark oxygen electrode (Section 11.6) and this can be used to calculate the phosphorylation:oxidation (P:O) ratio (Section 11.6.3).

An explanation for the mechanism of oxidative phosphorylation was proposed by P. Mitchell in 1961, and is now accepted. Mitchell's chemiosmotic theory proposes that an electrochemical gradient is set up across the inner mitochondrial membrane during the electron transport process. The FMN and ubiquinone function as hydrogen carriers, able to carry both H^+ and e^-, whereas the Fe-sulphur proteins and the cytochromes are able to carry only electrons. These carriers are arranged such that hydrogen and electron carriers alternate. A third hydrogen carrier, X, was also included to account for the third H^+ translocation site (Fig. 11.2). However, in 1975 Mitchell refined his theory and accepted that X really did not exist, but suggested that its function was actually carried out by a

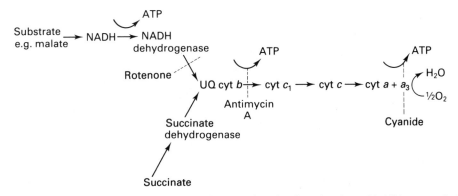

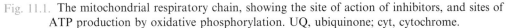

Fig. 11.1. The mitochondrial respiratory chain, showing the site of action of inhibitors, and sites of ATP production by oxidative phosphorylation. UQ, ubiquinone; cyt, cytochrome.

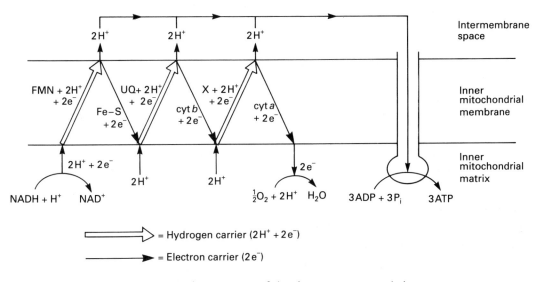

Fig. 11.2. Arrangement of the electron transport chain.

complex cyclic series of exchanges involving ubiquinone and cytochrome *b*, now commonly known as the Q-cycle.

The functioning of the electron transport chain thus ensures that, as electrons pass down the electron transport chain, protons are translocated from the inner matrix of the mitochondrion into the intermembrane space between the inner and outer mitochondrial membranes. A proton gradient is thus set up across the inner membrane and, as protons flow back across the inner membrane through a membrane-bound ATPase, ADP is phosphorylated to ATP.

Most mitochondria in tissues and carefully prepared samples of isolated mitochondria show respiratory control, in that in the absence of ADP the rate of electron transport slows down so NADH is not wasted. If ADP is added, then ATP synthesis can occur and the rate of electron transport increases (Section 11.6.3, Fig. 11.9). Such mitochondria would be said to be tightly coupled. In uncoupled mitochondria, electrons pass down the electron transport chain whether or not there is ADP to phosphorylate. Naturally uncoupled mitochondria are found in brown adipose tissue where they serve an important function of generating heat. Coupled mitochondria can be deliberately uncoupled by adding uncoupling agents such as 2,4-dinitrophenol. This renders the membrane permeable, such that the proton gradient is abolished, leaving no available power source to drive the ATPase. These phenomena can be demonstrated using a Clark oxygen electrode (Section 11.6).

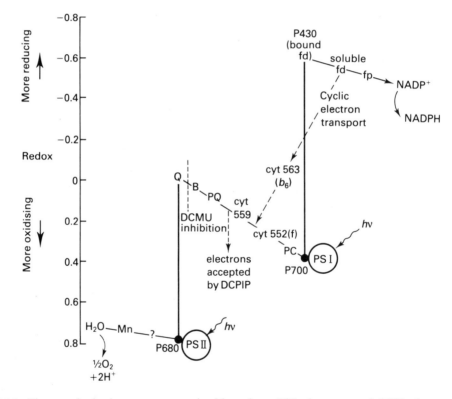

Fig. 11.3. Photosynthetic electron transport in chloroplasts. PSI, photosystem I; PSII, photosystem II; PQ, plastoquinone; cyt, cytochrome; Q,B, postulated intermediates; P680, P700, reaction centre chlorophylls; fd, ferredoxin; fp, flavoprotein (fd NADP oxidoreductase); ?, unknown component.

A second electron transport chain is found within the chloroplasts of higher plants. This photosynthetic electron transport chain (Fig. 11.3) is responsible for the trapping of energy from sunlight to form NADPH, which is then used in the

Calvin cycle for the fixation of atmospheric CO_2 to form carbohydrate. Within the photosynthetic electron transport chain, light energy is absorbed at two distinct sites such that an electron is passed from one carrier to another with a more negative redox potential (see Section 11.3 for a description of this term). This is an endergonic process requiring the expenditure of energy. Electrons travel from water to $NADP^+$ via photosystem II and photosystem I, producing oxygen and NADPH; this is known as non-cyclic electron transport and produces ATP by non-cyclic photophosphorylation. Electrons can also undergo cyclic electron transport around photosystem I, releasing no oxygen from water and producing no NADPH, but producing ATP by cyclic photophosphorylation.

Many aspects of photosynthetic electron transport and photophosphorylation are similar to mitochondrial electron transport and oxidative phosphorylation. As in mitochondria, when electrons or hydrogen atoms travel along the series of carriers a proton gradient becomes established. However, protons move inwards in response to photosynthetic electron transport, not outwards as in mitochondrial electron transport. Thus the interior of the chloroplast is considerably more acidic than the outside. The chloroplast ATPase is structurally very similar in its general shape and subunit structure to that found in the mitochondria, but is orientated differently because of the reversal of the direction of the proton gradient, such that the ATPase is on the outer rather than the inner face of the membrane. Photosynthetic uncouplers such as NH_4^+ are known to have effects similar to those of respiratory uncouplers. They speed up the rate of photosynthetic electron flow but abolish phosphorylation because they carry protons across the membrane. Inhibitors of photophosphorylation are also known that, like inhibitors of oxidative phosphorylation, combine with the ATPase at a specific site; one example is the antibiotic Dio-9.

11.1.3   The range of electrochemical techniques

Whereas at one time the only piece of electrochemical apparatus a biology or biochemistry undergraduate might ever actually use would be the pH electrode (Section 11.4), nowadays this is certainly not the case. Various ions other than H^+ may be measured using special electrodes termed ion-selective electrodes (Section 11.5). Most undergraduate students will also probably come into contact with the Clark oxygen electrode (Section 11.6), which has numerous biological applications, particularly in the study of chloroplast and mitochondrial activity. Biosensors (Section 11.8) are also becoming more common in clinical, industrial and environmental applications. Electrochemical detection of the effluent from HPLC columns (Section 10.4.7 and 11.7) is now a real alternative to the more classical ultraviolet detection techniques, and is the method of choice for the analysis of some important drugs (e.g. morphine) and biomolecules (e.g. neurotransmitters, such as adrenaline and noradrenaline).

Thus, since there are now such a great number of electroanalytical techniques available to the biologist, an understanding of their basic principles, and of the similarities and differences between various groups, is of paramount importance.

11.2   Principles of electrochemical techniques

11.2.1   Electrochemical cells and reactions

An electrochemical cell consists of two electrode–eloctrolyte systems termed half-cells, joined internally either by means of a common electrolyte or by means of a relatively concentrated solution of indifferent ions (often saturated or 4 M KCl or KNO_3) called a salt bridge.

It is convenient to consider electrochemical cells by the reactions that take place at only one of the two electrode surfaces (i.e. within one half-cell). Consider, for example, a copper wire dipping into a solution of $CuSO_4$. Within this system the metal electrode may throw off its ions into solution, leaving the electrons behind on the metal:

$$Cu_{solid} \rightarrow + \; Cu^{2+}_{(aqueous)} + \; 2e^-$$

The metal thus loses electrons – i.e. it is oxidised. This leaves the metal electrode negatively charged with respect to the surrounding solution. Alternatively, the metal ions in solution may combine with electrons at the metal surface (i.e. be reduced) to deposit metal atoms:

$$Cu^{2+}_{(aqueous)} + \; 2e^- \rightarrow Cu_{solid}$$

This leads to a deficit of electrons in the metal electrode leaving it positively charged with respect to the surrounding solution. Thus, a potential difference called the electrode potential is established across the electrode/electrolyte interface.

Many electrode reactions are of this type. There is, however, another type of half-cell in which both the oxidised and reduced forms of the substance exist in solution. For example, if a platinum wire is immersed in a solution of ferrous and ferric ions the equilibrium would be:

$$Fe^{3+} + \; e^- \rightleftharpoons Fe^{2+}$$

Although there is no fundamental difference between this half-cell and the $Cu/CuSO_4$ half-cell described above, an electrode in which the oxidised and reduced forms both exist in solution is called a redox electrode and its potential is known as a redox potential. Further consideration of such reactions is included in Section 11.3.

11.2.2 Schematic representation of electrochemical cells

To simplify the description of electrochemical cells, a standardised shorthand has evolved. Consider a simple electrochemical cell as shown in Fig. 11.4. This cell (the Daniell cell) consists of a strip of zinc and a strip of copper immersed in

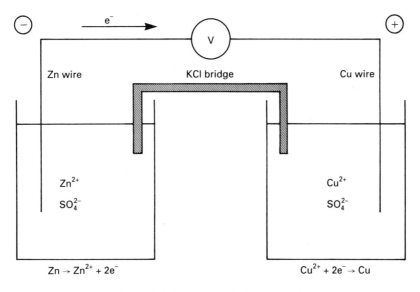

Fig. 11.4. Diagram of a Daniell cell.

solutions of $ZnSO_4$ and $CuSO_4$, respectively. If the Zn^{2+} and Cu^{2+} concentrations are approximately equal, then reactions will result in the oxidation of the zinc metal in one half-cell and the reduction of the copper ions in the other half-cell. If the two strips are then connected by a wire the solution in the first half-cell will show an increase in zinc ion concentration, whilst the solution in the second half-cell will show a depletion in cupric ion concentration. A salt bridge between the two compartments allows migration of ions between the two compartments but prevents gross mixing of the two solutions. Electrons will thus flow from the zinc anode to the copper cathode. Such an electrochemical cell can be described as follows:

$$- Zn \,|\, Zn^{2+} \,\|\, Cu^{2+} \,|\, Cu +$$

where the single vertical bars represent a phase boundary (electrode/electrolyte interface) and the double bars represent the KCl bridge. Such shorthand notation is in common usage.

11.2.3 A classification of electrochemical techniques

An electrochemical cell can be either a galvanic cell or an electrolytic cell (Fig. 11.5). A galvanic cell is one in which reactions occur spontaneously at the electrodes when they are connected externally by a conductor (such as the Daniell cell described in Section 11.2.2). Well-known examples of such cells include both non-rechargeable (e.g. Zn-MnO_2 cell) and rechargeable batteries. In contrast, an electrolytic cell is one in which reactions are effected by the imposition of an external voltage greater than the spontaneous potential of the cell. Well-known applications of electrolytic cells include the electrolytic synthesis of chlorine and aluminium, electrorefining of copper, and electroplating of silver and gold. (In

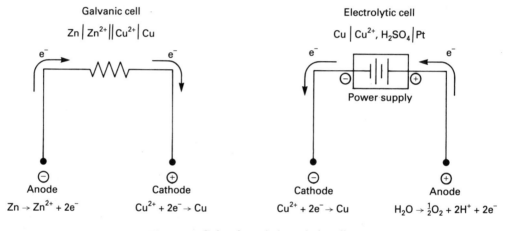

Fig. 11.5. Galvanic and electrolytic cells.

addition a rechargeable battery when it is being recharged is an electrolytic cell.)

It is unfortunately rather confusing that the anode in a galvanic cell is negatively charged, whereas that in the electrolytic cell is positively charged (Fig. 11.5). However, the reactions at the electrode surfaces are the same in both cases. In both types of cell, oxidation events occur at the anode and reduction events occur at the cathode.

Electroanalytical techniques can be subdivided into two main groups, according to whether the electrochemical cell is galvanic or electrolytic:

(i) **_Potentiometry_** involves the use of galvanic cells, with the measurement of the potential of an electrode without current flow.

(ii) ***Voltammetry*** involves the use of electrolytic cells, with the measurement of the current passing through an electrode at constant potential, or as the potential varies under potentiostatic control.

Potentiometry is discussed in Section 11.2.4. The commonest application of potentiometry is the pH electrode (Section 11.4), and in ion-selective and gas-sensing electrodes (Section 11.5). Electrolytic cells are discussed in Section 11.2.5. These are involved in all other forms of measurement (amperometric, polarographic and voltammetric). The commonest of these instruments likely to be encountered by biologists is the amperometric oxygen electrode (Section 11.6), though polaro-graphic and voltammetric applications such as those discussed in Section 11.2.5 are becoming more popular for the measurement of low concentrations of molecules, particularly in environmental and medical situations. One very useful application of voltammetry is in the field of HPLC (Section 10.4), where electrochemical detectors frequently offer better precision and higher selectivity than the more conventional spectrophotometric detectors (Section 11.7).

11.2.4 Introduction to potentiometry

Reference electrodes

Potentiometric measurements that involve an electrode which responds to a particular experimental situation by giving a spontaneous potential require a second, so-called, reference electrode of constant potential to be present so that the difference between the two can be measured. A reference electrode is necessary when pH electrodes, ion-selective electrodes and redox electrodes are used. His-torically, one of the most important reference electrodes has been the standard hydrogen electrode (SHE), which contains an inert metal electrode (e.g. platinum coated with platinum black) in a solution of a fixed concentration of hydrochloric acid (which supplies the H^+). Hydrogen gas at $10^5\,Pa$ (1450 lbf in.$^{-2}$) pressure is bubbled over the electrode, enabling the following equilibrium to be established:

$$\tfrac{1}{2}H_2 \rightleftharpoons H^+ + e^-$$

However, this electrode is highly inconvenient to use because the hydrogen has to be supplied at a constant pressure and must be oxygen free, and the platinum black is also readily contaminated. In practice, therefore, although oxidation reduction potentials (Section 11.3) are expressed relative to the SHE, other reference electrodes are used. Calomel electrodes (Fig. 11.6a) are commonly used as a reference. They consist of a solution of potassium chloride in contact with solid mercurous chloride (calomel) and mercury. This part of the circuit may be written as:

$$Hg \,|\, Hg_2Cl_2 \,|\, KCl \,\|\, \text{test solution}$$

The double lines indicating the presence of a salt bridge.

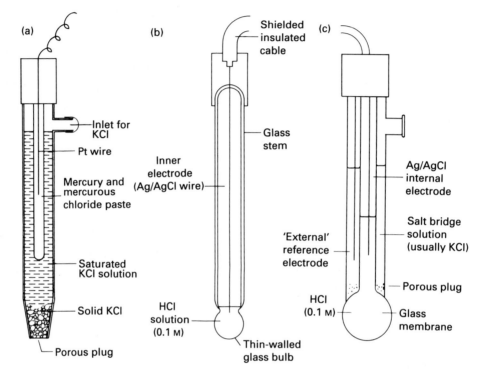

Fig. 11.6. Electrodes: (a) a calomel reference electrode; (b) glass electrode; (c) a combination electrode.

An alternative to the calomel electrode is the silver/silver chloride electrode. A deposit of silver chloride is present on metallic silver in a chloride solution such as KCl.

Any reference electrode must be in contact with the test solution via a liquid junction. These generally involve potassium chloride that slowly diffuses out of the electrode, giving electrical continuity. Unfortunately the liquid junction is likely to give an unknown junction potential, which cannot be eliminated completely. In using liquid junctions, care must be taken to ensure that it is the potassium chloride solution that diffuses out slowly, and not the test solution that diffuses in. Although outward diffusion does involve some contamination of the sample, this is not normally important, but, if either potassium or chloride is being measured, then a specially designed reference electrode called the double junction reference electrode must be used to prevent contamination.

There are several types of junction through which KCl diffuses: ceramic or fritted material, fibrous junctions and sleeve junctions. Fritted material is a collection of small particles pressed closely together, allowing some of the filling solution to leak through the gaps between the particles. Fibrous junctions can consist of woven fibres or of straight fibres, the latter giving an increased flow. The sleeve-type reference electrode has a narrow ring-shaped junction formed by the gap between an outer sleeve and the inner body of the electrode. The space between the sleeve and the electrode widens above the tip and forms a reservoir for the fluid. Flow occurs in some areas of the narrow ring junction but not in others. A sleeve junction is easier to clean than the other types (because the sleeve can be removed); it is also faster flowing, which means it is less likely to get clogged.

The Nernst equation

This equation is relevant to equipment that produces a potential, e.g. pH glass electrodes and ion-selective electrodes. Sections 11.3 and 11.4 discuss the Nernst equation more fully; however, essentially it describes electrode behaviour and can be expressed in the simplified form:

$$E = E_{constant} + 2.303 \frac{RT}{nF} \log_{10} A \tag{11.1}$$

or even better as:

$$E = E_{constant} + \frac{S}{n} \log_{10} C \tag{11.2}$$

where E is the total potential (V) developed between the sensing and reference electrodes, $E_{constant}$ is a constant potential that depends mainly on the reference electrode, $2.303RT/F\ (= S)$ is the Nernst factor or slope (R being the molar gas constant and T the absolute temperature), n is the number of charges on the ion, A is the activity of the ion, and C is the concentration of the ion.

Activity is an important physicochemical concept. It is the true measure of the ion's ability to affect chemical equilibria and reaction rates, and is its effective concentration in solution. In most biological situations the concentrations of ions are rather low and consequently the activity and concentration are equal. At higher concentrations, however, the activity becomes less than concentration.

The Nernst equation shows clearly that the electrode response depends on both temperature and the number of charges on the ion. At 25 °C the Nernst factor

2.303RT/F becomes 0.059, thus equation 11.2 becomes:

$$E = E_{\text{constant}} + \frac{0.059}{n} \log_{10} C \qquad\qquad (11.3)$$

There is thus a 59 mV change in potential for a 10-fold change in the activity of a monovalent ion, and a 29.5 mV change if the ion is divalent. An electrode is said to have Nernstian characteristics if it obeys Nernst's law. If the changes in potential relative to activity (slopes) are less than the theoretical values (in the operating range) this indicates either interference from other ions or an electrode malfunction.

11.2.5   Introduction to voltammetry

Voltammetric techniques involve reactions effected by an imposed external voltage greater than the spontaneous potential of the cell. The reactions occurring within such an electrolytic half-cell can be considered as reactions occurring at the surface of an electrode in which the electrode supplies electrons (reduction) or removes electrons (oxidation) in the same way as a chemical reagent may do so. Many biological substances are thus electro-active in that they can be relatively easily oxidised or reduced by applying a suitable potential to generate the necessary energy to drive the electrochemical reaction. Take, for example, a half-cell containing a solution of an electroreducible species; then, as the potential of the electrode is made progressively more negative, a point will be reached when reduction of the species will begin at the electrode and a small current will flow (E_1 in Fig. 11.7). As the potential is made even more negative, the current will increase dramatically as more reduction of the species occurs, until a potential is reached where the species is reduced at the electrode surface as quickly as it is able to reach the electrode surface by diffusion. At this point the maximum or *limiting current* is achieved, and further increases in potential will not affect the current (E_2 in Fig. 11.7).

The resulting plot (Fig. 11.7) of current against voltage (I/E) known as the polarographic wave has two important properties:

(i) the midpoint of the wave (called the half-wave potential, $E_{\frac{1}{2}}$) is characteristic for each species being reduced;

(ii) since the magnitude of the limiting current is controlled by the rate of diffusion of the reducible species, it will be directly proportional to the concentration of this species.

These features enable voltammetric techniques to be used both to identify and to quantify electro-active materials in solution. Unfortunately, voltammetric

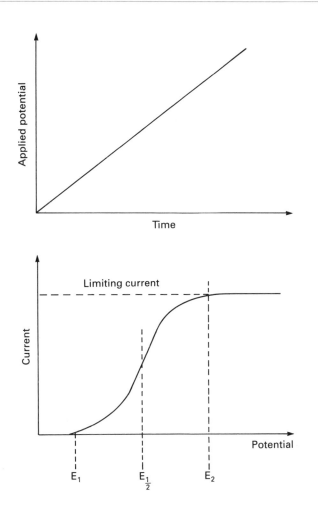

Fig. 11.7. Resultant polarographic wave as applied potential is increased linearly with time.

terminology remains rather confusing, and, although some authors do not use the term amperometry at all, a fairly standard classification includes:

(i) amperometry – measurement of the current passing through an electrode at constant potential;

(ii) voltammetry/polarography – measurement of the current while the potential varies under potentiostatic control.

Amperometry involves the reduction or oxidation of electro-active species, usually at the limiting current, such that the current is proportional to the concentration of the species of interest. The commonest application of amperometry is the Clark oxygen electrode (Section 11.6), where the reduction of oxygen at a platinum cathode gives rise to a current that is proportional to the oxygen tension in the solution, provided that an applied voltage maintains the platinum cathode at -0.6 V versus a silver anode. Amperometry may also be encountered by biologists using electrochemical detectors for the detection of solutes in the effluent from chromatography columns, especially HPLC (Section 11.7).

Polarography and voltammetry are used almost synonymously by some authors. Most modern texts, however, restrict the use of the term polarography to those techniques which use mercury, in the form of the dropping mercury electrode (DME), as the working electrode. This electrode consists of small droplets of mercury generated at the lower end of a glass capillary tube. The lifetime of each drop of mercury is usually arranged to be 0.5–5 s, the mass of each drop being around 5 mg. The DME is thus being continuously and reproducibly renewed; its past history is unimportant, giving rise to very reproducible responses. Polarography is used almost exclusively for the analysis of reducible species, usually metallic cations.

In contrast to polarography, voltammetric measurements involve no renewal of the working electrode surface. Instead a stationary hanging mercury drop electrode (HMDE) or a solid electrode made of platinum, gold, silver or carbon is employed. The advantage of this type of electrode over the DME is that a preconcentration step may be included in the measurement. The analyte is then usually stripped from the electrode surface during the potential scan; therefore the technique is often referred to as stripping voltammetry. Using these voltammetric techniques, oxidation reactions may be studied as readily as reduction reactions. Anodic stripping voltammetry involves a negative or oxidation current, whereas cathodic stripping voltammetry involves a positive or reduction current passing through the circuitry. These techniques are greatly superior to polarography in terms of sensitivity; however, the solid electrodes are generally more prone to contamination than is the mercury of the DME and, because they are not renewed, demonstrate poorer surface reproducibility.

Rather than apply a continuously increasing voltage, as in DC voltammetry, pulse voltammetry, differential pulse voltammetry, and AC voltammetry all produce more complex voltage changes, generally resulting in increased sensitivity over DC polarographic techniques. However, these are essentially advanced analytical techniques.

11.3   Redox reactions

11.3.1   Principles

Compounds capable of existing in an oxidised and a reduced form can be represented by the general equation:

A mixture of the reduced and oxidised form of a substance (e.g. Fe^{2+}/Fe^{3+} or $NADH/NAD^+$) is known as a redox couple. If an inert electrode (such as platinum) is put into a solution of a redox couple, the metal will become charged, and the potential difference set up between the metal and the solution can be compared with the steady potential produced by a reference electrode.

The scale of oxidation–reduction (redox) potentials is based on values obtained with the standard hydrogen electrode used as the reference electrode, and compared with the standard potential (E_o), which is produced by a platinum electrode dipping into 1 M concentrations of both the oxidised and reduced forms of a substance at 25 °C. Redox couples have positive or negative redox potentials according to whether they are more oxidising (positive) or reducing (negative) than the standard hydrogen electrode. The experimental potential measured (E) depends on the ratio of oxidised to reduced forms (i.e. their *relative* concentrations) and frequently on pH, but does not depend to a great extent on the actual concentrations involved. The measured redox potential E is related to the standard potential E_o by the Nernst equation:

$$E = E_o + 2.303 \frac{RT}{nF} \log \frac{[\text{oxid}]}{[\text{red}]}$$

where E is the measured redox potential of a couple of known composition (e.g. a mixture of 0.03 M oxidised form and 0.1 M reduced form), E_o is the standard redox potential with components of 1 M concentration at 25 °C and pH 0, [oxid] is the concentration of the oxidised form (oxidant = electron acceptor), [red] is the concentration of the reduced form (reductant = electron donor).

Additionally, if H^+ are involved in the reaction (i.e. the couple generates a pH change), then the equation must take this into account to become:

$$E = E_o + 2.303 \frac{RT}{nF} \log \frac{[\text{oxid}]}{[\text{red}]} + 2.303 \frac{RT}{nF} \log [\text{H}^+]^a$$

where a is the number of protons involved in the reaction.

In practice, the standard redox potential (E_o) used by chemists is little used by biologists because it involves pH 0 as one of the standard conditions. Instead

Table 11.1. *Standard redox potentials of interest to biologists (E_o' determined at pH 7 and 25 °C)*

E_o' (V)	Reaction	
	Oxidant + ne^-	→ Reductant
−0.42	$2H^+ + 2e^-$	→ H_2
−0.32	$NAD^+ + H^+ + 2e^-$	→ NADH
−0.22	$FAD + 2H^+ + 2e^-$	→ $FADH_2$
−0.19	Pyruvate + $2H^+ + 2e^-$	→ Lactate
−0.17	Oxaloacetate + $2H^+ + 2e^-$	→ Malate
−0.03	Fumarate + $2H^+ + 2e^-$	→ Succinate
+0.05	Ubiquinone + $2H^+ + 2e^-$	→ Ubiquinol
+0.08	Cytochrome $b(Fe^{3+}) + e^-$	→ Cytochrome b (Fe^{2+})
+0.25	Cytochrome c (Fe^{3+}) + e^-	→ Cytochrome c (Fe^{2+})
+0.29	Cytochrome a (Fe^{3+}) + e^-	→ Cytochrome a (Fe^{2+})
+0.30	$\frac{1}{2}O_2 + H_2O + 2e^-$	→ H_2O_2
+0.82	$\frac{1}{2}O_2 + 2H^+ + 2e^-$	→ H_2O

biologists adopt pH 7 as a standard condition, and consequently this standard redox potential is given the symbol E_o' to discriminate it from the E_o determined at pH 0. These two commonly used standard redox potentials can cause great confusion unless the pH is clearly stated and uniformity is maintained in comparisons of the potentials of different systems.

Table 11.1 includes some of the E_o' values of interest in biology. Note, for example, that the hydrogen ion E_o' is −0.42 V, whereas the standard value used by chemists (determined at pH 0) is by definition 0.00 V. Thus, a redox couple can theoretically be oxidised by (lose electrons to) a couple with a more positive E_o', and in turn will oxidise a couple with a more negative E_o'.

The free energy change in a coupled oxidation–reduction reaction is related to the redox potential of the couple and the number of electrons involved:

$$\Delta G^{o\prime} = -nF\Delta E_o'$$

where $\Delta G^{o\prime}$ is the standard free energy change, and $-nF\Delta E_o'$ is the potential difference between the two participating redox systems providing n is the same for each system.

If $\Delta E_o'$ is positive, then $\Delta G^{o\prime}$ will be negative and the coupled reaction will be exergonic, i.e. free energy will be released and the coupled reaction is thermodynamically favoured (this is often called a spontaneous reaction). If $\Delta E_o'$ is negative, $\Delta G^{o\prime}$ will be positive and the reaction will be termed endergonic; it will thus require an input of free energy to proceed in the direction stated (though the reverse reaction will be spontaneous).

It is difficult to predict the outcome of an interaction in a living cell of two redox couples whose E_0' values are very similar, since this may depend on the conditions within the cell. Factors that may influence the interaction include pH, the relative concentrations of the two molecules or ions, and the presence of chelating agents.

11.3.2   Applications of redox couples

Redox couples have numerous biological applications. Many are useful for investigating electron transport in that the reduced form of the redox couple can donate electrons to the electron transport chain and thus bypass the site of action of inhibitors such as rotenone and antimycin A, or can restart electron transport when an essential component has been removed. The precise site of action of inhibitor or the component can then be accurately pinpointed. Ascorbate is often used as an artificial electron donor, but is usually used in conjunction with another compound. With chloroplasts it has been found that ascorbate alone, or ascorbate with phenylenediamine, can replace water as electron donor to photosystem II, whereas ascorbate with 2,6-dichlorophenolindophenol donates an electron just before photosystem I, i.e. at a totally different site. With mitochondria, ascorbate is often used with cytochrome c or with N,N,N',N'-tetramethyl-p-phenylenediamine (TMPD); in both cases electrons are donated at cytochrome c.

Some redox couples are particularly useful in biochemical investigations because they change colour upon oxidation or reduction. Most of these oxidation–reduction indicators (redox dyes) are brightly coloured when oxidised and colourless when reduced, exceptions being the tetrazoliums and viologens. Examples are shown in Table 11.2. The rate of reduction of a redox dye as measured in a spectrophotometer can be used as the basis of an enzyme assay (Section 4.8.2); for example, the activity of succinate dehydrogenase isolated from mitochondria can be linked to methylene blue as follows:

$$\text{Succinate} + \text{methylene blue} \rightarrow \text{Fumarate} + \text{methylene blue}$$
$$\text{(oxidised} = \text{blue)} \qquad\qquad \text{(reduced} = \text{colourless)}$$

The rate of decolorisation of methylene blue, i.e. the rate of reaction, can thus be followed spectrophotometrically. However, oxygen must be excluded from the system or else the methylene blue will be reoxidised. The electron transport processes in chloroplasts, mitochondria and bacteria can also be studied using indicator dyes. Provided they have appropriate oxidation–reduction potential, the electrons are accepted by the indicators instead of being passed on to the next

Table 11.2. *Standard redox potentials of useful artificial redox couples (E_o' determined at pH 7 and 25 °C)*

E_o' (V)	Redox dye
−0.45	Methyl viologen
−0.36	Benzyl viologen
−0.08	TTC (2,3,5-triphenyltetrazolium chloride)
+0.01	Methylene blue
+0.08	PMS (phenazine methosulphate)
+0.22	DCPIP (2,6-dichlorophenol indophenol)
+0.36	Potassium ferricyanide

electron carrier in the chain. Unfortunately the vast majority of indicator dyes are not very specific and may receive electrons from several points in the electron transport chain. Care must be taken to interpret correctly the results of experiments using redox dyes because the dyes have been known to influence the nature of the reactions taking place and can inhibit enzymes or act as poisons for microorganisms. pH changes can also cause a change in colour, or in the ease of reduction. There is also the problem that the dye may not be readily able to cross membranes and may, therefore, not be able to reach the appropriate subcellular site. However, in the case of organelles or artificial vesicles, the fact that many dyes cannot penetrate the membrane can be exploited. Thus the extent of interaction of an electron transport component within the membrane and the external solution containing redox dyes indicates on which surface of the membrane the component is situated. The experiment can then be repeated with inverted vesicles, so that the other face of the membrane is then exposed to the dye. The subcellular localisation of enzymes can be ascertained by using carefully prepared tissue slices and staining them with dyes (a histochemical technique). Tetrazolium chloride is frequently used for this purpose because it gives an insoluble precipitate that does not readily diffuse from the site of formation.

11.4   The pH electrode

11.4.1   Principles

Perhaps the most convenient and accurate way of determining pH is by using a glass electrode. Developed in 1919, this device did not become popular until the 1930s, when reliable amplifiers became available. Nowadays, however, the pH electrode/pH meter is one of the most basic items of equipment found in biology laboratories.

The pH electrode depends on ion exchange in the hydrated layers formed on the glass electrode surface. Glass consists of a silicate network amongst which are metal ions coordinated to oxygen atoms, and it is the metal ions that exchange with H^+. The glass electrode acts like a battery whose voltage depends on the H^+ activity of the solution in which it is immersed. The size of the potential (E) due to H^+ is given by the equation:

$$E = 2.303 \frac{RT}{F} \log_{10} \frac{[H^+]_i}{[H^+]_o} \tag{11.4}$$

where $[H^+]_i$ and $[H^+]_o$ are the molar concentration of H^+ inside and outside the glass electrode.

In practice $[H^+]_i$ is fixed and is generally 10^{-1} because the electrode contains 0.1 M HCl. Since pH $= -\log_{10}[H^+]$, it follows that the developed potential is directly proportional to the pH of the solution outside the electrode. Glass electrodes are particularly useful because of the lack of interference from the components of the solution. On the whole these electrodes are not readily contaminated by molecules in solution, and if other ions are present they do not cause significant interference. However, at high pH they do respond to sodium. Inaccuracies also occur under very acid conditions.

A glass electrode (Fig. 11.6b) consists of a thin, soft glass membrane that is situated at the end of a hard glass tube, or sometimes an epoxy body. Also present in the glass electrode is an internal reference electrode of the silver/silver chloride (Ag/AgCl) type described in Section 11.2.4 surrounded by an electrolyte of 0.1 M HCl. This internal reference electrode gives rise to a steady potential. Thus, the varying potential of the glass electrode can be compared with a steady potential produced by an external reference electrode such as the standard calomel electrode by joining the internal and external reference electrodes to give:

	Glass electrode		Test solution	Reference electrode
Refence electrode (internal)	H⁺ (internal) i.e. 0.1 M HCl	Glass membrane	H⁺ (external) i.e. analyte	Refence electrode (external)

The external reference electrode can either be a separate probe (Fig. 11.6a) or built around the glass electrode giving a combination electrode (Fig. 11.6c). If a combination electrode is used, the level of the test solution must be high enough to cover the porous plug (liquid junction) but not so high as the level of the salt bridge solution (KCl) in the external electrode because it is essential for KCl to diffuse out slowly into the test solution.

Whatever reference electrode is used, the measured voltage is the result of the difference between that of the reference and the glass electrode. In practice, however, there are other potentials present in the system. These include the so-called asymmetry potential, which is poorly understood but which is present across the glass membrane even when the H^+ concentration is the same on both sides. Also included are the potentials due to the Ag/AgCl and to the liquid junction to the reference electrode, which gives a potential because the K^+ and Cl^- do not diffuse at exactly the same rate and therefore generate a small potential at the boundary between the sample and the KCl in the reference electrode.

The measured potential of a glass electrode is thus based on equation 11.4 but includes constants to account for the additional potentials within the device:

$$E = E^* + 2.303 \frac{RT}{F} \log_{10} \frac{[H^+]_i}{[H^+]_o} \tag{11.5}$$

where E^* includes the standard electrode potential of the glass electrode, and the constant junction potentials present in the system.

At 25 °C this becomes:

$$E = E^* + 0.059 \text{ pH} \tag{11.6}$$

where E^* now also includes a term to account for the internal H^+ concentration. As described in Section 11.2.4, at 25 °C, there is a 59 mV change for a 10-fold change in the activity of a monovalent ion; this means that a change of one pH unit produces a 59 mV change.

A pH electrode is used in conjunction with a pH meter. This records the potential due to the H^+ concentration, but is designed to take little current from the circuit. A large current flow would cause changes in the ion concentration and hence pH changes; this is prevented by having a high resistance present. The pH meter, glass electrode and reference calomel electrode are designed so that pH 7 gives a zero potential.

11.4.2   Operation of the pH electrode/meter

pH electrodes are available in a variety of shapes and sizes suitable for many different applications. These include electrodes for the measurement of the pH of blood, the mouth, flat moist surfaces such as isoelectric focusing gels, and equipment for field work. Intracellular pH can also be measured using miniature probes (microelectrodes). However, all of these devices rely on the same principle of measurement, and the vast majority will be operated in the same way.

It is important that the outer layer of glass on the glass electrode remains hydrated, and so it is normally immersed in solution. The thin glass membrane is fragile and care must be taken not to break or scratch it, or to cause a build-up of static electric charge by rubbing it. Many modern pH electrodes have a plastic casing surrounding the glass electrode to protect it from damage. Gelatinous and protein-containing solutions should not be allowed to dry out on the glass surface as they would inhibit response.

As can be seen from equation 11.5, the potential produced is dependent on the temperature (each pH unit change represents 54.2 mV at 0 °C and 61.5 mV at 37 °C). This effect is entirely predictable and can be compensated for. The further away from the pH 7 (the isopotential point, where temperature has no effect on potential) the more important it is that the temperature compensation is applied accurately because of the accumulation of errors. The pH meter will thus have a temperature compensation dial that must be correctly set before the meter is calibrated.

Calibration will necessitate the use of two solutions of widely differing pH. Usually calibration is first carried out with a pH 7 buffer, followed by a pH 4 buffer (if the sample is expected to be acidic) or a pH 9 buffer (if the sample is expected to be basic). Once the pH electrode is calibrated it can simply be immersed in the solution to be measured and a rapid and accurate estimate of pH can be made.

11.4.3   The pH-stat

The pH-stat is a form of automatic titrator that can be used to maintain a constant pH during a reaction that involves either the production or removal of H^+ from solution. The rate of reaction can thus be determined because a recorder draws a curve representing the volume of reagent added by the titrator, against time. A glass pH electrode, pH meter, recorder, controller, burette and magnetic stirrer are necessary; the burette can be an ordinary burette with a magnetic valve or, better still, a motor-driven burette syringe. A controller is necessary to break the current to the burette motor when the end point is reached. The best sort of titrators are arranged so that less and less reagent is added as the end point is approached. This avoids the danger of overshoot. The kind of pH glass electrode used needs to be very accurate and stable for kinetic work.

The pH-stat has some limitations in that, for example, the solution in the reaction vessel has to be stirred constantly, and this may cause denaturation of proteins and introduce atmospheric components that either may affect the pH or else may affect the reaction proceeding in the vessel. Other problems are the existence of an unknown junction potential and the tendency of the liquid from the burette tip (under the surface of the solution in the reaction vessel) to leak. The latter effect can be counteracted in part by making the density of the burette liquid lower than that of the reaction mixture.

11.5 Ion-selective and gas-sensing electrodes

11.5.1 Principles

The glass pH electrode is really a kind of ion-selective electrode (ISE) that is sensitive to H^+. Similar potentiometric electrodes have been developed which are responsive to other ions, for example Na^+, NH_4^+, Cl^-, and NO_3^-. The active material within these devices may be glass, an insoluble organic salt or an ion-exchange material. Glass is the active material within the pH electrode, but modified aluminosilicate glasses may also be used to produce a variety of monovalent cation-responsive electrodes (e.g. Na^+, Li^+ and NH_4^+). Insoluble inorganic salts such as silver sulphide may be used to produce electrodes responsive to Cu^{2+}, Pb^{2+}, and Cd^{2+}, whereas lanthanum fluoride may be used to produce electrodes responsive to F^-. Ion-exchange materials may be dissolved in a water immiscible solvent, then absorbed on to a Millipore filter to produce a liquid membrane or may be incorporated into PVC to give a solid membrane. The most frequently used electrodes of this type are those responsive to Ca^{2+}, K^+ and NO_3^-.

Ion-selective electrodes respond to the activity of a particular ion. However, if the instrument is calibrated with a standard of known concentration then, provided the ionic strengths of the solutions are similar, the concentration of the test solution will be recorded. To ensure that the ionic strengths are similar an ionic strength adjustor may be added. Ionic strength adjustors contain a high concentration of ions and sometimes pH adjustors and decomplexing agents or agents to remove species that interfere with the measurement. If, however, some of the ion is not free, but exists in a complex or an insoluble precipitate, then ion-selective electrodes will give a much lower reading than will a method that detects all of the ions present. Thus, atomic absorption spectrophotometry (Section 7.12) measures concentration and, for an ion such as calcium that readily forms calcium phosphate, it will give a significantly higher reading than that which would be obtained with an ISE. The ISE results, however, are significant because it is often the free ions that are responsible for clinical/biological effects.

An electrode may be ion selective, but not necessarily ion specific. Manufacturers' instructions will give information about this and will also mention chemicals that can poison the electrode. As with pH glass electrodes, ion-selective electrodes can be fouled by protein forming a surface film.

Many ions can be measured directly by the use of ISEs, or indirectly by titration. One form of indirect measurement is the use of the electrode as the end point indicator of a titration. The electrode can be sensitive either to the species being determined or to the titrant ion. Titrations are ten times more accurate than direct measurement because the procedure requires accurate measurement of a *change* in potential rather than the absolute *value* of the potential. This is an important point because ISEs are not intrinsically very accurate. For example, the determination of the concentration of calcium ions is best carried out by titrating the solution with EDTA, which is a strong complexing agent for calcium, using a calcium ISE. Since the electrode responds to the logarithm of the concentration of calcium ions in solution, as the EDTA is added a sharp end point is observed, giving a precision of 0.1% or better.

The response of ion-selective electrodes is (similar to the pH electrode) logarithmic, with 10-fold changes in ion activity giving equal increments on the meter scale. As with pH electrodes, the actual potential produced from an ISE is temperature dependent (Section 11.2.4), except at the isopotential point, which varies depending on the type of electrode. It is therefore important that temperature compensation is used. A reference electrode is also needed so that the varying potential of the ISE can be compared with the steady potential produced by a reference electrode (Section 11.2.4). If either K^+ or Cl^- are being measured, then a double junction reference electrode is needed to prevent contamination of the sample by the internal solution of the reference electrode.

Gas-sensing electrodes usually estimate the concentration of a gas by its interaction in a thin layer surrounding an ion-selective electrode, commonly a pH electrode. Carbon dioxide, ammonia, sulphur dioxide and nitrogen oxide can all be measured by their dissolution in a thin layer surrounding a pH electrode, and measuring the resultant pH of the layer.

The miniaturisation of ion-selective electrodes has been achieved by the modification of field effect transistors to respond to specific ions. Such ion-selective field effect transistors (ISFETs) are likely to have great clinical value. Multifunction ISFETs able to measure pH, Na^+, K^+, and Ca^{2+} are already available, and it is likely that such devices will become commonplace for the analysis of blood parameters either during surgery or in aftercare. ISFETs also make a suitable miniature transducer for incorporation into biosensors (Section 11.8).

11.5.2   Applications

ISEs are easy to use, economical, easily transportable, capable of continuous monitoring without hazard, and require little power. Because of these advantages, they are widely used, as shown in Table 11.3. Miniaturised electrodes called spearhead microelectrodes have been manufactured and are used to determine the ion contents of single cells, muscle and nerves.

Table 11.3. *Applications of ion-selective and gas sensing electrodes*

Ions or gas detected	Application
Na^+	Analysis of sea water, serum, soil, skin
K^+	Analysis of serum (often combined with Na^+ electrode)
Ca^{2+}	Analysis of serum, beer
Cl^-	Quick test for cystic fibrosis; food analysis
NO_3^-	Analysis of drinking water, fertilisers, microbial growth
NH_3	Analysis of solutions produced by Kjeldahl digestion of proteins (Section 4.4)
CO_2	Analysis of blood
Nitrogen oxides	Air pollution monitoring

Unlike measurements by atomic absorption spectrophotometry, ISEs respond over a wide range of concentrations, do not destroy the test sample, and are rapid in use. However, for clinical use where high precision is required, and where the normal range of blood cations is so small, ISEs are less commonly used than is atomic absorption or emission spectrophotometry.

11.6   The Clark oxygen electrode

11.6.1   Principles

In 1956 L.C. Clark Jr described a compact oxygen probe that has been found to be suitable in a great variety of biological applications. The electrode consists of a platinum cathode and a silver anode, both immersed in the same solution of saturated potassium chloride, and separated from the test solution by an oxygen-permeable membrane. When a potential difference of $-0.6\,V$ is applied across the electrodes such that the platinum cathode is made negative with respect to the silver anode, electrons are generated at the anode and are then used to reduce oxygen at the cathode. The oxygen tension at the cathode then drops and this acts as a sink so that more oxygen diffuses towards it to make up the deficit. Since the rate of diffusion of oxygen through the membrane is the limiting step in the reduction process, the current produced by the electrode is proportional to the oxygen tension in the sample. These electrode reactions may be summarised as:

$$\text{At silver anode} \qquad 4Ag + 4Cl^- \rightarrow 4AgCl + 4e^-$$
$$\text{At platinum cathode} \qquad O_2 + 4H^+ + 4e^- \rightarrow 2H_2O$$

There are many variants of the Clark electrode, the Rank electrode (produced by Rank Bros Ltd, Bottisham, Cambridge, England) being perhaps the most commonly used (Section 11.6.2). Probe-type Clark oxygen electrodes are also available and are of particular value in fermentation (so called pO_2 probes) and in environmental monitoring (Section 11.6.4). These probes generally need to be physically robust and in the case of fermenter probes need to be capable of withstanding the high temperatures necessary to sterilise the fermenter. Additionally, leaf disc electrodes capable of measuring gaseous oxygen concentrations have important biological applications (Section 11.6.5)

11.6.2   Operation of the Rank oxygen electrode (Clark electrode)

The Rank electrode (Fig. 11.8) allows a sample (usually about $3\,cm^3$) to be placed in an upper reaction vessel, which is separated from the electrode chamber by an oxygen-permeable membrane. Teflon ($12\,\mu m$ thick) is the usual choice of membrane, though Cellophane, polythene, silicone rubber and Cling Film have been used with varying degrees of success. Care must be taken to ensure that the membrane does not become contaminated, e.g. it should not be touched by hand; creasing and twisting can also cause problems. Thinner membranes give a quicker response but are more fragile. The membrane covers the electrodes and allows oxygen to diffuse towards them while preventing other reaction ingredients reaching the electrodes

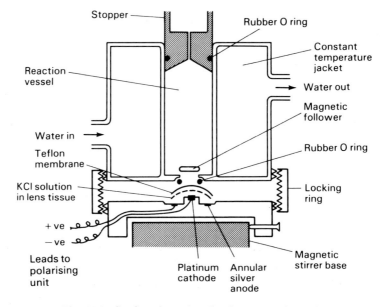

Fig. 11.8. Section through a Rank oxygen electrode.

and poisoning them. The electrodes are maintained in electrical continuity with each other via the potassium chloride solution. A square piece of microscope lens tissue is immersed in the potassium chloride to keep the solution in place, to provide some physical support for the thin membrane above it, and to make it easier to exclude air bubbles from the electrode compartment when the Teflon membrane is applied. However, the lens tissue square needs a 1–2 mm diameter hole in the centre to enable the platinum cathode to pass through the tissue. Electrodes should be clean and bright, and if they are contaminated they can be cleaned with dilute ammonium hydroxide. When the membrane has been changed, several minutes should be allowed for the electrode to give a steady response.

The oxygen electrode is mounted above a stirring motor, which is able to rotate a magnetic follower ('flea') when inserted into the reaction vessel. It is important that the contents of the reaction vessel are stirred as the platinum cathode reduces oxygen to produce the electrical current. An artificially low reading will be obtained if the magnetic stirrer should stop. However, when setting up an electrode (and after calibration), it is often useful to demonstrate the responsiveness of the electrode by switching off the magnetic stirrer and then restarting it after 10–15 s. A correctly set up electrode will show a reduction in current (oxygen tension) when the stirrer is switched off due to depletion of oxygen in the potassium chloride-filled electrode chamber. Resumption of stirring will result in a return of the current (oxygen tension in potassium chloride) to its previous level prior to the

Table 11.4. *Oxygen content of air-saturated water*

Temp. (°C)	Oxygen (μmol l^{-1})	Oxygen (p.p.m.)
15	305	9.8
20	276	8.8
25	253	8.1
30	230	7.5
35	219	7.0

stirrer being switched off. An electrode that does not respond in this way is unlikely to be of any value – the electrode should be disassembled and a fresh membrane applied. Since both the solubility and the rate of diffusion of oxygen are affected by temperature, then some form of temperature control is necessary for best results. Water from a thermocirculator may therefore be pumped through the water jacket surrounding the reaction vessel. During experimentation care should be taken to ensure that all reagents reach the required temperature before experimentation commences; if ice-cold reagents are pipetted into the reaction vessel and insufficient time is allowed for the reagents to reach the required temperature, incorrect results will be obtained. Calibration should also be carried out at the same temperature as that of the experiment.

The Rank oxygen electrode has a polarising module that enables the correct polarising voltage between the electrodes to be set, and also allows adjustment of the current for the 0% and 100% oxygen values so that output can be made directly to a chart recorder. One hundred per cent oxygen can be set with a gain (sensitivity) control using distilled water or appropriate reaction buffer that has stood in air at the temperature of the reaction vessel for several hours. The concentration of dissolved oxygen at that temperature and pressure can be found from various scientific tables. Some useful values are presented in Table 11.4. However, a more accurate calibration of the instrument can be obtained by using mitochondrial fragments to oxidise a known amount of NADH, the amount of this having been accurately determined by spectrophotometric means (Section 3.5). Zero per cent oxygen can be achieved by adding a few crystals of sodium dithionite, which removes oxygen from solution, by using yeast respiring sugars, or by bubbling nitrogen gas through the solution in the reaction vessel. At 0% oxygen, there should be no current flowing, but small faults in insulation of the platinum electrode may mean that a small leakage current is present.

The calibration solutions are removed from the reaction vessel (usually by a suction pump or Pasteur pipette) and the experimental samples pipetted in. The stopper is then put in position, so that oxygen from the air cannot enter. Air bubbles should be excluded from the reaction vessel whilst experimentation is in progress. Addition of small amounts of solutions such as inhibitors is best achieved using a Hamilton syringe to inject the addition through the small hole in the stopper. Many chemicals are adsorbed on to the surface of the membrane and reaction vessel; hence, it is important that the apparatus is thoroughly cleaned after each experiment. In addition, care must be taken when organic solvents are present in the reaction vessel because they may give an incorrect response. If the electrode is going to be reused without the membrane being changed, then it is essential that water is left in the reaction vessel to prevent the membrane from drying.

11.6.3   Applications of the Rank oxygen electrode

Due to their ability to give a continuous trace, oxygen electrodes have largely replaced manometric techniques in the study of reactions involving oxygen uptake and evolution.

Mitochondrial studies

The study of respiratory control and the effects of inhibitors on mitochondrial respiration and the measurement of phosphorylation : oxidation (P : O) ratios are best done by means of the oxygen electrode (Section 11.1.2). As shown in Fig. 11.9, a slow rate of respiration occurs with a mitochondrial suspension until ADP is added, whereupon ATP production starts and an enhanced rate of electron transport occurs. This produces an increased rate of loss of oxygen (the terminal electron acceptor in the mitochondrial electron transport chain) from solution as it is reduced to water (Fig. 11.1); the slope of the oxygen electrode trace thus increases (rate X in Fig. 11.9). Once the mitochondria have used up all of the available ADP (i.e. they have phosphorylated it to ATP) they can no longer phosphorylate and this causes a reduction in electron transport and in the rate of oxygen reduction; the slope of the oxygen electrode trace thus decreases (rate Y in Fig. 11.9). The respiratory control ratio (shown as the ratio of rates X:Y in Fig. 11.9) is a measure of the extent of coupling of respiration and phosphorylation. The P : O ratio, the number of molecules of ATP formed (P) per atom of oxygen consumed (O), can be calculated from the trace if the amount of ADP added and the amount of oxygen reduced (Fig. 11.9) are known. If an uncoupler such as 2,4-dinitrophenol (DNP) is added, the respiratory rate (electron transport rate) is not regulated by ADP.

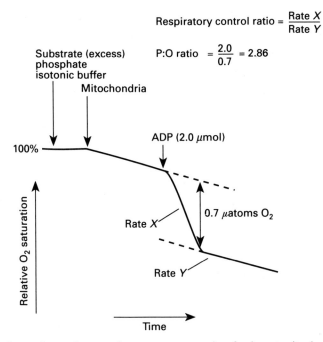

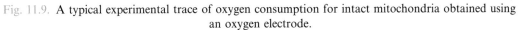

Fig. 11.9. A typical experimental trace of oxygen consumption for intact mitochondria obtained using an oxygen electrode.

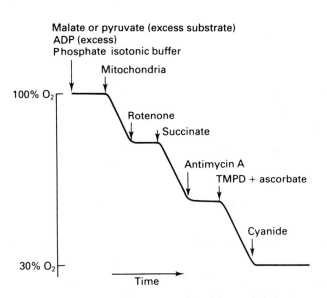

Fig. 11.10. Oxygen electrode trace showing the effect of inhibitors of electron transport and electron donors on mitochondrial respiration.

The sites of action of electron transport inhibitors can also be determined using an oxygen electrode. A typical result is shown in Fig. 11.10. Inhibition of respiration by rotenone can be overcome by use of succinate as an electron donor (it donates electrons below the site of action of rotenone on the electron transport chain). Antimycin A can still, however, inhibit respiration, which can be restarted by tetramethylphenylenediamine (TMPD) and ascorbate. However, even this rate can be inhibited by cyanide. These results enable the sites of action of the inhibitors to be placed as in Fig. 11.1.

Microorganism and chloroplast studies

Microorganisms that use oxygen as the terminal electron acceptor of respiratory electron transport can be studied using an oxygen electrode, and the effects of electron transport inhibitors determined. Respiration of yeast and other micro-organisms can be studied using different sugars. The most readily used sugars speed up the rate of respiration in starved yeast, which gives a steep slope on the oxygen electrode trace. The results also show that the sugar can enter the organism through the cell membrane. Oxygen evolution from cyanobacteria, algae, chloroplasts and chloroplast fractions enriched in photosystem II can be studied utilising a suitably illuminated Clark oxygen electrode. The oxygen content of the suspension medium must normally be reduced below 100% by bubbling nitrogen through the system. This ensures that the oxygen produced stays in solution and is recorded.

Enzyme assays

Enzymes are readily studied using a Clark oxygen electrode providing oxygen is involved in the reaction (Section 4.8.7). Glucose oxidase, D-amino acid oxidase and catalase are examples of enzymes whose properties can be studied in this way.

11.6.4   Probe-type Clark electrodes

Probe-type Clark oxygen electrodes have a variety of uses. These devices rely on the same principle of operation as the Rank electrodes; however, the cathode and retaining membrane are arranged at the end of a probe to enable insertion into a liquid phase. With such a configuration, however, problems may arise because of lack of stirring in the device.

Measurement of oxygen in bulk liquids

Oxygen concentrations are routinely monitored in fermentation processes, sewage and industrial waste treatment, and in inland, coastal and oceanic waters. This may involve a variant of the Clark electrode called a flush top sensor, which has a large cathode that gives a high current (but a negligible current at zero oxygen concentration). The equipment is rugged and easily manufactured, but to eliminate stirring effects the fluid must flow over the electrode. Oxygen solubility is different in fresh and salt water; hence such instruments have an adjustment for the degree of salinity of the sample. With respect to fermentation, dissolved oxygen probes (usually called pO_2 probes) need to be physically robust and capable of withstanding the high temperatures necessary to sterilise the fermenter. A popular electrode of this type is the Ingold pO_2 electrode, which may be autoclaved *in situ* within the fermenter and may operate maintenance-free for five to ten sterilisation/fermentation cycles or, in terms of time, for six to eight weeks. After this time, the internal electrolyte needs to be replaced before further use of the probe.

Clinical uses

An early clinical use of oxygen electrodes was in monitoring heart–lung machines during open-heart surgery. Because of their speed of response and ease of operation, oxygen electrodes have been used for testing patients who are being treated with oxygen. Some oxygen electrodes have been specially modified to be small enough to be inserted into a blood vessel, but frequently this is avoided because of the danger of infection or of blood clot formation. Often it is considered preferable to remove small samples of blood from a warmed earlobe or a finger-tip and measure the oxygen content of the blood in a small Clark-type pO_2 electrode.

11.6.5 The leaf disc electrode

Whilst the Rank oxygen electrode is ideally suited to many applications requiring a measurement of oxygen in aqueous samples, a leaf disc electrode such as the *Hansatech LD2* is of more use if gaseous oxygen measurements are required. Since the measurement of oxygen evolution is one of the easiest ways of following photosynthetic processes in leaves, this instrument has found many biological applications.

This device measures oxygen amperometrically using the same principle as the Rank electrode. However, instead of being a liquid-filled reaction vessel, the reaction chamber is designed to allow a leaf to be held in place and provided with saturating carbon dioxide (or bicarbonate as a source of carbon dioxide).

Illumination is usually provided by an array of light-emitting diodes (which produce little heat), and the oxygen emitted by the leaf during photosynthesis can be measured.

Calibration is slightly more complex than with the Rank electrode. A zero oxygen signal can be produced by passing nitrogen through the reaction chamber of the electrode. Once this is stopped, and air is passed into the chamber, the signal corresponding to 21% oxygen (i.e. the oxygen content of air) can be determined. However, in the closed chamber the *amount* of oxygen is related both to the oxygen concentration and the chamber volume. Thus, if the chamber volume is only $1 \, cm^3$ then it will contain $210 \, mm^3$ oxygen. In practice, because the leaf disc itself may reduce the *effective* volume of the chamber (i.e. that available to the oxygen), calibration involves injecting known volumes of air into the chamber and measuring the voltage response to obtain the effective volume of the chamber and thence a precise calibration of the electrode.

The leaf disc electrode has been used extensively for the study of the relationship between photosynthetic oxygen evolution under saturating carbon dioxide and the intensity of illumination, enabling calculation of quantum yield. The inclusion of probes to measure emitted fluorescence from the leaf disc at the same time as measurements of oxygen evolution are being made has resulted in a device that provides a variety of information useful to the plant physiologist or biochemist. Applications of those devices are diverse, ranging from studies of micropropagated plants to those of plants suffering from atmospheric pollution.

Although the leaf disc electrode was clearly designed for whole-leaf studies, photosynthetic rates of microalgae have also been studied in such electrodes. However, this demands that the algal suspension is first filtered, and the filter paper (with a covering of algal cells) used in the same way as a leaf in the electrode.

11.7   Electrochemical detectors for HPLC

11.7.1   Principles

Electrochemical detectors (ECDs) can be used in the same way as other detectors to detect and quantify analytes in the effluent from HPLC columns (Section 10.4.7). A variety of ECDs are commercially available, most working on the same principle of amperometric detection. Normally, when an ECD is used the potential is chosen so that it is sufficiently high to cause the reaction to occur. This information can be obtained from the polarographic wave of the compound determined as described in Section 11.2.5. The selected potential is then maintained at that level and a trace of current against time is recorded. As the effluent containing the analyte comes off the end of the HPLC column it flows through the detector and a peak is obtained, which may look very similar to those produced by ultraviolet or fluorescence detectors.

Not all compounds in the liquid give rise to a current at a particular chosen potential. This means that if two analytes have not been separated by chromatographic means, and one alone gives a current at the selected potential, then the ECD will give only one peak and it will be easy to make quantitative measurements despite the presence of the second analyte. In order to make use of this feature, it is important to know the reaction potentials of the various analytes in the mixture being injected on to the column. Such information can be found in the literature or, with modern detectors, by examining the compound under the same analytical conditions but at varying electrode potentials.

ECDs require an electrolyte to be present in the mobile phase so that a current may flow. This means that reversed-phase and ion-exchange chromatography (Sections 10.6.3 and 10.7), which both use aqueous solvents, can be coupled to ECDs if sufficient electrolytes are present. When the mobile phase is an organic solvent or a mixed organic/aqueous solvent, it is usual to add some inert salt such as potassium nitrate, ammonium thiocyanate or tetrabutylammonium perchlorate (which dissolves in pure organic solvents) to act as the electrolyte. ECDs are very sensitive to flow changes, hence they can be used only with pulse-free pumps (Section 10.4.5).

Whilst there are a great many configurations of ECD, they all comprise the same essential elements. One of the commonest detector types is the wall jet electrochemical detector (Fig. 11.11) in which the electrodes are mounted in a fluorocarbon block. The liquid flows under pressure up to the electrodes and through a jet to impinge on the surface of the working electrode. A high current efficiency ensues and the working electrode stays clean as the products of the electrochemical reaction are washed off in the flow. As well as the glassy carbon working electrode, there are two other electrodes present. One is the counter or auxiliary electrode, which is stainless steel and forms part of the inlet. Any deviation from the preset voltage of the working electrode can be corrected by the auxiliary electrode. The Ag/AgCl reference electrode gives a steady potential, and is in contact with the liquid as it leaves the wall jet flow cell.

11.7.2   Applications

ECDs can detect chemicals either by oxidation, in which the working electrode is kept at a positive potential, or by reduction, in which the working electrode is negative. At very high positive or negative potentials the current becomes large

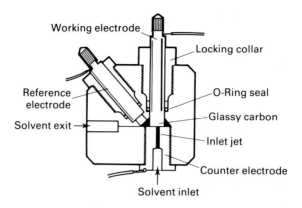

Fig. 11.11. A wall jet electrochemical detector.

owing to the solvent reacting; these are termed the anodic and cathodic limits of solvent

Most biological applications of ECDs involve the oxidation of the molecule

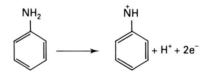

Fig. 11.12. Oxidation of an aromatic amine (aniline).

releasing electrons and giving rise to a current (see Fig. 11.12). Aromatic phenols, aromatic amines, heterocyclic nitrogen atoms and sulphur-containing molecules can all be detected by oxidation. Thus, drugs such as aspirin, paracetamol, morphine, nicotine and caffeine can be analysed by oxidation mode. ECD has become the method of choice for the analysis of catecholamines and neuro-transmitters, and ions such as sulphide, cyanide and iodide (Table 11.5).

Oxygenating compounds such as quinones, peroxides and amides can be detected by reduction. This approach is also suitable for aromatic nitro and nitroso groups and halogen compounds. From an electrochemical point of view, however, reductions are more difficult to carry out than oxidations because it is often difficult to obtain ideal experimental conditions. Solvents must therefore be thoroughly degassed, as oxygen will undergo electrochemical reactions, and meticulous care has to be taken to ensure that oxygen cannot re-enter the solution.

Advanced modern ECDs also offer the possibility of operating in a pulse mode, which is well suited to the detection of compounds such as carbohydrates that lack reproducible response in the traditional DC mode of operation.

Table 11.5. *Common applications of HPLC-ECD*

Anions
e.g. SCN^-, SO_3^{2-}, NO_2^-

Aromatic alcohols
Phenols, e.g. thyroxine
Catechols, e.g. adrenaline, dopamine, L-dopa
Morphine

Aromatic amines
Anilines e.g. benzidine
Sulphonamides

Indoles
Indole-3-derivatives, e.g. indole 3-acetic acid, melatonin
5-Hydroxyindoles, e.g. serotonin

Phenothiazines
e.g. promethazine

Purines
e.g. uric acid, xanthine, guanine

Thiols
e.g. cysteine, glutathione

Unsaturated alcohols
e.g. ascorbic acid

11.8   Biosensors

11.8.1   Introduction and principles

A biosensor is an analytical device consisting of a biocatalyst (enzyme, cell or tissue) and a transducer, which can convert a biological or biochemical signal or response into a quantifiable electrical signal. The biocatalyst component of most biosensors is immobilised on to a membrane or within a gel, such that the biocatalyst is held in intimate contact with the transducer, and may be reused.

Biosensors are already of major commercial importance, and their significance is likely to increase as the technology develops. This is because they can be made to respond specifically and with high sensitivity to a wide range of molecules including those of industrial, clinical and environmental importance. The best-developed systems are undoubtedly within the field of clinical medicine, where glucose responsive biosensors play a vital role in the measurement of blood glucose that is necessary for the management of diabetes.

Biosensors may be categorised as first-, second- or third-generation instruments according to the degree of intimacy between the biocatalyst and transducer. In first-generation instruments the two components (biocatalyst and transducer) may be easily separated and both may remain functional in the absence of the other. In second-generation instruments the two components interact in a more intimate fashion and removal of one of the two components affects the usual functioning of the other. In third-generation instruments the biochemistry and electrochemistry are even more closely linked and where the electrochemistry occurs at a semi-conductor the term biochip may be applied to describe such instruments.

11.8.2   First-generation instruments

The biocatalyst within a biosensor responds to the substrates in solution by catalysing a reaction. Glucose oxidase, for example, will catalyse the reaction:

$$Glucose + O_2 \rightarrow Gluconic\ acid + H_2O_2$$

By the use of first-generation technology, the rate of this reaction can be measured as follows:

(i) The rate of consumption of the substrate O_2 can be measured by its reduction at a platinum cathode polarised at $-0.6\,V$ versus the standard calomel electrode (i.e. with a Clark oxygen electrode).

(ii) The rate of production of the product H_2O_2 can be measured by its oxidation at a platinum anode polarised at $+0.7\,V$ versus the standard calomel electrode.

(iii) The rate of production of the product gluconic acid can be measured using a pH electrode to measure the associated decrease in pH.

The actual rate of reaction determined by these transducers will depend on the rate of enzyme reaction (usually following Michaelis–Menten kinetics (Section 4.7.2)), the rate of diffusion of substrate from the bulk phase into the enzyme layer, and the rate of diffusion of substrate within the enzyme layer. Which of these three rates is limiting depends on a number of factors of sensor design, but particularly on the amount of enzyme within the enzyme layer. However, although the theoretical kinetics are complex, in practice it is usual to find a rather linear relationship between rate of reaction and substrate concentration over a limited range of concentrations. Within this linear range or useful range it is relatively easy to calibrate the device, requiring just two standard solutions of known glucose concentration. Once the device is calibrated, the measurement of the rate of reaction by the biosensor can be used to quantify the concentration of glucose in solution.

This principle of operation was suggested by L.C. Clark Jr and C. Lyons in 1962, which led to the development of the Yellow Springs Instruments Model 23 in 1974, linking the enzyme glucose oxidase to a probe to measure H_2O_2. Immobilisation of the enzyme on a laminated membrane that covers the probe tip results in rapid response times of 40 s, and efficient reuse of the enzyme such that the membrane typically lasts two weeks before renewal becomes necessary. The analyser is able to measure the glucose content of whole blood, plasma or serum, requiring only 25 mm^3 sample.

Fig. 11.13. YSI Model 2700 SELECT Biochemistry Analyser. (Photograph courtesy of Clandon Scientific Ltd.)

The YSI model 23 is now unavailable, having been replaced by the more advanced YSI 2300 STAT Glucose and L-Lacate Analyser. The YSI 2700 SELECT Biochemistry Analyser (Fig. 11.13) essentially uses the same technology but allows the glucose oxidase membrane to be replaced with membranes containing other

Table 11.6. *Enzyme membranes available for analysers with H_2O_2 measuring probes as the transducer*

Analyte	Enzyme	Reaction
Alcohol	Alcohol oxidase	Ethanol + O_2 → H_2O_2 + Acetaldehyde
D-Glucose	Glucose oxidase	β-D-Glucose + O_2 → H_2O_2 + Gluconic acid
Lactose	Galactose oxidase	Lactose + O_2 → H_2O_2 + Galactose dialdehyde derivative
L-Lactate	L-Lactase oxidase	L-Lactate + O_2 → H_2O_2 + Pyruvate
Starch	⎰Amyloglucosidase ⎱Glucose oxidase	Starch + H_2O → β-D-Glucose β-D-Glucose + O_2 → H_2O_2 + Gluconic acid
Sucrose	⎰Invertase ⎱Mutarotase ⎱Glucose oxidase	Sucrose + H_2O → a-D-Glucose + β-D-Fructose a-D-Glucose → β-D-Glucose β-D-Glucose + O_2 → H_2O_2 + Gluconic acid

enzymes, enabling the analysis of a wide variety of substances of biological interest (see Table 11.6).

A great variety of other devices may be used as transducers. Many of the devices listed in Table 11.7 have been considered in previous sections. A photomultiplier is a device that can be used to detect the light emitted from enzymes such as luciferase (obtained from *Photinus pyralis*, the firefly) (Section 4.8.4). The photomultiplier is a sensitive photoelectric cell in which electrons emitted from a photocathode are accelerated to a second electrode where several electrons are liberated from each photoelectron. This amplification event may be repeated several times to enable the detection of very low levels of light. A thermistor is a semiconductor device in which the resistance decreases as the temperature increases. It can be used to measure very small temperature changes that take place as reactions proceed. Linking such transducers to appropriate enzymes allows biosensors to be constructed that are responsive to a wide variety of biologically important molecules (Table 11.8).

11.8.3   Second-generation instruments

There are a number of ways in which the degree of intimacy between biocatalyst and transducer can be increased. In sensors utilising oxidoreductases, second-generation instruments can be constructed by designing an electrode surface that is capable of capturing electrons, which are usually transferred in the oxidation or reduction reaction.

Table 11.7. *Types of transducer used in biosensors*

Type	Principle of operation/ cross-reference	Detectable species
Amperometric	Section 11.2.5	O_2, H_2O_2, I_2, NADH
Ion-selective electrode	Section 11.2.4	H^+, Na^+, Cl^-
Field effect transistor	Section 11.5.1	As ion-selective electrodes
Gas-sensing electrode	Section 11.5.1	CO_2, NH_3
Photomultiplier	Light emission	ATP (with luciferase present)
Thermistor	Heat of reaction	Universal

A good example of a commercially available second-generation instrument is provided by the ExacTech blood glucose meter (Fig. 11.14). In this device the rate of oxidation of glucose is measured not by the rate of disappearance of substrate or appearance of product, but by the rate of electron flow from glucose to an electrode surface. The reactions that occur in this device may be summarised as follows:

$$\text{Glucose} + \text{GO/FAD} \longrightarrow \text{Gluconic acid} + \text{GO/FADH}_2$$
$$\quad\;\; \text{(red)} \qquad \text{(ox)} \qquad\qquad \text{(ox)} \qquad\qquad \text{(red)}$$

$$\text{GO/FADH}_2 + 2M^+ \longrightarrow \text{GO/FAD} + 2M + 2H^+$$
$$\;\; \text{(red)} \qquad \text{(ox)} \qquad\qquad \text{(ox)} \quad\; \text{(red)}$$

At electrode:

$$2M \longrightarrow 2M^+ + 2e^-$$
$$\text{(red)} \qquad\; \text{(ox)}$$

where GO/FAD represents the FAD redox centre of glucose oxidase in its oxidised form, GO/FADH_2 represents the FAD redox centre of glucose oxidase in its reduced form, and M is a mediator, which in the ExacTech blood glucose meter is ferrocene. The electrons donated to the electrode surface then go to form a current that is proportional to the rate of oxidation of glucose, and hence proportional to the glucose concentration in the blood.

Devices of this type are far more suitable for miniaturisation. Whilst the ExacTech meter is itself only the size of a pen, devices using similar technology are now being produced that are so small they can be implanted under the skin to produce a blood glucose measuring system *in situ*. Work is ongoing to link such sensors to appropriate logic circuits and an insulin reservoir to provide diabetic patients with exactly the insulin they need throughout the day.

Fig. 11.14. The ExacTech blood glucose meter, showing test strip inserted. (Photograph courtesy of Medisense Britain Ltd.)

11.8.4   Third-generation instruments

Such instruments are essentially at the research level, rather than under commercial development. They involve the most intimate interactions of the biocatalyst and transducer. A glucose biosensor operating on the principle of the ExacTech meter (Section 11.8.3) but in which the enzyme was directly reduced at the electrode surface (obviating the need for a mediator) would be an example of such an instrument.

Table 11.8. *Examples of biosensors*

Analyte	Biocatalyst	Transducer	Immobilisation	Stability	Response time
Alcohol	Alcohol oxidase	O_2	Glutaraldehyde	>2 weeks	1–2 min
Arginine	*Streptococcus faecium*	NH_3	Physically entrapped	20 days	20 min
Cholesterol	*Nocardia erthyropolis*	O_2	Physically entrapped	4 weeks	35–70 s
D-Glucose	Glucose oxidase	O_2	Chemical	3 weeks	1 min
Glutamate	Glutamate decarboxylase	CO_2	Glutaraldehyde	1 week	10 min
NAD^+	NADase and *Escherichia coli*	NH_3	Dialysis membrane	1 week	5–10 min
Nitrate	*Azotobacter vinelandii*	NH_3	Physically entrapped	2 weeks	7–8 min
Penicillin	Penicillinase	H^+	Polyacrylamide	>2 weeks	15–30 s
Urea	Urease	NH_4^+	Polyacrylamide	>19 days	20–40 s

11.8.5   Cell-based biosensors

Whilst much of the pioneering work on biosensors has involved purified enzyme preparations as the biocatalyst component, there is a considerable amount of interest in the use of immobilised whole cells and tissues to produce biosensors. Such sensors are usually cheaper to produce than sensors depending on enzymes, as there is no requirement for complex biocatalyst isolation and purification procedures. However, cells contain many enzymes and care has to be taken to ensure selectivity of response (e.g. by adding specific enzyme or electron transport inhibitors to stop undesirable enzyme reactions). Cell-based electrodes may also have longer response times than do enzyme-based sensors, but a more serious problem is the time taken for cell-based sensors to return to baseline potential after use.

When cell-based sensors were first developed, they involved the use of harsh immobilisation procedures such as polyacrylamide gel entrapment, and the immobilised cells were not usually viable. However, the enzymes within them were still active. More recent immobilisation techniques have tended to use gentler physical methods such that cell viability is retained. The advantage of this is that such cells may be involved in converting substrate into product via a complex multienzyme pathway, without having to immobilise each of the enzymes and then provide them with expensive coenzymes.

A good example of a cell-based sensor is that using *Nocardia erythropolis* immobilised in polyacrylamide on agar on an oxygen electrode. The reaction carried out is:

$$\text{Cholesterol} + O_2 \xrightarrow[\text{Cholesterol oxidase}]{\textit{N. erythropolis}} \text{Cholest-4-en-3-one} + H_2O_2$$

The oxygen electrode measures the rate of oxygen uptake and this can be related to the cholesterol content of the biological sample (probably plasma).

More than one cell type can be incorporated into the electrode, thereby increasing the number of potential applications. Thus the biochemical oxygen demand (BOD) of organic matter in waste-water can be detected by a mixed culture of bacteria obtained from soil, because a single microorganism would be unable to use all of the organic compounds likely to be found in the sample.

It is also possible to combine an enzyme preparation with a cell type. For example purified NADase (from *Neurospora crassa*) plus *Escherichia coli* (high in nicotinamide deaminase activity) will produce the following reaction:

$$NAD^+ + H_2O \xrightarrow{\text{NADase}} \text{Nicotinamide} + \text{ADP-ribose}$$

$$\text{Nicotinamide} + H_2O \xrightarrow{\textit{E. coli}} \text{Nicotinic acid} + NH_3$$

The ammonia released can be detected by a gas-sensing electrode (transducer) to produce an NAD^+-sensitive biosensor.

Cell types other than microorganisms may also be utilised. A dopamine-sensitive biosensor has been produced by immobilising banana pulp in an oxygen electrode. Banana pulp is rich in the enzyme polyphenol oxidase, which causes the browning of the tissue. The pulp can thus be used as a rich source of the enzyme, and the enzyme found in banana has a high selectivity for the neuroactive agent dopamine.

11.8.6 Enzyme immunosensors

Several kinds of enzyme immunosensors have been developed. They combine the molecular recognition properties of antibodies with the high sensitivity of enzyme-based analytical methods. The enzyme is used as a marker as it reacts with its substrate, giving changes that can be detected by a transducer. There is a similarity between such methods and ELISA techniques (Section 2.6).

The enzyme immunosensor for IgG consists of an anti-IgG antibody bound to a membrane and linked to an oxygen electrode. Catalase is bound to the IgG, and so forms a label. This complex is then mixed with the sample that contains an unknown amount of unlabelled IgG. The labelled and unlabelled IgG compete for the antibody on a membrane. After exposure to test an IgG solution the sensor is rinsed to remove any non-specifically associated IgG and then immersed in a H_2O_2 solution as substrate for the catalase. The more unlabelled IgG present, the lower is the amount of catalase present and therefore the lower is the rate of oxygen evolution. A similar assay can be carried out for human chorionic gonadotrophin (hCG), a diagnostic hormone for pregnancy. Catalase is again used to label the hCG, and oxygen evolution is followed as before.

In attempts to construct enzyme immunosensors, bioluminescence, chemi-luminescence and fluorescence principles are often expoloited because of their great sensitivity. A luminescent immunoassay with catalase has been used to detect human serum albumin at only 1 $ng\,cm^{-3}$.

11.9 Suggestions for further reading

ANON. (1976) Classification and nomenclature of electroanalytical techniques (rules approved 1975). *Pure and Applied Chemistry* **45**(2), 81–97. (A fairly well accepted scheme of classifying electrochemical techniques.)

HALL, E.A.H. (1990) *Biosensors*. Open University Press, Milton Keynes. (An excellent introductory text to biosensor technology.)

HART, J.P. (1990) *Electroanalysis of Biologically Important Compounds*. Ellis Horwood Ltd, New

York. (An excellent introduction to principles and biological applications of electrochemistry.)

NICHOLLS D.G. and FERGUSON, S.J. (1992) *Bioenergetics 2*. Academic Press, London. (Excellent coverage of membrane bioenergetics and the chemiosmotic theory.)

PARVEZ, H., BASTART-MALSOT, M., PARVEZ, S., NAGATSU, T. and CARPENTIER, G. (Eds.) (1987) *Progress in HPLC*, vol. 2. VNU Science Press, Utrecht. (An advanced text detailing the diverse applications of HPLC-ECD; containing a very readable introductory chapter on the subject.)

RILEY, T. and TOMLINSON, C. (1987) *Principles of Electroanalytical Methods*. Wiley, Chichester. (An open learning text covering the basic theory and practical aspects of individual techniques.)

TURNER, A.P.F., KARUBE, I. and WILSON, G.S. (Eds.) (1987) *Biosensors: Fundamentals and Applications*. Oxford University Press, Oxford. (A definitive reference book on biosensor technology.)

WALKER, D. (1987) *The Use of the Oxygen Electrode and Fluorescence Probes in Simple Measurements of Photosynthesis*. Research Institute for Photosynthesis, Sheffield. (Contains much information on the Hansatech leaf disc electrode, together with useful coverage of Rank type electrodes and their applications.)

Index

575